HT46xx 单片机原理与实践

钟启仁 编著

北京航空航天大学出版社

内容简介

本书介绍 HT46 系列 8 位单片机的原理、开发和应用。讲述集成开发环境 HT3000 IDE 的特点和使用方法，特别是脱机（软件）仿真环境，可仿真虚拟外设（VPM），包括按键、电阻、三极管逻辑电路、LED 及字符型或点阵型 LCD 显示屏，屏幕样式可由用户定义，其程序设计、调试高效方便。书中有很多应用实例（含硬件图和程序清单），其软件有详尽的注释。本书附光盘 1 张，内含源程序代码以及相关资料。

本书既适合单片机初学者自学，也可供在校大学生和工程技术人员开发智能产品时学习和参考。

图书在版编目（CIP）数据

HT46xx 单片机原理与实践 / 钟启仁编著. —北京：北京航空航天大学出版社，2008.9
　ISBN　978-7-81077-863-3

Ⅰ. H…　Ⅱ. 钟…　Ⅲ. 单片微型计算机　Ⅳ. TP368.1

中国版本图书馆 CIP 数据核字（2008）第 083625 号

原书名《HT46xx 微控制器理论与实务宝典》。本书中文简体字版由台湾全华科技图书股份有限公司独家授权。仅限于中国大陆地区出版发行，不含台湾、香港、澳门。

北京市版权局著作权合同登记号图字：01-2006-0903

HT46xx 单片机原理与实践

钟启仁　编著

责任编辑　胡晓柏

＊

北京航空航天大学出版社出版发行

北京市海淀区学院路 37 号（100191）　发行部电话：010-82317024　传真：010-82328026
http://www.buaapress.com.cn　E-mail:bhpress@263.net
涿州市新华印刷有限公司印装　各地书店经销

＊

开本：787×1092　1/16　印张：30.25　字数：774 千字
2008 年 9 月第 1 版　2008 年 9 月第 1 次印刷　印数：4 000 册
ISBN 978-7-81077-863-3　定价：55.00 元（含光盘 1 张）

前　言

单片机（Microcontroller）历经了4位、8位、16位及32位等开发过程，被广泛地应用于各种领域，只要与操作接口有关的应用，都能发现它的踪迹。在国外，单片机的使用数量甚至成为评估收入与经济状况的指标之一。

近年来，台湾盛群半导体公司（Holtek）鉴于IC市场竞争越来越激烈，从消费性电子设计公司成功转型为专业单片机设计公司，专注于通用型与嵌入式单片机的开发。除了消费性、计算机外围、通信领域的嵌入式单片机外，还提供I/O、LCD、A/D、RF及A/D LCD等通用型单片机。盛群半导体公司的定位是以单片机为核心技术的IC设计公司，不同于其他单片机制造商。该公司的行销网络遍及全球，涵盖欧洲、北美、南美等地，其产品线广泛，不仅消费性产品用的单片机在德国获得飞利浦家电的采用，更是台湾最早推出符合工业标准规格单片机的设计公司之一。大陆市场部分也于近年开展，成立盛扬半导体公司，在I/O、LCD控制芯片以及Phone Controller市场均有所斩获。最值得一提的是该公司自行开发的开发工具，操作容易而且效能绝佳，具备绝佳的价格竞争优势，广泛获得欧美客户的采用。此外，该公司也与业界合作开发，除提供汇编语言外，也有C语言的编译器，可算是台湾提供IC开发工具上最为齐备的Design House！

目前，盛群半导体公司以提供8位OTP与Mask型单片机为主，未来则继续朝向可重复读/写的E^2PROM单片机发展，在技术层面上足以赶上欧美厂商。HT46xx系列单片机为盛群半导体公司所研发设计的A/D型8位单片机，被广泛应用于工业产品、家用电器、玩具等。由于它的高可靠性、故低障率、低成本、开发工具齐备等特点，在单片机的市场上早已占有一席之地。

本书主要针对HT46xx A/D型单片机的特性、功能、指令及相关的外围设备，编写了一系列的基本实验，如HT46xx内部的架构、基本功能特性、指令的应用都有详细的说明。本书共分为6章，各章内容如下：

第1章　HT46xx系列单片机简介：除了说明单片机特点之外，也介绍盛群半导体公司的HT46xx家族成员特性；并针对单片机未来发展趋势，提出笔者个人的一些浅见。

第2章　HT46xx系列系统体系结构：以循序渐进的方式，针对HT46xx内部硬件结构（包含存储器结构、I/O特性以及看门狗定时器、Timer/Event Counter、中断、I^2C传输接口、A/D转换接口、PWM接口等）做一番详尽的介绍。建议读者在阅读本章时，与第4章的基础实验相互搭配，这样才能增加对HT46xx系列单片机内部相关寄存器的印象，以免只是纸上谈兵而降低学习效果，失去学习的兴趣。

第3章　HT46xx指令集与开发系统：除了说明HT46xx系列单片机的指令之外，也介绍程序的编译流程与宏的写法。另外，所谓"工欲善其事，必先利其器"，盛群半导体公司提供了相当完善的开发工具,如HT-ICE以及完整的集成开发环境(HT-IDE3000V6)等。HT-IDE3000中的软硬件仿真功能VPM(Virtual Peripheral Manager)更能让用户在未接硬件电路(或没有ICE)的情况下,先行验证程序的功能。

第4章 基础实验篇:介绍几个基础实验,如跑马灯、LED、扫描式键盘、步进电机控制、Timer/Event Counter与WDT应用、外部中断、A/D、PWM、PFD、HALT Mode、I^{2}C接口等。希望通过这些基础实验,让读者对于HT46xx系列单片机的控制以及其内部各个单元,都能有初步的了解与认识。

第5章 进阶实验篇:介绍几个较深入的实验,如PWM直流电机控制、点矩阵控制、LCD控制、矩阵式与半矩阵式按键输入设置、I^{2}C Master-Slave数据传输等。相信通过这些实验,必定能够让读者对于HT46xx系列单片机的应用能有更深一层的了解。

第6章 实践应用篇:介绍几个实用、有趣的专题实验,如数字温度计、密码锁、猜数字游戏机、24小时定时时钟、逻辑笔、DTMF产生器、RS-232C串行接口传输等。希望读者能通过参与实验的过程中,学习到产品设计与开发的经验。

本书所有的例题程序及硬件电路,都经过实际的测试无误。读者可以直接编译之后烧录或是以ICE仿真,验证其正确性。所有实验都经过精心的安排与实际测试,每一个实验都有不同深度的学习。笔者想叮咛读者的是:虽然汇编语言不及VB或VC等高级语言来得人性化,但是在许多应用的场合,为了整体系统的效率(如RAM、ROM的需求,CPU的执行速度等),不得不使用汇编语言来编写程序,所以鼓励读者要多写程序、多除错,这样才能累积自己编写程序的经验。笔者经常告诉学生的是:"**程序一次写对,未必是好事;唯有从错误中学习,才是真正个人的经验累积**"。只要耐心研读,相信假以时日,您也可以成为单片机应用的佼佼者。书中的实验内容与顺序,都是经过刻意的安排。读者会发现越到后面的实验,大部分只是把之前使用的子程序加以重新组合而已,因此特别将几个常用的子程序列于附录中供读者参考,以便在需要之时可以快速查阅。

随书光盘中,除了有各个实验的源程序之外,也将实验中所使用的相关IC资料收录于其中,虽然是原文的内容,但却是IC制造厂商所提供最完整的资料。想要淋漓尽致地发挥IC的特性及功能,仔细阅读原厂的数据手册是不可缺少的必经过程。希望读者能够耐心研读,相信这对产品的设计、开发一定有所帮助。另外,由盛群半导体公司所提供的开发环境HT-IDE3000V6也一并收入到光盘中,不过在此还是鼓励读者多到网站www.holtek.com.tw上下载更新的程序版本,同时也可取得产品的最新资讯。

最后,衷心感谢台湾盛群半导体公司与家仑股份有限公司给予写作本书的机会以及在写作上提供的种种协助,尤其是盛群半导体公司产品推广部鲍惟圣经理在校稿过程中所提供的宝贵意见。企盼本书能带读者一窥单片机的世界,也期望读者能不吝于对本书的批评及指正!

钟启仁
于台湾风岗

目 录

第 1 章 HT46xx 系列单片机简介

1-1 单片机介绍及其未来趋势 2
1-2 HT46xx 单片机的特点介绍 5
1-3 HT46xx 家族介绍 9
1-4 HT46xx 硬件引脚功能描述 11
1-5 HT46xx 复位引脚($\overline{RES}$) 13
1-6 输入/输出引脚(PA、PB、PC、PD、PF) 14

第 2 章 HT46xx 系列系统体系结构

2-1 HT46xx 的内部体系结构 17
2-2 程序存储器结构 18
2-3 数据存储器结构 19
2-4 中断控制单元 24
2-5 定时器/计数器控制单元 26
2-6 输入/输出控制单元 29
2-7 PWM 输出接口 32
2-8 I^{2}C 串行接口 33
2-9 模/数转换器 38
2-10 WDT：看门狗定时器 41
2-11 复 位 42
2-12 省电模式 44

2-13 低电压复位 45

2-14 配置选项 46

第 3 章　HT46xx 指令集与开发工具

3-1 HT46xx 寻址模式与指令集（Instruction Set） 48

3-2 程序的编辑 67

3-3 HT-IDE3000 使用方式与操作 74

3-4 VPM 使用方式与操作 89

3-5 烧录器操作说明 93

第 4 章　基础实验篇

4-1 LED 跑马灯实验 97

4-2 LED 霹雳灯查表实验 101

4-3 七段显示器控制实验 107

4-4 指拨开关与七段显示器控制实验 111

4-5 按键控制实验 114

4-6 步进电机控制实验 118

4-7 4×4 键盘控制实验 125

4-8 喇叭发声控制实验 132

4-9 Timer/Event Counter 控制实验 137

4-10 Timer/Event Counter 中断控制实验 142

4-11 A/D 转换器控制实验 151

4-12 外部中断控制实验 156

4-13 PWM 接口控制实验 163

4-14 WDT 控制实验 168

4-15 "HALT Mode" 省电模式实验 173

4-16 I²C 串行接口控制实验 180

第5章 进阶实验篇

- 5-1 直流电机控制实验 197
- 5-2 马表——多颗七段显示器控制实验 201
- 5-3 静态点矩阵 LED 控制实验 207
- 5-4 动态点矩阵 LED 控制实验 211
- 5-5 LCD 字形显示实验 218
- 5-6 LCD 自建字形实验 234
- 5-7 LCD 与 4×4 键盘控制实验 240
- 5-8 LCD 之 DD/CG RAM 读取控制实验 244
- 5-9 LCD 的 4 位控制模式实验 252
- 5-10 比大小游戏实验 258
- 5-11 中文显示型 LCD 控制实验 266
- 5-12 半矩阵式（Half-Matrix）键盘与 LCD 控制实验 272
- 5-13 HT46xx I^{2}C Mater-Slave 传输实验 282

第6章 实践应用篇

- 6-1 专题一：数字温度计 302
- 6-2 专题二：密码锁 316
- 6-3 专题三：具记忆功能的密码锁（I^{2}C E^{2}PROM） 328
- 6-4 专题四：24 小时时钟 354
- 6-5 专题五：猜数字游戏机 367
- 6-6 专题六：逻辑测试笔 379
- 6-7 专题七：频率计数器（Counter）的制作 400
- 6-8 专题八：简易信号产生器的制作 411
- 6-9 专题九：复频信号（DTMF）产生器的制作 422
- 6-10 专题十：简易低频电压－频率转换器(VCO)的制作 .. 434
- 6-11 专题十一：简易声音调变器的制作 441
- 6-12 专题十二：RS-232 串行传输 450

附 录

- A　HT46xx 指令速查表 470
- B　HT46xx 家族程序存储器映射图 471
- C　HT46xx 家族数据存储器与特殊功能寄存器 472
- D　HT46xx 特殊功能寄存器速查表 473
- E　HT46xx 重置后的内部寄存器状态 474
- F　LCD 指令速查表 475
- G　本书常用子程序一览表 476

第1章

HT46xx 系列单片机简介

本章除了介绍单片机的特点及目前发展趋势之外，也将针对盛群半导体公司所生产的 A/D 类单片机——HT46xx 家族成员进行介绍。本章的内容包括：

1-1　单片机介绍及其未来趋势

1-2　HT46xx单片机的特点介绍

1-3　HT46xx家族介绍

1-4　HT46xx硬件引脚功能描述

1-5　复位引脚($\overline{\text{RES}}$)

1-6　输入/输出引脚(PA、PB、PC、PD、PF)

1-1 单片机介绍及其未来趋势

单片机(Microcontroller Unit, MCU)与微处理器(Microprocessor Unit, MPU)最基本的差别是单片机包括ROM或Flash存储器,并可编程设计、存储用户赋予的指令。越来越多的微处理器被应用在控制领域,由于嵌入式处理器(Embedded Processor)与Embedded Microcontroller功能相近(例如,数码照相机的影像控制芯片就导入了MCS-51与R3000芯片核心,也有厂商采用IBM公司的PowerPC MPU作为数字摄影机的内部处理器),因此"单片机"与"微处理器"已经越来越难以界定!单片机除了包括ROM或Flash 存储器的基本配备之外,近些年来单片机制造厂商更是将一些常用的外围元件,如A/D、D/A、Timer、PWM、串行端口等,集成到MCU内部,更扩展了单片机的应用领域。

在集成趋势发展之下,单片机核心集成多项功能以及提高存储器(RAM、ROM)容量,已经成为客户的基本需求,内置Flash存储器逐渐成为产品的主流。另外,将多媒体外围集成于单片机也是一个开发趋势,应用上包括数码照相机、PDA、打印机、影像处理设备与高速存取设备等。而单片机搭配上DSP(Digital Signal Processor),强化处理器运算效能,也是另一种技术导向。随着应用范围日益扩大,汽车已逐渐成为单片机应用的主流,例如安全气囊、雨刷等设备,都已逐步采用单片机来控制。高级车种上所采用的单片机数目也越来越多,从车体控制、安全气囊、电动舱到后视镜等,一般估计一辆汽车所使用的单片机大约在18个以上。在高价位的汽车,如BMW 7系列上,甚至使用高达80个以上的单片机。另外,IC卡也是颇具前途的应用,消费性产品也仍是各种单片机的主要应用领域。因此,单片机的应用领域十分广泛,从汽车、家电、IA、PC外围、显示器到通用市场,单片机均无所不在。

虽然2001年全球单片机市场有些下挫,但是依据SEMICO ReseaRCh公司(http://www.semico.com)统计,2002年8位单片机的产值达45亿美元,2003年则更上一层楼达到51亿美元,增长12.2%。从全球单片机市场来看,8位单片机占总出货量的60%,占单片机总产值的43%。表1-1-1是台湾工研院经资中心针对2002年全球各地MCU市场的调查报告,将其摘录供读者参考。

表1-1-1 2002年MCU的应用领域及全球各地区市场规模 单位:百万美元

应用领域	美洲	欧洲	日本	亚太	合计
消费性	131	234	1 357	1 151	2 873
车用	842	936	693	187	2 658
资讯	37	19	505	318	879
工业用	515	1 105	206	534	2 360
通信	84	131	122	140	477
航天与军事	47	56	0	9	112
合计	1 656	2 481	2 883	2 340	9359

注 资料来源:WSTS,IC Insights(2003/02);台湾工研院经资中心ITIS计划(2003/04)。

16位单片机的应用领域，主要是在工业控制应用与汽车市场，目前仍由欧美以及日系IC设计业者掌握。目前16位单片机与8位单片机还有相当幅度的价差，新的应用领域也仍在开发，根据业界统计，在2005年之前，8位的MCU仍是单片机产品的主流，而且未来必将朝向低电压、高效能发展。SEMICO ReseaRCh公司的市场调查报告指出，"相对于16位与32位，8位单片机虽然是属于比较低端的产品，但它的应用广泛，而且单片机并不需要先进的工艺水平，大部分8位单片机还是采用0.5 μs工艺水平，所以虽然它不再具有高度的成长性，但仍保有相当不错的获利能力"(请参考图1-1-1)。因此，虽然市场竞争激烈，仍有许多公司投入此市场(请参考表1-1-2)。

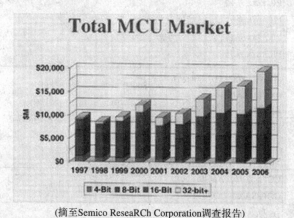

(摘至Semico ReseaRCh Corporation调查报告)
图1-1-1　4位、8位、16位和32位全球单片机市场预估

目前市面上单片机的种类繁多，如Freescale公司的68HC系列、Hitachi公司的H8系列、TI公司的MSP430系列、Microchip公司的PIC系列以及ADI公司的ADμC83x系列。由于工业自动化对单片机的需求以致于单片机发展迅速，台湾地区也有多家厂商加入单片机的制造行列，如盛群、民生、义隆、太欣、华邦、旺宏、茂硅、联电、硅成、联咏及凌阳等，产品线仍以4位及8位单片机为主。Freescale公司拥有8~32位的完整单片机产品线，但主力集中在8位与16位单片机，其中8位单片机的市场占有率为全球第一，16位排名第三。Freescale的8位单片机从过去的68HC11、68HC05转换到Flash的68HC08与68HCS08，68HCS08采用0.25μm工艺，频率则提升到20~40 MHz，多数应用于消费性电子产品。为了让越来越多消费性电子产品使用单片机，并且适应低功耗的需求，Freescale公司于2003初从3 V、5 V产品推进到2 V产品，低功耗特性将更适用于手持式设备。Freescale公司是全球目前汽车电子领域单片机市场占有率最高者之一，由于拥有齐全的产品线而且较早投入Flash单片机，所以具备技术成熟、产品稳定性高的优势，在以Flash工艺降低成本后，也逐渐提升了价格上的竞争力。

Microchip公司的单片机从1993年上市以来，迄今已累计超过20亿片的出货量。Microchip公司的单片机以8位为主，全球单片机市场占有率排名第二，主要用于汽车、消费性电子、计算机及PC外围、电信与通信、办公室自动化、工业控制等领域，其中汽车电子市场多在欧美地区，而亚太地区则以消费性电子为主。Microchip公司于2001年起将所有单片机产品线转为Flash型单片机，在功能上可让用户重复读/写，而且价格也较OTP型便宜10%~15%。除了已经集成的各种外围接口之外，目前Microchip公司正为单片机加入RF、Speech接口等功能。

表1-1-2　单片机主要供应商

公司	单片机型号		
	8 位	16 位	32 位
Freescale	M6801、M6802、M6809 M68HC11、M68HC12 Nitron 系列： 68HC908QT1、68HC908QT2 68HC908QT4、68HC908QY1 68HC908QY2、68HC908QY4	M68HC16 Flash 类 HC9S12A256、 HC9S12A128	M680X0 M6833X ColdFire 系列
Microchip	PIC12xx 系列 PIC16xx 系列 PIC17xx 系列 PIC18xx 系列		
Intel	80C51、80C151、80C251 8XC151SA/SB 8XC251SA/SB/SP/SQ	80960 系列 80296SA	386EX 486GX
TI	SE370CX、TMS370CX MSP430PX、MSP430CX TUS2140B、MSP430C1101		
NS	COPx 系列、Apollo COP8SAX、COP912		
DALLAS	DS87C550、DS80C310/320 DS87C520/530、DS83C530 DS80C323、DS1050		
Atmel	AT89SC、AT90SC T48Cx9x/M44Cx9x		
Zilog	Z86L97 系列 Z86L972、Z86L973、Z86L974		
NEC	78K0/Kx1	V850ES/Kx1	V850ES/Fx2
Toshiba	TMP87XX29U 系列 TMP88 系列 TMP86 系列	TMP91 系列 TMP93 系列 TMP95 系列	TMP94 系列
Mitsubishi	7630/7632、7532/36 7640、Slim740 系列	M16C/62、M16C/63 7902	M32R/D 系列 M32R/E 系列 M32R/I 系列
Epson	E0C88 系列		E0C33xx 系列
Hitachi		AE45x -B 系列	SH-1、SH-2、SH-3 SH3-DSP、SH-4
OKI	MSM80C 系列 MSM83C 系列 MSM85C 系列 MSM65 系列	MSM66 系列 ML66 系列	ML67 系列
Philips	80C51x2 系列	UBA2050/51 系列	
ST	ST62/63/7/7290/92 μPSD3200 系列	ST10	ST20

近一两年半导体产业并不景气，然而TI公司的单片机产品仍有相当幅度增长。MSP430超低功耗单片机家族，是以16位RISC处理器为基础所发展的超低功耗单片机，利用TI公司在高效能模拟和数字技术的优势，不但提供超低功耗，也让系统设计人员同时连接模拟信号、传感器和数字器件到同一电路中。MSP430家族阵容非常齐全，从0.49美元低端元件到高集成度产品，包括60 KB快闪存储器、2 KB RAM、高效能模/数转换器、数/模转换器以及其他多种高效能模拟与数字模块，全都集成至单片芯片内。MSP430提供极低的功耗和低于6μs快速启动能力，可在6 μs内从待命模式转换到正常操作模式，而待命模式最低只需0.7μA的电流，减少产品对电池的需求。

ADI公司于1999年开始涉足单片机市场，主要以8位单片机配合12~24位的ADC(Analog to Digital Converter)为主，其他如DAC(Digital to Analog Converter)及更大容量的Flash都集成到最新的产品中，应用领域主要在工业控制、测量仪器中的数据采集等。ADI的单片机以Intel 8051为核心，最新的ADμC83x系列内置的存储器已达62 K，更大的存储器容量将方便用户直接以C语言编写程序代码。

Atmel公司的AT89SC为8位单片机，可应用在智能卡上；AT90SC采用AVR CPU，也是8位RISC结构。日前推出MARC4，较流行的单片机产品如T48Cx9x/ M44Cx9x，它们的工作温度范围宽，适用于各种汽车及工业产品，如车辆引擎部分或车体电子系统(轮胎压力监视、冷却风扇控制、加热或车窗集成天线)的电子控制单元及工业传感器产品。

盛群半导体公司(Holtek)鉴于IC市场竞争将越来越激烈，近年来从消费性电子设计公司成功转型为专业单片机设计公司，专注于通用型与嵌入式单片机开发。除了消费性、计算机外围、通信领域的嵌入式单片机外，也提供I/O、LCD、A/D、RF及A/D LCD等通用型单片机。该公司的定位是以单片机为核心技术的IC设计公司，不同于台湾地区其他单片机制造商，其营销网络遍及全球，涵盖欧洲、北美、中南美洲等地，产品线广泛，其消费性产品用的单片机在德国获得飞利浦家电的采用。该公司是台湾地区最早推出符合工业规范的单片机设计公司。目前以提供8位的OTP与Mask型单片机为主，未来则朝向可重复读/写的E^2PROM 单片机发展，在技术层次上将足以赶上国外厂商。

盛群半导体公司的产品线相当完整，其主要产品请参考表1-1-3，8位单片机有十余项不同应用领域的专用产品，用户可以根据自己需求挑选最适用的单片机，以达到降低生产成本的最终目的。本书将以A/D类MCU——HT46xx家族为主，希望通过本书的介绍让读者能对此系列的单片机有所认识。由于盛群半导体公司MCU系列的兼容性很高，若能彻底了解HT46xx家族的微体系结构，想要再学习其他系列的单片机必定有事半功倍的效果。

1-2 HT46xx单片机的特点介绍

HT46xx系列是盛群半导体公司推出的8位A/D型单片机。此系列IC采用先进的COMS技术制造，因此具有低功耗、高执行速度的特性。其包括看门狗定时器、可编程计数器、ADC、PWM输出接口、I^2C-Bus接口、省电模式(HALT Mode)、低电压自动复位(Low Voltage Reset,

LVR)电路及双向I/O等强大功能,因此获得工业界的青睐。 HT46xx系列家族成员请参考表1-2-1。

表1-1-3 盛群半导体公司主要产品一览表

8位 MCU	Display Driver	Memory
Cost-Effective I/O 类 MCU 系列	RAM Mapping LCD Controller & Driver 系列	Mask ROM 系列
I/O 类 MCU 系列	Telephony LCD Driver 系列	OTP EPROM 系列
LCD 类 MCU 系列	VFD Controller & Driver 系列	SPI OTP EPROM 系列
A/D 类 MCU 系列	Dot Character VFD Controller & Driver	3-wire E^2PROM 系列
A/D with LCD 类 MCU 系列	OLED Driver 系列	I^2C E^2PROM 系列
R-F 类 MCU 系列		
Remote 类 MCU 系列		
Phone Controller MCU 系列		
Dot Matrix LCD MCU 系列		
Data Bank MCU 系列		
Voice MCU 系列		
Music MCU 系列		
Keyboard/Mouse/Joystick MCU 系列		
Remote Controller	**Power Management**	**Voice/Music**
Remote 类 MCU 系列	30 mA Regulator 系列	Voice MCU 系列
2^{12} Encoder/Decoder 系列	100 mA Regulator 系列	Music MCU 系列
3^9 Encoder 系列	300 mA Regulator 系列	Q-Voice^{TM} 系列
3^{12} Encoder/Decoder 系列	Negative Voltage Regulator 系列	Easy Voice^{TM} 系列
3^{18} Encoder/Decoder 系列	Voltage Detector 系列	Sound Effects 系列
Learning Encoder 系列	100 mA Step-up DC/DC Converter 系列	Piano 系列
TV Remote Controller 系列	Charge Pump DC/DC Converter 系列	
RFID 系列		
Doorbell 系列		
Computer	**Communication**	**Analog**
Keyboard/Mouse/Joystick MCU 系列	Phone Controller MCU 系列	D/A Converter 系列
Mouse 系列	Dual Mode Caller ID Phone Single Chip	Amplifier 系列
Keyboard 系列	Telecom Peripheral 系列	
Multimedia 系列	Basic Dialer 系列	
16位 Audio DSP 系列	IDD Lock Dialer 系列	
Miscellaneous		
Timepiece 系列		
Clinical Thermometer 系列		
Camera Peripheral 系列		
PIR Controller 系列		
Alphanumeric Recognition 系列		

第1章 HT46xx系列单片机简介

表1-2-1 HT46xx系列家族成员

型 号	V_{DD}/V	系统时钟/MHz	程序存储器	数据存储器	I/O	定时器/计数器 8位	定时器/计数器 16位	中断 Ext.	中断 Int.	栈	A/D	PWM
HT46R47 HT46C47	3.3~5.5	0.4~8	2 K×14	64×8	13	1	—	1	2	6	9位×4 ch	8位×1
HT46R22 HT46C22	2.2~5.5	0.4~8	2 K×14	64×8	19	1	—	1	3	6	9位×8 ch	8位×1
HT46R23 HT46C23	2.2~5.5	0.4~8	4 K×15	192×8	23	—	1	1	3	8	10位×8 ch	8位×2
HT46R24 HT46C24	2.2~5.5	0.4~8	8 K×16	384×8	40	—	2	1	4	16	10位×8 ch	8位×4

注：型号中R代表OTP型程序存储器；C代表Mask型程序存储器。

HT46xx的特点如下：

(1) 工作电压：2.2~5.5 V(f_{sys}=4 MHz)；4.5~5.5 V(f_{sys}=8 MHz)。

(2) 提供1个外部中断(External Interrupt)以及1~3个内部中断(Internal Interrupt)分别为A/D、Timer/Event Counter 和 I^{2}C Bus。

(3) 具有中断与7位预除功能的16位/8位定时器/事件计数器(Timer/Event Counter)。

(4) 提供PFD(Programmable Frequency Divider)，可作为音效的创建。

(5) 63个高级功能指令，是"类精简指令(RISC-like)"微体系结构。HT46xx 单片机1个指令周期只需要4个Clock的时间，而且除了改变PC(Program Counter)值的指令与查表指令需要2个指令周期之外，其余的指令都只需要1个指令周期的执行时间。

(6) 程序存储器(Program Memory)：2 K×14(HTx47、HTx22)、4 K×15(HTx23)、8 K×16(HTx24)。

(7) 数据存储器(Data Memory)：64×8(HTx47、HTx22)、192×8(HTx23)、384×8(HTx24)。

(8) 栈深度：6层(HTx47、HTx22)、8层(HTx23)、16层(HTx24)。

(9) 采用CMOS结构，具有强大的I/O驱动能力(工作在V_{DD}为5 V时，I/O端口的拉电流约为−10 mA；灌电流约为20 mA)。当频率为4 MHz，V_{DD}为5 V时，所需的电流约为2 mA；当进入省电模式(Power-Down Mode)时，只需要1 μA的电流(未启动看门狗定时器功能时)，可以说是相当省电。

(10) 模/数转换接口：HTx23与HTx24为8个通道，精度为10位；HTx47与HTx22则为6个通道，精度为9位。

(11) 具有看门狗WDT功能，使系统更加稳定(死机时，系统具有自动恢复的功能)。

(12) PWM输出接口与I^{2}C-Bus串行接口功能。

(13) 低电压自动复位电路：在电源电压不稳或电源需要连续开关的系统中，很可能会有复位不良的问题发生，使得设计者不得不在系统中再加上一些电路以克服此问题，这显然增加了系统的成本。HT46xx系列单片机将电源下降检测功能设计在单片机内部，提供用户多一项选择。

除了上述的特点之外，盛群半导体公司也提供了相当完善的开发工具，如HT-ICE以及完整的集成开发环境(HT-IDE3000V6)等，HT-IDE3000中的软硬件仿真功能(Virtual Peripheral

Manager；VPM)更能让用户在未接硬件电路的情况下，先行验证程序的功能。有了如此完整的开发环境，除了可以节省产品的开发时间之外，更可以使初学者在短时间之内了解HT系列单片机的特性及产品开发的技巧(请参考图1-2-1)。

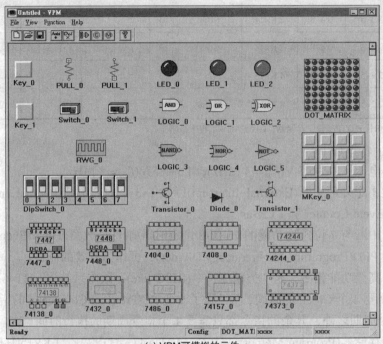
(a) VPM可模拟的元件

(b) HT-ICE(旧型)

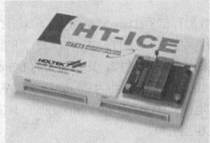
(c) HT-ICE+OTP Writer(新型)

(d) HT-写入器(RS-232接口)

(e) I/O接口卡

图1-2-1　HT46xx相关的开发工具

1-3　HT46xx家族介绍

　　HT46xx系列的家族成员约有8个,而且指令完全兼容,成员间主要的差异是程序存储器的大小与类型(Mask—HT46Cxx、OTP—HT46Rxx)、数据存储器的大小、I/O引脚总数以及外围设备的多少(如Timer、PWM、ADC)。用户可以根据所设计系统的实际需求,从中挑选出最适合自己需要的使用,请参考表1-2-1。

　　除此之外,同一型号的单片机也提供不同的引脚封装形式,让用户可以有多种选择,请参考图1-3-1。读者若需进一步了解更详细的资料,也可以到盛群半导体公司的网页上查询(http://www.holtek.com.tw)。各引脚的功能简述,请参考表1-3-1与1-4节。

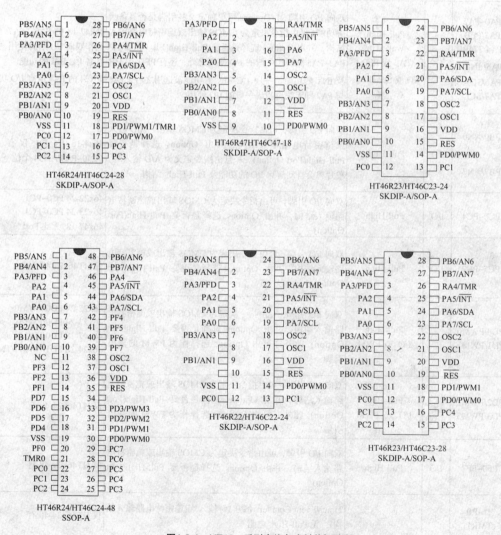

图1-3-1　HT46xx系列家族各式封装与引脚

表1-3-1　HT46xx系列家族各式封装的引脚定义

引脚名称	类型	Options	功能描述	附注
VDD	—	—	电源输入端(2.2~5.5 V) f_{SYS}=4 MHz, 2.2~5.5 V f_{SYS}=8 MHz, 4.5~5.5 V	
VSS	—	—	负电源输入端，通常接地	
OSC1 OSC2	I O	Crystal 或 RC	OSC1 及 OSC2 需根据 Options 选项的设定连接至 *RC* 电路或石英振荡器，以作为系统内部的工作时钟	
$\overline{RES}$	I	—	内为施密特电路结构，若为低电位，则单片机将停留于 RESET 状态	
PA0~PA2 PA3/PFD PA4/TMR PA5/$\overline{INT}$ PA6/SDA PA7/SCL	I/O	Pull-High Wake-Up PA3 或 PFD I/O 或 Serial Bus	双向 I/O 引脚，由 Options 可单独定义各引脚是否具备唤醒功能；可由指令决定为 CMOS 输出或施密特电路输入结构，并由 Options 选择是否要 Pull-High(Bit Option)。PA3~PA5 除可当一般的 I/O 使用之外，还有 PFD、TMR 及 $\overline{INT}$ 的功能。一旦 I²C-Bus 的功能被选用之后，PA6 与 PA7 即无法再当成一般的 I/O 使用	46x47-18 的 PA6、PA7 仅可当作双向 I/O 引脚使用，并不具备 I²C-Bus 的功能。 46x24 的 PA4 仅有 I/O 功能
PB0/AN0 ⋮ PB7/AN7	I/O	Pull-High	双向 I/O 引脚，可由指令决定为 CMOS 输出、A/D 输入或施密特电路输入结构，并由 Options 选择是否要 Pull-High(Port Option)。一旦由指令定义为 A/D 输入时，单片机会自动解除 I/O 的功能及 Pull-High 结构	46x4-18 的 PB 仅有 PA0/AN0~PA3/AN3 4 个位
PC0~PC4	I/O	Pull-High	双向 I/O 引脚，可由指令决定为 CMOS 输出或施密特电路输入结构，并由 Options 选择是否要 Pull-High(Port Option)	46x22-24 PC0~PC1 46x23-24 PC0~PC1 46x47-18 无此 Port
PC5~PC7	I/O	Pull-High	双向 I/O 引脚，可由指令决定为 CMOS 输出或施密特电路输入结构，并由 Options 选择是否要 Pull-High(Port Option)	仅 HT46x24-48 有
PD0/PWM0 PD1/PWM1	I/O	Pull-High I/O 或 PWM	双向 I/O 引脚，可由指令决定为 CMOS 输出或施密特电路输入结构，并由 Options 选择是否要 Pull-High(Port Option)。此外亦可由 Options 选择作为 PWM 的输出引脚	46x22-24 PD0/PWM0 46x23-24 PD0/PWM 46x47-18 PD0/PWM 46x24-28 PD1/PWM1/TMR1
PD2/PWM2 PD3/PWM3	I/O	Pull-High I/O 或 PWM	双向 I/O 引脚，可由指令决定为 CMOS 输出或施密特电路输入结构，并由 Options 选择是否要 Pull-High(Port Option)。此外亦可由 Options 选择作为 PWM 的输出引脚	仅 HT46x24-48 有
PF0~PF7	I/O	Pull-High	双向 I/O 引脚，可由指令决定为 CMOS 输出或施密特电路输入结构，并由 Options 选择是否要 Pull-High(Port Option)	仅 HT46x24-48 有
TMR0 TMR1	I	—	Timer/Event Counter 计数时钟输入端(施密特电路输入结构)，无 Pull-High 电阻	仅 HT46x24-48 有

注：Port Option是指必须定义整个Port的Pull-High功能；而Bit Option则表示可以定义单独Port的Pull-High功能。

1–4 HT46xx硬件引脚功能描述

1–4–1 电源引脚(V_{DD}、V_{SS})

HT46xx若工作在频率为4 MHz的环境下，直流电压范围为2.2~5.5 V；若工作在8 MHz的环境下，直流电压范围为4.5~5.5 V。工作频率为4 MHz、V_{DD}为5 V时所需的电流约为2 mA(若为RC振荡型则约为2.5 mA)；工作频率为8 MHz时所需的电流约为3 mA。V_{DD}为5 V时，当进入省电模式(HALT Mode)后，若WDT无效，只需要2 μA的电流(1 μA，当V_{DD}=3 V)；若WDT有效，也只需要10 μA的电流(5 μA，当V_{DD}=3 V)，可以说是相当省电。

1–4–2 振荡电路相关引脚(OSC1、OSC2)

HT46xx系列单片机可在RC或Crystal的振荡模式下工作(可由Options加以选择)，其电路连接方式请参考图1-4-1。在"Crystal Oscillator"振荡模式下，若f_{SYS}<1 MHz，则$C_1=C_2=300$ pF；否则C_1、C_2可以不用连接。

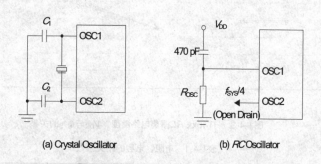

(a) Crystal Oscillator　　　　　　　　(b) RC Oscillator

图1-4-1　HT46R/Cxx振荡电路的连接方式

在RC Oscillator振荡模式下，注意R_{OSC}的值需介于30~750 kΩ之间。RC振荡模式的输出频率会受电源电压V_{DD}、电阻R_{OSC}、电容(470 pF)及温度的影响，请参考图1-4-2与图1-4-3。一般RC振荡器只要接到OSC1脚即可，而OSC2的输出为指令周期频率的方波，可以用来判断HT46xx单片机是否正常工作或者作为与外部逻辑电路同步之用。请读者注意的是OSC2为泄极开路式(Open Drain)的输出结构，如果必需使用此f_{SYS}/4的信号时，务必要记得接上上拉电阻。RC Oscillator的优点是结构简单，只要一个电阻和电容即可，因此成本极低；缺点就是振荡频率的准确度不高，容易受电源、元件、温度和湿度的影响，因此，RC振荡模式并不适用于对于工作频率要求相当精准及稳定的场合。

除此之外，RC振荡模式的工作频率还受芯片封装形式与电路板布线的影响；即使是同一型号的芯片，所产生的振荡频率也很难完全相同。但由于RC振荡模式的电路极为简单、成本相当低，在工作频率不需非常精确的场合，倒是一项节省成本的选择。由于RC振荡电路受电阻R_{OSC}与电容(470 pF)的影响很大，而一般电阻与电容的材质与其稳定度有相对的关系。特将其整理于表1-4-1，供读者参考。

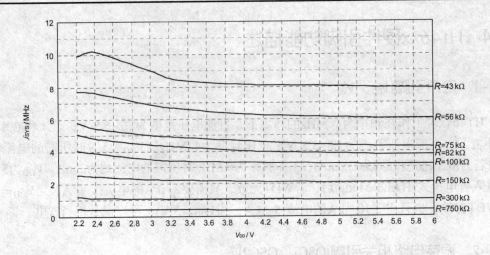

图1-4-2 HT46xx RC振荡电路 R_{OSC} 与 f_{SYS} 的关系

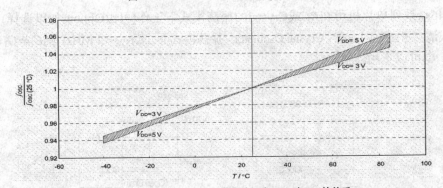

图1-4-3 HT46xx RC振荡电路温度、V_{DD} 与 f_{SYS} 的关系

表1-4-1 电阻、电容的相关系数

电 阻	碳素皮膜电阻	金属皮膜电阻	氧化金属皮膜电阻
一般误差	±5%	±1%	±5%以下
温度系数/（ppm/°C）	300	50	300
杂音指数/dB	−20~−10	−50~−10	−50~−10
电 容	陶瓷电容器	薄膜电容器	积层电容器
一般误差	10%	5%	20%
温度系数/（ppm/°C）	500	200	300
电感性	低	高	中

HT46xx单片机内部将OSC1引脚输入的时钟(即系统时钟 f_{SYS})以4个周期为单位称为指令周期(Instruction Cycle)。每一个指令周期可再细分为4个非重叠时段(T1~T4)，如图1-4-4所示，通常程序计数器PC是在T1加1，而在T4时依据PC值将指令从程序存储器调入内部的指令寄存器，至于指令的解码与执行则是在下一个指令周期的T1~T4完成。如此看来，好像执行一个指令需要花费两个指令周期的时间，其实不然。由于HT46xx单片机内部采用管线作业(Pipeline)的微体系结构，除了跳转指令与分支指令(即改变PC值的指令)之外，绝大多数指令的执行都只需花费一个指令周期的时间。请参考图1-4-5的例子。

第1章 HT46xx系列单片机简介

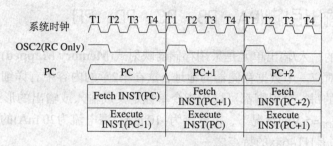

图1-4-4 Clock与指令周期的关系

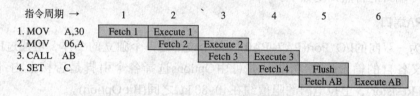

图1-4-5 HT46xx Pipeline微体系结构图例

首先在第1个指令周期时,依据PC的值在程序存储器提取指令(Fetch 1;"MOV A,30")。在第2个指令周期时,除了针对"MOV A,30"进行解码与执行之外(Execute 1),同时也将下一个指令调入指令寄存器(Fetch 2;"MOV 06,A")。在此执行期间,操作数(数据存储器)的读取是在T2进行的,而运算结果的存储则是在T4完成的。至于在执行分支及跳转指令(如CALL、JMP等)时,由于必须先将管线中的指令清空(Flush)以便存放新进的指令,所以必须用2个指令周期的时间才能完成。

1-5 复位引脚($\overline{RES}$)

HT46xx系统复位输入引脚内部为施密特触发(Schmitt Trigger)的形式。系统复位的主要目的,是让单片机回到已知的系统状态(请参考2-11节有关系统复位的内容)。当$\overline{RES}$引脚输入Low状态,HT46xx单片机随即进入复位状态,待$\overline{RES}$引脚回复High状态后,HT46xx单片机才恢复正常的工作。图1-5-1是HT46xx的$\overline{RES}$电路接法,其中电阻与电容值必须适当,以确保在$\overline{RES}$引脚回复到"High"状态之前,振荡电路已经稳定地输出了。

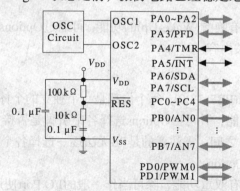

图1-5-1 HT46xx的$\overline{RES}$电路接法

1-6 输入/输出引脚(PA、PB、PC、PD、PF)

HT46xx系列的输入/输出单元是采用存储器映射式(**Memory Mapped**)的结构，也就是说每一个I/O端口都对应一个数据存储器的地址，这在第2章的内容中有详细的探讨，请读者自行参阅。HT46xx的每一个引脚都可以用指令单独定义成输入或输出的形式，其每一个引脚可输出20 mA(在V_{DD}=5 V的情况下，拉电流为–10 mA，灌电流为20 mA)的电流，如此强大的驱动能力，可以说是HT46xx的特点之一。

各I/O端口的特点归纳如下：

1. PA端口

PA为一双向的I/O Port(PA7~PA0)，每个位都是一个独立的个体，可通过PAC控制寄存器来定义各自的输入/输出功能，并可由Options选择各个引脚是否接上内部的上拉电阻(Pull-Up Resistor)，上拉电阻的阻值约在40~80 kΩ之间(Bit Option)。

PA的部分引脚具有双重功能，PA3可当成一般的I/O使用，也可以当成PFD(Programmable Frequency Divider)，通常用来驱动喇叭或Buzzer，需在配置选项(Options)选定功能。在PFD选项有效的情况下，若PA3被定义为输入模式(PAC3= "1")，则PA3仍可当成一般的输入引脚使用。若PAC3= "0"(输出模式)，此时如果设定PA3= "1"即开始输出PFD信号，其所产生的频率为定时器/计数器溢位频率的一半；设定PA3= "0"时将停止PFD信号的输出，并强迫使PA3的引脚呈现低电位的输出状态。

PA4也可当成Timer/Event Counter的输入控制(HT46x24除外)引脚。若选用计数模式时(Event Count)，此引脚即为计数脉冲的输入引脚；而当选用脉冲宽度测量模式时(Pulse Width Measurement)，待测的脉冲也是由此引脚输入。

PA5也可以作为外部中断(External Interrupt)的输入引脚，在外部中断有效的情况下，若由此引脚输入负脉冲，则代表要求CPU执行外部中断程序。

PA6与PA7尚可分别作为I^2C串行传输时的SCL(Serial Clock)与SDA(Serial Data)控制信号，此时PA6与PA7为NMOS泄极开路式(Open Drian)的输出引脚，请务必分别接上上拉电阻。详细内容请参考2-8节中"I^2C串行接口"的相关介绍。(注：HT46x47没有I^2C串行传输的功能。)

PA的每一个引脚均具有唤醒(Wake-Up)的功能，可通过Options加以选用，有关唤醒的相关介绍请参考2-12节的相关内容。

2. PB端口

PB为一双向的I/O Port(PB7~PB0)，每个位都是一个独立的个体，可通过PBC控制寄存器来定义各自的输入/输出功能，并可由Options选择各个引脚是否接上内部的上拉电阻，上拉电阻的阻值约在40~80 kΩ之间。需注意的是PB无法逐一选择各个引脚是否接上拉电阻，而是整个Port一起定义(Port Option)。

PB的每一个引脚都具有双重功能，除可当作一般的I/O Port使用外，也可作为模/数转换器的输入通道(AN0~AN7),详细内容请参考2-9节中关于"模/数转换"的介绍。请注意HT46x47的PB仅有4位(PB3/AN3~ PB0/AN0)。

3. PC端口

PC是一个单纯的双向I/O Port,其宽度只有5位(PC4~PC0;针对HT46x23-28、HT46x24-28系列),每一个引脚可以通过PCC控制寄存器单独定义其为输入或输出并可由Options选择是否接上内部的上拉电阻,上拉电阻的阻值约在40~80 kΩ之间。PC也无法逐一选择各个引脚是否要接上拉电阻,而是整个Port一起定义(Port Option)。请注意HT46x23-24与HT46x22的PC仅有2位(PC1~ PC0),HT46x47的PC并不存在,而HT46x24-48 PC的宽度为8位(PC4~PC7)。

4. PD端口

PD是一个双向的I/O Port,每一个引脚可以通过PDC控制寄存器单独定义其为输入或输出,请注意其宽度只有2位(PD1、PD0)。PD的每一个引脚都具有双重功能,除可当作一般的I/O Port使用外,也可作为PWM(Pulse Width Modulation)的输出引脚。当于Options选用PWM的功能后,若设定PD0(PD1)为"1",PWM信号即开始由PD0(PD1)输出;若设定为"0",则PWM信号停止输出,并强迫使PD0(PD1)引脚维持低电位。详细内容请参考2-7节中关于"PWM接口"的内容。请注意除了HT46x23/24-28的PD为2位、HT46x24-48的PD为4位外(PD0/PWM0~PD3/PWM3),其余的成员仅为1位(PD0/PWM)。

5. PF端口

PF是一个单纯的双向I/O Port,其宽度为8位(PF7~PF0);只有HT46x24-48具有此I/O Port,每一个引脚可以通过PDC控制寄存器单独定义其为输入或输出,并可由Options选择是否接内部的上拉电阻,上拉电阻的阻值约在40~80 kΩ之间。PD也无法逐一选择各个引脚是否接上拉电阻,而是整个Port一起定义(Port Option)。

6. TMR0和TMR1

TMR0和TMR1分别为Timer/Event Counter0与Timer/Event Counter1的输入控制(for HT46x24-48 only)引脚。若选用计数模式时,此引脚即为计数脉冲的输入引脚;而当选用脉冲宽度测量模式时,待测的脉冲也是由此引脚输入。

第 2 章

HT46xx 系列系统体系结构

本章以循序渐进的方式，对 HT46xx 的内部硬件体系结构做一番详尽的介绍。在第 1 章，针对其硬件引脚功能一一说明之后，本章对于存储器体系结构、I/O 特性、定时器/计数器(Timer/Event Conter)以及其他外围设备(如 ADC、PWM、PFD、WDT、I²C 串行接口)，都将有完整的说明。读者在阅读完本章之后，除了对 HT46xx 家族有更完整的认识之外，相信对于以后各章的硬件设计及程序编写必能更加得心应手。建议读者阅读本章时能够结合第 4 章的基础实验同步进行，将理论与实验相互结合学习，必定有事半功倍的效果。本章的内容包括：

- 2-1　HT46xx的内部体系结构
- 2-2　程序存储器结构
- 2-3　数据存储器结构
- 2-4　中断控制单元
- 2-5　定时器/计数器控制单元
- 2-6　输入/输出控制单元
- 2-7　PWM输出接口
- 2-8　I²C串行接口
- 2-9　模/数转换器
- 2-10　WDT：看门狗定时器
- 2-11　复　位
- 2-12　省电模式
- 2-13　低电压复位
- 2-14　配置选项

第 2 章 HT46xx 系列系统体系结构

2-1 HT46xx的内部体系结构

HT46x23的内部结构如图2-1-1所示,它主要是由程序存储器(Program ROM,4 096×15)、数据存储器(Data RAM,192×8)、特殊功能寄存器(TMRC、INTC0等)、输入/输出端口(PA、PB、PC、PD)、定时器/计数器(TMR)、8通道的模/数转换器、PWM输出接口与看门狗定时器WDT等组成。其中,程序存储器的宽度是15位,而数据存储器的宽度是8位。

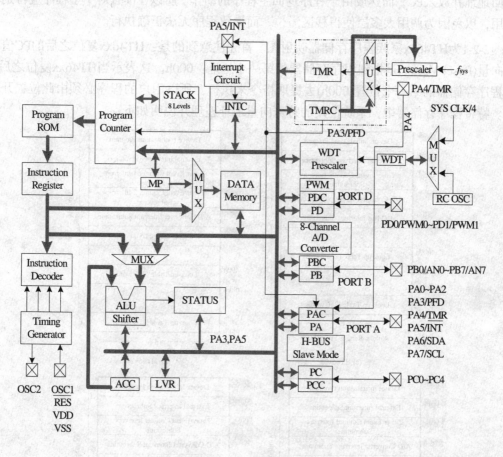

图2-1-1　HT46x23的内部结构图

关于程序的执行,首先是根据程序计数器PC值,在程序存储器中提取指令码(Op-Code)并存放到指令寄存器(Instruction Register),然后通过指令解码器(Instruction Decoder)解码之后配合时序产生电路(Timing Generator),产生一连串的硬件控制信号依序控制各个单元。

至于HT46xx家族中的其他成员,如HT46x22、HT46x24、HT46x47等,其结构都差不多。HT46R23/HT46C23算是家族中功能相当齐全的成员,因此本书将以HT46x23作为介绍的重点。

2-2 程序存储器结构

HT46x23的程序存储器大小为4 096(000h~FFFh)×15位。程序存储器主要是用来存放程序与表格数据，配合程序计数器PC来选择下次所要执行的指令地址。栈(Stack)则是另一个与程序执行流程相关的单元，它记录了程序的返回地址。值得提出说明的是：HT46x23的栈为8层，HT46x22、HT46x47的栈只有6层，而HT46x24为16层。当执行到CALL指令时，CPU会先将目前的PC值存放在栈寄存器，然后将所调用的子程序地址放入PC，如此便完成了调用子程序的跳转动作；而当执行到RET、"RET A,x"与RETI指令时，CPU则是从栈寄存器中取出返回地址并放入PC，而达成由子程序返回主程序的动作。所以，读者对于子程序应特别小心使用，以免因为调用太多层使得栈区溢位，而导致程序无法正确执行。

图2-2-1为HT46xx家族程序存储器映射图。首先注意到的是：HT46xx复位之后的PC值即复位向量(Reset Vector)是指到程序存储器的第一个地址：000h。这表示当HT46xx复位之后，会到程序存储器的第一个位置(000h)去提取指令来执行，所以用户的程序必须由此位置开始存放。特将程序存储器中，系统使用的特殊向量地址整理并说明如下：

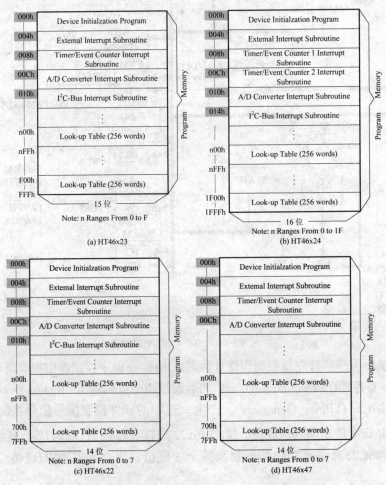

图2-2-1　HT46xx家族程序存储器映射图

第2章 HT46xx系列系统体系结构

(1) 000h(系统复位向量)

当单片机复位时,PC=000h,因此CPU会由此地址开始提取指令执行,所以用户的程序必须由此位置开始存放。

(2) 004h(外部中断向量)

如果在外部中断已经有效且栈寄存器还有空间的情况下,若产生外部中断,则CPU将跳至此地址执行程序。

(3) 008h(定时器/计数器中断向量)

在定时器/计数器中断已有效且栈寄存器还有空间的情况下,若产生定时器/计数器溢位的情形,则CPU将跳到此地址执行程序。(请注意:008h为HT46x24的Timer/Event Count 0中断向量。)

(4) 00Ch(A/D转换中断向量)

在A/D转换中断已有效且栈寄存器还有空间的情况下,若模/数转换器完成数据的转换时,则CPU将跳至此地址执行程序。(请注意:00Ch为HT46x24的Timer/Event Count 1中断向量。)

(5) 0010h(I^2C BUS中断向量)

在I^2C BUS中断已有效且栈寄存器还有空间的情况下,若I^2C的设备选择码(Device Select Code)相符或完成一个字节的数据传输时,则CPU将跳到此地址执行程序。(请注意:010h为HT46x24的A/D转换中断向量,其I^2C BUS中断向量在014h。HT46x47无I^2C BUS串行传输功能。)

由以上的说明中可以发现HT46xx的一项特点:即使中断已经有效,HT46xx在进入中断服务之前,会先检查栈寄存器是否仍有空间存放返回地址。如果栈寄存器已经放满了,则将暂时不执行目前的中断服务子程序,待栈寄存器有空间存放返回地址时再去执行。如此的设计,可以排除部分因为栈溢位(Stack Overflow)而导致程序无法正常执行的状况。

如前所述,程序存储器除了存放程序代码之外,也可用来存放表格数据,例如七段显示器的显示码、点矩阵的字形码及字串数据等;也可配合特殊的查表指令(TABRDC [m]、TABRDL [m]),则每笔数据的宽度可达15位(HT46x23)、16位(HT46x24)或14位(HT46x22、HT46x47)。至于查表指令的用法,请读者参考第3章的指令说明。

2-3 数据存储器结构

HT46x23的数据存储器为224×8位,其中有192×8位可以当成一般的数据存储器使用。图2-3-1(a)即为HT46x23数据存储器的结构图,其中00h~27h为特殊功能寄存器,因其内含值牵涉到HT46x23的内部运作与外围设备的特性,使用时要特别小心;40h~FFh则是一般用途寄存器,用户可以自由使用;在00h~3Fh间还有几个保留的地址,这些空间是保留给将来新型号的单片机使用,当读取这些保留区时将读回00h。

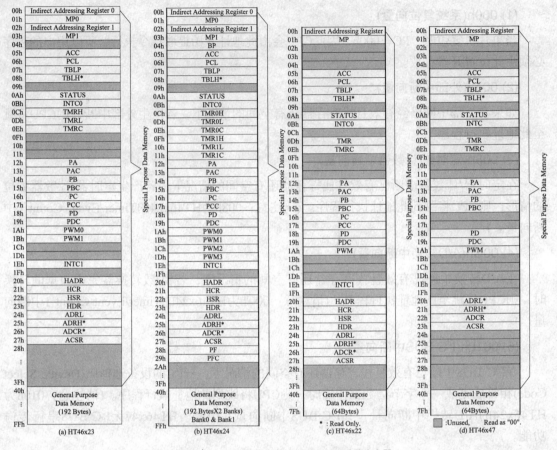

图2-3-1 HT46xx数据存储器与特殊功能寄存器

　　HT46x24的一般用途寄存器共有两个Banks，可以通过BP(Bank Pointer)控制寄存器进行切换，所以其一般用途寄存器为384×8位(参考图2-3-1(b))。至于HT46xx家族其他成员的数据存储器结构，请参考图2-3-1(c)与图2-3-1(d)。以下将以HT46x23为主体，说明每一个特殊功能寄存器的功能。其实这些特殊功能寄存器可说是单片机的精髓，想要对芯片操控自如或发挥其最大的效能，一定要先彻底了解这些寄存器的特性。但对初学者而言，刚开始就要记住这么多不同功能的寄存器(而且各个位又有不同的控制功能)，难免会有"雾里看花、越看越花"的感觉，导致对单片机的学习失去信心及兴趣。在此强烈建议初学者，结合第4章的实验内容，以循序渐进、边做边学的方式逐一了解各个特殊功能寄存器的作用，如此方能坐收事半功倍的效果。

2-3-1 IAR0和MP0

　　◇ IAR0(00h)：间接寻址寄存器(Indirect Addressing Register 0)；
　　◇ MP0(01h)：数据存储器指位器0(Memory Pointer 0)。

　　IAR0是一个8位的间接寻址寄存器，它主要是配合MP0寄存器(Memory Pointer 0；01h)来完成间接存取数据存储器的工作。在HT46R23/C23内部实际上并没有IAR0寄存器，但指令中若以IAR0作为操作数时，其实HT46x23的动作是将MP0寄存器内所存放的内容当成数据存储器的地址，然后针对该地址内的数据执行所指定的运算，此即所谓"间接寻址法(Indirect Addressing Mode)"。下面用例子来说明：

例：将40h~4Fh(共16字节)数据存储器内容清除为0。

```
1.        INCLUDE  "HT46R23"
2.        MOV      A,40h        ;将欲清除的寄存器起始地址
3.        MOV      MP0,A        ;存入MP0(01h)中
4.        MOV      A,16         ;设定A=16
5. LOOP:  CLR      IAR0         ;清除MP0所指定的寄存器地址
6.        INC      MP0          ;MP0加1,指向下一个地址
7.        SDZ      ACC          ;ACC=ACC-1
8.        JMP      LOOP         ;若ACC≠0,继续执行下一个地址
9.        ...
```

请读者注意的是，每一个特殊功能寄存器都占据数据存储器的一个地址，理当以这些地址来做数据的存取。但是光看地址实在很难联想起该寄存器的功能及作用，因此盛群半导体公司在设计开发工具时，特别将特殊功能寄存器的名称与其在数据存储器中所占地址的对应关系定义在HT46R23.INC文件中。用户只要在程序开头先以INCLUDE指令将此定义文件载入，就可以直接使用特殊功能寄存器的名称来编写程序，增加程序的可读性及编写时的方便性，而本例中第一行指令的目的就在此。请特别注意第5行"CLR IAR0"指令，其动作并不是将00h的寄存器地址内容清除。如前述的说明：指令中若以IAR0作为操作数时，HT46xx的动作是将MP0寄存器内所存放的内容当成数据存储器的地址，然后针对该地址内的数据执行所指定的运算。当第1次执行第5行时，MP0=40h，所以40h的地址内容被清0；同理，当(ACC)-1≠0，第2次执行第5行时，MP0=41h，所以41h的地址内容被清0；其余以此类推。此段程序中利用ACC(累加器)作为计数器，控制所要清除的寄存器位置数目；MP0则当成位置指针，记录所要清除的存储器地址。

2-3-2　IAR1和MP1

◇ IAR1(02h)：间接寻址寄存器1(Indirect Addressing Register 1)；
◇ MP1(03h)：数据存储器指位器1(Memory Pointer 1)。

HT46x23提供了两组寄存器作为间接寻址之用，IAR0与MP0是其中一组，IAR1与MP1是另外一组。指令中若以IAR1作为操作数时，HT46xx的动作是将MP1寄存器内所存放的内容当成数据存储器的地址，然后针对该地址内的数据执行所指定的运算。至于详细的运用方式，请读者参阅2-3-1小节中的说明。

注意：除了HT46x23与HT46x24之外，其他成员之间接寻址寄存器只有一组，分别称为IAR(Indirect Addressing Register)与MP(Memory Pointer)。

2-3-3　BP

HT46x24的一般用途寄存器(40h~FFh)共有两个Banks，可以通过BP(Bank Pointer；04h)控制寄存器进行切换(见图2-3-1(b))。BP(for HT46x24 only)用来选择地址40h~FFh所对应的

Bank，对于00h~3Fh的地址并无任何作用。图2-3-2为HT46x24的BP控制寄存器描述。

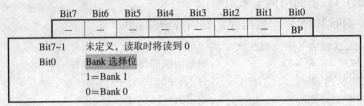

图2-3-2　HT46x24的BP控制寄存器

2-3-4　ACC累加器

ACC(Accumulator；05h)累加器是所有寄存器中唯一可以执行常数值运算的寄存器，所以凡是立即数(Immediate Data)的运算以及寄存器间的数据传输，都必须通过ACC寄存器来完成。

2-3-5　PCL程序计数器的低8位

HT46x23的程序计数器为12位，而其中的低8位可以当成一般的寄存器作为指令的操作数。程序计数器的内含值代表CPU下一个要执行指令的存放地址，因此，通过PCL(06h)的变化可以达到程序跳转的目的，这就是一般所谓的"计算式跳转(Computational Jump)"。请看以下的应用程序：

```
1.  ADDM   A,PCL      ;将 PCL 与 ACC 相加,结果存回 PCL
2.  JMP    CASE0      ;若 ACC=0,跳至 CASE0
3.  JMP    CASE1      ;若 ACC=1,跳至 CASE1
4.  JMP    CASE2      ;若 ACC=2,跳至 CASE2
5.  ……
```

如上例，是根据不同的ACC值控制CPU跳转至不同的程序地址(CASE0(A=0)、CASE1(A=1)、CASE2(A=2)、…)，有点类似C程序语言中SWITCH与CASE指令。但是要注意的是，由于芯片内的算术与逻辑运算电路均只有8位，因此当"ADDM A,PCL"加法指令产生进位时，并不会将进位累进至程序计数器的高4位，也就是说计算式跳转仅限于同一个程序页的范围。所以，在运用此类计算式跳转时，请特别注意有进位发生时的情况。

2-3-6　TBLP表格指位器

HT46xx提供两个查表指令："TABRDL [m]"和"TABRDC [m]"，方便用户在查表时的使用，TBLP(07h)代表所要查询数据所在的地址。以HT46x23为例，程序存储器共有4 096个地址(000h~FFFh)，但TBLP寄存器只有8位，因此，HT46xx将程序存储器以256个位置为单位称为一个程序页(Program Page)。当使用"TABRDC [m]"指令时，依据TBLP所指示的地址将目前程序页的内容读出，并将低8位存入数据存储器m地址中；当使用"TABRDL [m]"指令时，根据TBLP所指示的地址将程序最后一页(F00h~FFFh)的内容读出,并将低8位存入数据存储器m地址中。总结查表位置与TBLP的关系如表2-3-1所列。

表2-3-1 查表位置与TBLP的关系(HT46x23)

指令	查表数据所在地址											
	*11	*10	*9	*8	*7	*6	*5	*4	*3	*2	*1	*0
TABREC [m]	P11	P10	P9	P8	@7	@6	@5	@4	@3	@2	@1	@0
TABRDL [m]	1	1	1	1	@7	@6	@5	@4	@3	@2	@1	@0

表2-3-1中的*11~*0代表读取数据所在位置,@7~@0则是TBLP寄存器的内含值,P11~P8则代表目前程序计数器高4位的值。

至于HT46x22与HT46x47,程序存储器共有2 048个地址(000h~7FFh),其查表位置与TBLP的关系如表2-3-2所列。

表2-3-2 查表位置与TBLP的关系(HT46x22、HT46x47)

指令	查表数据所在地址										
	*10	*9	*8	*7	*6	*5	*4	*3	*2	*1	*0
TABREC [m]	P10	P9	P8	@7	@6	@5	@4	@3	@2	@1	@0
TABRDL [m]	1	1	1	@7	@6	@5	@4	@3	@2	@1	@0

表2-3-2中的*10~*0是代表读取数据的所在位置,@7~@0则是TBLP寄存器的内含值,P10~P8则代表目前程序计数器高3位的值。

HT46x24的程序存储器共有8 192个地址(0000h~1FFFh),其查表位置与TBLP的关系如表2-3-3所列,其中*12~*0是代表读取数据的所在位置,@7~@0则是TBLP寄存器的内含值,P12~P8则代表目前程序计数器高5位的值。

表2-3-3 查表位置与TBLP的关系(HT46x24)

指令	查表数据所在地址												
	*12	*11	*10	*9	*8	*7	*6	*5	*4	*3	*2	*1	*0
TABREC [m]	P12	P11	P10	P9	P8	@7	@6	@5	@4	@3	@2	@1	@0
TABRDL [m]	1	1	1	1	1	@7	@6	@5	@4	@3	@2	@1	@0

2-3-7 TBLH表格数据的高位

HT46x23的程序存储器为4 096×15位、HT46x24为8 192×16位、HT46x22与HT46x47为2 048×14位,然而所提供的查表指令("TABRDL [m]"及"TABRDC [m]")则只是将8位的查表结果存放在地址m的数据存储器。为了解决表格数据为15位(或16位、14位)而数据存储器只有8位的情形,HT46xx是将读回的低8位置于指令所指定的寄存器(m)中,而其他的位则固定地存于TBLH(08h)寄存器内。提醒读者的是,TBLH是一个只读的寄存器,无法用"TABRDL [m]"和"TABRDC [m]"以外的指令来更改其值。因此,最好避免在主程序与中断服务子程序中同时使用查表指令,如果无法避免,最好在查表指令之前先禁止中断的发生,待TBLH的值存放到适当寄存器之后再将其有效,以免发生TBLH在中断服务子程序中被破坏的情形。

2-3-8 STATUS状态寄存器

状态寄存器(Status Register；0Ah)由反应指令执行结果的状态标志位(C、AC、Z、OV)和系统状态标志位(PDF、TO)组成，如图2-3-3所示。

Bit7	Bit6	Bit5	Bit4	Bit3	Bit2	Bit1	Bit0
—	—	TO	PDF	OV	Z	AC	C

Bit7~6	未定义，读取时将读到0
Bit5	TO：看门狗定时器(WDT)状态位 1：看门狗定时器溢位时 0：Power-on、执行"CLR WDT"或"HALT"指令之后
Bit4	PDF：省电状态位(Power-Down Flag) 1：执行"HALT"指令后 0：Power-on 或执行"CLR WDT"指令之后
Bit3	OV：溢位标志位(Overflow Flag) 1：若执行运算时造成 Bit7 进位而 Bit6 没有进位；或 Bit6 进位而 Bit7 没有进位，则表示发生溢位,OV 被设定为 1 0：若执行运算时造成 Bit7 与 Bit6 都有进位；或 Bit7 与 Bit6 都无进位，则表示未发生溢位,OV 被设定为 0
Bit2	Z：零标志位(Zero Flag) 1：当执行算数/逻辑指令后的结果等于 0 时 0：当执行算数/逻辑指令后的结果不为 0 时
Bit1	AC：辅助进/借位标志位(Auxiliary Carry Flag) 加法指令(ADD、ADC、ADDM、ADCM) 1：执行加法指令后，低 4 位(Low Nibble)有进位时 0：执行加法指令后，低 4 位(Low Nibble)无进位时 减法指令(SUB、SBC、SUBM、SBCM) 1：执行减法指令后，低 4 位(Low Nibble)无借位时 0：执行减法指令后，低 4 位(Low Nibble)有借位时
Bit0	C：进/借位标志位(Carry Flag) 加法指令(ADD、ADC、ADDM、ADCM) 1：执行加法指令后产生进位时 0：执行加法指令后没有进位时 减法指令(SUB、SBC、SUBM、SBCM) 1：执行减法指令后无借位时 0：执行减法指令后有借位时

图2-3-3 HT46xx的状态寄存器

2-4 中断控制单元

◇ INTC0(0Bh)：中断控制寄存器0(Interrupt Control Register 0)；
◇ INTC1(1Eh)：中断控制寄存器1(Interrupt Control Register 1)。

HT46x23提供了4种不同的中断：外部中断、定时器/计数器中断、A/D转换中断以及I^2C BUS中断，而其结构是属于可屏蔽中断(Maskable Interrupt)，也就是当有中断要求产生时，CPU不一定会跳到相关的中断向量地址去执行中断服务子程序(Interrupt Service Routine；ISR)，需视内部相关控制位的设定而定，而INTC0与INTC1就是用来控制这些中断是否有效的开关，如图2-4-1所示。

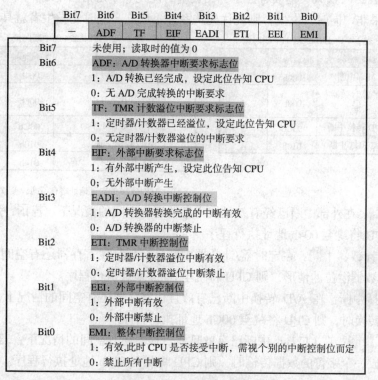

(a) HT46x23的INTC0控制寄存器

(b) HT46x23的INTC1控制寄存器

图2-4-1　HT46x23的INTC0/INTC1控制寄存器

　　INTC0与INTC1中断控制寄存器的各个位可以分为两大类。其一是用来反映是否有中断要求发生的状态标志位，如HIF、ADF、TF与EIF；另一类则是用来控制CPU是否接受中断的控制位，如EHI、EADI、ETI、EEI与EMI。当中断发生时，相对应的状态标志位会被设定为1，待进入中断程序执行完中断服务子程序后，HT46x23会自动清除该标志位，代表该次的中断要求事件已经处理完毕。

　　EMI位可视为HT46x23是否接受中断的控制总枢纽，当EMI位为0时，表示HT46x23将不接受任何中断(不论EHI、EADI、ETI、EEI位的设定为何)；反之，当EMI位为1时，则需视EHI、EADI、ETI、EEI位的设定状况再决定是否接受中断。各个中断的向量地址与优先级(Priority)

如表2-4-1和表2-4-2所列，请读者注意：在表中的优先顺序是指当中断"同时"发生时的优先级关系；如果非"同时"发生，则此优先关系并不成立，此时先发生者就具有高优先级。

表 2-4-1　HT46x23 的中断向量

中断来源	向量地址	优先级
外部中断	004h	1
定时器/计数器中断	008h	2
A/D 转换中断	00Ch	3
I²C BUS 中断	010h	4

表 2-4-2　HT46x24 的中断向量

中断来源	向量地址	优先级
外部中断	004h	1
定时器/计数器 0 中断	008h	2
定时器/计数器 1 中断	00Ch	3
A/D 转换中断	010h	4
I²C BUS 中断	014h	5

注：HT46x24 的向量地址略有不同，参考图 2-2-1。

外部中断：在外部中断已经有效且栈寄存器还有空间的情况下，若 $\overline{INT}$ 引脚出现 1→0 的变化，则 CPU 将跳至 004h 地址执行程序。

定时器/计数器中断：在定时器/计数器中断已有效且栈寄存器还有空间的情况下，若产生定时器/计数器溢位的情形，则 CPU 将跳至 008h 地址执行程序。

A/D 转换中断：在 A/D 转换中断已有效且栈寄存器还有空间的情况下，若模/数转换器完成数据的转换时，则 CPU 将跳至 00Ch 地址执行程序。

I²C Bus 中断：在 I²C Bus 中断已有效且栈寄存器还有空间的情况下，若 I²C 的设备选择码相符或完成一个字符的数据传输时，则 CPU 将跳至 010h 地址执行程序。

由以上的说明中可以发现HT46xx中断机制的一项特点，即使中断已经有效，HT46xx在进入中断服务之前，会先检查栈寄存器是否仍有空间存放返回地址。如果栈寄存器已经放满了，则将暂时不执行目前的中断服务子程序，待栈寄存器有空间存放返回地址时再去执行。如此设计，可以排除部分因为栈溢位而导致程序无法正常执行的状况。不过，HT46xx只有在进入中断之前，才会进行检查栈寄存器是否仍有空间的动作，所以在中断服务子程序中用户必须留意，不能有太多调用子程序的动作，否则仍会造成栈溢位的状况。中断服务子程序的最后一个指令可以是RET或RETI指令，不同的是：RETI指令在返回主程序之前会先将EMI位设定为1(中断有效)，而RET指令则不会。

再次提醒读者：一旦进入中断，HT46xx会首先将EMI位清除(中断有效)，如此的设计是为了避免发生巢状中断(Interrupt Nesting)的情形，因此在中断程序执行的过程中若仍有其他的中断发生，HT46xx只会将其对应的标志位(HIF、ADF、TF或EIF)设定为1，但并不会进入其向量位置执行程序。所以，如果读者在应用上允许中断程序执行中仍有其他中断发生的需求(请参考6-12节)，就必须在中断服务子程序中再次将EMI位设定为1。

2-5　定时器/计数器控制单元

◇ TMRH(0Ch)：定时器/计数器高8位(Time/Event Counter High Byte)；
◇ TMRL(0Dh)：定时器/计数器低8位(Time/Event Counter Low Byte)；
◇ TMRC(0Eh)：TMR控制寄存器(Time/Event Counter Control Register)；

◇ TMR1H(0Fh)：Time/Event Counter 1 High Byte(for 46x24 only);
◇ TMR1L(10h)：Time/Event Counter 1 Low Byte(for 46x24 only);
◇ TMR1C(11h)：Time/Event Counter 1 Control Register(for 46x24 only).

HT46x23内部有一个16位的上数型定时器/计数器(Up-Counter)，它最大的功能是存储定时的时间(或者说是计数时钟脉冲的总数)。其内部电路结构如图2-5-1所示。

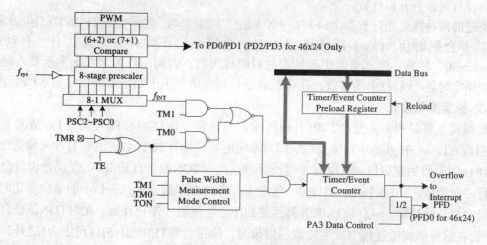

图2-5-1　HT46x23 Timer/Event Counter内部电路结构

TMRH与TMRL分别代表此计数器的高8位和低8位。计数的信号来源、计数分频可以由TMRC寄存器控制，TMRC各位功能说明如图2-5-2所示。

Bit7	Bit6	Bit5	Bit4	Bit3	Bit2	Bit1	Bit0
TM1	TM0	—	TON	TE	PSC2	PSC1	PSC0

Bit7~6	TM1 TM0：工作模式选择位(TMR Operation Mode) 00：未使用 01：计数模式,计数时钟脉冲由 TMR 引脚输入 10：计时模式,计数时钟脉冲由内部电路产生 11：脉宽测量模式
Bit5	未使用；读取时的值为 0
Bit4	TON：定时/计数控制位(Enable/Disable Counting) 1：开始计数 0：停止计数
Bit3	TE：定时/计准位控制位(Define TMR Active Edge) 1：当计数时钟脉冲由 Low→High 时加 1 0：当计数时钟脉冲由 High→Low 时加 1
Bit2~0	PS2~PSC0：TMR 计数分频位(Define Prescaler) 000：$f_{INT}=f_{SYS}$　　　100：$f_{INT}=f_{SYS}/16$ 001：$f_{INT}=f_{SYS}/2$　　101：$f_{INT}=f_{SYS}/32$ 010：$f_{INT}=f_{SYS}/4$　　110：$f_{INT}=f_{SYS}/64$ 011：$f_{INT}=f_{SYS}/8$　　111：$f_{INT}=f_{SYS}/128$

图2-5-2　HT46x23的TMRC控制寄存器

如图2-5-1所示，当写入数值到TMRL寄存器时，将会根据TMRL的值更新Preload Register的Low Byte；当写入数值到TMRH寄存器时，Preload Register将根据TMRH与TMRL的值重新设定。

定时器/计数器有3种工作模式，由TM1及TM0位选择。

(1) 计数模式

计数时钟脉冲由芯片的TMR引脚输入,主要用来记录外部事件的发生次数,可用TE位选择计数器是在正边沿或负边沿时加1,并以TON位控制是否开始计数。

(2) 计时模式

由芯片系统时钟脉冲(fSYS)经过8位分频后,作为计数的时钟脉冲来源(fINT),并以TON位控制计数器是否开始计数。

上述的两种模式当计数器由FFFFh→0000h(产生溢位)时,除了会将TF(INT0.5)位设定为1之外,并会同时由Preload Register载入起始数值继续计数;亦即其具有"自动载入(Auto-reload)"功能,除非是要更改计数的时间或次数,否则用户无须每次重新载入起始数值。如果定时器/计数器中断有效,在计数器溢位时也将产生中断请求的信号要求CPU服务。

(3) 脉宽测量模式

此模式主要是用来测量由TMR引脚所输入的脉宽。在TON与TE位均为1的情况下,当TMR引脚有Low→High的电位变化时,即启动定时器开始计数(计数的时钟脉冲来源为fINT),一直到TMR引脚回复到Low电位时才停止计数,此时TON位会自动清0;也就是说只会对脉宽进行一次测量。在TON=1,TE=0的情况下,当TMR引脚有High→Low的电位变化时,即启动定时器开始计数,一直到TMR引脚回复到High电位时才停止计数,此时TON位也会自动清0。若在脉宽的测量过程中发生计数器溢位时,除了会将TF(INT0.5)位设定为1之外,并会同时由Preload Register载入起始数值继续计数。如果定时器/计数器中断被有效,在计数器溢位时也将产生中断请求的信号要求CPU服务。

上述3种计数模式在TON=0(Timer/Count Counter 停止计数)的情况下,载入数值到Preload Register时会同时更新Timer/Count Counter的数值;但是在TON=1时更新Preload Register,则Timer/Count Counter会继续计数,待溢位产生时才会更新数值。另外需注意的是:当读取Timer/Count Counter时(如"MOV A,TMRH"),系统会暂时将计数时钟脉冲移开。这可能造成计数数值产生误差,用户在设计程序时要注意到这一点。

定时器/计数器溢位的信号可以用来产生PFD(Programmable Frequency Divider),其所产生的频率为定时器/计数器溢位频率的一半(需由Options选择此功能,此时PFD信号将由RA.3输出)。

注意:HT46x22与HT46x47计数单元的控制模式和HT46x23完全相同,唯一的差异是定时器/计数器的宽度仅有8位(称为TMR)。HT46x24的计数单元共有两组,其TMR0的相关寄存器分别称为TMR0H、TMR0L与TMR0C,控制方式与前述的TMR完全相同。

HT46x24 TMR0计数单元的控制模式与HT46x23完全相同,但其有另外一组计数单元TMR1,控制方式也是十分相似,其内部电路结构如图2-5-3所示,可看出TMR1与TMR(TMR0 for HT46x24)相比,只是少了一组Prescaler而已。图2-5-4描述了HT46x24的TMR1C控制寄存器。

第 2 章 HT46xx 系列系统体系结构

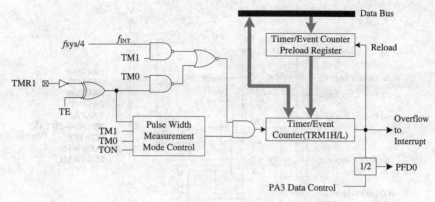

图2-5-3 HT46x24 Timer/Event Counter 1内部电路结构

Bit7	Bit6	Bit5	Bit4	Bit3	Bit2	Bit1	Bit0
TM1	TM0	—	TON	TE	—	—	—

Bit7~6　TM1 TM0：工作模式选择位
　　　　00：未使用
　　　　01：计数模式，计数时钟脉冲由 TMR 引脚输入
　　　　10：定时模式，计数时钟脉冲由内部电路产生
　　　　11：脉宽测量模式
Bit5　　未使用；读取时的值为 0
Bit4　　TON：定时/计数控制位
　　　　1：开始计数
　　　　0：停止计数
Bit3　　TE：定时/计准位控制位
　　　　1：当计数时钟脉冲由 Low→High 时加 1
　　　　0：当计数时钟脉冲由 High→Low 时加 1
Bit2~0　未使用；读取时的值为 0

图2-5-4 HT46x24的TMR1C控制寄存器

2-6 输入/输出控制单元

2-6-1 PA和PAC

◇ PA(12h)：输入/输出寄存器(Input/Output Register；Port A)；
◇ PAC(13h)：输入/输出控制寄存器(Port A Control Register)。

　　HT46xx系列单片机的输入/输出端口采用寄存器映射式I/O(Memory-Map I/O)的结构，即其每一个I/O端口都对应一个寄存器的位置。PA为一双向的I/O端口(PA7~PA0)，每一个位都是一个独立的个体，可通过PAC控制寄存器来定义各自的输入/输出功能，并可由Options选择其是否接内部的上拉电阻。上拉电阻的阻值在40~80 kΩ之间。

　　HT46x23的每一支I/O引脚，均具有如图2-6-1所示的结构。除了负责将输出数据锁存的数据锁存器(Data Bit Latch；DB_Latch)之外，还有一个用来选择输入或输出功能的控制寄存器(Control Bit Latch；CB_Latch)。在使用I/O端口之前，必须通过指令先将PC_Latch控制寄存器做适当的设定，然后才能将数值正确地写到I/O端口上。

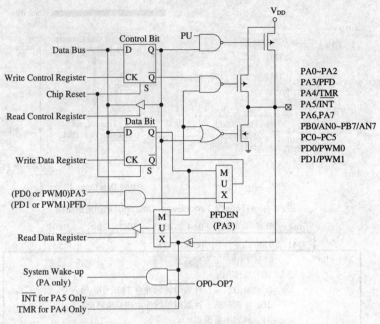

图2-6-1　HT46x23 I/O端口内部电路结构

例：将PA的高4位设为输出模式、低4位为输入模式。
1.　　INCLUDE　"HT46R23"
2.　　MOV　A,00001111B　　　;Acc=00001111B
3.　　MOV　PAC,A　　　　　　;存入 PA 控制寄存器中

如前所述：所有的I/O端口都可以被单独定义为输入或输出。当CB_Latch控制寄存器写入1时，表示其对应的引脚被定义成输入功能；反之，若控制寄存器写入0，则表示对应的引脚被定义为输出功能。RCR(Read Control Register)与WCR(Write Control Register)分别代表CB_Latch的读取与写入信号，当执行I/O控制寄存器的读写指令时，HT46Rxx内部的控制电路会自动产生这些时序信号来控制硬件做适当的动作，也就是说除了可直接设定I/O端口的输出/输入之外，还可读回目前I/O端口的设定状态。

WDR(Write Data Register)与RDR(Read Data Register)分别代表DB_Latch的写入与读取信号，即I/O端口读/写控制信号。通过定义之后，由BC_Latch的Q输出来控制2-To-1多工器(MUX)。若定义为输入(Q=1)，当读取I/O端口的数据时，则直接读取I/O引脚上的状态；若定义为输出(Q=0)，当数据写到I/O端口时，则先送到DB_Latch后再直接送到I/O引脚上。如果在定义为输出(Q=0)模式而又去读取I/O端口，此时读到的值是DB_Latch上的值(即最后一次写到DB_Latch上的状态)，而非I/O引脚上的状态，这一点必须加以注意！

另外，若以"SET [m],i"、"CLR [m],i"、"CPL [m]"、"CPLA [m]"指令对I/O端口控制时，要特别注意此类指令会有"先读再写(Read-Modify-Write)"的动作。例如，"CLR PA,3"是先将整个PA(8位)的值读进CPU，执行位运算后再将结果写到PA上。但PA有一些引脚是双向I/O引脚(如PA5)，假设当执行"CLR PA,3"指令时PA5是输入模式，则PA5引脚上的状态会先被读入再写到DB_Latch上，覆盖原先DB_Latch上的数据。因此，只要PA5一直是输入模式就没有问题；一旦PA5切换为输出模式，则PD_Latch上的数据是不可预知的！

PA的部分引脚具有双重功能，请参考图2-6-1。PA3可当成一般的I/O使用，也可以当成PFD(Programmable Frequency Divider)，需在配置选项(Options)时选定该功能。在PFD选项有效的情况下，若PA3被定义为输入模式(PAC3＝1)，则PA3仍可当成一般的输入引脚使用。若PAC3＝0(输出模式)，此时如果设定PA3＝1即开始输出PFD信号，其所产生的频率为定时器/计数器溢位频率的一半；设定PA3＝0时将停止PFD信号的输出，并强迫使PA3的引脚呈现低电位的输出状态。

PA4也可当成Timer/Event Counter的输入控制，若选用计数模式时，此引脚即为计数脉冲的输入引脚；而当选用脉宽测量模式时，待测的脉冲也由此引脚输入。

PA5也可以作为外部中断的输入引脚，在外部中断有效的情况下，若由此引脚输入一负边沿的脉冲，则代表要求CPU执行外部中断程序。

PA6与PA7还可分别作为I^2C串行传输时的SCL(Serial Clock)与SDA(Serial Data)控制信号，此时PA6与PA7为NMOS漏极开路式(Open Drian)的输出引脚，请务必分别接上上拉电阻。详细内容请参考2-8节中关于"I^2C串行接口"的说明。

PA的每一个引脚均具有唤醒(Wake-Up)功能，可通过Options来选用，有关唤醒的相关说明请参考2-12节中有关"省电模式"的内容。

2-6-2 PB和PBC

- ◇ PB(14h)：输入/输出寄存器(Input/Output Register；Port B)；
- ◇ PBC(15h)：输入/输出控制寄存器(Port B Control Register)。

同PA一样，它是一个8位双向的I/O端口(PB7~PB0)，每一个引脚可以通过PBC控制寄存器单独定义其为输入或输出。PB的每一个引脚都具有双重功能，除可当一般的I/O端口使用之外，也可作为模/数转换器的输入通道(A0~A7)，详细内容请参考2-9节中关于"模/数转换器"的说明。

2-6-3 PC和PCC

- ◇ PC(16h)：输入/输出寄存器(Input/Output Register；Port C)；
- ◇ PCC(17h)：输入/输出控制寄存器(Port C Control Register)。

PC是一个单纯的双向I/O端口，其宽度只有5位(PC4~PC0)，每一个引脚可以通过PCC控制寄存器单独定义其为输入或输出(注意：HT46xx家族在I/O端口的位数目上略有不同，请读者参阅图2-6-1)。

2-6-4 PD和PDC

- ◇ PD(18h)：输入/输出寄存器(Input/Output Register；Port D)；
- ◇ PDC(19h)：输入/输出控制寄存器(Port D Control Register)。

PD是一个双向的I/O端口，每一个引脚可以通过PDC控制寄存器单独定义其为输入或输出，请注意其宽度只有2位(PD1、PD0)。PD的每一个引脚都具有双重功能，除可当一般的I/O端口使用之外，也可作为PWM(Pulse Width Modulation)的输出引脚。当Options选用PWM的功能后，若设定PD0(PD1)为1，PWM信号即开始由PD0(PD1)输出；若设定为0，则PWM信号停止输出,并强迫使PD0(PD1)引脚维持低电位。详细内容请参考2-7节中关于"PWM接口"的说明。

注意：HT46x24-28的PD1又可当成TMR1计数单元的信号输入引脚；而HT46x24-48的PD共有4个位(PD3~PD0),每一位都可当成PWM的输出引脚(PWM3~PWM0)。

2-6-5 PF和PFC

◇ PF(28h)：输入/输出寄存器(Input/Output Register；Port F) for 46x24-48 only；
◇ PFC(29h)：输入/输出控制寄存器(Port F Control Register)for 46x24-48 only。

PF是一个单纯的双向I/O端口,其宽度为8位(PF7~PF0),每一个引脚可以通过PFC控制寄存器单独定义其为输入或输出(注意：HT46xx家族在I/O端口的位数目上略有不同,请读者参阅图2-6-1)。

2-7 PWM输出接口

◇ PWM0(1Ah)：PWM0周期控制寄存器(PWM0 Duty Cycle)；
◇ PWM1(1Bh)：PWM1周期控制寄存器(PWM1 Duty Cycle)；
◇ PWM2(1Ch)：PWM2周期控制寄存器(PWM0 Duty Cycle) for 46x24 only；
◇ PWM3(1Dh)：PWM3周期控制寄存器(PWM1 Duty Cycle) for 46x24 only。

HT46x23提供两个通道的PWM输出,必须在Options中选用此功能,此时PWM信号将由PD0、PD1输出,PWM计数器的时钟脉冲来源为f_{SYS},而其Duty Cycle与调变周期(Modulation Cycle)分别由PWM0与PWM1寄存器控制。当Options选用PWM的功能后,若设定PD0(PD1)为1,PWM信号即开始由PD0(PD1)输出；若设定为0,则PWM信号停止输出,并强迫使PD0(PD1)引脚维持在低电位状态。HT 46x23提供两种PWM模式：(6+2)Mode与(7+1)Mode,必须在Options中予以指定,以下说明这两种模式的不同。

(1) (6+2)Mode

此时,PWM周期被分割成4个调变时段(Modulation Cycle 0~3),每个调变时段为64个工作时钟脉冲,而PWM寄存器的控制位区分为DC(PWM7~ PWM2)与AC(PWM1~ PWM0)两部分,每一个调变时段的Duty Cycle关系如表2-7-1所列。

表2-7-1 (6+2)Mode各调变时段的Duty Cycle

参 数	AC(0~3)	Duty Cycle
Modulation Cycle i i＝0~3	i＜AC	$\frac{DC+1}{64}$
	i≥AC	$\frac{DC}{64}$

请参考如图2-7-1所示的图例。

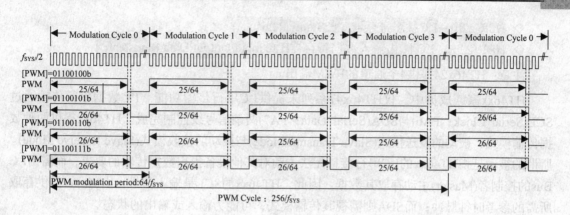

图 2-7-1　(6+2)PWM Mode 图例

注意：HT46x47仅有(6+2)Mode的PWM控制方式。

(2) (7+1)Mode

此时PWM周期被分割成两个调变时段(Modulation Cycle 0~1)，每个调变时段为128个工作时钟脉冲，而PWM寄存器的控制位区分为DC(PWM7~PWM1)与AC(PWM0)两部分，每一个调变时段的Duty Cycle关系如表2-7-2所列。

表2-7-2　(7+1)Mode各调变时段的Duty Cycle

参　数	AC(0~1)	Duty Cycle
Modulation Cycle i	i＜AC	$\dfrac{DC+1}{128}$
i＝0~1	i≥AC	$\dfrac{DC}{128}$

请参考如图2-7-2所示的图例。

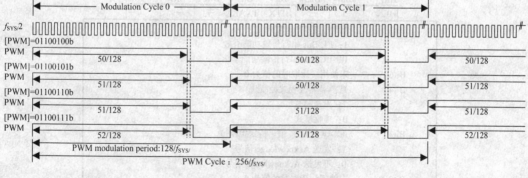

图 2-7-2　(7+1)PWM Mode 图例

2-8　I²C串行接口

◇ HADR(20h)：I²C地址寄存器(I²C Slave Address Register)；
◇ HCR(21h)：I²C控制寄存器(I²C Control Register)；

◇ HSR(22h)：I²C状态寄存器(I²C Status Register)；
◇ HDR(23h)：I²C输出/输入数据寄存器(I²C Input/Output Data Register)；

注意：HT46x24提供4个通道的PWM输出。

HT46xx家族成员中，仅HT46x47系列未提供I²C串行传输功能。I²C数据串行传输通过SCL(Serial Clock；PA6)与SDA(Serial Data；PA7)两条信号线控制完成。HT46xx提供两种数据传输模式：被动的发送模式(Slave Transmit Mode)与被动的接收模式(Slave Receive Mode)。所谓的被动是指HT46xx的I²C串行接口无法主动对其他设备提出数据传输的要求，而需由I²C Bus的控制者(Master)主动存取其数据。因此，HT46xx的SCL是输入信号，由Master提供存取所需的参考时钟脉冲；而SDA则需视其传输模式，可能为输入或输出的状态。

I²C Bus上连接的设备有各自的地址，在数据存取开始之前需由I²C Bus Master先送出所欲存取的设备地址，而HADR寄存器就是HT46xx的设备地址。当Bus Master所送出的地址与HADR的内容相吻合时，HT46xx的I²C串行传输接口才会与Bus Master继续进行数据存取所需的交握程序；若地址不吻合，I²C串行传输接口将不予理会。HADR寄存器的格式如图2-8-1所示。

Bit7	Bit6	Bit5	Bit4	Bit3	Bit2	Bit1	Bit0
			Slave Address				—

图2-8-1　HT46xx的HADR寄存器

Slave Address为7位，HADR.0并未定义。若要使用I²C串行传输接口功能，首要的任务就是要在此寄存器设定地址，以便I²C Bus Master通过此接口来存取HT46xx的数据。

HCR是用来控制I²C串行传输接口的寄存器，I²C串行传输接口功能的启用、传输模式等，都由HCR寄存器设定。请参阅图2-8-2，当HT46xx是负责提供数据让Bus Master读取时，需将HTX位设定为1(Transmit Mode)；反之，若HT46xx是接收从Bus Master所送出的数据，则需将HTX位设为0(Receive Mode)

Bit7	Bit6	Bit5	Bit4	Bit3	Bit2	Bit1	Bit0
HEN	—	—	HTX	TXAK	—	—	—

Bit7	HEN：I²C Bus 有效位
	1：I²C Bus 串行传输功能有效
	0：I²C Bus 串行传输功能禁止
Bit6~5	未使用；读取时的值为0
Bit4	HTX：传输模式控制位
	1：发送模式
	0：接收模式
Bit3	TXAK：ACK 信号传送控制位
	1：发送 Acknowledge 信号
	0：不发送 Acknowledge 信号
Bit2~0	未使用；读取时的值为0

图2-8-2　HT46xx的HCR控制寄存器

在I²C串行传输的过程中，数据是以字节为单位由最高位(MSB)开始循序从SDA线送出，负责接收的设备在收完最后一位之后，如果要继续读取下一个字节的数据，就必须送出ACK(ACK＝0)信号通知发送端继续送出下一笔数据；否则就别送出ACK(ACK＝1)信号。发送端就是以ACK信号的存在与否来判定要继续送出下一笔数据或是结束此次的传输。TXAK位就是当HT46xx在接收模式(Slave Receive Mode)时，是否要送出ACK信号的控制位。

HDR是I^2C串行传输中数据接收与发送的寄存器，在开始发送之前，必须先将欲发送的数据写入HDR寄存器。相反的，在接收模式时，I^2C串行接口会将接收到的数据存放在HDR寄存器，不过要提醒读者的是在开始接收数据之前，必须先对HDR寄存器进行一次无效的读取(Dummy Read)。

HSR则是反映I^2C串行接口目前状态的寄存器，其结构如图2-8-3所示。如前所述，I^2C Bus上连接的设备有各自的地址，在数据存取开始之前需由I^2C Bus Master送出所欲存取的设备地址，当Bus Master所送出的地址与HADR的内容相吻合时，HASS位与HIF(INTC1.4)会被设定为1，同时发出I^2C Bus中断的请求信号并告知CPU。

在I^2C串行传输的过程中，数据是以字节为单位，由最高位(MSB)开始循序从SDA线送出(或接收)，用户可通过HCF位得知目前发送的状态，HCF＝0表示8位的数据已发送(或接收)完成，此时会设定HIF(INTC1.4)为1，同时发出I^2C Bus中断的请求信号告知CPU；若HCF＝1则代表数据正在发送中。I^2C Bus Master在送出设备地址时，会同时送出要对该设备进行读或写的控制命令，而此命令会由HT46xx的I^2C串行接口记录在SRW位，用户通过此位决定该工作在发送模式(SRW＝1)或是接收模式(SRW＝0)。HBB则是反应I^2C串行接口是否处于忙碌状态的位。

如前所述，I^2C串行传输时发送端是以ACK信号的存在与否来判定要继续送出下一笔数据或是结束此次的传输。当HT46xx在发送模式时，接收端是否送出ACK信号会由I^2C接口记录在RXAK位，如果RXAK＝0，表示接收端要继续读取下一笔数据，用户就必须以程序控制送出下一笔数据(写入HDR寄存器)；若RXAK＝1，表示接收端未送出ACK信号，HT46xx可以结束此次数据传输。

Bit7	Bit6	Bit5	Bit4	Bit3	Bit2	Bit1	Bit0
HCF	HASS	HBB	—	—	SRW	—	RXAK

Bit7	HCF：I^2C Bus 传输状态标志位 1：1字节数据正进行传输中 0：1字节的数据传输已经完成
Bit6	HASS：I^2C Bus 地址比对状态标志位 1：地址比对结果与设备吻合 0：地址比对结果与设备不吻合
Bit5	HBB：I^2C Bus 忙碌标志位 1：I^2C 接口正处于忙碌状态 0：I^2C 接口不处于忙碌状态
Bit4~3	未使用；读取时的值为0
Bit2	SRW：I^2C Bus 读/写状态标志位 1：Bus Master 要读取设备的数据 0：Bus Master 要写入数据到设备
Bit1	未使用；读取时的值为0
Bit0	RXAK：I^2C Bus 接收 ACK 信号状态标志位 1：数据接收设备未送出 ACK 信号 0：数据接收设备送出 ACK 信号

图2-8-3　HT46xx的HSR状态寄存器

如果要运用HT46xx的I²C串行接口功能,请读者务必注意以下几项要点,并请参考图2-8-4所示的流程图。

① 在 HADR 寄存器中写入 7 位设备地址(Device Address);
② 将 HEN(HCR.7)位设定为 1,启动 I²C 串行接口;
③ 设定 EIH(INTC1.0)为 1, I²C 串行接口的中断功能有效。

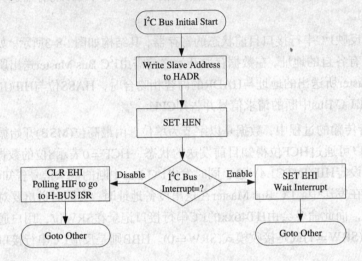

图2-8-4 使用I²C串行接口功能的程序初始化流程

由图2-8-4可知当启动I²C串行接口功能时,除了以中断方式来进行数据的传输之外,还可以用轮询(Polling)的方式来处理;但基于让单片机的运作更有效率的考虑,仍建议读者以中断的方式来处理。

HT46xx的I²C串行接口在两种情况都会产生中断请求信号要求CPU服务,其一是当地址比对吻合时,其二则是完成8位数据的发送(或接收)时。因此,一旦进入I²C中断服务子程序后,必须先由HASS(HSR.6)位分辨中断发生的原因。若HASS=1,表示是地址比对吻合所造成的中断,接着就需要依照SRW(HSR.2)位的状态设定要工作在发送模式(SRW=1)或是接收模式(SRW=0)。如果是发送模式,就接着将要发送的数据写入HDR寄存器即可;若是接收模式,请务必紧接着进行读取HDR的动作,请注意此时读到的并非接收到的数据,而只是让I²C串行接口正常动作所需要的"Dummy Read"。

若HASS=0,表示是8位数据传输完毕的中断,可再依据HTX判别是发送完成(HTX=1)还是接收完成(HTX=0)所造成的。若HTX=0,表示发送端串行送出的8位数据已经完整地接收,此时可由HDR寄存器读取数据;若HTX=1,表示I²C串行接口已串行送出8位数据,此时必须检查RXAK位,判别接收端是否送出ASK信号,以决定是否要继续发送数据,请参考图2-8-5的流程图。

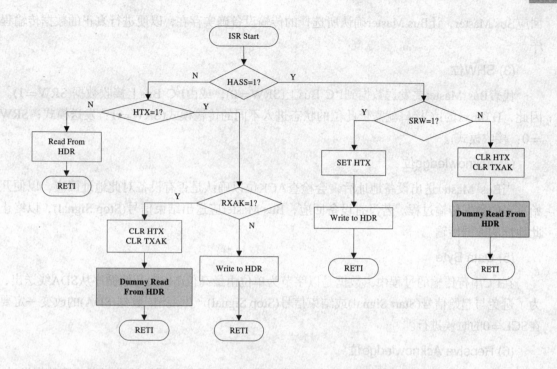

图2-8-5　使用I^2C Bus中断服务子程序的流程

为使读者能更深入了解I^2C串行接口的传输控制方式，请参考图2-8-6的时序图及以下的说明。

(1) Start Signal(起始信号)

这是由I^2C Bus Master所送出的信号，用于通知连接在I^2C Bus上的所有设备，准备接收地址信息及读/写命令。HT46xx的I^2C接口检测到起始信号时，会设定HBB位，表示进入忙碌状态。所谓起始信号是指SCL＝1时，SDA由1→0的状态。

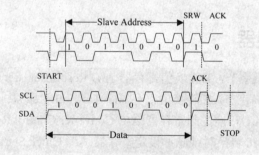

图2-8-6　I^2C Bus时序图

(2) Slave Address(设备地址)

在起始信号之后，Bus Master必须送出设备地址(7位)以选定数据传输的对象。若HT46xx的I^2C接口检测到Bus Master所送出的地址与自己的地址(HADR)相同时，就会产生I^2C Bus中断并设定HASS位，将第8个位存入SRW位，并在第9个位时间于SDA送出ACK信号(低电位)

回应Bus Master，让Bus Master确认所选择的传输设备确实存在，以便进行真正的数据传输程序。

(3) SRW位

代表Bus Master要发送数据到I²C Bus上(SRW=0)，或由I²C Bus上接收数据(SRW=1)。因此，HT46xx的I²C接口就需依此位的状态进入不同的传输模式(SRW=1：发送模式；SRW=0：接收模式)。

(4) Acknowledge位

当Bus Master送出设备地址后，会检查ACK位以确认是否有设备对此地址回应，以便开始真正的数据传输过程。若没有设备回应，Bus Master应送出结束信号(Stop Signal)，以终止此地址的数据传输。

(5) Data Byte

在I²C串行传输的过程中，数据是以字节为单位由最高位(MSB)开始循序从SDA线送出，为了避免与起始信号(Start Signal)或结束信号(Stop Signal)产生混淆，数据(SDA)的改变一定要在SCL=0的时候进行。

(6) Receive Acknowledge位

当接收端接收到8个位的数据之后，必须让发送端知道是否要继续进行下一笔数据发送。因此接收端必须在第9个位时间于SDA送出ACK信号(低电位)告知发送端，当HT46xx工作在发送模式时，就必须检查RXAK位以决定是否要继续发送数据；而工作在接收模式时，就必须根据TXAK的设定送出ACK信号，让发送端判断是否需要继续送出数据。

(7) Stop Signal(结束信号)

这是由I²C Bus Master所送出的信号，用于通知连接在I²C Bus上的所有设备结束目前的数据发送。当HT46xx的I²C接口检测到结束信号时，会清除HBB位，表示I²C Bus的忙碌状态已经结束。所谓结束信号是指SCL=1时，SDA由0→1的状态。

2-9 模/数转换器

- ◇ ADRL(24h)：低字节A/D转换值寄存器(A/D Result Register Low Byte)；
- ◇ ADRH(25h)：高字节A/D转换值寄存器(A/D Result Register High Byte)；
- ◇ ADCR(26h)：A/D转换控制寄存器(A/D Converter Control Register)；
- ◇ ACSR(27h)：A/D转换时钟脉冲设定寄存器(A/D Clock Setting Register)。

HT46x23、HT46x24提供8个通道(A7~A0)的模/数转换功能，转换器的分辨率为10位，模拟输入信号由PB(PB7~PB0)的引脚输入，转换的结果(D9~D0)则存放在ADRH与ADRL寄存器中，请参考表2-9-1的数据存放格式。

表2-9-1　HT46x23与HT46x24的A/D转换结果存放格式

寄存器	Bit7	Bit6	Bit5	Bit4	Bit3	Bit2	Bit1	Bit0
ADRL	D1	D0	—	—	—	—	—	—
ADRH	D9	D8	D7	D6	D5	D4	D3	D2

转换器的转换速度可由ACSR寄存器设定，请参考图2-9-1的描述。

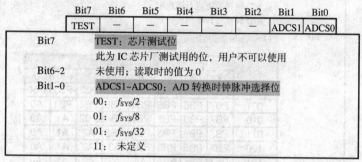

图2-9-1　HT46xx的ACSR控制寄存器

请参考图2-9-2，依时序图所示，A/D转换器完成一次转换约需花费76个T_{AD}的时间(此即转换时间：Conversion Time)，而T_{AD}所指的就是转换的时钟脉冲周期。以$f_{SYS}=4$ MHz为例，若选择ADCS1~ADCS0＝01，则此时的$T_{AD}=2$ μs，转换时间＝152 μs。不过请读者注意在原厂数据手册的一项限制：$T_{AD} \geq 1$ μs，也就是说HT46x23的A/D转换最短的转换时间为76 μs，如果所选择的转换时钟脉冲(ADC Clock Source)小于1 μs，则并不保证转换结果的正确性。

ADCR为A/D转换控制寄存器，虽然HT46x23有8个通道的模拟输入，但因为内部仅有一组A/D转换器，因此一次仅能进行单一通道的数据转换。有关转换的通道以及转换的开始与否，均由ADCR寄存器控制，请参考图2-9-3。

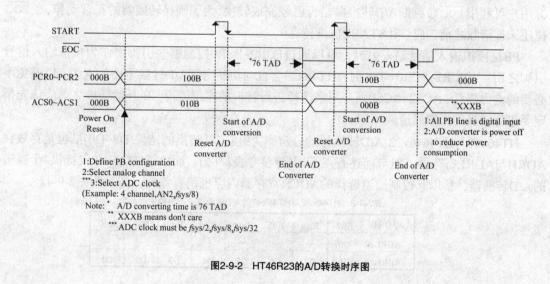

图2-9-2　HT46R23的A/D转换时序图

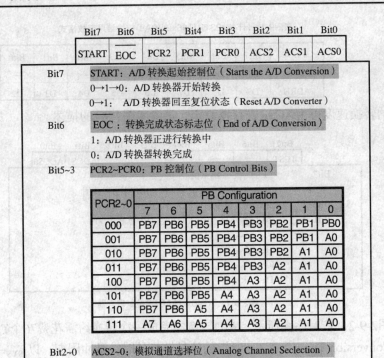

图2-9-3 HT46x23、46x22与46x24的ADCR控制寄存器

ADCR寄存器中的START位，是控制A/D转换器停滞于复位状态或开始进行转换的控制开关，当START由0→1，则令A/D转换器回到复位状态；当START由0→1→0，则要求A/D转换器开始针对选择的模拟通道进行转换，当转换完成时A/D转换接口会自动将$\overline{EOC}$位清除为0。用户可利用$\overline{EOC}$位判断A/D转换器是否已经完成转换，为了确保转换器的正常动作，在$\overline{EOC}$位还未被清除之前，应该让START位维持在0。

PB是模拟输入信号输入引脚，可以通过PCR2~PCR0控制哪些引脚要作为模拟输入信号引脚之用。当PCR2~PCR0＝000时，HT46x23会自动切断内部A/D转换电路的电源，以避免不必要的电力损耗。ACS2~ACS0则是用来选择进行转换的模拟通道，在开始转换之前应该先指定要进行转换的模拟通道。

HT46x47与HT4646x22 A/D转换器的分辨率为9位，其转换的结果(D8~D0)虽也是存放在ADRH与ADRL寄存器，但与前述有点不同(请参考表2-9-2)。而由于HT46x47仅提供4个通道的A/D转换器，所以其控制通道选择的ADCR寄存器内容也稍有差异，请参阅图2-9-4。

表2-9-2 HT46x22与HT46x47的A/D转换结果存放格式

寄存器	Bit7	Bit6	Bit5	Bit4	Bit3	Bit2	Bit1	Bit0
ADRL	D0	—	—	—	—	—	—	—
ADRH	D8	D7	D6	D5	D4	D3	D2	D1

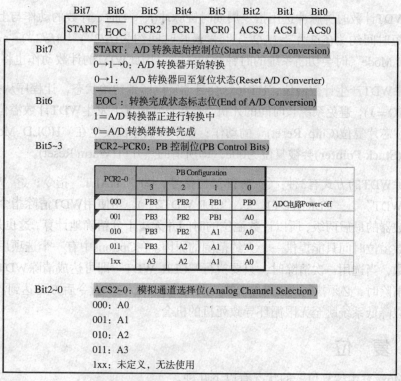

图2-9-4　HT46x47的ADCR控制寄存器

2-10　WDT：看门狗定时器

看门狗计数器最主要的功能，是避免程序因不可预期的因素而造成系统长时间的瘫痪(如跳至无限循环)。它的计数时钟脉冲可以来自HT46xx内部专属的RC自振式振荡器(Free-Running On-Chip RC Oscillator)，或指令周期时钟脉冲，请参考图2-10-1。

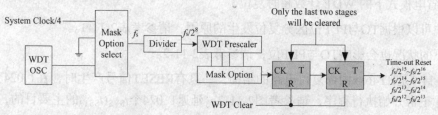

图2-10-1　HT46xx的WDT内部结构

看门狗计时的时钟脉冲来源、预分频器(WDT Prescaler)、WDT清除模式以及是否使用WDT的功能，都必须由Options中的选项来设置。

如图2-10-1所示，当WDT计数时钟脉冲选用内部RC自振式振荡器(WDT OSC)时，计数的时钟脉冲周期约为65μs(工作于5 V时)，配合预分频器的选用，其计数溢位的最短时间约为300~600 ms，最长时间约为2.3~4.7 s。此时即使单片机已进入"睡眠模式(HALT Mode)"，WDT仍继续计数。由于计数时钟脉冲是由内部的RC电路产生，其受温度及工作电压的影响甚大，使用时应特别注意。

当WDT计数时钟脉冲选用指令周期时钟脉冲($f_{SYS}/4$)时,计数的动作与上述相同,计数溢位的最短时间约为$(f_S/2^{12})^{-1}$~$(f_S/2^{13})^{-1}$s;最长时间约为$(f_S/2^{15})^{-1}$~$(f_S/2^{16})^{-1}$s。当单片机进入"HOLD Mode"时会切断系统的时钟脉冲(f_{SYS}),因此WDT的计数动作也将随之停止。

如果WDT产生计数溢位,HT46xx会自动复位回到初始状态,让程序从头开始执行(此时会设定TO=1),避免系统长时间的死机。若在正常情况下发生WDT计数溢位,此时系统会自动产生"芯片复位(Chip Reset)"的动作;如果计数溢位发生在"HOLD Mode"时,则只有PC与SP(Stack Pointer)会被复位为00h,即所谓的热开机(Warm Reset)。

清除WDT的方式有3种:① 外部复位信号;② "HALT"指令;③ WDT清除指令(如"CLR WDT"、"CLR WDT1"、"CLR WDT2")。当使用WDT清除指令时,仅会清除看门狗计数器的最后两级,所以计数溢位的时间也并无法十分精确地计算,这也是上述有关最短、最长计数溢位时间只能提供一个大约范围的原因。在Options中有一个选项用来选择看门狗的清除次数,当选用一次清除时,只要执行"CLR WDT"即可达成清除WDT的目的;但当选择两次清除时,必须执行"CLR WDT1"与"CLR WDT2"指令后方可达到清除WDT的效果,如此可以降低系统跳至无限循环导致死机的机会。

2-11 复位

HT46xx发生"复位"的状况有以下几种。

◇ 系统开机复位;

◇ 正常工作状况下的$\overline{RES}$复位(即$\overline{RES}$引脚有Low状态);

◇ 省电模式下的$\overline{RES}$复位;

◇ 正常工作状况下的WDT计数超时复位;

◇ 省电模式下的WDT计数超时复位。

用户可以根据TO与PDF位区分复位发生的原因,请参考表2-11-1。

以下的状况也会影响TO与PDF位,请参考表2-11-2。

图2-11-1是HT46xx内部复位电路的结构,一旦有RESET信号产生时,在1 024个t_{SYS}后,单片机才正常开始执行程序,请参考图2-11-2。延迟1 024个t_{SYS}(t_{SST})的主要目的,是为了确保内部的振荡电路能够开始且稳定地振荡,此为SST定时器(System Set-Up Timer)的功能。

表2-11-1 Reset后,TO与PDF位的状态

TO	PDF	RESET 起因
0	0	$\overline{RES}$ during Power-up
0	0	正常工作状况下的$\overline{RES}$复位
0	1	省电模式下的$\overline{RES}$复位
1	0	正常工作状况下的WDT计数溢位复位
1	1	省电模式下的WDT计数溢位复位

表2-11-2 影响TO与PDF位的事件

事 件	TO	PDF
Power-Up	0	0
WDT 计数溢位	0	U[(1)]
"HALT" 指令	0	1
"CLR WDT" 指令[(2)]	0	0
"CLR WDT1"与"CLR WDT2"指令[(3)]	0	0

注:(1)U 表示维持原来的状态;
(2)当选择 WDT 为一次清除模式时;
(3)当选择 WDT 为两次清除模式时,且两个指令都执行过后。

第 2 章 HT46xx 系列系统体系结构

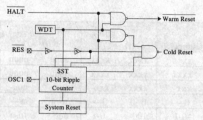
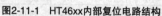

图2-11-1 HT46xx内部复位电路结构

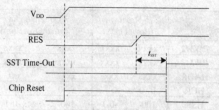

图2-11-2 HT46xx复位时序

表2-11-3是HT46xx复位之后内部寄存器的状态,由于3种复位状况会有不同的设定值,为方便读者查阅,故一并将其整理于表中。总结系统复位的几个重点,整理如下:

◇ PC＝000h,表示CPU提取的第一个指令是在程序存储器的第0个位置,所以要将程序由此地址放起。

◇ 所有的中断被禁止,若要使用中断功能必须以适当指令将相关的中断控制位使能。

◇ WDT会清0,并于复位状况解除后立即开始计数的动作。

◇ Timer/Event Counter处于停止计数状态,即TON＝0。

◇ 所有的输入/输出端口被定义为输入状态。

◇ SP指向栈寄存器的最上层。

2-11-3 HT46x23、HT46x24复位后的内部寄存器状态

Register	Reset (Power On)		WDT Time-Out (Normal Operation)		$\overline{\text{RES}}$ Reset (Normal Operation)		$\overline{\text{RES}}$ Reset (HALT)		WDT Time-Out (HALT)*	
TMRL	xxxx	xxxx	xxxx	xxxx	xxxx	xxxx	xxxx	xxxx	uuuu	uuuu
TMRH	xxxx	xxxx	xxxx	xxxx	xxxx	xxxx	xxxx	xxxx	uuuu	uuuu
TMRC	00-0	1000	00-0	1000	00-0	1000	00-0	1000	uu-u	uuuu
TMR1L	xxxx	xxxx	xxxx	xxxx	xxxx	xxxx	xxxx	xxxx	uuuu	uuuu
TMR1H	xxxx	xxxx	xxxx	xxxx	xxxx	xxxx	xxxx	xxxx	uuuu	uuuu
TMR1C	00-0	1---	00-0	1---	00-0	1---	00-0	1---	uu-u	u---
Program Counter	000H		000H		000H		000H		000H	
BP	----	---0	----	---0	----	---0	----	---0	----	---u
MP0	xxxx	xxxx	uuuu	uuuu	uuuu	uuuu	uuuu	uuuu	uuuu	uuuu
MP1	xxxx	xxxx	uuuu	uuuu	uuuu	uuuu	uuuu	uuuu	uuuu	uuuu
ACC	xxxx	xxxx	uuuu	uuuu	uuuu	uuuu	uuuu	uuuu	uuuu	uuuu
TBLP	xxxx	xxxx	uuuu	uuuu	uuuu	uuuu	uuuu	uuuu	uuuu	uuuu
TBLH	-xxx	xxxx	-uuu	uuuu	-uuu	uuuu	-uuu	uuuu	-uuu	uuuu
STATUS	--00	xxxx	--1u	uuuu	--uu	uuuu	--01	uuuu	--11	uuuu
INTC0	-000	0000	-000	0000	-000	0000	-000	0000	-uuu	uuuu
INTC1	---0	---0	---0	---0	---0	---0	---0	---0	---u	---u
PA	1111	1111	1111	1111	1111	1111	1111	1111	uuuu	uuuu
PAC	1111	1111	1111	1111	1111	1111	1111	1111	uuuu	uuuu
PB	1111	1111	1111	1111	1111	1111	1111	1111	uuuu	uuuu
PBC	1111	1111	1111	1111	1111	1111	1111	1111	uuuu	uuuu
PC	---1	1111	---1	1111	---1	1111	---1	1111	---u	uuuu
PCC	---1	1111	---1	1111	---1	1111	---1	1111	---u	uuuu
PD	----	--11	----	--11	----	--11	----	--11	----	--uu
PDC	----	--11	----	--11	----	--11	----	--11	----	--uu
PF	1111	1111	1111	1111	1111	1111	1111	1111	uuuu	uuuu
PFC	1111	1111	1111	1111	1111	1111	1111	1111	uuuu	uuuu
PWM0	xxxx	xxxx	xxxx	xxxx	xxxx	xxxx	xxxx	xxxx	uuuu	uuuu
PWM1	xxxx	xxxx	xxxx	xxxx	xxxx	xxxx	xxxx	xxxx	uuuu	uuuu

续表

Register	Reset (Power On)		WDT Time-Out (Normal Operation)		$\overline{RES}$ Reset (Normal Operation)		$\overline{RES}$ Reset (HALT)		WDT Time-Out (HALT)*	
PWM2	xxxx	xxxx	xxxx	xxxx	xxxx	xxxx	xxxx	xxxx	uuuu	uuuu
PWM3	xxxx	xxxx	xxxx	xxxx	xxxx	xxxx	xxxx	xxxx	uuuu	uuuu
HADR	xxxx	xxx-	xxxx	xxx-	xxxx	xxx-	xxxx	xxx-	uuuu	uuu-
HCR	0--0	0---	0--0	0---	0--0	0---	0--0	0---	u--u	u---
HSR	100-	-0-1	100-	-0-1	100-	-0-1	100-	-0-1	uuu-	-u-u
HDR	xxxx	xxxx	xxxx	xxxx	xxxx	xxxx	xxxx	xxxx	uuuu	uuuu
ADRL	xx--	----	xx--	----	xx--	----	xx--	----	uuuu	----
ADRH	xxxx	xxxx	xxxx	xxxx	xxxx	xxxx	xxxx	xxxx	uuuu	uuuu
ADCR	0100	0000	0100	0000	0100	0000	0100	0000	uuuu	uuuu
ACSR	1---	--00	1---	--00	1---	--00	1---	--00	u---	--uu

注：* 代表热复位(Warm Reset)；u 代表状态没改变；x 代表未知状态。

注意：由于HT46xx家族的部分寄存器位数不尽相同，故本表主要是以HT46x23为主，而表中粗斜字体部分代表仅HT46x24所拥有的寄存器，各成员的详细复位数据还是请读者参阅原厂的数据手册。

2-12 省电模式

大多数单片机的应用场合都是以电池为主要的供电设备，为了达到省电的目的，HT46xx提供了省电模式的功能，一旦系统进入省电模式，HT46xx单片机大约只消耗数μA的电流，请参考表2-12-1。

表2-12-1 HT46x23省电模式的电流消耗比较

工作模式	状态		Typ.	Max.	单位
	V_{DD}/V	测试条件			
Operating Current	3	$f_{SYS}=4$ MHz	0.6	1.5	mA
	5	$f_{SYS}=4$ MHz	2	4	mA
Standby Current WDT Enable	3	No Load, System HALT	—	5	μA
	5		—	10	μA
Standby Current WDT Disable	3	No Load, System HALT	—	1	μA
	5		—	2	μA

执行HALT指令后，系统随即进入省电模式，单片机并将执行以下的工作：

◇ 关闭系统时钟脉冲(但若WDT计数时钟脉冲选用内部RC自振式振荡器(WDT OSC)时，WDT仍会继续计数)；

◇ 所有内部数据存储器的内容维持不变；

◇ 清除WDT，并重新开始计数(若WDT计数时钟脉冲选用WDT OSC)；

◇ 清除TO，并设定PDF=1。

进入省电模式后，I/O端口会保持在执行HALT指令之前的状态。所以，若要达到更佳的省电效果，最好在进入省电模式之前也一并将外围的负载元件一起关闭，以减少电流的损耗。要注意的是，若选用WDT OSC作为WDT计数时钟脉冲，则在进入省电模式的后它仍旧会继续计数；所以如果希望系统能长时间地停留在省电状态，就必须先把看门狗定时器关闭，否则每隔一段时间(视WDT预分频器的设定状况而定),系统就会自动被唤醒。让芯片由省电模式重新回复工作的方式有4种,分别是:

① 外部硬件复位(External Reset):即在$\overline{RES}$引脚输入Low电位,至于复位后的寄存器状态,请参考2-11节中的说明。

② WDT计时溢位复位:注意在使用此种方式唤醒时,必须在Options中选用WDT OSC作为WDT计数时钟脉冲。

上述的两种方式都是以"复位"的方式来唤醒单片机,不同的是WDT计时溢位所产生的复位,是所谓的"热复位(Warm Reset)",此时仅有SP及PC寄存器被重新设定为0h,单片机的内部电路均维持原来的状态。

③ 中断唤醒:如果在省电模式中,有中断要求发生,致使状态标志位(EIF、TF、ADF、HIF)由0变为1,将使单片机脱离省电模式。此时,若相对的中断有效且栈寄存器还有空间存放返回地址,CPU将跳至对应的中断向量去提取指令执行;否则CPU将执行HALT(即进入省电模式)的下一行指令。

④ PA有1→0的准位变化发生:使用此种方式唤醒的前提是在Options中必须选用Wake-Up功能,唤醒后的CPU将执行HALT的下一行指令。

当唤醒的动作发生时,单片机会延迟1 024个系统时钟脉冲(f_{SYS})周期后才重新执行工作。若是以中断方式唤醒而又需跳至中断向量提取指令,则又会多延迟1个时钟脉冲周期。在需要实时反应的应用中,请读者务必特别留意。

2-13 低电压复位

HT46xx系列单片机提供了低电压复位电路(LVR),用于监测单片机电源电压的变化,欲使用此项自动复位功能,必须在Options中加以选用。选用LVR功能后,若芯片工作电压范围为0.9~3.3 V(如电池正处于充电状态),而且在原状态维持1 ms以上,则LVR会自动将单片机复位,请参考图2-13-1的图例。

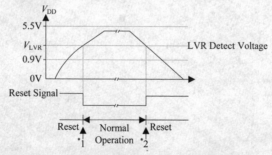

1:为确保振荡电路已处于稳定振荡的状态,在Reset信号解除后会由SST产生1 024个时钟脉冲的延迟,然后单片机才进入正常工作的状态。
2:由于低电压状态必须维持1 ms以上,LVR才会产生系统复位信号,因此在检测到低电压后约1 ms的时间,单片机才会进入复位状态。

图2-13-1 HT46xx系列单片机低电压复位

2-14 配置选项

配置选项(Options)主要是让用户在程序烧录(OTP Type)阶段，对于HT46xx的系统功能及状态加以设定的一些控制位，在程序执行时并无法以任何指令来改变这些位值，芯片要能如预期的正常工作，必须选定适当的选项。Options是提供用户对芯片多重应用的一种解决方案，请读者参考表2-14-1中对于各位的功能说明。

表2-14-1　HT46xx Options功能说明

No.	配置选项(Options)功能说明
1	选择振荡形式，决定是 *RC* 振荡器或是石英振荡器
2	WDT 计数时钟脉冲来源，有 3 种模式选择：① 芯片内部的 *RC* 振荡器；② 指令周期时钟脉冲；③ 关闭 WDT 计数功能
3	"CLR WDT" 次数选择，定义 WDT 的清除方式： "单次清除(One Time)"：只要执行 "CLR WDT" 指令即可将 WDT 清除； "两次清除(Two Times)"：必须执行 "CLR WDT1" 与 "CLR WDT2" 后才可将 WDT 清除
4	"唤醒功能"选项：可以指定 PA(PA7~PA0)的哪些引脚具有将单片机由省电模式中唤醒的功能
5	上拉电阻(Pull-High)选项：选择当 I/O 端口定义为输入模式时是否接上内部上拉电阻(PA7~PA0 可以个别地选定是否要 Pull-High，而其他 I/O 端口则无法个别地选定，必须整个端口一起定义)
6	选择 PA3 为一般的准位(Level)输出或是 PFD 输出的形式
7	PWM 相关选项：① (7+1)Mode 或 (6+2)Mode； ② PD0 为准位(Level)输出或是 PWM0 的输出； ③ PD1 为准位(Level)输出或是 PWM1 的输出； ④ PD2 为准位(Level)输出或是 PWM2 的输出(HT46x24 only)； ⑤ PD3 为准位(Level)输出或是 PWM3 的输出(HT46x24 only)
8	WDT 超时周期相关选项，有 4 种选择：计数时钟脉冲/2^{12}、计数时钟脉冲/2^{13}、计数时钟脉冲/2^{14} 或计数时钟脉冲/2^{15} 另外，还须选择 WDT 的计数时钟脉冲来源为 On-Chip OSC 或指令周期时钟脉冲(如果计数时钟脉冲来源为指令周期时钟脉冲，一旦进入 "HALT" Mode 后，WDT 将不再计数)
9	低电压复位选项：选择是否使用 LVR 功能
10	I^2C Bus 选项：指定 PA7 与 PA6 是一般的 I/O 端口或具备 I^2C BUS 功能

第 3 章

HT46xx 指令集与开发工具

所谓"工欲善其事，必先利其器"，本章除了将说明 HT46xx 的指令之外，也将介绍程序的编译流程与宏的写法。另外，盛群半导体公司为了让用户对其产品的使用能更加得心应手，提供了一套相当完整的开发环境——HT-IDE3000，其集成了编辑、编译、ICE 仿真、虚拟硬件仿真器以及烧录器等功能，建议读者随时浏览盛群半导体公司的网站并下载最新的 IDE 的软件。本章将针对这些功能的操作方式给出详细的说明，主要的内容包括：

3-1　HT46xx寻址模式与指令集(Instruction Set)

3-2　程序的编辑

3-3　HT-IDE3000使用方式与操作

3-4　VPM使用方式与操作

3-5　烧录器操作说明

3-1 HT46xx寻址模式与指令集(Instruction Set)

所谓寻址模式是指CPU查找操作数的方式(或者说是途径)，HT46xx提供5种不同的寻址模式。

① 立即寻址：即运算的常数值直接跟在运算码之后。

```
例如：    MOV    A,50h      ;Acc=50h
         AND    A,55h      ;Acc=Acc AND 55h
```

② 直接寻址：即操作数的地址直接跟在运算码之后。

```
例如：    MOV    A,[50h]    ;Acc=数据存储器地址50h的内容
         ADD    A,[55h]    ;Acc=Acc AND数据存储器地址55h的内容
```

③ 间接寻址：即操作数的地址是存放在特殊的寻址寄存器中(MP0、MP1)。

```
例如：    MOV    A,50h      ;Acc=50h
         MOV    MP0,A
         CLR    IAR0       ;清除MP0(=50h)所指定的数据存储器，
                           ;即执行结果[50h]=00h
```

④ 特殊寄存器寻址：即针对某一特殊寄存器做运算。

```
例如：    CLR    WDT        ;清除看门狗计时器
```

⑤ 指针寻址：此寻址方式主要是执行查表动作。

```
例如：    MOV    A,80h      ;Acc=80h
         MOV    TBLP,A     ;TBLP=80h
         TABRDC [50h]      ;读取目前程序页地址80h的内容,并将低8
                           ;位存放到数据存储;器(地址50h),其余
                           ;位则放置在TBLH寄存器
         TABRDL ACC        ;读取程序存储器最末页地址80h的内容
                           ;(F80h),并将低8;位存放到Acc寄存器，
                           ;其余则放置在TBLH寄存器
```

HT46xx为"类精简指令集"的体系结构，因此具有指令少、指令解码速度快的特性。HT46xx总共有63个指令，依其功能可分为9大类。

① 算术运算指令：包括ADD、ADDM、ADC、ADCM、SUB、SUBM、SBC、SBCM以及DAA指令。

② 逻辑运算指令：报括AND、OR、XOR、ANDM、ORM、XORM、CPL以及CPLA指令。

③ 递增递减指令：包括INCA、INC、DECA以及DEC指令。

④ 循环移位指令：包括RRA、RR、RRCA、RRC、RLA、RL、RLCA以及RLC指令。

⑤ 数据传送指令：即MOV相关指令。

⑥ 位运算指令：即CLR [m].i与SET [m].i指令。

⑦ 分支跳转指令：包括JMP、SZ[m]、SZA、SZ、SNZ、SIZ、SDZ、SIZA、SDZA、CALL以及RET指令。

⑧ 查表专用指令：包括TABRDC与TABRDL指令。

⑨ 其他功能指令：包括NOP、CLR [m]、SET [m]、CLR WDT、CLR WDT1、CLR WDT2、SWAP、SIZA、SWAPA以及HALT指令。

关于指令的执行时间，除了有跳转功能的分支指令(条件成立时的跳转)占2个指令周期之外，其余的指令都只占1个指令周期；而每个指令周期(T_{INT})等于4倍的振荡器振荡周期(4/f_{SYS})。举例来说，若HT46x23工作在4 MHz的频率，则一个指令周期的时间即为：

$$T_{INT}=4\times(4\text{ MHz})^{-1}=1\,\mu\text{s}$$

为方便读者写程序时查阅方便，特将HT46xx的指令集整理于表3-1-1中。若读者想进一步了解指令的动作，请参考本节后续对于个别指令的详细说明(介绍的顺序依照英文字母由小至大排列)。 指令说明之中，各运算符号所代表的意义如下：

"x"　　=8位的常数值(立即寻址)。
"m"　　=8位的数据存储器地址(直接寻址)，为了方便有时候称为"寄存器m"或"m寄存器"。
"Acc"　=累加器(Accumulator)，或称为"Acc寄存器"。
"i"　　=0~7，代表位的位置。
"Addr"=程序存储器地址，HT46x23为12位，HT46x22、HT46x47为11位，HT46x24为13位。
"ˇ"　　=标志位受影响。

表3-1-1　HT46xx指令速查表

助记符		指令功能描述	指令周期	受影响的标志					
				C	AC	Z	OV	PDF	TO
ADC	A,[m]	累加器A、寄存器[m]与进位标志C相加，结果存至累加器A	1	ˇ	ˇ	ˇ	ˇ		
ADCM	A,[m]	寄存器[m]、累加器A与进位标志C相加，结果存至[m]	1①	ˇ	ˇ	ˇ	ˇ		
ADD	A,[m]	累加器A与寄存器[m]相加，结果存至累加器A	1	ˇ	ˇ	ˇ	ˇ		
ADD	A,x	累加器A与常数x相加，结果存至累加器A	1	ˇ	ˇ	ˇ	ˇ		
ADDM	A,[m]	寄存器[m]与累加器A相加，结果存至[m]	1①	ˇ	ˇ	ˇ	ˇ		
AND	A,[m]	累加器A与寄存器[m]执行AND运算，结果存至累加器A	1			ˇ			
AND	A,x	累加器A与常数x执行AND运算，结果存至累加器A	1			ˇ			
ANDM	A,[m]	寄存器[m]与累加器A执行AND运算，结果存至[m]	1①			ˇ			
CALL	Addr	调用子程序指令(PC=Addr)	2						
CLR	[m],i	将寄存器[m]的第i位清除为0 (i=0~7)	1①						
CLR	[m]	将寄存器[m]内容清除为0	1①						
CLR	WDT	清除看门狗定时器	1					ˇ	ˇ
CLR	WDT1	看门狗时器清除指令1	1					②	②
CLR	WDT2	看门狗时器清除指令2	1					②	②
CPL	[m]	对寄存器[m]内容取补数，再将结果回存至[m]	1①			ˇ			
CPLA	[m]	对寄存器[m]内容取补数，再将结果存至A	1			ˇ			
DAA	[m]	累加器A的内容转成BCD码后存至[m]	1①	ˇ					
DEC	[m]	寄存器[m]−1，结果存至寄存器[m]	1①			ˇ			
DECA	[m]	寄存器[m]−1，结果存至累加器A	1			ˇ			
HALT		进入省电模式	1					ˇ	ˇ

助记符	指令功能描述	指令周期	C	AC	Z	OV	PDF	TO
INC [m]	寄存器[m]+1，结果存至[m]	1①			■			
INCA [m]	寄存器[m]+1，结果存至累加器 A	1			■			
JMP Addr	跳转至地址 Addr(PC=Addr)	2						
MOV A,[m]	将寄存器[m]内容放入累加器 A	1						
MOV [m],A	将累加器 A 内容放入寄存器[m]	1①						
MOV A,x	将常数 x 放入寄存器[m]	1						
NOP	不动作	1						
OR A,[m]	累加器 A 与寄存器[m]执行 OR 运算，结果存至累加器 A	1			■			
OR A,x	累加器 A 与常数 x 执行 OR 运算，结果存至累加器 A	1			■			
ORM A,[m]	寄存器[m]与累加器 A 执行 OR 运算，结果存至[m]	1①			■			
RET	子程序返回指令(PC=Top of Stack)	2						
RET A,x	子程序返回指令(PC=Top of Stack)，并将常数 x 放入累加器 A	2						
RETI	中断子程序返回指令(PC=Top of Stack)，并设定 EMI flag=1	2						
RL [m]	寄存器[m]内容左移 1 位	1①						
RLA [m]	寄存器[m]内容左移 1 位后，将结果存至累加器 A	1						
RLC [m]	寄存器[m]内容连同进位标志 C 一起左移 1 位	1①	■					
RLCA [m]	寄存器[m]内容连同进位标志 C 一起左移 1 位后，将结果存至 A	1	■					
RR [m]	寄存器[m]内容右移 1 位	1①						
RRA [m]	寄存器[m]内容右移 1 位后，将结果存至累加器 A	1						
RRC [m]	寄存器[m]内容连同进位标志 C 一起右移 1 位	1①	■					
RRCA [m]	寄存器[m]内容连同进位标志 C 一起右移 1 位后，将结果存至 A	1	■					
SBC A,[m]	累加器 A 与寄存器[m]、进位标志 C 相减，结果存至 A	1	■	■	■	■		
SBCM A,[m]	累加器 A 与寄存器[m]、进位标志 C 相减，结果存至[m]	1①	■	■	■	■		
SDZ [m]	将寄存器[m]−1 结果存至寄存器[m]，若结果为 0 则跳过下一行	1①						
SDZA [m]	将寄存器[m]−1 结果存至累加器 A，若结果为 0 则跳过下一行	1						
SET [m],i	将寄存器[m]的第 i 位设定为 1 (i=0~7)	1①						
SET [m]	将寄存器[m]内容设定为 FFh	1①						
SIZ [m]	将寄存器[m]+1 结果存至寄存器[m]，若结果为 0 则跳过下一行	1①						
SIZA [m]	将寄存器[m]+1 结果存至累加器 A，若结果为 0 则跳过下一行	1						
SNZ [m],i	若寄存器[m]的第 i(i=0~7)位不为 0 则跳过下一行	1①						
SUB A,x	累加器 A 与常数 x 相减，结果存至累加器 A	1	■	■	■	■		
SUB A,[m]	累加器 A 与寄存器[m]相减，结果存至累加器 A	1	■	■	■	■		
SUBM A,[m]	累加器 A 与寄存器[m]相减，结果存至[m]	1①	■	■	■	■		
SWAP [m]	将寄存器[m]的高低四位互换	1①						
SWAPA [m]	将寄存器[m]的高低四位互换后的结果存至累加器 A	1						
SZ [m]	若寄存器[m]内容为 0 则跳过下一行	1②						
SZ [m],i	若寄存器[m]的第 i(i=0~7)位为 0 则跳过下一行	1②						
SZA [m]	将寄存器[m]内容存至累加器 A，若为 0 则跳过下一行	1②						
TABRDC [m]	依据 TBLP 读取程序存储器(目前页)的值并存放至 TBLH 与[m]	2①						
TABRDL [m]	依据 TBLP 读取程序存储器(最末页)的值并存放至 TBLH 与[m]	2①						
XOR A,[m]	累加器 A 与寄存器[m]执行 XOR 运算，结果存至累加器 A	1			■			
XOR A,x	累加器 A 与常数 x 执行 XOR 运算，结果存至累加器 A	1			■			
XORM A,[m]	累加器 A 与寄存器[m]执行 XOR 运算，结果存至[m]	1①			■			

第3章 HT46xx指令集与开发工具

说明：i=某个位(0~7)；x=8位常数；[m]=数据存储器位置；Addr=程序存储器位置；■=标志受影响；□=标志不受影响。

① 若有载入数值至PCL寄存器，则执行时间增加1个指令周期。
② 若条件成立，跳转到下一行指令执行时，则执行时间增加1个指令周期。
③ 执行"CLR WDT1"与"CLR WTD2"后，TO和PDF标志会被清0；若只是单独执行任何一个指令，则TO及PDF标志并不受影响。

指 令：	ADC A,[m]	影响标志				
		TO	PDF	OV	Z	AC	C
动 作：	Acc←Acc+[m]+C			✓	✓	✓	✓

功能说明： Acc的内容值、地址m的数据存储器内容值与C(进位标志位)相加，并将运算结果存入Acc寄存器。
范　例： ADC A,[40h]
执行前： 地址40h的数据存储器内容为1Fh，寄存器Acc内容为01h，C=1。
执行后： 地址40h的数据存储器内容为1Fh，寄存器Acc内容为21h。
注　意： "m"代表地址m的数据存储器，为方便说明，以后一律以寄存器m(或m寄存器)称之。

指 令：	ADCM A,[m]	影响标志				
		TO	PDF	OV	Z	AC	C
动 作：	[m]←Acc+[m]+C			✓	✓	✓	✓

功能说明： Acc的内容值、寄存器m的内容值与C(进位标志)相加，并将运算结果存入寄存器m。
范　例： ADCM A,[40h]
执行前： 寄存器40h内容为1Fh，寄存器Acc内容为01h，C=0。
执行后： 寄存器40h内容为20h，寄存器Acc内容为01h。

指 令：	ADD A,[m]	影响标志				
		TO	PDF	OV	Z	AC	C
动 作：	Acc←Acc+[m]			✓	✓	✓	✓

功能说明： Acc的内容值与寄存器m的内容相加，并将运算结果存入Acc寄存器。
范　例： ADD A,[40h]
执行前： 寄存器40h内容为30h，寄存器Acc内容为20h。
执行后： 寄存器40h内容为30h，寄存器Acc内容为50h。

指 令：	ADD A,x	影响标志				
		TO	PDF	OV	Z	AC	C
动 作：	Acc←Acc+x			✓	✓	✓	✓

功能说明： Acc的内容值与常数x相加，并将运算结果存入Acc寄存器。
范　例： ADD A,40h
执行前： 寄存器Acc内容为20h。
执行后： 寄存器Acc内容为60h。

指　令：	ADDM A,[m]	影响标志
		TO　PDF　OV　Z　AC　C
动　作：	[m]←Acc+[m]	

功能说明：Acc 的内容值与寄存器 m 的内容相加，并将运算结果存入 m 寄存器。
范　　例：ADDM A,[40h]
执 行 前：寄存器 40h 内容为 30h，寄存器 Acc 内容为 20h。
执 行 后：寄存器 40h 内容为 50h，寄存器 Acc 内容为 20h。

　　　　　ADD(M)、ADC(M)指令通常在 8 位以上的加法运算中配合使用，下面是将两笔 16 位数据相加的范例(WORD2＝WORD1＋WORD2)：

范　　例：
```
        ……                          ;数据存储器定义区
        WORD1       DW    ?         ;保留 2Byte 的数据空间
        WORD2       DW    ?         ;保留 2Byte 的数据空间
        ……                          ;程序区
        MOV         A,WORD1[0]      ;取得 Low Bytes
        ADDM        A,WORD2[0]      ;Low Byte 相加
        MOV         A,WORD1[1]      ;取得 High Bytes
        ADCM        A,WORD2[1]      ;High Byte 相加
```
执 行 前：[WORD1]＝5566h，[WORD2]＝33AAh，Acc＝36h。
执 行 后：[WORD1]＝5566h，[WORD2]＝8910h，Acc＝55h。
注　　意：在 HT 的 Assembly Language 中，m[N]代表 m＋N 的数据存储器地址。

指　令：	AND A,[m]	影响标志
		TO　PDF　OV　Z　AC　C
动　作：	Acc←Acc "AND" [m]	

功能说明：Acc 的内容值与寄存器 m 的内容执行"AND"运算，并将结果存入 Acc 寄存器。
范　　例：AND A,[40h]
执 行 前：寄存器 40h 内容为 88h，寄存器 Acc 内容为 08h。
执 行 后：寄存器 40h 内容为 88h，寄存器 Acc 内容为 08h。

指　令：	AND A,x	影响标志
		TO　PDF　OV　Z　AC　C
动　作：	Acc←Acc "AND" x	

功能说明：Acc 的内容值与常数 x 执行"AND"运算，并将结果存入 Acc 寄存器。
范　　例：AND A,F0h
执 行 前：寄存器 Acc 内容为 88h。
执 行 后：寄存器 Acc 内容为 80h。

指　令：	ANDM A,[m]	影响标志
		TO　PDF　OV　Z　AC　C
动　作：	[m]←Acc "AND" [m]	

第 3 章　HT46xx 指令集与开发工具

功能说明：以 Acc 的内容值与寄存器 m 的内容执行 "AND" 运算，并将结果存入 m 寄存器。
范　　例：ANDM A,[40h]
执 行 前：寄存器 40h 内容为 88h，寄存器 Acc 内容为 08h。
执 行 后：寄存器 40h 内容为 08h，寄存器 Acc 内容为 08h。

指　令：	CALL Addr	影响标志					
		TO	PDF	OV	Z	AC	C
动　作：	PC←Addr,Stack←PC+1						

功能说明：调用 Addr 所在地址的子程序，此时 PC 值已经加 1(指向下一个指令地址)，先将此 PC 值存放到栈寄存器中(Stack Memory)之后，再将 Addr 放到 PC，达到跳转的目的。
范　　例：CALL READ_KEY
执 行 前：PC=110h，READ_KEY=300h，Stack=000h。
执 行 后：PC=300h，READ_KEY=300h，Stack=111h。

指　令：	CLR [m].i	影响标志					
		TO	PDF	OV	Z	AC	C
动　作：	[m].i←"0"						

功能说明：将寄存器 m 的第 i 位清 "0"。
范　　例：CLR [40h].2
执 行 前：[40h]=66h。
执 行 后：[40h]=60h。

指　令：	CLR [m]	影响标志					
		TO	PDF	OV	Z	AC	C
动　作：	[m]←"00h"						

功能说明：将寄存器 m 清为 "00h"。
范　　例：CLR [40h]
执 行 前：[40h]=66h。
执 行 后：[40h]=00h。

指　令：	CLR WDT	影响标志					
		TO	PDF	OV	Z	AC	C
动　作：	WDT 及 WDT 计时器←00h,PDF 及 TO←0	0	0				

功能说明：将 WDT 与 WDT 计时器予以清除，同时 TO 及 PDF 位也会被清除为 "0"。
范　　例：CLR WDT
执 行 前：WDT 计时器=86h，TO= "1"，PDF= "1"。
执 行 后：WDT 计时器=00h，TO= "0"，PDF= "0"。

指　令：	CLR WDT1	影响标志					
		TO	PDF	OV	Z	AC	C
动　作：	WDT 及 WDT 计时器←00h,PDF 及 TO←0	0*	0*				

53

功能说明: 当在 Options 选项中选择两个指令清除 WDT 时才有用。必须配合"CLR WDT2"指令,必须在"CLR WDT1"与"CLR WDT2"指令均执行过的情况下,才可以将 WDT 与 WDT 计时器予以清除,同时 TO 及 PDF 位也会被清 0;否则并不会影响 TO 及 PDF 位。

范　　例: `CLR　WDT1`
　　　　　`CLR　WDT2`

执 行 前: WDT 计时器=86h,TO=1,PDF=1。

执 行 后: WDT 计时器=00h,TO=0,PDF=0。

指　令:	CLR WDT2	影响标志				
		TO	PDF	OV	Z	AC	C
动　作:	WDT 及 WDT 计时器←00h,PDF 及 TO←0	0*	0*				

功能说明: 当在 Options 选项中选择两个指令清除 WDT 时才有用。必须配合"CLR WDT1"指令,必须在"CLR WDT1"与"CLR WDT2"指令均执行过的情况下,才可以将 WDT 与 WDT 计时器予以清除,同时 TO 及 PDF 位也会被清 0;否则并不会影响 TO 及 PDF 位。

范　　例: `CLR　WDT1`
　　　　　`CLR　WDT2`

执 行 前: WDT 计时器=86h,TO=1,PDF=1。

执 行 后: WDT 计时器=00h,TO=0,PDF=0。

指　令:	CPL [m]	影响标志					
		TO	PDF	OV	Z	AC	C
动　作:	[m]←/[m]				✓		

功能说明: 将寄存器 m 取 1 的补数(1's Complement)。

范　　例: `CPL [40h]`

执 行 前: [40h]=55h。

执 行 后: [40h]=AAh。

指　令:	CPLA [m]	影响标志					
		TO	PDF	OV	Z	AC	C
动　作:	Acc←/[m]				✓		

功能说明: 将寄存器 m 取 1 的补数(1's Complement),并将结果存入 Acc 寄存器。

范　　例: `CPLA [40h]`

执 行 前: [40h]=55h,Acc=88h。

执 行 后: [40h]=55h,Acc=AAh。

指　令:	DAA [m]	影响标志					
		TO	PDF	OV	Z	AC	C

动作:	If Acc.3~Acc.0>9 or AC=1 Then [m].3~[m].0←Acc.3~Acc.0+6,AC1=/AC Else 　　　[m].3~[m].0←Acc.3~Acc.0+6,AC1=1 and If Acc.7~Acc.4>9 or C=1 Then [m].7~[m].4←Acc.7~Acc.4+6+AC1,C=0 Else 　　　[m].7~[m].4←Acc.7~Acc.4+AC1,C=C

功能说明： 将 Acc 寄存器内容调整为 BCD 格式，并将结果存入 m 寄存器。

范　　例： `ADD  A,05h`
　　　　　`DAA  [40h]`

执 行 前： [40h]=55h，Acc=89h。

执 行 后： [40h]=94h，Acc=89h。

"DAA"在 BCD(Binary Coded Decimal)的加法运算上是极为重要的调整指令，若能善加利用可以省去很多麻烦。例如，要求得 1+2+3+…+10(答案=55)的结果，程序写法如下：

范　　例：
```
      #INCLUDE    HT46R23.INC     ;载入寄存器定义文件
      ……                          ;数据存储器定义区
      COUNT   DB    ?             ;保留 1Byte 的数据空间
      ……                          ;程序区
      MOV     A,10
      MOV     COUNT,A             ;设定 COUNT=10,因为有 10 笔数据相加
      CLR     ACC                 ;Acc=0
NEXT: ADD     A,COUNT             ;累加
      DAA     ACC                 ;进行 BCD 调整
      SDZ     COUNT               ;10 笔数据加完了吗
      JMP     NEXT                ;未加完,继续累加
      NOP                         ;10 笔数据加完了
      ……
```

上列的程序执行到"NOP"时，Acc寄存器=55h。如果读者将"DAA ACC"调整指令舍去，执行的结果为Acc寄存器=37h(十进制的55)。

指　令：	DEC [m]	影响标志
		TO PDF OV Z AC C
动　作：	[m]←[m]−1	

功能说明： 将 m 寄存器内容减 1，结果存回 m 寄存器。

范　　例： `DEC [40h]`

执 行 前： [40h]=55h，Acc=89h。

执 行 后： [40h]=54h，Acc=89h。

指　令：	DECA [m]	影响标志
		TO PDF OV Z AC C
动　作：	Acc←[m]−1	

功能说明：将 m 寄存器内容减 1，结果存回 Acc 寄存器。
范　　例：DECA [40h]
执 行 前：[40h]=55h，Acc=89h。
执 行 后：[40h]=55h，Acc=54h。

指　令：	HALT	影响标志					
		TO	PDF	OV	Z	AC	C
动　作：	PC←PC+1,PDF←1,TO←0	0	1				

功能说明："HALT" 指令将使单片机进入省电模式(Power-Down Mode)；此时系统频率将被关闭，因此程序将停止执行。不过在进入省电模式之前，会先清除 WDT 以及 WDT 计数器，并将 PDF 设定为 1、TO 清 0。进入省电模式后，数据存储器以及寄存器的内容将保持不变。

范　　例：HALT
执 行 前：PC=100h，PDF=0，TO=0。
执 行 后：PC=101h，PDF=1，TO=0。

指　令：	INC [m]	影响标志					
		TO	PDF	OV	Z	AC	C
动　作：	[m]←[m]+1						

功能说明：将 m 寄存器内容加 1，结果存回 m 寄存器。
范　　例：INC [40h]
执 行 前：[40h]=55h，Acc=89h。
执 行 后：[40h]=56h，Acc=89h。

指　令：	INCA [m]	影响标志					
		TO	PDF	OV	Z	AC	C
动　作：	Acc←[m]+1						

功能说明：将 m 寄存器内容加 1，结果存回 Acc 寄存器。
范　　例：INCA [40h]
执 行 前：[40h]=55h，Acc=89h。
执 行 后：[40h]=55h，Acc=56h。

指　令：	JMP Addr	影响标志					
		TO	PDF	OV	Z	AC	C
动　作：	PC←Addr						

功能说明：将欲跳转的目的地址(Addr)放到 PC，达到跳转的功能。
范　　例：JMP START
执 行 前：PC=110h，START=000h。
执 行 后：PC=000h，START=000h。

指　令：	MOV A,[m]	影响标志					
		TO	PDF	OV	Z	AC	C
动　作：	Acc←[m]						

功能说明：将 m 寄存器的内容复制到 Acc 寄存器。
范　　例：MOV A,[40h]
执 行 前：[40h]=55h，Acc=89h。

第3章 HT46xx 指令集与开发工具

执 行 后：[40h]=55h，Acc=55h。

指 令：	MOV [m],A		影响标志				
动 作：	[m]←Acc		TO	PDF	OV	Z	AC	C

功能说明：将 Acc 寄存器的内容复制到 m 寄存器。
范　　例：MOV [40h],A
执 行 前：[40h]=55h，Acc=89h。
执 行 后：[40h]=89h，Acc=89h。

指 令：	MOV A,x		影响标志				
动 作：	Acc←x		TO	PDF	OV	Z	AC	C

功能说明：将立即数(常数 x)传送到 Acc 寄存器。
范　　例：MOV A,40h
执 行 前：Acc=89h。
执 行 后：Acc=40h。

指 令：	NOP		影响标志				
动 作：	PC←PC+1		TO	PDF	OV	Z	AC	C

功能说明：不执行任何运算，但是 PC 值会加 1。虽然是不执行任何运算，但执行 NOP 指令仍是会耗费一个指令周期的时间。因此，除了用来填补程序空间之外，有时候也当作延迟时间之用。

范　　例：NOP
执 行 前：PC=89h。
执 行 后：PC=8Ah。

指 令：	OR A,[m]		影响标志				
动 作：	Acc←Acc"OR"[m]		TO	PDF	OV	Z	AC	C

功能说明：Acc 的内容值与寄存器 m 的内容执行"OR"运算，并将结果存入 Acc 寄存器。
范　　例：OR A,[40h]
执 行 前：寄存器 40h 内容为 80h，寄存器 Acc 内容为 08h。
执 行 后：寄存器 40h 内容为 80h，寄存器 Acc 内容为 88h。

指 令：	OR A,x		影响标志				
动 作：	Acc←Acc"OR"x		TO	PDF	OV	Z	AC	C

功能说明：Acc 的内容值与常数值 x 执行"OR"运算，并将结果存入 Acc 寄存器。
范　　例：OR A,40h
执 行 前：寄存器 Acc 内容为 08h。
执 行 后：寄存器 Acc 内容为 48h。

指 令：	ORM A,[m]		影响标志				
动 作：	[m]←Acc"OR"[m]		TO	PDF	OV	Z	AC	C

功能说明：Acc 的内容值与寄存器 m 的内容执行 "OR" 运算，并将结果存入 m 寄存器。
范　　例：ORM A,[40h]
执 行 前：寄存器 40h 内容为 80h，寄存器 Acc 内容为 08h。
执 行 后：寄存器 40h 内容为 88h，寄存器 Acc 内容为 08h。

指　令：	RET	影响标志					
		TO	PDF	OV	Z	AC	C
动　作：	PC←Stack						

功能说明：将栈寄存器所存放的 PC 值取回，并存入 PC。
范　　例：RET
执 行 前：PC=066h, Stack=888h。
执 行 后：PC=888h, Stack=888h。

指　令：	RET A,x	影响标志					
		TO	PDF	OV	Z	AC	C
动　作：	PC←Stack,Acc←x						

功能说明：将栈寄存器所存放的 PC 值取回，并存入 PC；同时将常数值 x 放入 Acc 寄存器。
范　　例：RET A,58h
执 行 前：PC=066h, Stack=888h, Acc=34h。
执 行 后：PC=888h, Stack=888h, Acc=58h。

指　令：	RETI	影响标志					
		TO	PDF	OV	Z	AC	C
动　作：	PC←Stack,EMI←"1"						

功能说明：将栈寄存器所存放的 PC 值取回，并存入 PC；并把 EMI 位设定为 1。RETI 通常是中断服务子程序的最后一个指令，因为在进入中断服务子程序时，单片机会自动清除 EMI 位，以防止其他中断再发生。因此，利用 RETI 指令在返回主程序的同时，将中断重新使能。

范　　例：RETI
执 行 前：PC=066h, Stack=888h, EMI=0。
执 行 后：PC=888h, Stack=888h, EMI=1。

指　令：	RL [m]	影响标志					
		TO	PDF	OV	Z	AC	C
动　作：	[m].(i+1)←[m].i for i=0~6 [m].0←[m].7						

功能说明：将寄存器 m 的内容左移，并将结果存回 m 寄存器，如下所示：

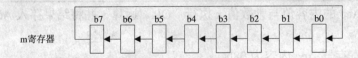

范　　例：RL [40h]
执 行 前：[40h]=44h。
执 行 后：[40h]=88h。

第3章 HT46xx 指令集与开发工具

指 令：	RLA [m]	影响标志
动 作：	Acc.(i+1) ← [m].i for i=0~6 Acc.0 ← [m].7	TO PDF OV Z AC C

功能说明： 将寄存器 m 的内容左移，并将结果存回 Acc 寄存器，如下所示：

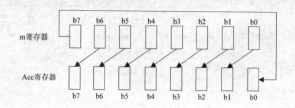

范　例： RLA [40h]
执行前： [40h]=44h, Acc=66h。
执行后： [40h]=44h, Acc=88h。

指 令：	RLC [m]	影响标志
动 作：	[m].(i+1) ← [m].i for i=0~6 [m].0 ← C, C ← [m].7	TO PDF OV Z AC C

功能说明： 将寄存器 m 的内容伴随进位标志(C)一起左移，并将结果存回 m 寄存器，如下所示：

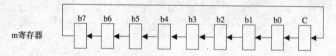

范　例： RLC [40h]
执行前： [40h]=41h, C=1。
执行后： [40h]=83h, C=0。

指 令：	RLCA [m]	影响标志
动 作：	Acc.(i+1) ← [m].i for i=0~6 Acc.0 ← C, C ← [m].7	TO PDF OV Z AC C

功能说明： 将寄存器 m 的内容伴随进位标志(C)一起左移，并将结果存回 Acc 寄存器，如下所示：

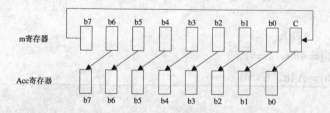

范　例： RLCA [40h]
执行前： [40h]=41h, Acc=66h, C=1。
执行后： [40h]=41h, Acc=83h, C=0。

指　令：	RR [m]	影响标志
		TO　PDF　OV　Z　AC　C
动　作：	[m].i ← [m].(i+1) for i=0~6 [m].7 ← [m].0	

功能说明：将寄存器 m 的内容右移，并将结果存回 m 寄存器，如下所示：

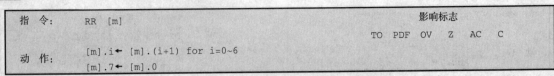

范　例： RR [40h]
执行前： [40h]=45h。
执行后： [40h]=A2h。

指　令：	RRA [m]	影响标志
		TO　PDF　OV　Z　AC　C
动　作：	Acc.i ← [m].(i+1) for i=0~6 Acc.7 ← C	

功能说明：将寄存器 m 的内容右移，并将结果存回 Acc 寄存器，如下所示：

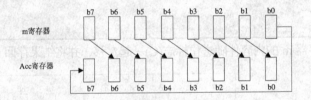

范　例： RRA [40h]
执行前： [40h]=45h，Acc=66h。
执行后： [40h]=45h，Acc=A2h。

指　令：	RRC [m]	影响标志
		TO　PDF　OV　Z　AC　C
动　作：	[m].i ← [m].(i+1) for i=0~6 [m].7 ← C,C ← [m].0	✓

功能说明：将寄存器 m 的内容伴随进位标志(C)一起右移，并将结果存回 m 寄存器，如下所示：

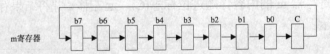

范　例： RRC [40h]
执行前： [40h]=48h，C=1。
执行后： [40h]=A4h，C=0。

指　令：	RRCA [m]	影响标志
		TO　PDF　OV　Z　AC　C
动　作：	Acc.i ← [m].(i+1) for i=0~6 Acc.7 ← C,C ← [m].0	✓

功能说明：将寄存器 m 的内容伴随进位标志(C)一起右移，并将结果存回 Acc 寄存器，如下所示：

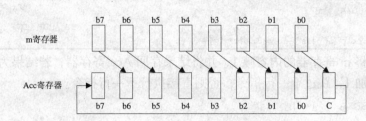

范　例：	RRCA [40h]						
执行前：	[40h]=48h，Acc=66h，C=1。						
执行后：	[40h]=48h，Acc=A4h，C=0。						

指　令：	SBC A,[m]	影响标志					
		TO	PDF	OV	Z	AC	C
动　作：	Acc←Acc+/[m]+C			✓	✓	✓	✓

功能说明：将 Acc 寄存器与 m 寄存器的内容相减(考虑进位标志)，并将结果存回 Acc 寄存器。

范　例： SBC A,[40h]

执行前： Acc=50h，[40h]=30h，C=1(状况一)。

执行后： Acc=20h，[40h]=30h，C=1。

执行前： Acc=50h，[40h]=30h，C=0(状况二)。

执行后： Acc=1Ah，[40h]=30h，C=1。

指　令：	SBCM A,[m]	影响标志					
		TO	PDF	OV	Z	AC	C
动　作：	[m]←Acc+/[m]+C			✓	✓	✓	✓

功能说明：将 Acc 寄存器与 m 寄存器的内容相减(考虑进位标志)，并将结果存回 m 寄存器。

范　例： SBCM A,[40h]

执行前： Acc=50h，[40h]=30h，C=1(状况一)。

执行后： Acc=50h，[40h]=20h，C=1。

执行前： Acc=50h，[40h]=30h，C=0(状况二)。

执行后： Acc=50h，[40h]=1Ah，C=1。

指　令：	SDZ [m]	影响标志					
		TO	PDF	OV	Z	AC	C
动　作：	Skip if ([m]−1)=0,[m]←[m]−1						

功能说明：将 m 寄存器的内容减 1，并将结果存回 m 寄存器，若结果为 0，就将 PC 值再加 1，跳过下一行指令；否则直接执行下一行指令。

范　例： SDZ [40h]

　　　　　 RR　[41h]　　　　;[40h]−1≠0

　　　　　 RL　[41h]　　　　;[40h]−1=0

执行前： [40h]=50h，[41h]=21h(状况一)。

执行后： [40h]=4Ah，[41h]=21h。

执行前： [40h]=01h，[41h]=21h(状况二)。

执行后： [40h]=00h，[41h]=42h。

指　令：	SDZA [m]	影响标志
动　作：	Skip if ([m]−1)=0, Acc←[m]−1	TO　PDF　OV　Z　AC　C

功能说明：将 m 寄存器的内容减 1，并将结果存回 Acc 寄存器，若结果为 0，就将 PC 值再加 1，跳过下一行指令；否则直接执行下一行指令。

范　　例：　SDZA　[40h]
　　　　　　RR　　[41h]　　　　；[40h]−1≠0
　　　　　　RL　　[41h]　　　　；[40h]−1=0

执 行 前：[40h]=50h，[41h]=21h，Acc=38h(状况一)。
执 行 后：[40h]=50h，[41h]=21h，Acc=4Ah。
执 行 前：[40h]=01h，[41h]=21h，Acc=38h(状况二)。
执 行 后：[40h]=01h，[41h]=42h，Acc=00h。

指　令：	SET [m]	影响标志
动　作：	[m]←"FFh"	TO　PDF　OV　Z　AC　C

功能说明：将寄存器 m 的内容设定为 FFh。

范　　例：　SET　[40h]
执 行 前：[40h]=86h。
执 行 后：[40h]=FFh。

指　令：	SET [m].i	影响标志
动　作：	[m]←"1"	TO　PDF　OV　Z　AC　C

功能说明：将寄存器 m 的第 i 位设定为 1。

范　　例：　SET　[40h].3
执 行 前：[40h]=60h。
执 行 后：[40h]=68h。

指　令：	SIZ [m]	影响标志
动　作：	Skip if ([m]+1)=0, [m]←[m]+1	TO　PDF　OV　Z　AC　C

功能说明：将 m 寄存器的内容加 1，并将结果存回 m 寄存器，若结果为 0，就将 PC 值再加 1，跳过下一行指令；否则直接执行下一行指令。

范　　例：　SIZ　[40h]
　　　　　　RR　　[41h]　　　　；[40h]+1≠0
　　　　　　RL　　[41h]　　　　；[40h]+1=0

执 行 前：[40h]=50h，[41h]=21h(状况一)。
执 行 后：[40h]=51h，[41h]=21h。
执 行 前：[40h]=FFh，[41h]=21h(状况二)。
执 行 后：[40h]=00h，[41h]=42h。

指　令：	SIZA　[m]			影响标志				
				TO	PDF	OV	Z	AC　C
动　作：	Skip if ([m]+1)=0, Acc←[m]+1							

功能说明：　将 m 寄存器的内容加 1，并将结果存回 Acc 寄存器，若结果为 0，就将 PC 值再
　　　　　　加 1，跳过下一行指令；否则直接执行下一行指令。

范　　例：　SIZA　　[40h]
　　　　　　RR　　　[41h]　　　　　　　；[40h]−1≠0
　　　　　　RL　　　[41h]　　　　　　　；[40h]−1=0

执 行 前：　[40h]=50h, [41h]=21h, Acc=38h(状况一)。
执 行 后：　[40h]=50h, [41h]=21h, Acc=4Ah。
执 行 前：　[40h]=FFh, [41h]=21h, Acc=38h(状况二)。
执 行 后：　[40h]=01h, [41h]=42h, Acc=00h。

指　令：	SNZ　[m].i			影响标志				
				TO	PDF	OV	Z	AC　C
动　作：	Skip if [m].i≠"0"							

功能说明：　判断寄存器 m 的第 i 位是否为 1，若是 1 就将 PC 值再加 1，跳过下一行指令；
　　　　　　则直接执行下一行指令。

范　　例：　SNZ　　[40h].7
　　　　　　RR　　　[41h]　　　　　　　；[40h].7="0"
　　　　　　RL　　　[41h]　　　　　　　；[40h].7="1"

执 行 前：　[40h]=60h, [41h]=21h(状况一)。
执 行 后：　[40h]=60h, [41h]=21h。
执 行 前：　[40h]=80h, [41h]=21h(状况二)。
执 行 后：　[40h]=80h, [41h]=42h。

指　令：	SUB　A,[m]			影响标志				
				TO	PDF	OV	Z	AC　C
动　作：	Acc←Acc+/[m]+1					✓	✓	✓　✓

功能说明：　将 Acc 寄存器与 m 寄存器的内容相减(不考虑进位标志)，并将结果存回 Acc 寄
　　　　　　存器。

范　　例：　SUB　A,[40h]
执 行 前：　Acc=50h, [40h]=30h。
执 行 后：　Acc=20h, [40h]=30h。

指　令：	SUBM　A,[m]			影响标志				
				TO	PDF	OV	Z	AC　C
动　作：	[m]←Acc+/[m]+C					✓	✓	✓　✓

功能说明：　将 Acc 寄存器与 m 寄存器的内容相减(不考虑进位标志)，并将结果存回 Acc 寄
　　　　　　存器。

范　　例：　SUBM　A,[40h]
执 行 前：　Acc=50h, [40h]=30h。
执 行 后：　Acc=50h, [40h]=20h。

SUB(M)、SBC(M)指令通常在 8 位以上的减法运算中配合运用,以下是将两笔 16 位数据相减的范例(WORD2＝WORD1－WORD2):

```
范    例:  ......                ;数据存储器定义区
           WORD1   DW   ?        ;保留 2Byte 的数据空间
           WORD2   DW   ?        ;保留 2Byte 的数据空间
           ......                ;程序区
           MOV    A,WORD1[0]     ;取得 Low Bytes
           SUBM   A,WORD2[0]     ;Low Byte 相减
           MOV    A,WORD1[1]     ;取得 High Bytes
           SBCM   A,WORD2[1]     ;High Byte 相减
```

执 行 前: [WORD1]＝5566h,[WORD2]＝33AAh,Acc＝36h。
执 行 后: [WORD1]＝5566h,[WORD2]＝21BCh,Acc＝55h。
注   意: 在 HT 的 Assembly Language 中,m[N]代表 m＋N 的数据存储器地址。

指　令:	SUB A,x	影响标志					
		TO	PDF	OV	Z	AC	C
动　作:	Acc←Acc＋/x＋1			✓	✓	✓	✓

功能说明: 将 Acc 寄存器与常数 x 相减,并将结果存回 Acc 寄存器。
范   例: SUB A,40h
执 行 前: Acc＝50h。
执 行 后: Acc＝10h。

指　令:	SWAP [m]	影响标志					
		TO	PDF	OV	Z	AC	C
动　作:	[m].3~[m].0↔[m].7~[m].4						

功能说明: 将寄存器 m 的 High Nibble(高 4 位;b7~b4)与 Low Nibble(低 4 位;b3~b0)互换,并将结果存回 m 寄存器,如下所示:

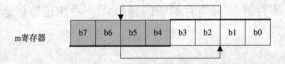

范   例: SWAP [40h]
执 行 前: [40h]＝28h。
执 行 后: [40h]＝82h。

指　令:	SWAPA [m]	影响标志					
		TO	PDF	OV	Z	AC	C
动　作:	Acc.3~Acc.0←[m].3~[m].0 Acc.7~Acc.4←[m].7~[m].4						

功能说明: 将寄存器 m 的 High Nibble(高 4 位;b7~b4)与 Low Nibble(低 4 位;b3~b0)互换,并将结果存回 Acc 寄存器,如下所示:

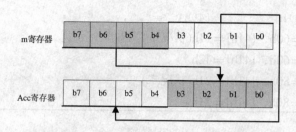

范　例： SWAPA [40h]
执行前： [40h]=28h, Acc=66h。
执行后： [40h]=28h, Acc=82h。

指　令：	SZ [m]	影响标志					
动　作：	Skip if [m]="00h"	TO	PDF	OV	Z	AC	C

功能说明： 判断寄存器 m 的内容是否为 00h, 若是就将 PC 值再加 1, 跳过下一行指令; 否则直接执行下一行指令。

范　例： SZ [40h]
　　　　 RR [41h]　　　　　;[40h]≠"00h"
　　　　 RL [41h]　　　　　;[40h]="00h"
执行前： [40h]=60h, [41h]=21h(状况一)。
执行后： [40h]=60h, [41h]=21h。
执行前： [40h]=00h, [41h]=21h(状况二)。
执行后： [40h]=00h, [41h]=42h。

指　令：	SZA [m]	影响标志					
动　作：	Skip if [m]="00h", Acc←[m]	TO	PDF	OV	Z	AC	C

功能说明： 将寄存器 m 的内容复制到 Acc 寄存器, 并检查寄存器 m 的内容是否为 00h, 若是就将 PC 值再加 1, 跳过下一行指令; 否则直接执行下一行指令。

范　例： SZA [40h]
　　　　 RR [41h]　　　　　;[40h]≠"00h"
　　　　 RL [41h]　　　　　;[40h]="00h"
执行前： [40h]=60h, [41h]=21h, Acc=88h(状况一)。
执行后： [40h]=60h, [41h]=21h, Acc=60h。
执行前： [40h]=00h, [41h]=21h, Acc=88h(状况二)。
执行后： [40h]=00h, [41h]=42h, Acc=00h。

指　令：	SZ [m].i	影响标志					
动　作：	Skip if [m].i="0"	TO	PDF	OV	Z	AC	C

功能说明： 判断寄存器 m 的第 i 位是否为 0, 若是 0 就将 PC 值再加 1, 跳过下一行指令; 否则直接执行下一行指令。

范　例： SZ [40h].7
　　　　 RR [41h]　　　　　;[40h].7="0"

```
                    RL  [41h]           ;[40h].7="1"
```
执 行 前： [40h]=60h，[41h]=21h(状况一)。
执 行 后： [40h]=60h，[41h]=42h。
执 行 前： [40h]=80h，[41h]=21h(状况二)。
执 行 后： [40h]=80h，[41h]=21h。

指 令：	TABRDC [m]	影响标志					
		TO	PDF	OV	Z	AC	C
动 作：	[m]←ROM Code (Low Byte) TBLH←ROM Code (Low Byte)						

功 能 说 明： TBLP(8Bit)寄存器的内容为地址，到程序存储器的"目前页"提取数据，并将低位组数据存入寄存器 m，高位组数据存入 TBLH 寄存器(C：Current Page)。

范　　例： `TABRDC [40h]`

执 行 前： [40h]=50h，TBLP=30h，TBLH=F4h，[170h]=5566h，[F70h]=3388h(HT46x23程序存储器的"最末页")，PC=105h(所以"目前页"为 100h~1FFh)。

执 行 后： [40h]=66h，TBLP=30h，TBLH=55h，[170h]=5566h，[F70h]=3388h(HT46x23程序存储器的"最末页")，PC=106h。

指 令：	TABRDL [m]	影响标志					
		TO	PDF	OV	Z	AC	C
动 作：	[m]←ROM Code (Low Byte) TBLH←ROM Code (Low Byte)						

功 能 说 明： TBLP(8Bit)寄存器的内容为地址，到程序存储器的"最末页"提取数据，并将低位组数据存入寄存器 m，高位组数据存入 TBLH 寄存器。(L：Last Page)

范　　例： `TABRDL [40h]`

执 行 前： [40h]=50h，TBLP=30h，TBLH=F4h，[170h]=5566h，[F70h]=3388h(HT46x23程序存储器的"最末页")，PC=105h(所以"目前页"为 100h~1FFh)。

执 行 后： [40h]=88h，TBLP=30h，TBLH=33h，[170h]=5566h，[F70h]=3388h(HT46x23程序存储器的"最末页")，PC=106h。

指 令：	XOR A,[m]	影响标志					
		TO	PDF	OV	Z	AC	C
动 作：	Acc←Acc "XOR" [m]				✓		

功 能 说 明： Acc 的内容值与寄存器 m 的内容执行"XOR"运算，并将结果存入 Acc 寄存器。

范　　例： `XOR A,[40h]`

执 行 前： 寄存器 40h 内容为 55h，寄存器 Acc 内容为 AAh。
执 行 后： 寄存器 40h 内容为 55h，寄存器 Acc 内容为 FFh。

指 令：	XOR A,x	影响标志					
		TO	PDF	OV	Z	AC	C
动 作：	Acc←Acc "XOR" x				✓		

功 能 说 明： Acc 的内容值与常数值 x 执行"XOR"运算，并将结果存入 Acc 寄存器。

范　　例： `XOR A,0FFh`

执 行 前： 寄存器 Acc 内容为 55h。
执 行 后： 寄存器 Acc 内容为 AAh。

第 3 章　HT46xx 指令集与开发工具

指　令：	XORM A,[m]	影响标志
动　作：	[m]←Acc "XOR" [m]	TO　PDF　OV　Z　AC　C

功能说明：　Acc 的内容值与寄存器 m 的内容执行 "XOR" 运算，并将结果存入 m 寄存器。
范　　例：　XORM A,[40h]
执 行 前：　寄存器 40h 内容为 55h，寄存器 Acc 内容为 AAh。
执 行 后：　寄存器 40h 内容为 FFh，寄存器 Acc 内容为 AAh。

3-2　程序的编辑

3-2-1　汇编语言格式

　　为了缩短程序验证与产品开发的时间，目前市面上许多厂家生产的单片机都配备有C语言编译器(C-Complier)的开发环境。盛群半导体公司也不例外，Holtek C-Complier就是仿效ANSI 标准的C编辑器，但是由于单片机本身硬件体系结构的限制，因此并不完全兼容。若着眼于程序代码的长度与执行效率，还是应使用汇编语言作为单片机开发的主要语言。因此，在编写HT46xx程序时，必须先了解汇编语言的格式，那么剩下的，便是熟用芯片所提供的指令集来编写程序，如此便完成汇编语言程序的编写；至于功力的高低，就视个人所付诸的心血与努力了。对Holtek C-Complier程序语言的体系结构及语法有兴趣的读者，可以在HT–IDE的 "Help Menu" 中参考 "Holtek C Programmer's Guide"，本章以汇编语言为主进行介绍。

　　汇编语言格式的每一行指令可以分成4个字节，如下所述：

[Name：] Op-Code　[operand1[,operand2]]　[;Comment]

1. 标记栏(Name Field)

　　标记栏是指某一列指令中的第一个字节，是用来代表该列指令所在的实际程序存储器地址，因此程序设计者只要在程序中以标记代表该指令地址，而不需自己算出跳转目地的实际存储器地址，这对程序设计者而言是非常方便的。待程序完成，再将整个程序交给编译器翻译成机器码时，编译器便会自动帮助用户算出该标记的所在实际存储器地址。标记栏一定要在每列指令的开头处，且不可有空格。并非每列指令都一定要有标记，标记也可以独立占用一列。Holtek 汇编语言中的标记栏可由 "A" ~ "Z"、"a" ~ "z"、"0" ~ "9"、"?"、"_"、"@" 符号组成，但是第一个字符不得为数字，而且编译器只识别前面31个字符。

2. 指令栏(OP-code Field)

　　指令栏是一个指令列的主体，它的位置是在标记栏之后，空一格以上的位置，它指出了要CPU做什么事，如传送指令、加减运算指令、位设定指令等。而此栏除了可以写CPU所要做的指令之外，另外伪指令(Pseudo Instruction，也称编译指引(Assembly Directive))也写在这个字节中，如EQU、ORG、IF、END等伪指令；常用的伪指令请参考3-2-2小节。

3. 操作数栏(Operand Field)

操作数栏是在指令栏之后空一格以上的位置开始，操作数栏指出了指令运算的对象，因此根据指令类型的不同，操作数的个数可能是一个、两个或没有。

4. 注解栏(Comment Field)

注解栏其实并不属于程序的一部分，它是保留给程序设计者，对某一行指令做文字注解来说明程序的功能，以增加程序的可读性。而习惯上注解栏是写在操作数栏之后，因此在编译时，编译器将不会管注解栏识别字符分号";"之后的文字。注解的文字叙述，可以写在任何地方，如果能在程序中加上适当的注解，将能更有效地维护程序。

3-2-2 常用的HT46xx伪指令(Assembly Directives)

伪指令给用户在编写程序时提供了极大的方便，Holtek编译器(HASMW32)提供许多的伪指令。详细的内容读者可以查阅HT-IDE "Help Menu"中的"Assembly Language"，以下只摘录其中较常使用的指令做介绍。

```
指  令：.CHIP
格  式：.CHIP description-file
```

功能：Holtek编译器(HASMW32)可以编译全系列的8位单片机汇编语言程序，因此必须以".CHIP"指令指定编译之后所产生的是哪一个单片机的机器码。

范例：.CHIP HT46R23   ;产生 HT46R23 的机器码

```
指  令：.SECTION
格  式：name.SECTION [align] [combine] 'class'
```

功能：Holtek编译器(HASMW32)的工程管理方式中，是将整个Project视为一段程序的总和每一个程序段就是一个Section，其后的参数意义如下：

name：定义此Section的名称；

align：指定此Section要放到寄存器的哪一个位置，有以下几种不同的存放方式(若未指明，其Default设定为"BYTE")：

　　BYTE：此Section可放至寄存器的任何一个位置；
　　WORD：此Section必须放至寄存器的偶数位置；
　　PARA：此Section必须放至寄存器地址可被16整除位置；
　　PAGE：此Section必须放至寄存器地址可被256整除位置。

combine：指定此Section要如何与名称(name)、align相同的Section结合，有以下选择：

　　at Addr：指定此Section必须放在Addr的地址；
　　common：指定此Section可以与其他Section重叠。

class：指定存储器的种类，ROM(程序存储器)或DATA(数据存储器)。

范例：MY_DATA .SECTION 'DATA'   ;== DATA SECTION ==

　　　MY_CODE .SECTION AT 0 'CODE'   ;== PROGRAM SECTION ==

```
指  令：ORG
格  式：ORG <expression>
```

第3章 HT46xx 指令集与开发工具

功能：设定程序存储器(或数据存储器)的起始值。

范例：
```
        ORG     000h
        JMP     START           ;此指令将存放于寄存器地址 000h
        ORG     08h
START:  NOP                     ;此指令将存放于寄存器地址 008h
        :                       ;此指令将存放于寄存器地址 009h
```

指　令：END
格　式：END

功能：表示程序结束。

范例：
```
START:  NOP
        :
        END
```

当编译器发现 END 指令后，即停止编译。因此，就算继续在后面写一些指令，也不会产生任何的机器码或错误消息。

指　令：PROC、ENDP
格　式：name PROC、name ENDP

功能：定义程序模组的起始与结束。

范例：
```
DELAY   PROC
        MOV     DEL1,A
DEL_1:  MOV     A,30
        MOV     DEL2,A          ;SET DEL2 COUNTER
DEL_2:  MOV     A,110
        MOV     DEL3,A          ;SET DEL3 COUNTER
DEL_3:  SDZ     DEL3            ;DEL3 DOWN COUNT
        JMP     DEL_3
        SDZ     DEL2            ;DEL2 DOWN COUNT
        JMP     DEL_2
        SDZ     DEL1            ;DEL1 DOWN COUNT
        JMP     DEL_1
        RET
DELAY   ENDP
```

指　令：EQU
格　式：name EQU expression

功能：设定编译时的常数的值。

范例：
```
        FOUR    EQU     4           ;定义 FOUR=4
        TEMP    EQU     12h         ;TEMP=16
        MAX     EQU     8822h
        :
```

指　令：LOW、HIGH
格　式：LOW expression、HIGH expression

功能：取得常数的高(HIGH)、低(LOW)位组。

范例：
```
MAX      EQU      8822h
HI_BYTE  EQU      HIGH MAX    ;HI_BYTE=88h
LO_BYTE  EQU      LOW MAX     ;LO_BYTE=22h
 :
```

指　令：INCLUDE
格　式：#INCLUDE <include_file>

功能：将其他源程序文件载入本程序中，让程序精简并增加可读性。读者可以自己编写 INCLUDE 文件，在系统中也提供了几个文件(HT46R23.INC、HT46C22.INC 等)供用户当作范例使用。其放置在"X:\HIDE-3000V6\Include"当中，X 为用户目前安装的路径。

范例：#INCLUDEHT46R23.INC

指　令：PUBLIC 与 EXTERN
格　式：PUBLIC name1 [,name2 [,...]]
　　　　EXTERN name1:type [,name2:type [, ...]]

功能：声明变量由程序内部定义或者是程序外部定义。Holtek 编译器(HASMW32)是以工程来设定用户程序，故在同一工程中两个不同文件要互相调用、参数相互传递则需使用 PUBLIC 与 EXTERN。在程序中这两个功能可用在任何地方与声明数次。

范例：
```
PUBLIC   START,LOOP_1,LABEL_1
EXTERN   START: BIT,LOOP_1: BYTE,LABEL_1: WORD
```

指　令：MACRO、LOCAL 与 ENDM
格　式：
```
name    MACRO [dummy-parameter [,,,]]
            LOCAL dummy-name [,,,,]
        statements
ENDM
```

功能：宏指令的定义，name 为宏指令的名称，紧接着在 MACRO 之后是此宏指令的参数。LOCAL 则定义此宏指令中所使用到的标号，由 LOCAL 指令之后一直到 ENDM 指令之前为宏指令的程序主体。由 LOCAL 所定义的标号将只是一个暂时的伪名称，当宏展开时 HASMW32 将以程序中唯一的符号—"??digit"取代这些标号(digit 代表 0000~FFFF 的数值)。读者在使用宏指令时，一定要记得将宏内的标号声明成 LOCAL 形式，否则如果程序中参用一次以上的宏指令，HASMW32 就会产生错误的消息。

范例：
```
DIFF_X_Y  MACRO   X,Y,DIFF
          LOCAL   PLUS
          MOV     A,X
          SUB     A,Y
          SZ      C
          JMP     PLUS
          CPL     ACC
          INCA    ACC
PLUS:     MOV     DIFF,A              ENDM
```

```
指 令: $
格 式: 无
```

功能: 目前程序计数器(Program Counter)的数值。

范例: JMP $

让程序跳转至目前的程序计数器地址，此范例中则是变成在原地跳转，可以利用此方式来制造一个死循环。其语法与"Label：JMP Label"意义相同。

范例: SDZ   DEL   ;Jump Here
       JMP $-1   ;Jump to Last Line
       NOP

有时候在程序中不希望有太多的标号，如上例的小循环，那么可以使用$+n 或$-n 来取代目的地址。

```
指 令: DBIT、DB、DW、DUP
格 式: [name]    DB     [value1 [,value2...]]
       [name]    DW     [value1 [,value2...]]
       [name]    DBIT
       [name]    DB     repeated-count DUP(?)
       [name]    DW     repeated-count DUP(?)
```

功能: 保留一个位(DBIT)、一个字节(DB)或两个字节(DW)的寄存器空间，其中 name 代表该地址的标号，"value1、…"代表在保留的寄存器地址中存入的数值。注意：如果保留的是数据存储器空间，则不可以指定存入的数值。此时的"value1…"必须以"?"取代。而 DUP 代表此类型(BYTE 或 WORD)的寄存器空间要保留几组。

范例: MY_DATA   .SECTION 'DATA'   ;== DATA SECTION ==
      BUF1   DB   ?   ;Reserved 1 Byte
      BUF2   DW   ?   ;Reserved 2 Bytes
      FLAG_1   DBIT   ;Reserved 1 Bit
      ARRAY_BYTE   DB   20 DUP(?)   ;Reserved 20×1 Byte
      ARRAY_WORD   DW   10 DUP(?)   ;Reserved 10×2 Byte
        ⋮
      MY_CODE   .SECTION AT 0 'CODE'   ;== PROGRAM SECTION ==
      TAB_1   DW   1,2,4,8,16,32,64,128,256
      TAB_2   DW   'ABCDEFG'

```
指 令: OFFSET
格 式: OFFSET name、OFFSET label
```

功能: 取得变量 name 的数据存储器地址或标号的地址。

范例: MY_DATA   .SECTION 'DATA'   ;== DATA SECTION ==
      BUF   DB   10 DUP (?)   ;Reserved 1 Byte
      COUNT   DB   ?
      MY_CODE   .SECTION AT 0 'CODE'   ;== PROGRAM SECTION ==
            MOV   A,10
            MOV   A,OFFSET BUF1
            MOV   MP0,A   ;CLEAR ARRAY BUF BY INDIRECT
            CLR   IAR0   ;ADDRESSING MODE

```
        INC     MP0
        SDZ     COUNT
        JMP     $-3
        MOV     A,OFFSET STR1
        MOV     TBLP,A                   ;LOAD STRING1 START ADDRESS
        TARBRDL ACC                      ;Acc='M' CHARACTER
        ORG     LASTPAGE
STR1:   DC      'MISS!',STR_END          ;DEFINE STRING DATA 1
STR2:   DC      'BINGO!',STR_END         ;DEFINE STRING DATA 2
```

指　令：DC
格　式：[label:] DC expression1 [,expression2 [,...]]

功能：保留程序存储器空间，其中 Label 代表该地址的标号；"value1…"代表在保留的寄存器地址中存入的数值。至于程序存储器空间是几个位，Holtek 编译器(HASMW32)会由 ".CHIP" 指令所指定的单片机型号加以判定。

范例：
```
MY_CODE    .SECTION AT 0 'CODE'          ;== PROGRAM SECTION ==
ORG        LASTPAGE
STR1:      DC      'MISS!',STR_END       ;DEFINE STRING DATA 1
STR2:      DC      'BINGO!',STR_END      ;DEFINE STRING DATA 2
```

保留字

编写程序时为增加其可读性，一定都会使用一些字符将变量名称改为用户习惯的字符串，但其中仍需注意有些字符是不能使用的，编写程序时请特别注意，其中包括下列几个大项：

◇ 寄存器：A、WDT、WDT1 和 WDT2。
◇ 指令：

ADC	CPL	JMP	RLC	SDZ	SWAP
ADCM	CPLA	MOV	RLCA	SDZA	SWAPA
ADD	DAA	**NOP**	RR	SET	SZ
ADDM	DEC	OR	RRA	SIZ	SZA
AND	DECA	ORM	RRC	SIZA	TABRDC
ANDM	HALT	RET	RRCA	SNZ	TABRDL
CALL	INC	RETI	SBC	SUB	XOR
CLR	INCA	RLA	SBCM	SUBM	XORM

◇ 伪指令与运算符号：

$	.CHIP	EXTERRN	LOCAL	ORG
*	DB	HIGH	LOW	PAGE
+	DBIT	IF	MACRO	PROC
—	DW	IFDEF	MOD	PUBLIC
.	ELSE	IFE	.NOLIST	ROMBANK
/	END	IFNDEF	.NOLISTINCLUDE	RAMBANK
=	ENDIF	INCLUDE	NOLISTMACRO	SHL
?	ENDM	.LIST	NOT	SHR
[ ]	ENDP	LISTINCLUDE	OFFSET	XOR
AND	EQU	.LISTMACRO	OR	

3-2-3 程序的编写

所谓的编译，就是从软件程序的编写与程序编译，到最后的软件程序与硬件电路能够配合，达到预先所设想的功能，图3-2-1呈现其每个步骤及过程。

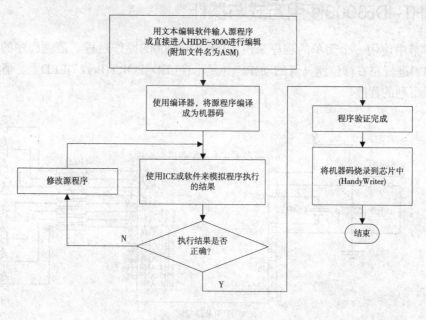

图3-2-1　编译流程

3-2-4 程序的编译

当程序设计者完成程序的设计之后，首先须以文本编辑软件来编写程序(或是直接在HT-IDE所提供的环境下进行编辑)，接着就是利用编译程序，将此源程序的文本内容转换成为单片机所能接受的机器码，而此转换的过程便称之为编译。因此，程序设计者在完成编写程序之后，只要执行HT-IDE的编译功能，便会自动将源程序内的指令转换成相对应的机器码。如此一来，程序设计者即可利用完成编译后所产生的相关文件进行调试、模拟与验证的工作。

程序编译无误后，只是语法上没有问题而已，并不代表它就一定能达到设计者所期望的工作。因为可能还有语意或逻辑上的错误，需要进一步的测试与调试，才能使程序达到预期的执行结果。一般在软、硬件配合的设计时，必须以"在线模拟器"(In-Circuit Emulator，一般简称为ICE)来完成程序与硬件最后的调试及验证工作，但此开发设备并非一般用户都能随时拥有。所幸盛群半导体公司为了推广其单片机的市场，提供了用户相当方便的开发环境，读者可以在其网页(www.holtek.com.tw)上免费地下载一套功能完备的开发软件—HT-IDE3000V6，其中的软件模拟器Simulator提供一般程序调试时所需的各种工具，如单步执行(Single step)、断点设定(Break point)等，唯一的缺点是看不到接上硬件之后，程序实际执行的情况。而HT-IDE3000V6的软硬件模拟功能(Virtual Peripheral Manager，VPM)恰可弥补此项缺点，VPM可以让用户在计算机的屏幕上先预览接上硬件之后的情形。读者可以随时上网下载最新的HT-IDE版本。在3-3节中，将以实例来说明程序的开发过程。

3-3　HT-IDE3000使用方式与操作

在本节中使用一个简单的程序来说明HT-IDE3000的操作过程。范例程序的动作要求如下：让LED进行左右移位跑马灯的动作，并显示在LED_PORT(Port 7)LED上。请参考图3-3-1的电路及下列的程序。

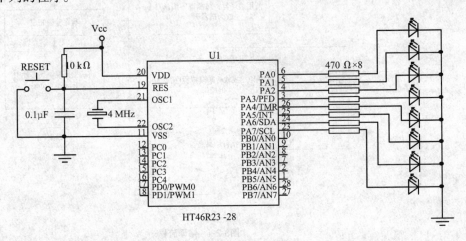

图3-3-1　LED跑马灯实验电路

1. 程序 3-3-1 LED 跑马灯实验

```
1     ; PROGRAM : 3-1.ASM   (3-1.PRJ)            BY STEVEN
2     ; FUNCTION: LED SCANNING DEMO PROGRAM      2002.DEC.07.
3     #INCLUDE         HT46R23.INC
4                      CHIP HT46R23
5     ;----------------------------------------------------------------
6     MY_DATA     .SECTION   'DATA'              ;== DATA SECTION ==
7     DEL1        DB         ?                   ;DELAY LOOP COUNT 1
8     DEL2        DB         ?                   ;DELAY LOOP COUNT 2
9     DEL3        DB         ?                   ;DELAY LOOP COUNT 3
10    ;----------------------------------------------------------------
11    LED_PORT    EQU        PA                  ;DEFINE LED_PORT
12    LED_PORTC   EQU        PAC                 ;DEFINE LED_PORT CONTROL REG.
13
14    MY_CODE     .SECTION   AT 0 'CODE'         ;== PROGRAM SECTION ==
15            ORG        00H                     ;HT-46RXX RESET VECTOR
16    MAIN:
17            CLR        LED_PORTC               ;CONFIG LED_PORT AS O/P MODE
18            CLR        LED_PORT                ;SET INITIAL LED STATE
19            SET        C                       ;SET CARRY FLAG (STATUS.0)
20    RIGHT:
21            RRC        LED_PORT                ;SHIFT RIGHT
22            MOV        A,100
23            MOV        DEL1,A                  ;SET DELAY FACTOR
24            CALL       DELAY                   ;DELAY 100*1ms
25            SNZ        LED_PORT.0              ;IS ALL LEDs HAVE BEEN LIT?
```

```
26          JMP       RIGHT                    ;NO. CONTINUE RIGHT SHIFT.
27  LEFT:
28          RLC       LED_PORT                 ;SHIFT LEFT
29          MOV       A,200
30          MOV       DEL1,A                   ;SET DALAY FACTOR
31          CALL      DELAY                    ;DELAY 200*1ms
32          SNZ       LED_PORT.7               ;IS ALL LEDs HAVE BEEN LIT?
33          JMP       LEFT                     ;NO. CONTINUE LEFT SHIFT.
34          JMP       RIGHT                    ;REPEAT THE RIGHT PROCESS.
35  ;********************************************************************
36  ;                    Delay about DEL1*1ms
37  ;********************************************************************
38  DELAY   PROC
39          MOV       A,03
40          MOV       DEL2,A                   ;SET DEL2 COUNTER
41  DEL_2:  MOV       A,110
42          MOV       DEL3,A                   ;SET DEL3 COUNTER
43  DEL_3:SDZ         DEL3                     ;DEL3 DOWN COUNT
44          JMP       DEL_3
45          SDZ       DEL2
46          JMP       DEL_2                    ;DEL2 DOWN COUNT
47          SDZ       DEL1
48          JMP       DELAY                    ;DEL1 DOWN COUNT
49          RET
50  DELAY   ENDP
51          END
```

程序说明

7~9   依序定义变量地址。

11~12 定义 LED_PORT 为 PA，定义 LED_PORTC 为 PAC。

15   声明寄存器地址由 000h 开始。

17   将 LED_PORT 定义成输出模式。

20~26 LED_PORT 内容设为 0 并将进位标志(CARRY FLAG)设为 1。进位标志的目地在于执行循环动作时包括了进位标志，带进位位右移动作会使得进位标志的 1 带进位位左移至 LED_PORT 的 BIT 7。

27~34 对 LED_PORT 进行带进位位右移的动作，由进位标志开始带进位位右移一圈，检查 LED_PORT.0 是否为 1，成立表示带进位位右移动作已经结束。程序中 DEL1 是用来有效地控制 DELAY 子程序的延迟时间(DEL1×1 ms)。

25~31 对 LED_PORT 进行带进位位左移的动作，由进位标志开始带进位位左移一圈，检查 LED_PORT.7 是否为 1，成立表示带进位位左移动作已经结束。重新回到 RIGHT 再进行带进位位右移的动作。不断反复执行即可让 LED 持续地变化。

38~50 延迟子程序，其延迟时间由 DEL1 的内含值决定，大约为 DEL1×1 ms。至于其计算方式，请参考实验 4-1 的说明。

请读者注意第3行"#INCLUDE HT46R23.INC"指令的用意是将定义文件HT46R23.INC载入到程序中，该定义文件由盛群半导体提供，当用户完成HT-IDE3000安装的程序之后会同时建立许多的定义文件。在第2章的内容中已经介绍许多寄存器，如MP0、MP1、IAR1、IAR0、TMRC等，相信读者早就忘记这些寄存器所对应的地址了吧？如果在编写程序的过程中，能

HT46xx 单片机理论与实践宝典

直接使用这些寄存器的名称不是更方便、也更容易看懂吗？ "#INCLUDE HT46R23.INC" 这行指令的目的就在此。HT46R23.INC 文件的内容如下，请读者参考：

```
1   ;ht46r23.INC
2   ; This file contains the definition of registers for
3   ;     Holtek ht46r23 microcontroller.
4   ; Generated by Cfg2IncH V1.0.
5   ; Do not modify manually.
6
7       IAR0        EQU     [00H]
8       R0          EQU     [00H]    ;old style declaration, not recommended for use
9       MP0         EQU     [01H]
10      IAR1        EQU     [02H]
11      R1          EQU     [02H]    ;old style declaration, not recommended for use
12      MP1         EQU     [03H]
13      ACC         EQU     [05H]
14      PCL         EQU     [06H]
15      TBLP        EQU     [07H]
16      TBLH        EQU     [08H]
17      STATUS      EQU     [0AH]
18      INTC0       EQU     [0BH]
19      TMRH        EQU     [0CH]
20      TMRL        EQU     [0DH]
21      TMRC        EQU     [0EH]
22      PA          EQU     [012H]
23      PAC         EQU     [013H]
24      PB          EQU     [014H]
25      PBC         EQU     [015H]
26      PC          EQU     [016H]
27      PCC         EQU     [017H]
28      PD          EQU     [018H]
29      PDC         EQU     [019H]
30      PWM0        EQU     [01AH]
31      PWM1        EQU     [01BH]
32      INTC1       EQU     [01EH]
33      HADR        EQU     [020H]
34      HCR         EQU     [021H]
35      HSR         EQU     [022H]
36      HDR         EQU     [023H]
37      ADRL        EQU     [024H]
38      ADRH        EQU     [025H]
39      ADCR        EQU     [026H]
40      ACSR        EQU     [027H]
41      C           EQU     [0AH].0
42      AC          EQU     [0AH].1
43      Z           EQU     [0AH].2
44      OV          EQU     [0AH].3
45      PDF         EQU     [0AH].4
46      TO          EQU     [0AH].5
47      EMI         EQU     [0BH].0
48      EEI         EQU     [0BH].1
49      ETI         EQU     [0BH].2
50      EADI        EQU     [0BH].3
51      EIF         EQU     [0BH].4
52      TF          EQU     [0BH].5
53      ADF         EQU     [0BH].6
54      TE          EQU     [0EH].3
55      TON         EQU     [0EH].4
56      PA0         EQU     [012H].0
57      PA1         EQU     [012H].1
58      PA2         EQU     [012H].2
59      PA3         EQU     [012H].3
60      PA4         EQU     [012H].4
61      PA5         EQU     [012H].5
62      PA6         EQU     [012H].6
```

```
63    PA7         EQU     [012H].7
64    PB0         EQU     [014H].0
65    PB1         EQU     [014H].1
66    PB2         EQU     [014H].2
67    PB3         EQU     [014H].3
68    PB4         EQU     [014H].4
69    PB5         EQU     [014H].5
70    PB6         EQU     [014H].6
71    PB7         EQU     [014H].7
72    PC0         EQU     [016H].0
73    PC1         EQU     [016H].1
74    PC2         EQU     [016H].2
75    PC3         EQU     [016H].3
76    PC4         EQU     [016H].4
77    PC5         EQU     [016H].5
78    PC6         EQU     [016H].6
79    PC7         EQU     [016H].7
80    PD0         EQU     [018H].0
81    PD1         EQU     [018H].1
82    EHI         EQU     [01EH].0
83    HIF         EQU     [01EH].4
84
85    LASTPAGE    EQU     0F00H
```

如果使用其他型号的单片机，就必须载入其他的定义文件，读者可以在安装完HT-IDE3000后，到"x:/HT-IDE3000V6/Include/"的子目录下查看由原厂所提供的定义文件("x"代表安装的磁盘位置)。用户也可以建立自己的定义文件，以提升程序的管理与使用的方便性。

2. HT-ICE 操作说明

(1) 软件需求

执行HT-IDE3000时可以配合盛群半导体公司所推出的HT-ICE(见图3-3-2)，这套ICE功能与IC本身的功能相同，让个人计算机能够成为一套HT46xx系列单片机的开发系统。若读者在经济能力上无法购买ICE，HT-IDE3000也提供软硬件模拟器，稍后也将一并介绍。

(a) 新型ICE+Programmer　　　　　　(b) 旧型ICE(目前已不提供)

图3-3-2　HT-ICE的外观

(2) 起始设置

读者可由本书所附赠的光盘中取得HI-IDE3000，或上盛群半导体公司网站(www.holtek.com.tw)取得HT-IDE的最新版本。软件安装完成后读者可在"程序"中看到一个新增的文件夹——HOLTEK HT-IDE3000，表示软件已安装完成。其中提供多种的开发工具，本书将只介绍有关编译与模拟的功能。执行HT-IDE之后会出现如图3-3-3所示的画面。

图3-3-3　进入HT-IDE的画面

此时如果用户已经接上HT-ICE，将直接进入开发环境，否则将会出现如图3-3-4所示的错误消息，请读者直接单击"否(N)"按钮进入开发环境，如图3-3-5所示。

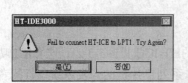

图3-3-4　未接上HT-ICE的错误消息

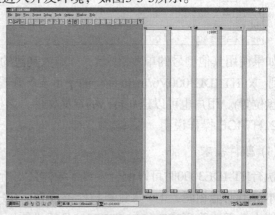

图3-3-5　开发环境画面

由于HT-IDE3000的开发环境是以工程(Project)的方式来管理用户的文件，所以要开发一个新的应用程序，首先必须建立一个Project，实际上就是供HT-IDE3000管理用的文件。现在就请读者由Project菜单下选择NEW命令，如图3-3-6所示。

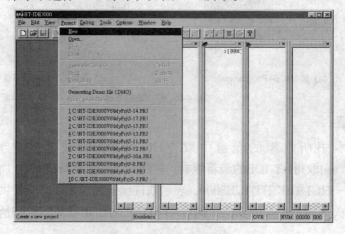

图3-3-6　由Project菜单下选择New命令

第 3 章　HT46xx 指令集与开发工具

此时会出现如图3-3-7所示的对话框,此对话框要求用户输入工程的名称(Project Name)与所使用的单片机型号(Micro Controller)。读者可以指定自己的工程名称,至于单片机型号就根据程序3-3-1指定为HT46R23。设定完成之后就单击OK,然后又会弹出如图3-3-8所示的对话框提供相关的选项,说明如下:

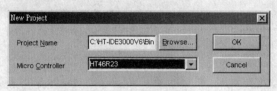

图3-3-7　选择NEW命令后出现的对话框

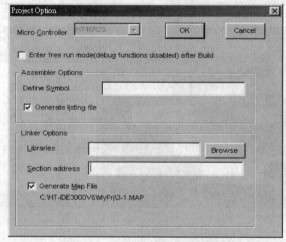

图3-3-8　单击OK后出现的对话框

① Enter free run mode(debug functions disabled)after Build:当工程创建之后进入全速执行模式,此时无法使用Debug模式下的功能。除非已经确定程序完全无误,否则不建议勾选此选项。

② Generate listing file:编译以后会产生列表文件,列表文件通常会提供一些值得参考的信息,建议读者勾选此选项。

③ Libraries:指定函数库的所在地址及文件名,程序3-3-1未使用任何函数库,所以不需指定。

④ Section address:指定程序段的地址,在本例中不需指定;HD-IDE3000会依程序中的SECTION命令设定地址。

⑤ Generate Map File:产生标号(Name、Lable)的对应地址,这些信息将存放在xxx.MAP文件中(如本例为3-1.MAP)。

设定完成之后单击OK按钮,又会弹出Options的对话框,如图3-3-9所示,有关Options选项中各项参数所代表的意义请读者参阅2-14节的说明。本节中的应用范例程序3-1较为简单,读者不需更改任何的设置,只要使用HT-IDE3000的默认设置即可。但若他日有其他的应用,就需根据实际的需要设置Options选项的内容。单击OK按钮以后,就真正进入HT-IDE3000的开发环境,如图3-3-10所示。

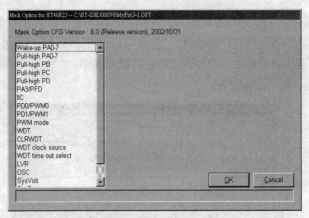

图3-3-9 Options对话框

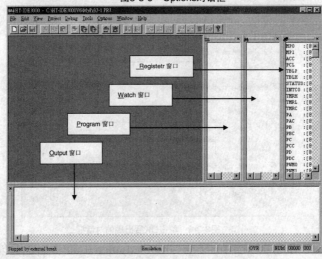

图3-3-10 完成Options编辑后出现的窗口

① Register窗口：此窗口用于显示所有寄存器的内容。
② Watch窗口：显示"DATA SECTION"中所定义的变量所在地址及其内含值。
③ Program窗口：显示程序存储器的内容。
④ Output窗口：当编译或创建时，所有的输出消息会在此窗口显示。

因为目前的工程中还未包含任何程序，因此在此出现空白或"@"符号。

接下来就开始输入或编写程序，读者可以直接使用HT-IDE3000提供的编辑器或自己熟悉的文本编辑软件(但是要存成纯文本文件)来进行程序的编写。

如果以HT-IDE3000编辑器来编写程序，请在File菜单中执行New命令(见图3-3-11)，然后就会立即弹出如图3-3-12所示的编辑窗口，用户可以开始进行程序编辑的工作。如果读者之前已经创建程序文件，可在File菜单中执行Open命令将其打开，现在就请读者将本书所附的程序3-1打开。

第 3 章　HT46xx 指令集与开发工具

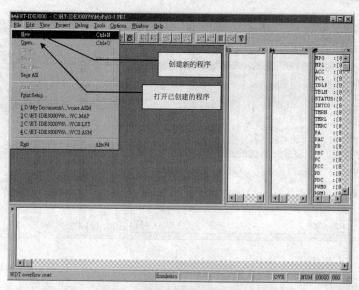

图3-3-11　在File菜单中执行New命令

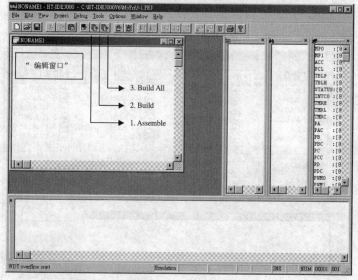

图3-3-12　程序编辑窗口

在此先介绍HT-IDE3000在调试模式时的3个命令：

① Assemble：编译程序，按下此按钮将进行程序编译，并产生调试或烧录时所需的文件。

② Build：建立工程管理，此时会建立程序与工程文件中的相关链接。

③ Build All：建立工程管理，此时会建立工程文件中所有程序的相关链接。

将程序3-1打开之后，程序的内容就呈现在编辑窗口，请注意此时只是将程序打开观看或进行编辑而已，还未将程序加入Project中管理。请读者执行Project菜单下的Edit命令(见图3-3-13)，接着就会出现如图3-3-14所示的Project编辑窗口。

81

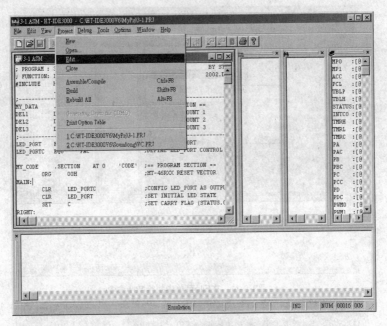

图3-3-13 载入程序后的编辑窗口

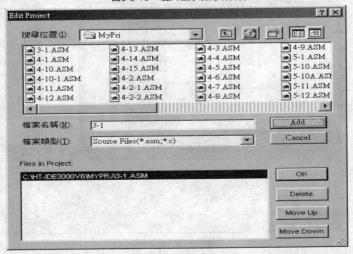

图3-3-14 Project的编辑窗口

现在可以选择程序3-1将其加入(Add)到此Project中。一个Project中可以包括数个文件,但是其中对于寄存器(包含程序存储器与数据存储器)的分配与变量名称必须妥善安排,也就是说要灵活运用"SECTION"伪指令,通常在多人合作开发的大型应用程序中,会使用一个工程来同时管理数个文件。对于初学者或小型的程序开发,还是建议读者以一个工程管理一个程序的形式较为单纯。

请读者单击图3-3-12所介绍的Build或Build All命令来进行编译与链接的动作,此时编译的相关消息会出现在Output窗口(见图3-3-15),如果程序有语法错误,则会在Output窗口显示行号及错误的原因。

第3章　HT46xx指令集与开发工具

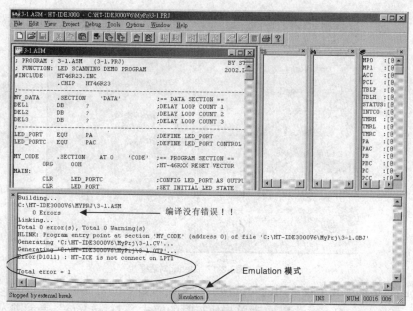

图3-3-15　进行编译与链接的窗口

读者发现程序编译并没有错误，但是却出现一行"HI-ICE is not connect on LPT1"的错误消息。这是因为LPT1上没有接上HI-ICE，而用户又选择"Emulation" Mode的原因。"Emulation" Mode一定要接上HT-ICE才可以操作HT-IDE3000的相关功能，而其操作方式与"Simulation" Mode完全相同，所以这里就将操作模式切换为软件模拟的"Simulation" Mode，在"Simulation" Mode下介绍HT-IDE3000相关功能的操作方式。请读者执行Option菜单下的Debug命令(见图3-3-16)，此时将弹出如图3-3-17所示的对话框。

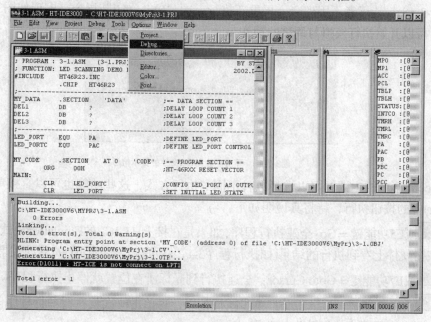

图3-3-16　执行Option菜单下的Debug命令

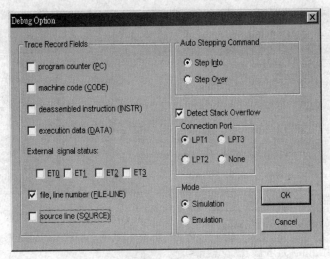

图3-3-17　Debug对话框

在Debug对话框中几个较常用的勾选项目意义说明如下：

① Auto Stepping Command：自动单步执行时的模式，"Step Into"表示进入子程序执行自动单步；"Step Over"表示不进入子程序执行自动单步。

② Detect Strack Overflow：建议勾选此选项，当栈寄存器溢位时，HT-IDE3000会产生警告消息提醒用户。

③ Connection Port：选择HT-ICE的链接端口，如果读者要使用HT-ICE，可要谨慎勾选此项。

④ Mode：选择HT-IDE3000的操作模式，没有接上HT-ICE的读者请勾选Simulation。

单击OK之后会产生一个警示信息，这只是告知用户在Simulation模式下不会产生.COD文件(见图3-3-18)，请读者不用理会。再重新执行一次"Build All"指令，出现如图3-3-19所示的画面，可以开始执行模拟与调试的动作。

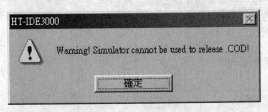

图3-3-18　不会产生.COD文件的警示信息

在Debug菜单栏中读者可以选择数种不同的执行方式，这些功能也可以直接单击工具栏(Tool Bar)中的按钮执行，特将其说明如下：

① ▆或F5功能键＝Go：全速执行程序，如果是在接上HT-ICE的Emulation模式下执行，此时应该在硬件上看到执行的结果(如执行程序3-1，将可以观察到LED的移动效果)。

② ▆或F7功能键＝Goto Cursor：全速执行到光标所在位置就停止执行程序。

③ ▆或F8功能键＝Step Into：单步执行程序，每单击一次就执行一行程序。如果调用子程序，则进入子程序执行单步的动作。

④ 或F10功能键＝Step Over：单步执行程序，每单击一次就执行一行程序。如果调用子程序，则不会进入子程序执行单步的动作，而是等子程序执行完毕之后再停下来。

⑤ 或Shift+F10功能键＝Step Out：当单步执行程序进入子程序时，可以选择此功能，那么当执行到返回指令时(如"RET"、"RETI"、"RET A,x")，程序就停止执行。

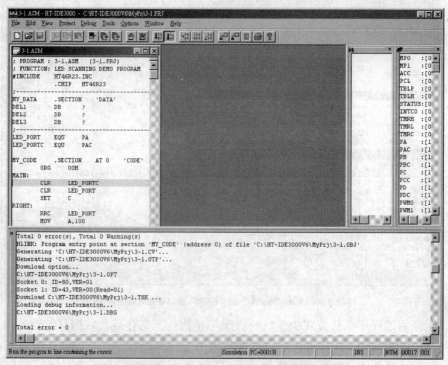

图3-3-19 Simulation模式下一切OK的显示画面

在程序全速执行的过程中，有以下几种方式让程序停止执行：

① 或Alt+F5功能键＝Stop：停止程序执行。

② 或F4功能键＝Reset：系统复位，程序停止执行，存储器的值不会改变，但寄存器的值将还原到Reset的状态。

③ ＝Power-on Reset：电源开机复位，程序停止执行，存储器的值不会改变，但寄存器的值将还原至Power-on Reset的状态。

另外在全速执行之前，也可以先设定断点(Breakpoints)，让程序执行到某一行程序即停止，以便观察寄存器与存储器的变化，分析程序的正确性，相关的命令如下：

① ＝Toggle Breakpoints：将光标移至欲设断点的位置单击此命令或快速双击，即可完成断点的设定，此时断点位置将以咖啡色光棒表示(见图3-3-20)。再次单击 或快速双击，可以取消此断点。

② ＝Clear All Breakpoints：清除所有断点。

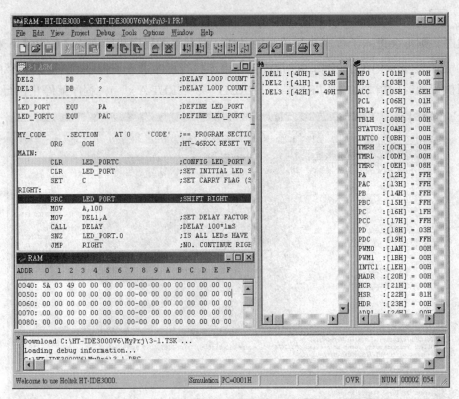

图3-3-20 可由"Watch"或"RAM"窗口观察数据存储器的变化

在以上介绍的程序执行方式中，只要程序停止执行，寄存器及数据存储器的内含值就会呈现在相对应的窗口内，用户可以通过这些数值的变化判断程序的正确性，如图3-3-20所示。如果用户觉得单步执行时，每单击一次只执行一个指令太过于麻烦，还可以选择"Debug→Stepping"，即自动单步执行功能，此时程序会全速执行但是执行速度较慢。除此之外，每执行一行就会将寄存器与存储器的变化显示在相对应窗口内。至于自动单步的执行速度，可以在弹出的对话框中设置，共有"FAST~5 Second"7种不同的速度选择，如图3-3-21所示。

图3-3-21 自动单步执行的速度调整对话框

在程序编写过程中，通常都会以惯用的名称来取代存储器的地址(例如程序3-1中的DEL1、DEL2与DEL3)，如果直接由RAM窗口来观察，用户可能不知道这些变量的对应地址。此时就可以在Watch窗口内直接输入变量名称，这样就可以看到其所对应的地址与内含值(参见图3-3-20)。

注意：请记得在变量名称前加一个"."符号，然后按下Enter键即可。

另外，在一些应用的场合中，经常需要知道程序的执行时间以便决定单片机的系统时钟脉冲，HT-IDE3000提供了一个统计执行指令周期的功能，当用户有上述的需求时可以加以应用。例如，用户想知道程序3-1中第39～47行的执行时间，就必须先计算出其指令周期，此时就可以通过"View→Cycle Count"来完成。请读者按照下列步骤操作：首先，在第39与47行分别设定断点，接着打开"Cycle Count"对话框，单击"🔄"(或F4功能键)让系统复位后再单击"⬇"(或F5功能键)全速执行程序，此时窗口的变化如图3-3-22所示。在"Accumulative Cycle Count"窗口上用户可以观察到"Cycle Count：8"的显示情况，这代表程序由第1行执行到第39行共花费了8个指令周期的时间，但这并不是我们所要的。单击"Accumulative Cycle Count"窗口中的Reset后，可以将Cycle Count归0。再单击"⬇"(或F5功能键)全速执行程序，此时PC停留在第47行，而"Accumulative Cycle Count"窗口变化为如图3-3-23所示的情况，这就代表由第39行执行到47行共耗费了1 003个指令周期的时间。

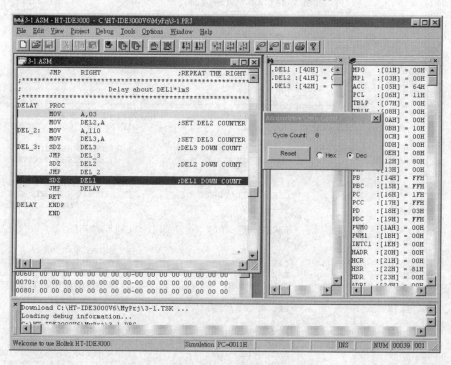

图3-3-22 "Cycle Count"功能范例

至于"Accumulative Cycle Count"窗口中的Hex与Dec选项，表示选择指令周期数的数字显示模式为十六进制或十进制，读者可依自己的习惯勾选。不过"Cycle Count"的计数值最大只到65 535，如果超过将会显示Overflow。

除了之前介绍的显示窗口之外，用户还可以打开其他窗口，观察单片机其他资源的变化情形，请读者选择"Window→Stack"可以监视栈寄存器的变化(见图3-3-24)，避免Stack Overflow的现象发生；Trace List窗口则记录着最近执行的指令，提高调试的效率。

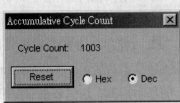

图3-3-23 第39~47行的执行时间　　　　图3-3-24 监视栈寄存器的Stack窗口

在"Project"(见图3-3-8)选项中勾选了"Generate listing file"与"Generate Map File"两项功能。究竟其文件内容是什么，请读者参阅清单3-3-1和清单3-3-2。

清单3-3-1　Listing File格式

```
File: 3-1.ASM      Holtek Cross-Assembler  Version 2.86      Page 1
   1  0000         ; PROGRAM : 3-1.ASM    (3-1.PRJ)                  BY STEVEN
   2  0000         ; FUNCTION: LED SCANNING DEMO PROGRAM             2002.DEC.07.
   3  0000         #INCLUDE   HT46R23.INC
   4  0000                    .CHIP HT46R23
   5  0000         ;--------------------------------------------------------------
   6  0000         MY_DATA    .SECTION    'DATA'      ;== DATA SECTION ==
   7  0000 00      DEL1       DB       ?              ;DELAY LOOP COUNT 1
   8  0001 00      DEL2       DB       ?              ;DELAY LOOP COUNT 2
   9  0002 00      DEL3       DB       ?              ;DELAY LOOP COUNT 3
  10  0003         ;--------------------------------------------------------------
  11  0003         LED_PORT   EQU      PA             ;DEFINE LED_PORT
  12  0003         LED_PORTC  EQU      PAC            ;DEFINE LED_PORT CONTROL REG.
  13  0003
  14  0000         MY_CODE    .SECTION AT 0 'CODE'    ;== PROGRAM SECTION ==
  15  0000                    ORG      00H            ;HT-46RXX RESET VECTOR
  16  0000         MAIN:
  17  0000 1F13               CLR      LED_PORTC      ;CONFIG LED_PORT AS OUTPUT MODE
  18  0001 1F12               CLR      LED_PORT       ;SET INITIAL LED STATE
  19  0002 300A               SET      C              ;SET CARRY FLAG (STATUS.0)
  20  0003         RIGHT:
  21  0003 1B92               RRC      LED_PORT       ;SHIFT RIGHT
  22  0004 0F64               MOV      A,100
  23  0005 0080   R           MOV      DEL1,A         ;SET DELAY FACTOR
  24  0006 2010               CALL     DELAY          ;DELAY 100*1ms
  25  0007 3812               SNZ      LED_PORT.0     ;IS ALL LEDs HAVE BEEN LIT?
  26  0008 2803               JMP      RIGHT          ;NO. CONTINUE RIGHT SHIFT.
  27  0009         LEFT:
  28  0009 1A92               RLC      LED_PORT       ;SHIFT LEFT
  29  000A 0FC8               MOV      A,200
  30  000B 0080   R           MOV      DEL1,A         ;SET DALAY FACTOR
  31  000C 2010               CALL     DELAY          ;DELAY 200*1ms
  32  000D 3B92               SNZ      LED_PORT.7     ;IS ALL LEDs HAVE BEEN LIT?
  33  000E 2809               JMP      LEFT           ;NO. CONTINUE LEFT SHIFT.
  34  000F 2803               JMP      RIGHT          ;REPEAT THE RIGHT PROCESS.
  35  0010         ;***************************************************************
  36  0010         ;                    Delay about DEL1*1ms
  37  0010         ;***************************************************************
  38  0010         DELAY PROC
  39  0010 0F03               MOV      A,03
  40  0011 0080   R           MOV      DEL2,A         ;SET DEL2 COUNTER
  41  0012 0F6E     DEL_2:    MOV      A,110
  42  0013 0080   R           MOV      DEL3,A         ;SET DEL3 COUNTER
  43  0014 1780   R DEL_3:    SDZ      DEL3           ;DEL3 DOWN COUNT
  44  0015 2814               JMP      DEL_3
  45  0016 1780   R           SDZ      DEL2           ;DEL2 DOWN COUNT
  46  0017 2812               JMP      DEL_2
  47  0018 1780   R           SDZ      DEL1           ;DEL1 DOWN COUNT
  48  0019 2810               JMP      DELAY
  49  001A 0003               RET
  50  001B         DELAY ENDP
  51  001B         END
       0 Errors
```

列表文件大致分成5个字节，第1栏为行号，第2栏为所占存储器地址，请注意DEL1~DEL3的地址均为00h。列表文件是在Assemble过程所产生的文件，而HT-IDE3000是在Build All(或Build)阶段才由Linker安排数据存储器地址，因此这里的00h并非是真正的地址，而是一个代号。所以在第4栏中的R(Relocatable)是代表该地址还未被正确指定。第3栏为指令所对应的机器码，第5栏则是源程序。编译过程中若发生语法错误，也会在列表文件中显示出来。

清单3-3-2　Map File格式

```
Holtek (R) Cross Linker Version 7.35
Copyright (C) HOLTEK Semiconductor Inc. 2002-2003. All rights reserved.

Input Object File: C:\HT-IDE3000V6\MyPrj\3-1.OBJ

Input Library File: C:\HT-IDE3000V6\LIB\MATH6.LIB

    Start      End        Length     Class     Name
    0000h      001ah      001bh      CODE      MY_CODE (C:\HT-IDE3000V6\MyPrj\3-1.OBJ)
    0040h      0042h      0003h      DATA      MY_DATA (C:\HT-IDE3000V6\MyPrj\3-1.OBJ)


    Indepentent Local Sections


    Start      End        Length     Class     Name
    0043h      0043h      0000h      ILOCAL    DELAY (C:\HT-IDE3000V6\MyPrj\3-1.OBJ)

ROM Usage Statistics
    Size       Used       Percent
    1000h      001bh      0%


RAM Usage Statistics
    Size       Used       Percent
    00c0h      0003h      1%


Call Tree


HLINK:Program entry point at section 'MY_CODE' (address 0) of file 'C:\HT-IDE3000V6\MyPrj\3-1.OBJ'

Total 0 error(s), Total 0 Warning(s)
```

MAP文件由Linker产生，提供了程序存储器以及数据存储器使用状况的相关消息，请读者直接参考文件中的说明。

3-4　VPM使用方式与操作

以上的模拟过程中，如果没有接上HT-ICE以及硬件电路，则无法看到LED的灯号变化情形，而只能由PA的数值变化来分析程序的正确与否，其实这对有经验的工程师而言已经是绰绰有余，但对初学者来说就有点隔靴搔痒，没什么感觉！HT-IDE3000提供的"Virtual Peripherals Manager"恰可解决此一窘境！在Tools菜单下还有几个选项(见图3-4-1)，在此稍做说明：

① Mask Option：可以更改Options中的选项。
② Diagnose：提供HT-ICE自我检测功能。
③ Handy Writer：配合烧录器进行OTP型单片机的程序烧录工作。
④ Library Manager：函数库管理工具。
⑤ Voice：有关Voice ROM的数据管理工具。

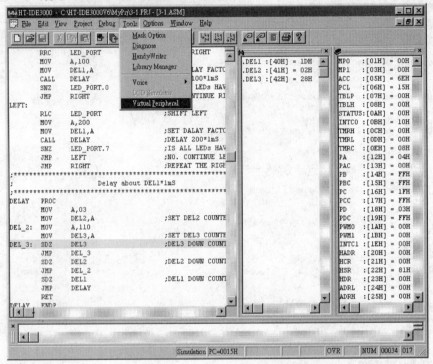

图3-4-1　Tools菜单下的选项

读者选择进入"Virtual Peripherals"选项，准备进行虚拟的硬件模拟(在进入VPM模拟之前，请确定HT-IDE3000是在Simulation的模式之下)。此时会弹出如图3-4-2所示的窗口，接下来必须在此窗口放置与连接要模拟的电路。首先介绍工具栏上所提供的工具。

① ：加入元件，HT-IDE3000 VPM目前提供大约20余种模拟元件，单击 会弹出零件的对话框(见图3-4-3)让用户挑选。

② ：删除元件，单击 (或是Delete键)可以删除电路中的元件。

③ ：定义元件的引脚连接状态。

④ ：进入电路绘制模式(Config Mode)，只能绘制电路，无法模拟。

⑤ ：切换VPM模式，在电路绘制模式下单击 则进入电路模拟模式；若在电路模拟模式下单击 ，则恢复电路绘制模式。

第 3 章　HT46xx 指令集与开发工具

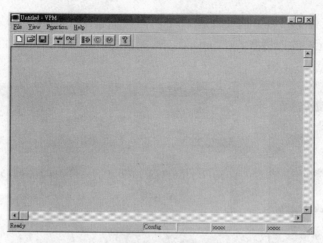

图3-4-2　VPM管理窗口

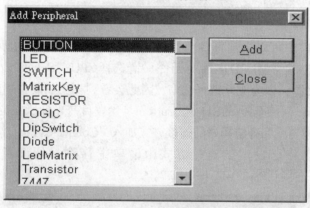

图3-4-3　VPM模拟元件选用对话框

现在就请读者开始绘制图3-3-1的电路,准备执行程序3-3-1的模拟。首先单击 按钮,加入8个LED与电阻,如图3-4-4所示。

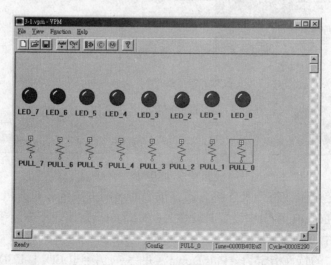

图3-4-4　电路绘制

将光标移到元件上，右击该元件后可设定元件的特性(Config)；若快速双击该元件(或单击)，就可设定元件的连接方式(Connect)。现在请先设定LED_0的特性(见图3-4-5)，然后再设定其连接方式(见图3-4-6)。

图3-4-5　LED特性窗口

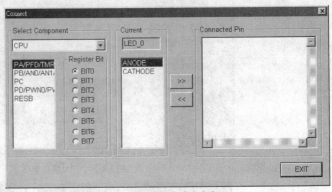

图3-4-6　LED连接方式设定窗口

在LED特性窗口中可以指定其颜色，而连接方式设定窗口中显示几项参数："Select Component"表示选择要与LED连接的元件，请选择CPU "Select Component"下方则是Component的相关资源，因为刚才选择CPU，因此会看到相关的硬件引脚，请结合图3-3-1的电路，将LED_0的Anode接至PA的BIT 0。Current是代表目前光标所在位置的元件名称，选择好对应的引脚之后单击将选择好的连接方式加入"Connectes Pin"清单，如图3-4-7所示。若单击 << ，可以将已设定的连接方式由清单中移除。请读者重复上述过程，将LED_1~LED_7依序连接到PA的BIT 1~ BIT 7。

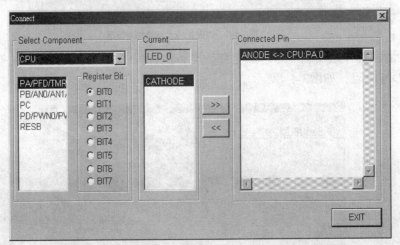

图3-4-7　LED连接方式设置

接着开始设定电阻的特性及连接方式，按照前述的方法依序打开PULL_0电阻的Config与Connect窗口，在Config窗口中设置其特性为"Pull Down"(代表一端连接至地，见图3-4-8)，并在Connect窗口中设置将另一端接到LED_0的CATHOD(见图3-4-9)。重复上述的程序，依序将PULL1~PULL_7的特性指定为"Pull Down"并一一连接到LED_1~LED_7的CATHOD。如此便完成了图3-3-1的模拟电路绘制，模拟电路如图3-4-10所示。

第 3 章　HT46xx 指令集与开发工具

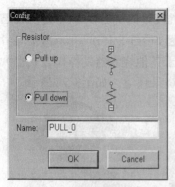

图3-4-8　Resistor特性窗口

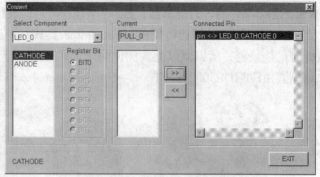

图3-4-9　Resistor连接方式设置窗口

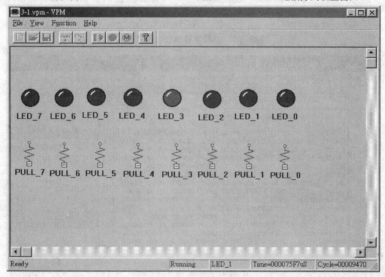

图3-4-10　模拟电路的执行

　　接下来就是一边执行程序，一边观察LED显示的变化。请读者单击 进入电路模拟模式，然后在HT-IDE3000窗口单击 (或F5功能键)全速执行程序，此时VPM会将程序的输出反映到VPM窗口中，如图3-4-10所示，LED将会左、右来回显示。一旦进入电路模拟模式之后，不管是单步执行或全速执行，程序的输出都会在VPM的窗口显示出来。此时如果再单击 ，又可以回到电路绘制模式，使用户可以再次修改电路。在VPM窗口的右下角所呈现的数值则分别代表程序执行的时间与指令周期数。

　　或许用户已经发现，此时LED轮流点亮的速度会受Windows操作系统的影响，例如用户若稍微移动一下鼠标，整个速度就慢了下来。另外，在DELAY子程序上的时间控制也不十分精确，如果硬要挑剔，"无法真实反应执行的时间"应该算是VPM模拟环境的唯一缺点。

3-5　烧录器操作说明

　　经过HT-ICE或VPM的模拟验证之后，可以开始把程序烧录到OTP型单片机，让整个系统成为一个Stand alone的成品。如图3-5-1所示为盛群半导体公司所提供的各种烧录器，

HT-Writer是RS-232接口的烧录器；HandyWriter则是列表机接口的烧录设备(不过现在已不提供)；目前盛群半导体公司已将ICE与Programmer集成在一起如图3-5-1(a)所示。由于笔者手边只有此型的烧录器，所以就以此为例说明烧录的程序，至于其他烧录器的操作方式也大致相同。请读者由HT-IDE3000选择Tools→HandyWriter进入其烧录环境，如图3-5-2所示。

(a) 新型 ICE+Programmer　　　(b) 旧型HandyWriter　　　(c) HT-Writer

图3-5-1　盛群半导体公司的各种烧录器

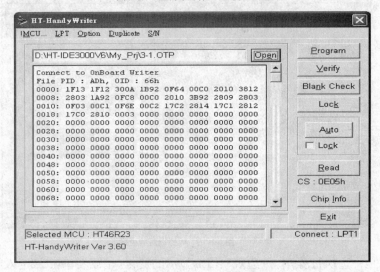

图3-5-2　Programmer操作窗口

首先由Open选择要烧录的文件(.OTP文件)，其次由"！MCU"指定单片机的型号，接下来可以进入Option检查Options中的选项是否正确。最后就可以单击右边的功能键进行烧录或检查的动作，特将其说明如下：

① Program：开始进行程序的烧录。

② Verify：验证烧录是否正确无误。

③ Blank Check：检查是否为空白 IC。

④ Lock：可以对烧录完成的 IC 进行锁定的动作，防止 IC 内部的程序码被读出。

⑤ Auto：自动执行 Blank Check、Program、Verify 三项动作，如果勾选其下方的 Lock 选项，那么在执行完上述三项动作之后，会自动对芯片进行锁定的动作。

⑥ Read：读回芯片内的数据并存回 PC 的缓冲区，其下方的 CS 为检查和(Check Sum)。读回的数据也可以存成文件，其附加文件名为 OTP。

⑦ Chip Info：读回芯片的 Power-on ID、Software ID、ROM size 及 Option size，并加以显示，如图 3-5-3 所示。

⑧ Exit：离开 HandyWriter 烧录环境。

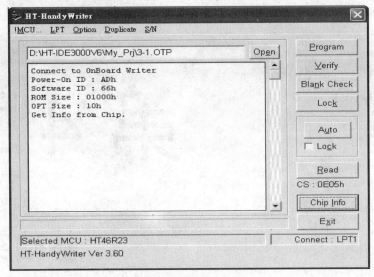

图3-5-3 Chip Information

此外，烧录器还提供Duplicate与S/N功能。S/N顾名思义就是设定芯片的序号，方便日后对于芯片功能的分辨与产品的追溯管理。选择S/N→Setup命令后，将出现如图3-5-4所示的对话框，用户可以指定起始序号，写入所指定地址的低8位。当选择S/N→Enable命令启动此项功能后，所设定的序号会出现在HandyWriter操作窗口的右下角，而且每烧录一颗IC，序号就会自动加1。Duplicate功能则是要复制多颗IC时相当方便的工具，选择Duplicate→Setup，出现如图3-5-5所示对话框，依照实际需要选择要执行的动作。当选择Duplicate→Enable启动此项功能后，只要检测到芯片插入HandyWriter上的活动式IC引脚座，就会自动执行刚刚在Setup功能中选定的工作。

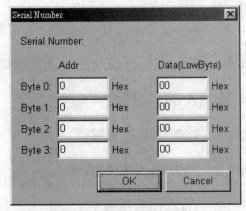

图3-5-4 Serial Number对话框

图3-5-5 Duplicate对话框

第 4 章

基础实验篇

本章将介绍几个简单的实验，如跑马灯、LED、扫描式键盘、步进电机控制、Timer/Event Counter、PFD、外部中断、PWM 接口、A/D 转换器、I^2C 串行接口与 WDT 应用等。希望通过这些基础实验让读者对 HT46xx 的控制及其内部各单元能有初步的了解与认识。由于 *RC* 振荡方式的振荡频率的准确度较差，容易受电源、元件、温度和湿度的影响，所以实验中的电路均采用 Crystal 振荡的方式；如果读者仍想采用成本较低的 *RC* 振荡方式，就必须注意延迟时间的调整，以求获得正确的实验结果。本章的实验内容包括：

- 4-1 　LED 跑马灯实验
- 4-2 　LED 霹雳灯查表实验
- 4-3 　七段显示器控制实验
- 4-4 　指拨开关与七段显示器控制实验
- 4-5 　按键控制实验
- 4-6 　步进电机控制实验
- 4-7 　4×4 键盘控制实验
- 4-8 　喇叭发音控制实验
- 4-9 　Timer/Event Counter 控制实验
- 4-10　Timer/Event Counter 中断控制实验
- 4-11　A/D 转换器控制实验
- 4-12　外部中断控制实验
- 4-13　PWM 接口控制实验
- 4-14　WDT 控制实验
- 4-15　"HALT Mode" 省电模式控制实验
- 4-16　I^2C 串行接口控制实验

4-1 LED跑马灯实验

4-1-1 目 的

本实验将利用带进位位左移(RLC)、带进位位右移(RRC)的指令,让8个LED达到循序点亮、来回移动的效果。

4-1-2 学习重点

通过本实验,读者对于HT46xx的I/O 端口定义、特性以及延迟子程序(DELAY)的时间计算都应有透彻的了解。

4-1-3 电路图

如图4-1-1所示,当连接至LED的输出引脚输出信号为High时,则LED为顺向偏压(Forward Bias),所以LED会亮;反之,当连接至LED输出引脚的输出状态为Low时,则并没有电流流过LED,所以LED不会亮。程序中利用带进位位左移(RLC)、带进位位右移指令(RRC)并搭配不同的延迟时间,让8个LED达到循序点亮、来回移动的效果。读者可以通过降低电阻值(470 Ω)来提高LED的亮度,但是要特别留意HT46xx系列单片机原厂数据手册中,对于I/O端口所提供电流的限制:每一个单独I/O引脚的电流驱动能力为拉电流20 mA,灌电流10 mA。

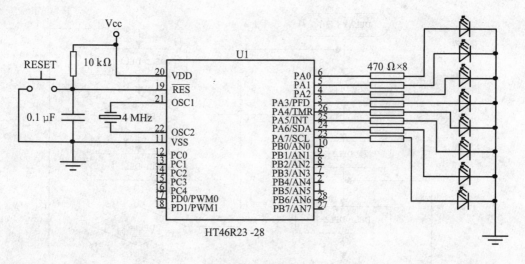

图4-1-1 LED控制电路

4-1-4 流程图及程序

1. 流程图(见图 4-1-2)

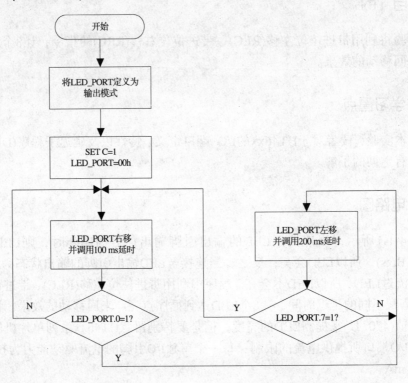

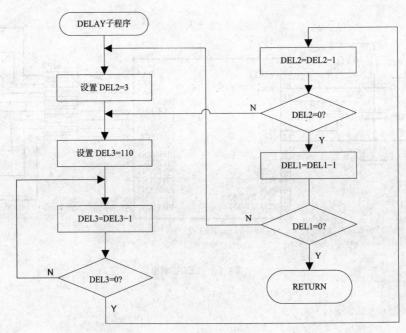

图4-1-2 流程图

2. 程序 4-1-1 LED 跑马灯实验

```
1    ; PROGRAM : 4-1.ASM   (4-1.PRJ)                    BY STEVEN
2    ; FUNCTION: LED SCANNING DEMO PROGRAM               2002.DEC.07.
3    #INCLUDE        HT46R23.INC
4                    CHIP HT46R23
5    ;----------------------------------------------------------------
6    MY_DATA       .SECTION    'DATA'           ;== DATA SECTION ==
7    DEL1          DB          ?                ;DELAY LOOP COUNT 1
8    DEL2          DB          ?                ;DELAY LOOP COUNT 2
9    DEL3          DB          ?                ;DELAY LOOP COUNT 3
10   ;----------------------------------------------------------------
11   LED_PORT      EQU         PA               ;DEFINE LED_PORT
12   LED_PORTC     EQU         PAC              ;DEFINE LED_PORT CONTROL REG.
13
14   MY_CODE       .SECTION  AT 0 'CODE'        ;== PROGRAM SECTION ==
15           ORG         00H                    ;HT-46RXX RESET VECTOR
16   MAIN:
17           CLR         LED_PORTC              ;CONFIG LED_PORT AS O/P MODE
18           CLR         LED_PORT               ;SET INITIAL LED STATE
19           SET         C                      ;SET CARRY FLAG (STATUS.0)
20   RIGHT:
21           RRC         LED_PORT               ;SHIFT RIGHT
22           MOV         A,100
23           MOV         DEL1,A                 ;SET DELAY FACTOR
24           CALL        DELAY                  ;DELAY 100*1ms
25           SNZ         LED_PORT.0             ;IS ALL LEDs HAVE BEEN LIT?
26           JMP         RIGHT                  ;NO. CONTINUE RIGHT SHIFT.
27   LEFT:
28           RLC         LED_PORT               ;SHIFT LEFT
29           MOV         A,200
30           MOV         DEL1,A                 ;SET DALAY FACTOR
31           CALL        DELAY                  ;DELAY 200*1ms
32           SNZ         LED_PORT.7             ;IS ALL LEDs HAVE BEEN LIT?
33           JMP         LEFT                   ;NO. CONTINUE LEFT SHIFT.
34           JMP         RIGHT                  ;REPEAT THE RIGHT PROCESS.
35   ;***************************************************************;
                            Delay about DEL1*1ms
36   ;***************************************************************
DELAY         PROC
37           MOV         A,03
38           MOV         DEL2,A                 ;SET DEL2 COUNTER
39   DEL_2: MOV         A,110
40           MOV         DEL3,A                 ;SET DEL3 COUNTER
41   DEL_3: SDZ         DEL3                    ;DEL3 DOWN COUNT
42           JMP         DEL_3
43           SDZ         DEL2                   ;DEL2 DOWN COUNT
44           JMP         DEL_2
45           SDZ         DEL1                   ;DEL1 DOWN COUNT
46           JMP         DELAY
47           RET
48   DELAY ENDP
49           END
```

HT46xx 单片机原理与实践

程序说明

7~9　　依序定义变量地址。

11~12　定义 LED_PORT 为 PA，定义 LED_PORTC 为 PAC。

15　　　声明存储器地址由 000h 开始(HT46xx Reset Vector)。

17　　　将 LED_PORT 定义成输出模式。

18~19　LED_PORT 内容设为 0 并将进位标志位(CARRY FLAG)设为 1。

20~26　对 LED_PORT 进行带进位位右移的动作，由进位标志位开始带进位位右移一圈，检查 LED_PORT.0 是否为 1，成立表示带进位位右移动作已经结束。程序中 DEL1 是用来控制 DELAY 子程序的延迟时间(DEL1×1 ms)。

27~33　对 LED_PORT 进行带进位位左移的动作，由进位标志位开始带进位位左移一圈，检查 LED_PORT.7 是否为 1，成立表示带进位位左移动作已经结束。重新回到 RIGHT 再进行带进位位右移的动作。不断地反复执行即可让 LED 持续地变化。

38~50　DELAY 子程序，其延迟时间由 DEL1 的内含值决定，大约为 DEL1×1 ms。至于其计算方式，请参考后续的说明。

由于人类的眼睛有视觉暂留的特性，因此在点亮一个LED之后，必须加入适当的延迟，人的肉眼方能观察到灯号的变化。延迟程序是一般应用中经常会使用到的一个子程序，在此特别将DELAY子程序的时间计算加以说明。首先，用户必须知道每一个指令执行所需的时间，在HT46xx系列指令当中，凡是会改变PC值的指令需费时2个指令周期，而其他指令则只需1个指令周期。因此，如RET A,x、RET、RETI、CALL、JMP等指令需2个指令周期；而SDZ、SDZA、SIZ、SIZA、SNZ或SZ则需视其是否发生跳转的情况而定。如果条件不成立，则直接执行下一行指令，所以只要1个指令周期；但若条件成立，则需跳过下一行指令，此时则需2个指令周期。至于1个指令周期究竟是多少时间，就需视外部接的振荡器频率而定。其关系为：一个指令周期=$4/f_{SYS}$，其中f_{SYS}为外部振荡器的振荡频率；因此当外部振荡频率为4 MHz时，一个指令周期即为$1\mu s$。

下面来分析一段程序。

程　　式				指令执行周期			循环
1:	DELAY:	MOV	A,03 ;	1		×DEL1	
2:		MOV	DEL2,A ;	1		×DEL1	
3:	DEL_2:	MOV	A,110 ;	1	×3	×DEL1	
4:		MOV	DEL3,A ;	1	×3	×DEL1	
5:	DEL_3:	SDZ	DEL3 ;	(109×1+2)	×3	×DEL1	A B C
6:		JMP	DEL_3 ;	109×2	×3	×DEL1	
7:		SDZ	DEL2 ;	(2×1+2)		×DEL1	
8:		JMP	DEL_2 ;	2×2		×DEL1	
9:		SDZ	DEL1 ;			(DEL1-1)+2	
10:		JMP	DELAY ;			(DEL1-1)×2	
11:		RET	;	2			

此程序由3个循环组成，循环次数分别由DEL1、DEL2与DEL3寄存器控制，请读者参考DELAY子程序的流程图。首先观察由DEL3控制的小循环(A)，由于在3~4行已将DLE3设为110，所以第5行会执行110次。其中有109次减1不为0，需费时109×1μs；只有1次减1为0，需费时2μs(因为结果为0，必须跳过第6行→PC值改变，需费时2个指令周期)。3~8行为中循环(B)，由DEL2控制；当DEL2减1不等于0时，就会再重复执行一次小循环。最后，当由DEL1控制的大循环(C)减1不为0时，就会再重复执行一次中循环。因此，只要将各个循环所需执行的执行周期加起来，就可以得知其延迟的时间，其指令周期数为(1 006×DEL1+1)Cycles。读者只要在调用DELAY子程序前，先设置DEL1的值，就可以获得所需的延迟时间，其延迟时间范围约为1~256 ms。

值得提醒读者的是，当DEL1设置为0时，DELAY子程序所执行的指令周期数为257 537(1 006×256+1)Cycles。这是因为SDZ指令是先将存储器内容减1再做判断(0−1＝0FFh)，所以循环总共执行了256次。

4-1-5 动动脑＋动动手

- 试着改变 LED 跑动的速度(即缩短延迟的时间)，观察在什么延迟时间以下就无法再分辨出 LED 是逐一被点亮的？

- 改写 DELAY 子程序，使时间延迟的范围为 10~2 560 ms。

- 原程序的执行状态是 LED 先带进位位右移 1 次后再带进位位左移 1 次，试着改写程序，使 LED 的显示状态变成先带进位位右移 2 次之后再带进位位左移 3 次。

- 将 RRC 与 RLC 指令改成 RR 与 RL，并重新改写程序 4-1-1 使其完成与原来相同的动作。

4-2 LED霹雳灯查表实验

4-2-1 目 的

本实验主要介绍HT46x23的查表指令，并利用查表方式让8个LED达到各式各样的灯号变化效果。

4-2-2 学习重点

通过本实验，读者应了解HT46xx的查表指令TABRDC [m]和TABRDL [m]指令的差异。另外，也会介绍一种通过改变PCL而达到查表动作的程序写法，即计算式跳转(Computational Jump)。这对于随机(Random)查询表格中的任意数据相当方便，读者也应该彻底了解其原理。

4-2-3 电路图（见图4-2-1）

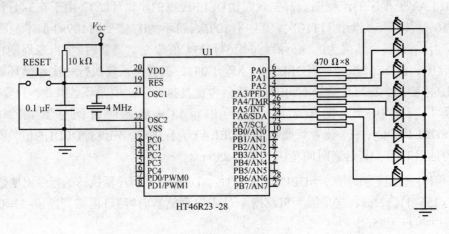

图4-2-1　霹雳灯控制电路

在4-1节中，是以HT46xx的带进位位左移、带进位位右移指令让8个LED达到循序点亮、来回移动的效果；如果想产生其他各种灯号的变化，查表法是最方便的方式。只要事先将所要的灯号变化数据依序建表，再由程序控制在适当的时机送出，就可以让LED展现不同的显示效果。

4-2-4 流程图及程序

1. 流程图(见图 4-2-2)

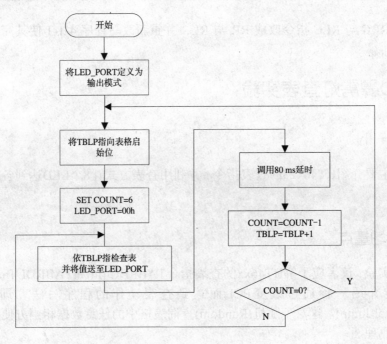

图4-2-2　流程图

2. 程序 4-2-1　LED 霹雳灯查表实验

```
1    ;PROGRAM : 4-2 PROGRAM  (4-2.PRJ)            2002.DEC.30.
2    ;FUNCTION: DISPLAY 1 PORT LED PILI LIGHT             BY STEVEN
3    #INCLUDE       HT46R23.INC
4                   .CHIPHT46R23
5    ;------------------------------------------------------------
6    MY_DATA        .SECTION    'DATA'        ;== DATA SECTION ==
7    DEL1           DB          ?             ;DELAY LOOP COUNT 1
8    DEL2           DB          ?             ;DELAY LOOP COUNT 2
9    DEL3           DB          ?             ;DELAY LOOP COUNT 3
10   COUNT          DB          ?
11   ;------------------------------------------------------------
12   LED_PORT       EQU         PA            ;DEFINE LED PORT
13   LED_PORTC      EQU         PAC           ;DEFINE LED PORT CONTROL REG.
14
15   MY_CODE        .SECTION    AT 0 'CODE'   ;== PROGRAM SECTION ==
16          ORG     00H                       ;HT-46RXX RESET VECTOR
17          CLR     LED_PORTC                 ;CONFIG PA AS OUTPUT MODE
18   MAIN:
19          MOV     A,TAB_PILI
20          MOV     TBLP,A                    ;INITIAL POINTER START ADDRESS
21          MOV     A,06
22          MOV     COUNT,A                   ;SET DATA COUNT OF TAB_PILI
23   LOOP:
24          TABRDC  LED_PORT                  ;READ TABLE AND SEND TO PA
25          MOV     A,80
26          MOV     DEL1,A                    ;SET DELAY FACTOR
27          CALL    DELAY                     ;DELAY 80*1ms
28          INC     TBLP                      ;UPDATE TABLE POINTER
29          SDZ     COUNT                     ;COUNT DOWN BY 1
30          JMP     LOOP                      ;JUMP IF NOT TABLE END
31          JMP     MAIN                      ;RESTART
32   TAB_PILI:
33          DC      000000010000001B          ;TABLE FOR MDK02
34          DC      000000001000010B          ;"1" FOR TURN ON
35          DC      000000000100100B
36          DC      000000000011000B
37          DC      000000000100100B
38          DC      000000001000010B
39   ;****************************************************************
40   ;              Delay about DEL1*10ms
41   ;****************************************************************
42   DELAY  PROC
43          MOV     A, 30
44          MOV     DEL2,A                    ;SET DEL2 COUNTER
45   DEL_2: MOV     A,110
46          MOV     DEL3,A                    ;SET DEL3 COUNTER
47   DEL_3: SDZ     DEL3                      ;DEL3 DOWN COUNT
48          JMP     DEL_3
49          SDZ     DEL2                      ;DEL2 DOWN COUNT
50          JMP     DEL_2
51          SDZ     DEL1                      ;DEL1 DOWN COUNT
52          JMP     DELAY
53          RET
54   DELAY  ENDP
55          END
```

程序说明

7~10　　依序定义变量地址。

12~13　　定义 LED_PORT 为 PA。

16　　声明存储器地址由 000h 开始(HT46x23 Reset Vector)。

HT46xx 单片机原理与实践

行号	说明
17	将 LED_PORT 定义成输出模式。
19~20	将 TBLP 指向表格(TAB_PILI)起始地址。
21~22	将计数器 COUNT 设定成 6。
24	依 TBLP 指示的值到目前的程序页读取数据,并送到 LED_PORT 显示。
25~27	调用 DELAY 子程序,延迟 80 ms。
29~31	判断 COUNT−1 是否等于 0,成立则重新开始(MAIN);反之则回到 LOOP 显示下一笔数据。
32~38	LED 显示数据建表区。
42~54	DELAY 子程序,延迟时间的计算请参考 4-1 节。

HT46xx 提供两个查表指令(TABRDL [m]及 TABRDC [m])方便用户在查表时使用,TBLP 寄存器则代表所要查询的数据所在地址。以 HT46x23 为例,程序存储器共有 4 096 个地址(000h~FFFh),但 TBLP 寄存器只有 8 个位,因此 HT46xx 将程序存储器以 256 个位置为单位称为一个程序页。当使用 TABRDC [m]指令时,依据 TBLP 所指示的地址将目前程序页的内容读出,并将低 8 位存入数据存储器 m 地址中(其余位则存入 TBLH 寄存器)。而若使用 TABRDL [m]指令,则根据 TBLP 所指示的地址将程序最后一页(F00h~FFFh)的内容读出,并将低 8 位存入数据存储器 m 地址中(其余位则存入 TBLH 寄存器)。总结查表位置与 TBLP 的关系如表 4-2-1 所列。

表 4-2-1 查表位置与 TBLP 的关系(HT46x23)

指令	查表数据所在地址											
	*11	*10	*9	*8	*7	*6	*5	*4	*3	*2	*1	*0
TABREC [m]	P11	P10	P9	P8	@7	@6	@5	@4	@3	@2	@1	@0
TABRDL [m]	1	1	1	1	@7	@6	@5	@4	@3	@2	@1	@0

表 4-2-1 中的*11~*0 代表读取数据的所在位置,@7~@0 则是 TBLP 寄存器的内含值,P11~P8 代表目前程序计数器高 4 位的值。通常我们都习惯把查表的数据置于程序之后,以增加程序的可读性,但是万一表格有跨页(Cross Page)的情形时,就无法正确地取得数据,这点请读者特别注意。为避免此问题,建议读者把数据放在程序存储器的最末页(HT46x23:F00h~FFFh,HT46x22、HT46x47:700h~7FFh),然后利用 TABRDL [m]指令来读取数据。以程序 4-2-1 为例,可将程序改写成如下所示:

3. 程序 4-2-2 LED 霹雳灯查表实验(TABRDL)

```
1   ;PROGRAM : 4-2-1.ASM  (4-2-1.PRJ)              2002.DEC.30.
2   ;FUNCTION: DISPLAY 1 PORT LED PILI LIGHT (LAST PAGE)   BY STEVEN
3   #INCLUDE   HT46R23.INC
4              .CHIPHT46R23
5   ;------------------------------------------------ -----------------
6   MY_DATA    .SECTION   'DATA'         ;== DATA SECTION ==
7   DEL1       DB         ?              ;DELAY LOOP COUNT 1
8   DEL2       DB         ?              ;DELAY LOOP COUNT 2
9   DEL3       DB         ?              ;DELAY LOOP COUNT 3
10  COUNT      DB         ?
11  ;--------------------------------------------------------------
12  LED_PORT   EQU        PA             ;DEFINE LED PORT
13  LED_PORTC  EQU        PAC            ;DEFINE LED PORT CONTROL REG.
14
15  MY_CODE    .SECTION   AT 0 'CODE'    ;== PROGRAM SECTION ==
```

```
16              ORG         00H                     ;HT-46RXX RESET VECTOR
17              CLR         LED_PORTC               ;CONFIG PA AS OUTPUT
18      MAIN:
19              MOV         A,TAB_PILI
20              MOV         TBLP,A                  ;INITIAL POINTER START ADDRESS
21              MOV         A,06
22              MOV         COUNT,A                 ;SET DATA COUNT OF TAB_PILI
23      LOOP:
24              TABRDL      LED_PORT                ;LOAD FROM LAST PAGE
25              MOV         A,80
26              MOV         DEL1,A                  ;SET DELAY FACTOR
27              CALL        DELAY                   ;DELAY 80*1ms
28              INC         TBLP                    ;TABLE POINTER+1
29              SDZ         COUNT                   ;END OF TABLE?
30              JMP         LOOP                    ;NO. NEXT ENTRY.
31              JMP         MAIN                    ;RESTART
32      ;****************************** ******************************
33      ;                Delay about DEL1*10ms
34      ;****************************** ******************************
35      DELAY PROC
36              MOV         A, 30
37              MOV         DEL2,A                  ;SET DEL2 COUNTER
38      DEL_2:  MOV         A,110
39              MOV         DEL3,A                  ;SET DEL3 COUNTER
40      DEL_3:  SDZ         DEL3                    ;DEL3 DOWN COUNT
41              JMP         DEL_3
42              SDZ         DEL2                    ;DEL2 DOWN COUNT
43              JMP         DEL_2
44              SDZ         DEL1                    ;DEL1 DOWN COUNT
45              JMP         DELAY
46              RET
47      DELAY ENDP
                ORG         LASTPAGE                ;ORG 0F00H
56      TAB_PILI:
57              DC          000000010000001B        ;TABLE FOR MDK02
58              DC          000000001000010B        ;"1" FOR TURN ON
59              DC          000000000100100B
60              DC          000000000011000B
61              DC          000000000100100B
48              DC          000000001000010B
49              END
```

程序4-2-2与程序4-2-1的差异有两点：其一是改用TABRDL [m]指令来读取数据；其二是利用ORG LASTPAGE伪指令将数据存放在程序存储器的最后一页。

不管是TABRDL [m]还是TABRDC [m]指令，在循序(Sequential)读取表格数据时，可以直接利用递增或递减指令来更改指针(INC TBLP、DEC TBLP)。但是，如果想随机(Random)读取表格内的任意一笔数据，使用上述指令就显得有些笨拙。

所幸，HT46xx的程序计数器(PCL)可以当成一般寄存器拿来运算，若能利用此特性，也可作为查表的另一项选择，尤其在随机读取表格数据时更显其功效。

此种查表方式主要是通过PCL(即PC的低8位)值的改变与"RET A,x"指令来完成。请参考以下的程序范例。

4. 程序 4-2-3　LED 霹雳灯查表实验(ADDM A, PCL)

```
1   ;PROGRAM : 4-2-2.ASM   (4-2-2.PRJ)                    2002.DEC.30.
2   ;FUNCTION: DISPLAY 1 PORT LED PILI LIGHT (ADDM A,PCL)    BY STEVEN
3   #INCLUDE    HT46R23.INC
4   .CHIP       HT46R23
5   ;------------------------------------------------------------------
```

```
6   MY_DATA     .SECTION    'DATA'           ;== DATA SECTION ==
7   DEL1        DB          ?                ;DELAY LOOP COUNT 1
8   DEL2        DB          ?                ;DELAY LOOP COUNT 2
9   DEL3        DB          ?                ;DELAY LOOP COUNT 3
10  COUNT       DB          ?
11  INDEX
12  ;--------------------------------------------------------------
13  LED_PORT    EQU         PA               ;DEFINE LED PORT
14  LED_PORTC   EQU         PAC              ;DEFINE LED PORT CONTROL REG.
15  MY_CODE     .SECTION    AT 0 'CODE'      ;== PROGRAM SECTION ==
16  ORG         00H                          ;HT-46RXX RESET VECTOR
17  CLR         LED_PORTC                    ;CONFIG PA AS OUTPUT
18  MAIN:
19              CLR         INDEX            ;INITIAL POINTER START ADDRESS
20              MOV         A,06
21              MOV         COUNT,A          ;SET DATA COUNT OF TAB_PILI
22  LOOP:
23              MOV         A,INDEX          ;LOAD INDEX TO Acc
24              CALL        TRANS_PILI       ;LOOK-UP TABLE
25              MOV         LED_PORT,A       ;SEND TO LED PORT
26              MOV         A,80
27              MOV         DEL1,A           ;SET DELAY FACTOR
28              CALL        DELAY            ;DELAY 80*1ms
29              INC         INDEX            ;TABLE INDEX+1
30              SDZ         COUNT            ;END OF TABLE?
31              JMP         LOOP             ;NO. NEXT ENTRY.
32              JMP         MAIN             ;RESTART
33  ;****************************************************************
34  ;              Delay about DEL1*10ms
35  ;****************************************************************
36  DELAY       PROC
37              MOV         A,30
38              MOV         DEL2,A           ;SET DEL2 COUNTER
39              MOV         A,110
40              MOV         DEL3,A           ;SET DEL3 COUNTER
41  DEL_3:      SDZ         DEL3             ;DEL3 DOWN COUNT
42              JMP         DEL_3
43              SDZ         DEL2             ;DEL2 DOWN COUNT
44              JMP         DEL_2
45              SDZ         DEL1             ;DEL1 DOWN COUNT
46              JMP         DELAY
47              RET
48  DELAY       ENDP
49  ;****************************************************************
50  ;     LOOK-UP TABLE BY Acc AS INDEX, AND RETURN DATA IN Acc
51  ;****************************************************************
52  TRANS_PILI              PROC
53              ADDM        A,PCL            ;PCL=PCL+Acc
54              RET         A,10000001B      ;RETURN FROM HERE IF Acc=0
55              RET         A,01000010B      ;RETURN FROM HERE IF Acc=1
56              RET         A,00100100B      ;RETURN FROM HERE IF Acc=2
57              RET         A,00011000B      ;RETURN FROM HERE IF Acc=3
58              RET         A,00100100B      ;RETURN FROM HERE IF Acc=4
59              RET         A,01000010B      ;RETURN FROM HERE IF Acc=5
60  TRANS_PILI              ENDP
61              END
```

如本例中的TRANS_PILI子程序，当跳到此子程序执行"ADDM A,PCL"指令(第54行)时，此刻PC值已经指向"RET A,10000001B"指令(第55行)。因此，若Acc＝0，则执行"ADDM A,PCL"后，就接着执行"RET A,10000001B"回到主程序，并在Acc载入表中第一笔显示值"10000001B"；同理，如果Acc＝1，则执行"ADDM A,PCL"(第54行)后，会接着执行"RET A,01000010B"(第56行)回到主程序，此时在Acc载入表中第二笔显示值"01000010B"，其余以此类推。

此种查表方式是利用改变PCL而达到改变程序流程的目的,一般称之为"计算式跳转(Computational Jump)"。往后的实验中经常会使用这种程序技巧,读者一定要把其原理搞清楚。

另外,应注意的是:由于HT46xx的算数逻辑单元(ALU)只有8位,当执行"ADDM A,PCL"指令时,系统只会将低8位的运算结果放置在PC<7:0>(即PCL),其余位维持不变(也就是说,如果Acc与PCL相加时既使有进位产生,此进位是不会累进至PC的高位组)。因此,如果想以"ADDM A,PCL"指令达到查表目的,千万要记得查表的数据一定不能有跨页的情况。这种程序技巧不只用来作为随机查询表格的利器,也可用来作为控制程序流程的方式,如下例就是依据寄存器Acc的数值来控制程序跳到不同的执行地址。
例题:

```
ADDM    A,PCL       ;PCL=Acc+PCL
JMP     EQU_0       ;Jump to EQU_0 if Acc=0
JMP     EQU_1       ;Jump to EQU_1 if Acc=1
JMP     EQU_2       ;Jump to EQU_2 if Acc=2
JMP     EQU_3       ;Jump to EQU_3 if Acc=3
……
```

4-2-5 动动脑 + 动动手

- 改写程序,让 LED 显示图案的速度越来越快。
- 在程序 4-1-1 第 32 行指令前插入 "ORG 1FDh" 指令,重新编译并执行程序,程序是否正常执行?是什么原因导致此结果?请读者利用单步执行的功能观察 PC 与 Stack 的变化情形。
- 在程序 4-2-3 第 53 行指令前插入 "ORG 1FDh" 指令,重新编译并执行程序,程序是否正常执行?是什么原因导致此结果?请读者利用单步执行的功能观察 PC 与 Stack 的变化情形。

4-3 七段显示器控制实验

4-3-1 目 的

本实验以HT46x23来控制一个共阴极(Common Cathod)的七段显示器,并采用查表法的方式使七段显示器重复依序地显示0~9的数字。

4-3-2 学习重点

通过本实验,读者应熟悉七段显示器的控制与HT46xx的查表方式与应用。

4-3-3 电路图(见图4-3-1)

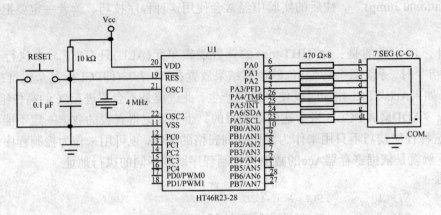

图4-3-1 七段显示器控制电路

七段显示器是在一般电子电路中广泛被使用到的一种元件。它由8个发光二极管组合而成，依照其内部的构造可分为共阴极和共阳极两种。共阳极七段显示器将所有的LED阳极连接在一起，输入Low时LED才会亮，其内部构造如图4-3-2所示。如果需要显示0~F的数字，只要让适当位置的LED发亮即可。如表4-3-1所列为共阳极与共阴极七段显示器的0~F字形表。

表4-3-1 共阳极/共阴极七段显示器的字形表

字形	共阳极七段显示器		共阴极七段显示器	
	RB(7,…,0) (h,g,…,b,a)	HEX	RB(7,…,0) (h,g,…,b,a)	HEX
0	11000000	C0h	00111111	3Fh
1	11111001	F9h	00000110	06h
2	10100100	A4h	01011011	5Bh
3	10110000	B0h	01001111	4Fh
4	10011001	99h	01100110	66h
5	10010010	92h	01101101	6Dh
6	10000010	82h	01111101	7Dh
7	11111000	F8h	00000111	07h
8	10000000	80h	01111111	7Fh
9	10011000	98h	01100111	67h
A	10001000	88h	01110111	77h
B	10000011	83h	01111100	7Ch
C	10100111	A7h	01011000	58h
D	10100001	A1h	01011110	5Eh
E	10000110	86h	01111001	79h
F	10001110	8Eh	01110001	71h

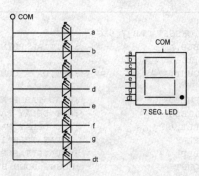

图4-3-2 共阳极七段显示器内部结构图

4-3-4 流程图及程序

1. 流程图(见图 4-3-3)

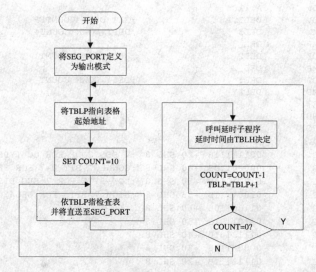

图4-3-3 流程图

2. 程序 4-3 七段显示器控制实验

```
1    ;PROGRAM : 4-3.ASM  (4-3.PRJ)                  2002.DEC.30.
2    ;FUNCTION: 7 SEGMENT LED DEMO PROGRAM          BY STEVEN
3    #INCLUDE    HT46R23.INC
4                .CHIPHT46R23
5    ;--------------------------------------------------------------
6    MY_DATA     .SECTION    'DATA'       ;== DATA SECTION ==
7    DEL1        DB      ?                ;DELAY LOOP COUNT 1
8    DEL2        DB      ?                ;DELAY LOOP COUNT 2
9    DEL3        DB      ?                ;DELAY LOOP COUNT 3
10   COUNT       DB      ?
11   ;--------------------------------------------------------------
12   SEG_PORT    EQU         PA           ;DEFINE 7-SEGMENT LED PORT
13   SEG_PORTC   EQU         PAC          ;DEFINE SEG_PORT CONTROL REG.
14   MY_CODE     .SECTION    AT 0 'CODE'  ;== PROGRAM SECTION ==
15         ORG       00H                  ;HT-46RXX RESET VECTOR
16         CLR       SEG_PORTC            ;CONFIG PA AS OUTPUT MODE
```

```
17      MAIN:
18              MOV     A,OFFSET TAB_7_SEG
19              MOV     TBLP,A                  ;INITIAL POINTER START ADDRESS
20              MOV     A,10
21              MOV     COUNT,A                 ;SET DATA COUNT OF TAB_7_SEG
22      LOOP:
23              TABRDL  SEG_PORT                ;READ TABLE AND SEND TO PA
24              MOV     A,TBLH                  ;GET DELAY FACTOR
25              MOV     DEL1,A                  ;SET DELAY FACTOR
26              CALL    DELAY                   ;DELAY TBLH*10mS
27              INC     TBLP                    ;UPDATE TABLE POINTER
28              SDZ     COUNT                   ;COUNT DOWN BY 1
29              JMP     LOOP                    ;JUMP IF NOT TABLE END
30              JMP     MAIN                    ;RESTART
31      ;****************************************************************
32      ;                  Delay about DEL1*10ms
33      ;****************************************************************
34      DELAY   PROC
35              MOV     A,30
36              MOV     DEL2,A                  ;SET DEL2 COUNTER
37      DEL_2:  MOV     A,110
38              MOV     DEL3,A                  ;SET DEL3 COUNTER
39      DEL_3:  SDZ     DEL3                    ;DEL3 DOWN COUNT
40              JMP     DEL_3
41              SDZ     DEL2                    ;DEL2 DOWN COUNT
42              JMP     DEL_2
43              SDZ     DEL1                    ;DEL1 DOWN COUNT
44              JMP     DELAY
45              RET
46      DELAY   ENDP
47              ORG     LASTPAGE
48      TAB_7_SEG:
49              DC      00111111B +20 SHL 8     ;7 SEG CODE FOR CC TYPE LED
50              DC      00000110B +40 SHL 8     ;CONNECT BIT 0~7 FOR SEGMENT A~H
51              DC      01011011B +50 SHL 8     ;LOW-BYTE FOR 7 SEG DISPLAY CODE
52              DC      01001111B +55 SHL 8     ;HIGH 7 BITs FOR DELAY FACTOR
53              DC      01100110B +60 SHL 8
54              DC      01101101B +50 SHL 8
55              DC      01111101B +40 SHL 8
56              DC      00000111B +30 SHL 8
57              DC      01111111B +20 SHL 8
58              DC      01100111B +95 SHL 8
59              END
```

程序说明

7~10   依序定义变量地址。

12~13   定义 SEG_PORT 为 PA。

16   声明存储器地址由 000h 开始(HT46xx Reset Vector)。

17   将 SEG_PORT 定义成输出模式。

19~20   将 TBLP 指向表格(TAB_PILI)起始地址。

21~22   将计数器 COUNT 设置成 10(因为要显示 0~9 共 10 个数字)。

23   依 TBLP 指示的值到程序存储器最末页(Last Page)读取数据,并送到 SEG_PORT 显示,此时在 TBLH 寄存器中为各个数字显示时间的控制常量。

24~26   调用延时子程序,延时时间由 TBLH 的值决定。

27	将 TBLP 加 1，指向下一笔数据。
28~30	判断 COUNT-1 是否等于 0，成立则重新开始(MAIN)；反之则回到 LOOP 显示下一笔数据。
34~46	DELAY 子程序，延时时间的计算请参考 4-1 节。
48~58	七段显示码数据与延时时间常量建表区。

本实验采用查表法来读取所要输出的字形数据,此法须先将输出的字形数据建立表格,且数据的排列顺序要事先安排好,如此才可读取到正确的字形。比较特别的是HT46x23的程序存储器为4 096×15位，HT46x22与HT46x47为2 048×14位、HT46x24为8 192×16位，然而所提供的查表指令(TABRDL [m]及TABRDC [m])则只是将8位的查表结果存放在地址m的数据存储器。为了解决表格数据为15位(或14位、16位)而数据存储器只有8位的情形,HT46xx将读回的低8位置于指令所指定的寄存器(m)中,而其他的位则固定地存于TBLH寄存器内。程序中的表格数据除了七段显示器的字形码之外,也利用SHL伪指令(Assebmly Directive)将控制显示时间的常量集成在一个存储器位置中,因此通过TABRDL SEG_PORT指令执行查表动作后,除了将显示码(Bit7~0)送到SEG _PORT显示数值之外,其余位(Bit14~8)也一并读到TBLH寄存器中,再以TBLH的值来控制DELAY子程序,用于达到控制显示时间的目的。提醒读者的是,TBLH是一个只读的寄存器,无法用TABRDL [m]及TABRDC [m]以外的指令来更改其值。因此,最好避免在主程序与中断服务子程序中同时使用查表令,如果无法避免,最好在查表指令之前先禁止中断的发生,待TBLH的值存放到适当寄存器之后再将其使能,以免发生TBLH在中断服务子程序中被破坏的情形。

4-3-5 动动脑 + 动动手

- 改写程序，让显示的数值由递增改成递减。
- 改写程序，让显示的速度随着显示的数值增加而减慢。
- 改写程序，让显示的速度随着显示的数值增加而变快。

4-4 指拨开关与七段显示器控制实验

4-4-1 目 的

本实验以指拨开关为输入设备，控制七段显示器以不同的速度显示0~9的数字。

4-4-2 学习重点

通过本实验，读者应熟悉HT46xx I/O端口的输入控制方式，并对查表法的应用更加得心应手。

4-4-3 电路图

如图4-4-1所示的硬件设计方式,当指拨开关(DIP Switch)拨到ON时,相对应的输入引脚读入为Low;反之若拨至OFF时,则读入为High。由于HT46xx的I/O端口具备Pull-High功能,上拉电阻的阻值在40~80 kΩ之间,所以电路图中接在指拨开关的电阻实际上是可以不用接的,但请读者务必在Options(配置选项)中选择PA的Pull-High功能。

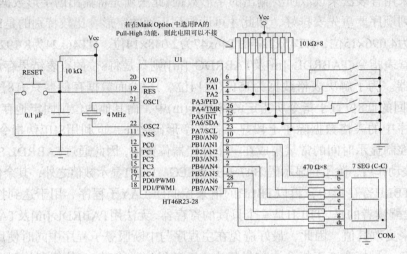

图4-4-1 指拨开关与七段显示器控制电路

4-4-4 流程图及程序

1. 流程图(见图4-4-2)

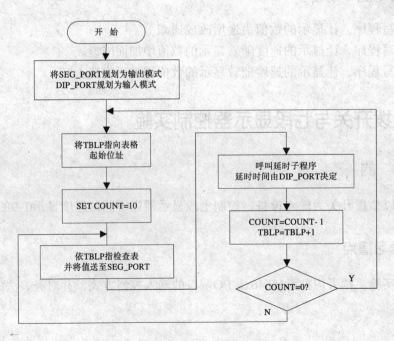

图4-4-2 流程图

2. 程序 4-4 指拨开关与七段显示器控制实验

```
1    ;PROGRAM : 4-4.ASM  (4-4.PRJ)                        2002.DEC.30.
2    ;FUNCTION: 7 SEGMENT LED WITH SPEED CONTROL BY DIP_SW   BY STEVEN
3    #INCLUDE       HT46R23.INC
4                   .CHIPHT46R23
5    ;------------------------------------------------------------------
6    MY_DATA        .SECTION    'DATA'       ;== DATA SECTION ==
7    DEL1           DB      ?                ;DELAY LOOP COUNT 1
8    DEL2           DB      ?                ;DELAY LOOP COUNT 2
9    DEL3           DB      ?                ;DELAY LOOP COUNT 3
10   COUNT          DB      ?
11   ;------------------------------------------------------------------
12   SEG_PORT       EQU     PA               ;DEFINE 7-SEG PORT
13   SEG_PORTC      EQU     PAC              ;DEFINE 7-SEG PORT CONTROL REG.
14   DIP_PORT       EQU     PB               ;DEFINE DIP PORT
15   DIP_PORTC      EQU     PBC              ;DEFINE DIP PORT CONTROL REG.
16
17   MY_CODE        .SECTION  AT 0 'CODE'    ;== PROGRAM SECTION ==
18          ORG     00H                      ;HT-46RXX RESET VECTOR
19          SET     DIP_PORTC                ;CONFIG DIP_PORT AS INPUT MODE
20          CLR     SEG_PORTC                ;CONFIG SEG_PORT AS OUTPUT MODE
21   MAIN:
22          MOV     A,OFFSET TAB_7_SEG
23          MOV     TBLP,A                   ;INITIAL POINTER START ADDRESS
24          MOV     A,10
25          MOV     COUNT,A                  ;SET DATA COUNT OF TAB_7_SEG
26   LOOP:
27          TABRDL  SEG_PORT                 ;READ TABLE AND SEND TO SEG_PORT
28          MOV     A,DIP_PORT               ;GET DELAY FACTOR FROM DIP_SW
29          MOV     DEL1,A                   ;SET DELAY FACTOR
30          CALL    DELAY                    ;DELAY DIP_PORT*10ms
31          INC     TBLP                     ;UPDATE TABLE POINTER
32          SDZ     COUNT                    ;COUNT DOWN BY 1
33          JMP     LOOP                     ;JUMP IF NOT TABLE END
34          JMP     MAIN                     ;RESTART
50   ;******************************************************************
51   ;              Delay about DEL1*10ms
52   ;******************************************************************
53   DELAY  PROC
54          MOV     A,30
55          MOV     DEL2,A                   ;SET DEL2 COUNTER
56   DEL_2: MOV     A,110
57          MOV     DEL3,A                   ;SET DEL3 COUNTER
58   DEL_3: SDZ     DEL3                     ;DEL3 DOWN COUNT
59          JMP     DEL_3
60          SDZ     DEL2                     ;DEL2 DOWN COUNT
61          JMP     DEL_2
62          SDZ     DEL1                     ;DEL1 DOWN COUNT
63          JMP     DELAY
64          RET
65   DELAY  ENDP
66          ORG     LASTPAGE
67   TAB_7_SEG:
68          DC      000000000111111B         ;7 SEG CODE FOR CC TYPE LED
69          DC      000000000000110B         ;CONNECT BIT 0~7 FOR SEGMENT A~H
70          DC      000000001011011B         ;LOW-BYTE FOR 7 SEG DISPLAY CODE
71          DC      000000001001111B
72          DC      000000001100110B
73          DC      000000001101101B
74          DC      000000001111101B
75          DC      000000000000111B
76          DC      000000001111111B
77          DC      000000001100111B
78          END
```

程序说明

7~10	依序定义变量地址。
12~15	定义 SEG_PORT 为 PA，定义 PB_SW 为 PB.0。
18	声明存储器地址由 000h 开始(HT46xx Reset Vector)。
17	将 SEG_PORT 定义成输出模式，PB_SW 为输入模式。
22~23	将 TBLP 指向表格(TAB_PILI)起始地址。
24~25	将计数器 COUNT 设置成 10(因为要显示 0~9 共 10 个数字)。
27	依 TBLP 指示的值到程序存储器最末页读取数据，并送到 SEG_PORT 显示。
28~30	由 DIP_PORT 读入控制数字显示时间的常量，调用延时子程序，延时时间由 DIP_PORT 的值决定。
31	将 TBLP 加 1，指向下一笔数据。
32~34	判断 COUNT−1 是否等于 0，成立则重新开始(MAIN)；反之则回到 LOOP 显示下一笔数据。
55~67	DELAY 子程序，延时时间的计算请参考 4-1 节。
36~46	七段显示码数据建表区。

本实验利用指拨开关读入的数值，控制DELAY子程序的延时时间。因此，每一个数字的显示时间大约可由10 ms变化至2 560 ms。关于DELAY子程序的时间计算在4-1节中已有完整的介绍。

4-4-5 动动脑 + 动动手

- 改写程序，当 DIP_PORT.7 设置为 1 时就停止计数。
- 改写程序，当 DIP_PORT.0 设置为 1 时就只显示奇数(1、3、5、7、9、1、…)；而若 DIP_PORT.0 设置为 0 时只显示偶数(0、2、4、6、8、0、…)。

4-5 按键控制实验

4-5-1 目 的

本实验以按键控制七段显示器的显示；每按一次按键，七段显示器的显示就加1；否则显示器上的数值维持不变。

4-5-2 学习重点

通过本实验，读者应了解按键弹跳现象的排除方式。

4-5-3 电路图

首先由电路图4-5-1中可知，若按键未被按下，则对应的I/O引脚应呈现高电位；反之，若按下按键，则应该读到低电位的信号。本实验主要利用I/O(PB0)引脚电位的高、低来控制七段显示器是否该加1。由于HT46xx的I/O端口具备Pull-High功能，上拉电阻的阻值在40~80 kΩ之间，所以电路图中接于按钮开关的电阻实际上是可以不用接的，但请读者务必在Options中选择PB的Pull-High功能。

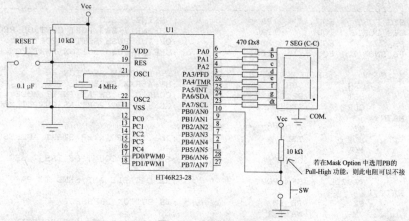

图4-5-1 按键控制电路

4-5-4 流程图及程序

1. 流程图(见图 4-5-2)

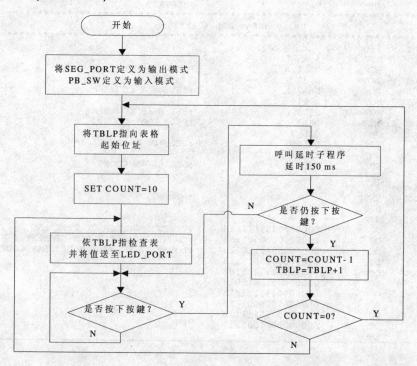

图4-5-2 流程图

2. 程序 4-5 按键控制实验

```
1    ;PROGRAM : 4-5.ASM  (4-5.PRJ)                    2002.DEC.30.
2    ;FUNCTION: 7 SEGMENT LED WITH PUSH BOTTON CONTROL  BY STEVEN
3    #INCLUDE    HT46R23.INC
4                .CHIPHT46R23
```

```
5   ;----------------------------------------------------------------------
6   MY_DATA     .SECTION    'DATA'          ;== DATA SECTION ==
7   DEL1        DB      ?                   ;DELAY LOOP COUNT 1
8   DEL2        DB      ?                   ;DELAY LOOP COUNT 2
9   DEL3        DB      ?                   ;DELAY LOOP COUNT 3
10  COUNT       DB      ?
11  ;----------------------------------------------------------------------
12  SEG_PORT    EQU     PA                  ;DEFINE 7-SEG PORT
13  SEG_PORTC   EQU     PAC                 ;DEFINE 7-SEG PORT CONTROL REG.
14  PB_SW       EQU     PB.0                ;DEFINE PUSH BOTTON PORT
15  PB_SWC      EQU     PBC.0               ;DEFINE PUSH BOTTON PORT CON REG.
16
17  MY_CODE     .SECTION    AT 0 'CODE'     ;== PROGRAM SECTION ==
18          ORG     00H                     ;HT-46RXX RESET VECTOR
19          SET     PB_SWC                  ;CONFIG PB_SW AS INPUT MODE
20          CLR     SEG_PORTC               ;CONFIG SEG_PORT AS OUTPUT MODE
21  MAIN:
22          MOV     A,OFFSET TAB_7_SEG
23          MOV     TBLP,A                  ;INITIAL POINTER START ADDRESS
24          MOV     A,10
25          MOV     COUNT,A                 ;SET DATA COUNT OF TAB_7_SEG
26  LOOP:
27          TABRDL  SEG_PORT                ;READ TABLE AND SEND TO PA
28  WAIT_1:
29          SZ      PB_SW                   ;IS BUTTON PRESSED?
30          JMP     WAIT_1                  ;NO.
31          CALL    DELAY                   ;DELAY, SKIP TO SEE BOUNCING!!
32          SZ      PB_SW
33          JMP     WAIT_1
34          INC     TBLP                    ;UPDATE TABLE POINTER
35          SDZ     COUNT                   ;COUNT DOWN BY 1
36          JMP     LOOP                    ;JUMP IF NOT TABLE END
37          JMP     MAIN                    ;RESTART
38  ;**********************************************************************
39  ;                   Delay about DEL1(15)*10ms
40  ;**********************************************************************
41  DELAY   PROC
42          MOV     A,15
43          MOV     DEL1,A
44  DEL_1:  MOV     A,30
45          MOV     DEL2,A                  ;SET DEL2 COUNTER
46  DEL_2:  MOV     A,110
47          MOV     DEL3,A                  ;SET DEL3 COUNTER
48  DEL_3:  SDZ     DEL3                    ;DEL3 DOWN COUNT
49          JMP     DEL_3
50          SDZ     DEL2                    ;DEL2 DOWN COUNT
51          JMP     DEL_2
52          SDZ     DEL1                    ;DEL1 DOWN COUNT
53          JMP     DEL_1
54          RET
55  DELAY   ENDP
56          ORG     LASTPAGE
57  TAB_7_SEG:
58          DC      000000000111111B        ;7 SEG CODE FOR CC TYPE LED
59          DC      000000000000110B        ;CONNECT BIT 0~7 FOR SEGMENT A~H
60          DC      000000010011011B        ;LOW-BYTE FOR 7 SEG DISPLAY CODE
61          DC      000000010001111B
62          DC      000000011100110B
63          DC      000000011011101B
64          DC      000000011111101B
65          DC      000000000000111B
66          DC      000000011111111B
67          DC      000000011100111B
68          END
```

程序说明

7~10   依序定义变量地址。

行号	说明
12~15	定义 SEG_PORT 为 PB，定义 DIP_PORT 为 PA。
18	声明存储器地址由 000h 开始(HT46xx Reset Vector)。
17	将 SEG_PORT 定义成输出模式，DIP_PORT 为输入模式。
22~23	将 TBLP 指向表格(TAB_PILI)起始地址。
24~25	将计数器 COUNT 设定成 10(因为要显示 0~9 共 10 个数字)。
27	依 TBLP 指示的值至程序存储器最末页读取数据，并送到 SEG_PORT 显示。
29~30	等待用户按下按键。
31	调用延时子程序，延时 150 ms 的弹跳时间。
32~33	检查按键是否仍处于被按下的状态。
34	将 TBLP 加 1，指向下一笔数据。
35~37	判断 COUNT−1 是否等于 0，成立则重新执行程序(JMP MAIN)；反之则回到 LOOP 显示下一笔数据。
41~55	DELAY 子程序，延时时间的计算请参考 4-1 节。
57~67	七段显示码数据建表区。

程序中利用"SZ PB_SW"指令检测按键是否按下，以决定是否更新七段显示器上的字形。如果将"CALL DELAY"指令由程序中删除，此时程序看起来仍旧合理，但是按下按键之后，发现显示器的显示值并非加1，而是毫无规则地乱加一通，有时加2、有时加3……这是由按键本身的"弹跳现象"所造成的，请参考图4-5-3。

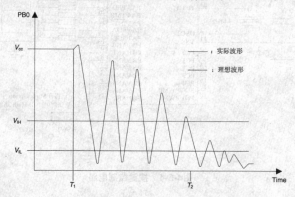

图4-5-3 波形图

理论上如果在时间T_1按下按键，那么应该在PB0测得如图4-5-3中虚线的信号。可是因为按键内部为机械式的接点，所以实际上会得到类似图4-5-3中的实线波形。图4-5-3中另外标示了V_{IH}及V_{IL}两个电压，分别代表HT46xx把输入当成High的最低电位与当成Low的最高电位(当$V_{DD}=5$ V时，$V_{IH}=0.7V_{DD}$，$V_{IL}=0.3V_{DD}$)。虽然用户只按了一次按键，但是HT46xx的PB0却接收到好几次的High、Low变化，由于指令执行的速度相当快(μs级)，因此HT46xx会误以为用户按了好几次按键，这就是为什么显示器所显示的值会乱跳，而非逐次加1的原因。解决弹跳现象，可从硬件或软件两方面着手，硬件的解决方式一般采用R-S正反器；而软件的解决方式，基本上就是等信号稳定后(T_2)再做按键事件的处理。为了减少硬件的成本，通常只要有微处理机的场合，大都采用软件来解决弹跳现象。至于按键上的信号需多久的时间才会稳定，当然需视其本身的材质与特性而定，一般在15~50 ms之间。以上的程序就是以软件的方式来解决弹跳现象,其做法是当检测到SW1(即PB0)变Low时,等延时150 ms之后再去执行指针更新及显示的动作。之所以要延时如此久的原因,是要让按键持续一直按着不放时,七段显示器的累加效果得以显现。

4-5-5 动动脑+动动手

- 📖 修改程序,使按键每按一次,七段显示器的显示值就减1。
- 📖 本实验的程序动作是当按键每按一次,就将七段显示器的显示值加1。若按键不放开,则显示器的值就每隔150 ms自动加1。请修改范例程序,使得按键即使按着不放,在每隔0.5 s之后显示值仍会自动加1。
- 📖 续上例,如果按键每按下一次只能加1,除非用户放开再按下,否则就不能再累加。请改写程序达成此要求。
- 📖 参考图4-5-4,写一个程序完成以下的动作:每按一次SW1,七段显示器的显示值就减1;每按一次SW2,七段显示器的显示值就加1。

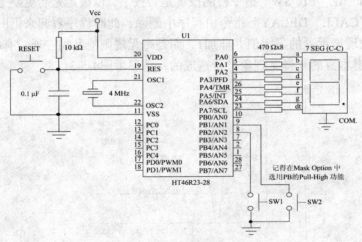

图4-5-4 按键控制电路

4-6 步进电机控制实验

4-6-1 目 的

本实验以程序控制步进电机的转速与转向。

4-6-2 学习重点

通过本实验,读者应了解步进电机的工作原理。

4-6-3 电路图(见图4-6-1)

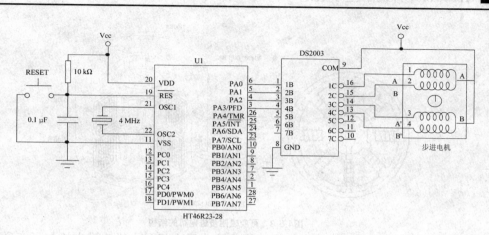

图4-6-1 步进电机控制电路

步进电机是一种单体运动设备,在开回路系统(Open Loop System)中使用步进电机更能达到精确位置与速度控制,而且设计过程极为简单,因此被广泛应用,如光驱、磁盘驱动器、打印机、绘图仪等。步进电机之所以广泛地当作驱动设备,主要有下列特性:

◇ 步进电机主要是以寸动方式来运转,可利用脉冲式电流来驱动,而脉冲数的多少则是决定电机转动的步进角大小。

◇ 只要将脉冲依序加到各相绕组,则转子旋转的速度和脉冲频率成正比,同时步进电机也较易瞬间启动或急速停止。

◇ 使用开回路系统控制,而不需复杂的回路控制。

步进电机依其结构可分为三大类,分别为:

◇ 永磁步进电机(Permanent Magnet Motor):结构如图4-6-2所示,转轴(Rotor)由永久磁铁组成,此类电机体系结构简单、成本低,因此适于大量制造。其特点为转速低、转矩小、步进角度(Step Angle)大,所以适于转速低、扭力低且不需精密角度控制的应用上。

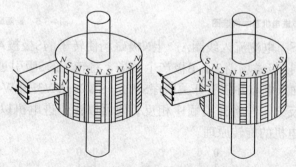

图4-6-2 永磁步进电机的结构

◇ 可变磁阻步进电机(Variable Reluctance Motors):结构如图4-6-3所示,此类电机内部并无永久磁铁,因此具有体积小的优点。

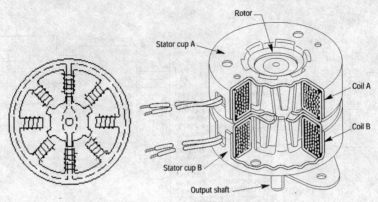

图4-6-3 可变磁阻步进电机的结构

◇ 混合式步进电机(Hybrid Motors)：混合式步进电机是永磁步进电机与可变磁阻步进电机的综合体，也是目前在工业上应用最为广泛的步进电机。

如图4-6-4所示，步进电机中间的转子(Rotor)由永久磁铁组成，一边为N极另一边为S极定子(Stator)。在步进电机线圈中流动的电流称为励磁电流。对于四相步进电机而言，其定子有四组线圈A、A'、B及B'，各线圈的C端共接电池正极，另一端则通过开关接电池负极，如图4-6-5所示。

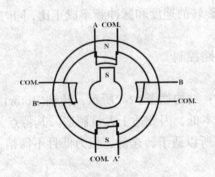

图4-6-4 步进电机基本构造图

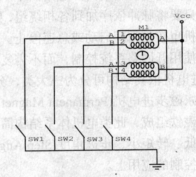

图4-6-5 绕阻励磁电路

当开关SW1按下，电流流入线圈A产生N极磁场使转子的S极被A极吸引过来，而SW1放掉立即按下SW2则A极磁场消失，B极产生磁场，将转子的S吸引过来，则转子就会顺时针旋转。若将四极加入电流并依照开关顺序SW1→SW2→SW3→SW4，则步进电机就会以顺时针方向旋转；反之，若将开关顺序相反，使可以让步进电机以逆时针方向旋转。如图4-6-6所示为步进电机的旋转原理。

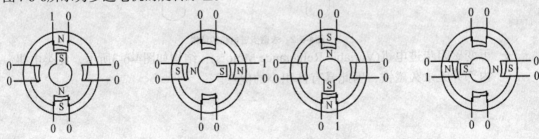

图4-6-6 步进电机旋转原理

由于步进电机各相线圈的励磁场顺序会影响转动方向,且励磁的脉冲可决定转动的速率,因此励磁脉冲频率不能太高,否则会使转子产生失步现象,也就是线圈受到励磁,转子还来不及转动,则又要去励磁,使得转子无法转换。步进电机依照定子线圈的相可分单相、双相、三相、四相和五相;因定子线圈上产生正负两个磁极方式的不同,又分为单极驱动与双极驱动。以一个四相步进电机的单极励磁驱动电路为例,单极励磁依各相之间励磁顺序的不同,可分为单相励磁、双相励磁及单–双相励磁三种。

- ◇ 单相励磁:每次励磁一相线圈,步进角为 θ,本身消耗电力小而且角度精确度高,但转距小相对阻尼效果差,振动现象也大。因此除非有特别要求角度精确度,否则一般较少使用,如图4-6-7(a)所示。
- ◇ 双相励磁:每次励磁二相励磁线圈,步进角为 θ,转距大,在稳定的操作区内使用相对阻尼效果较好,是目前使用率较高的一种励磁方式,如图4-6-7(b)所示。
- ◇ 单—双相励磁:这种方法是单相励磁和双相励磁的混合方式,其最大优点在于每一步的角度为原来的一半,所以精度提高一倍,且能很平滑运转,如图4-6-7(c)所示。

STEP	A	B	A'	B'
1	1	0	0	0
2	0	1	0	0
3	0	0	1	0
4	0	0	0	1

STEP	A	B	A'	B'
1	1	1	0	0
2	0	1	1	0
3	0	0	1	1
4	1	0	0	1

STEP	A	B	A'	B'
1	1	0	0	0
2	1	1	0	0
3	0	1	0	0
4	0	1	1	0
5	0	0	1	0
6	0	0	1	1
7	0	0	0	1
8	1	0	0	1

(a) 单相励磁　　　　　　　　(b) 双相励磁　　　　　　　　(c) 单—双相励磁

图4-6-7　步进电机的励磁方式

如图4-6-6所示,四相步进电机因为定子上有四组相对线圈,分别提供90°相位差,当步进电机为单极励磁时,送入一个脉冲则转子转动一步即停止,这时转子所旋转的角度称为步进角,其步进角的计算公式如下:

步进角=360°/(相数×转子步数)=360°/寸动数

以四相50齿为例,其步进角 $\theta = 360°/(4 \times 50) = 1.8°$;若步进电机走200步则正好是 $(200 \times 1.8° = 360)$ 一圈。图4-6-8是200-Step步进电机内部结构的示意图。

本实验采用单相励磁的方式来驱动步进电机,一般在步进电机的驱动电路中使用4个达灵顿晶体管电路来接受单片机的控制。图4-6-9所示的驱动电路由4个达灵顿晶体管组成,将控制信号加到晶体管的基极,以控制相组的导通或截止,若晶体管进入饱和区则相组就会受到励磁。

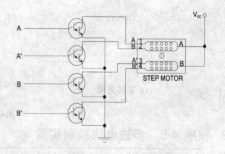

图4-6-8　200-Step步进电机　　　　　图4-6-9　步进电机驱动电路

4-6-4 流程图及程序

1. 流程图(见图 4-6-10)

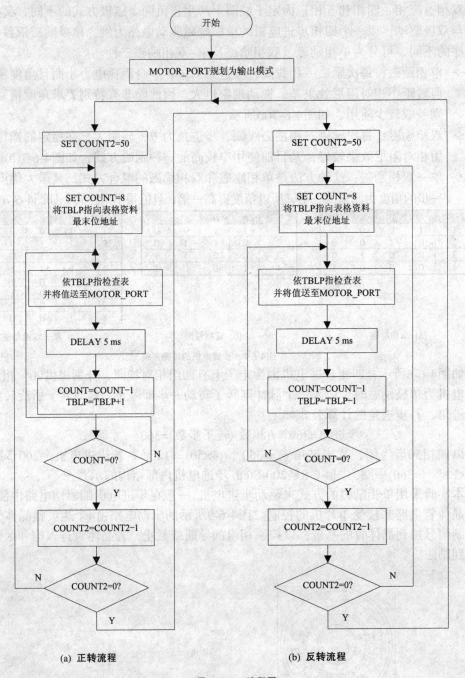

(a) 正转流程　　　　　　(b) 反转流程

图4-6-10　流程图

2. 程序 4-6　步进电机控制实验

```
1  ;PROGRAM : 4-6.ASM  (4-6.PRJ)              2002.DEC.30.
2  ;FUNCTION: STEP MOTOR HALF STEP CONTROL DEMO PROGRAM BY STEVEN
```

```
3       #INCLUDE    HT46R23.INC
4                   .CHIPHT46R23
5       ;--------------------------------------------------------------
6       MY_DATA     .SECTION    'DATA'          ;== DATA SECTION ==
7       DEL1        DB      ?                   ;DELAY LOOP COUNT 1
8       DEL2        DB      ?                   ;DELAY LOOP COUNT 2
9       DEL3        DB      ?                   ;DELAY LOOP COUNT 3
10      COUNT       DB      ?                   ;PHASE COUNT
11      COUNT2      DB      ?                   ;8 PHASE COUNT
12      ;--------------------------------------------------------------
13      MOTOR_PORT  EQU         PA              ;DEF. MOTOR PORT
14      MOTOR_PORTC EQU         PAC             ;DEF. MOTOR PORT CONTROL REG
15      MY_CODE     .SECTION    AT 0 'CODE'     ;== PROGRAM SECTION ==
16              ORG         00H                 ;HT-46RXX RESET VECTOR
17              CLR         MOTOR_PORTC         ;CONFIG PA AS OUTPUT MODE
18      MAIN:
19              MOV         A,50                ;CW MOTION
20              MOV         COUNT2,A            ;SET STEP COUNT
21      LOOP_1:
22              MOV         A,TAB_CW
23              MOV         TBLP,A              ;SET TABLE POINTER
24              MOV         A,8
25              MOV         COUNT,A             ;SET TABLE COUNT
26      LOOP_2:
27              TABRDL      MOTOR_PORT          ;GET EXCIATED CODE SEND TO PA
28              CALL        DELAY               ;FOR DELAY
29              INC         TBLP                ;NEXT PHASE CODE
30              SDZ         COUNT               ;8 PHASE OVER?
31              JMP         LOOP_2              ;NO, NEXT PHASE
32              SDZ         COUNT2              ;1 TURN?
33              JMP         LOOP_1              ;NO, NEXT STEP PHASE
34
35              MOV         A,50                ;CCW MOTION
36              MOV         COUNT2,A            ;SET STEP COUNT
37      LOOP_3:
38              MOV         A,TAB_CW+7
39              MOV         TBLP,A              ;SET START TABLE POINTER
40              MOV         A,8
41              MOV         COUNT,A             ;SET PAHSE COUNT
42      LOOP_4:
43              TABRDL      MOTOR_PORT          ;GET EXCIATED CODE SEND TO PA
44              CALL        DELAY               ;FOR DELAY
45              DEC         TBLP                ;NEXT PHASE CODE
46              SDZ         COUNT               ;8 PHASE OVER
47              JMP         LOOP_4              ;NO, NEXT PHASE
48              SDZ         COUNT2              ;1 TURN?
49              JMP         LOOP_3              ;NO, NEXT STEP PAHSE
50              JMP         MAIN                ;YES, RESTART.
51      ;****************************************************************
52      ;                   Delay about DEL1(5)*1ms
53      ;****************************************************************
54      DELAY   PROC
55              MOV         A,5
56              MOV         DEL1,A
57      DEL_1:  MOV         A,3
58              MOV         DEL2,A              ;SET DEL2 COUNTER
59      DEL_2:  MOV         A,110
60              MOV         DEL3,A              ;SET DEL3 COUNTER
61      DEL_3:  SDZ         DEL3                ;DEL3 DOWN COUNT
62              JMP         DEL_3
63              SDZ         DEL2                ;DEL2 DOWN COUNT
64              JMP         DEL_2
65              SDZ         DEL1                ;DEL1 DOWN COUNT
66              JMP         DEL_1
67              RET
68      DELAY   ENDP
69              ORG         LASTPAGE
```

```
70      TAB_CW:
71          DC      000000000000001B        ;STEP MOTOR HALF STEP CONTROL
72          DC      000000000000011B        ;EXCIATION TABLE
73          DC      000000000000010B
74          DC      000000000000110B
75          DC      000000000000100B
76          DC      000000000001100B
77          DC      000000000001000B
78          DC      000000000001001B
79          END
```

程序说明

行	说明
7~11	依序定义变量地址。
13~14	定义 MOTOR_PORT 为 PA。
16	声明存储器地址由 000h 开始(HT46xx Reset Vector)。
17	将 MOTOR_PORT 定义成输出模式。
19~20	设置循环计数器 COUNT2 为 50，将步进电机连续激发四相(4×50)，让步进电机朝某一特定方向转动，在此是让步进电机带进位位右移。
22~25	将 TBLP 指向励磁表的起始地址，并将 COUNT 设置为 8，至于 COUNT 的作用是什么，下面将会有详细的说明。
27	进行查表动作，取得欲激发步进电机的数值，并送至 MOTOR_PORT。
28	调用 DELAY 子程序，延时 5 ms。
29~31	将 TBLP 加 1 后，判断 COUNT−1 是否等于 0，即判断是否已将 8 个相位全部激发完毕，若不成立表示需继续激发下个相位；反之，则根据 COUNT2 的内容值重新激发 8 个相位或进行下面的反转动作。
32~33	判断步进电机是否已被连续激发 8 个相位 50 次(即 400 个 Half Steps)，成立则进行反转动作，反之则重新激发 8 个相位使步进电机正转。
35~36	重新设置 COUNT2 为 50，准备进行步进电机反转的动作。
38~39	让步进电机往反方向转动方法，就是将原来激发步进电机的励磁顺序颠倒，因此必须将 TBLP 指向励磁表的最后一笔数据，将激发步进电机的顺序由后到前。
40~41	将 COUNT 设置为 8。
43	进行查表动作，取得欲激发步进电机的数值，并送到 MOTOR_PORT。
44	调用 DELAY 子程序，延时 5 ms。
45~47	将 TBLP 递减后，判断 COUNT−1 是否等于 0，即判断是否已将 8 个相位全部激发完毕，若不成立表示需继续激发下一个相位；反之，则根据 COUNT2 的内容值重新激发 8 个相位或回复正转动作。
48~50	判断步进电机是否已被连续激发 4 个相位 50 次，成立则重新进行正转动作，反之则重新激发 8 个相位，使步进电机正转。如此重复执行，即可看见步进电机一正一反来回旋转。
54~68	DELAY 子程序，延迟时间的计算请参考 4-1 节。
70~78	步进电机单-双相励磁信号建表区。

步进电机的控制程序并不困难，其实也只是查表法的运用而已。要提醒读者的是：因为步进电机是一种机械设备，因此需要一点时间才能将转子移动到励磁的位置。但究竟需要多少时间，则需视电机及其驱动电流大小而定。本例中在每次送出励磁信号之后，都给

予 10 ms 的延时，读者可以试着减短 DELAY 子程序的延时时间，并观察其结果。

4-6-5 动动脑 + 动动手

- 改写程序，让步进电机的速度越转越快。
- 改写程序，让步进电机的速度越转越慢。
- 在 PA 上加上 DIP Switch(见图 4-6-11 的电路)，并改写程序，使其可以由 DIP Switch 的设置来控制电机的转速与转向。

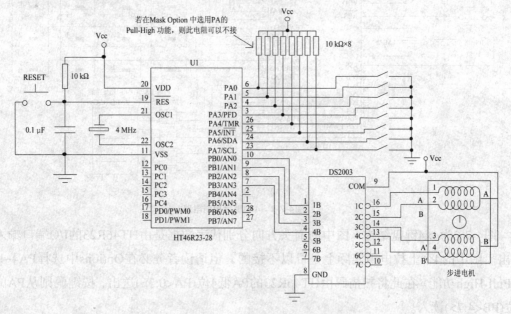

图4-6-11　修改后的步进电机控制电路

4-7 4×4键盘实验控制实验

4-7-1 目　的

本实验控制七段显示器上显示4×4键盘的按键值。

4-7-2 学习重点

通过本实验，读者应了解4×4键盘的扫描方式及工作原理。4×4键盘是经常使用到的输入设备，读者务必了解其原理及控制程序的写法。

4-7-3 电路图

在4-5节中介绍了单一按键的输入控制方式，虽然按键的检测相当方便，但是每增加一个按键，就必须额外占用一个I/O引脚，这在需要多按键输入的场合是极不经济的。在实际的应用上，如果需要4个以上的按键，一般就会采用矩阵式的键盘，以减少I/O引脚的浪费。如图4-7-1所示，以4×4矩阵式键盘为例，说明其工作原理。

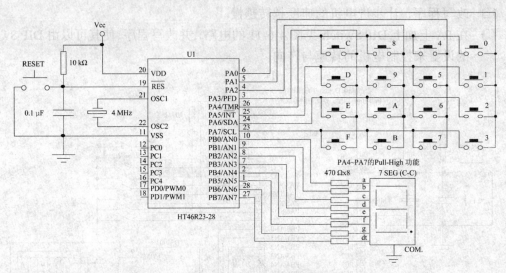

图4-7-1 4×4键盘控制电路

图4-7-2为4×4键盘结构，图中的箭头方向分别代表数据是由HT46R23的I/O端口输入或输出。图中的4个上拉电阻实际上是可以不接的，但请读者务必在Options中选择PA4~PA7的Pull-High功能。在此将扫描码由HT46R23的PA低4位(PA<0:3>)送出，按键码则从PA的高4位(PB<4:7>)读入。

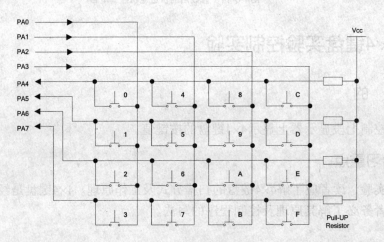

图4-7-2 4×4键盘

所按下的键值与扫描码、按键码间的关系分析如下：

① 若PA0~PA3送出1111b的扫描码，则无论是否有按下按键，由PA4~PA7所读回的

按键码都是 1111b。

② 若 PA0~PA3 送出 0000b 的扫描码，若没有按下按键，则由 PA4~ PA7 所读回的按键码为 1111b；若有按键被按下，则此时读回的按键码与被按按键的所在行数有关，如表 4-7-1 所列。

表4-7-1　被按按键与读回的按键码的关系

PA4~PA7 读回值	被按按键所在行数	可能的键值
0111b	0	0,1,2,3
1011b	1	4,5,6,7
1101b	2	8,9,A,B
1110b	3	C,D,E,F

由此可知如果扫描码为0000b，虽然可以检测是否按下按键以及是哪一行的按键被按下，但是却无法分辨是按下该行的哪一个键。

③ 依上述的推论，必须逐列扫描(送0)，然后根据读回的按键码与所送出的扫描码的关系，才能区分出究竟是哪一个键被按下。其关系如表4-7-2所列。

表4-7-2　扫描码与按键码的关系

扫描码	按键码			
	0111b (PA4~PA7)	1011b (PA4~PA7)	1101b (PA4~PA7)	1110b (PA4~PA7)
0111b(PA.0~3)	"0"	"1"	"2"	"3"
1011b(PA.0~3)	"4"	"5"	"6"	"7"
1101b(PA.0~3)	"8"	"9"	"A"	"B"
1110b(PA.0~3)	"C"	"D"	"E"	"F"

按键的检测由第0~3列周而复始不断地扫描，如果扫描过程中没有按下按键，则所读回的按键码都为1111b；若有键被按下，当扫描到该列时就会传回被按按键的行数对应值。举例来说，若A键被按下：

① 扫描第 0 列(由 PA0~PA3 送出扫描码为 0111b)，由 PA4~PA7 读回的按键码为 1111b，表示本列没有键被按下。

② 扫描第 1 列(由 PA0~PA3 送出扫描码为 1011b)，由 PA4~PA7 读回的按键码为 1111b，表示本列也没有键被按下。

③ 扫描第 2 列(由 PA0~PA3 送出扫描码为 1101b)，由 PA4~PA7 读回的按键码为 1101b，表示本列第 2 行的按键被按下。

接下来就是如何将扫描码与按键码转换为对应的按键值，一般可采取查表的方式，虽然较为简单，但是因为要建表，所以会占用较多的存储器。另一种方法，是直接找出行、列值与按键值的关系，以本实验的按键安排为例，如果A键被按下，可以由上述的扫描码与按键码得知是第2列、第2行被按下。由于行与行之间的按键值相差4，所以只要把行数乘上4再加上列数就可以得到其按键值，如2(第2列)×4＋2(第2行)＝0Ah。在本实验中的 READ_KEY 子程序就是采用此方法来求取按键值；亦即该行若无按键按下，就将按键值加1。所以，若该列各行都没有按键被按下，按键值被加4，这就如同乘4的作用。因为往后的实验经常会使用4×4键盘，所以请读者务必参考程序中的说明，彻底了解键盘的译码动作。

4-7-4 流程图及程序

1. 流程图(见图 4-7-3)

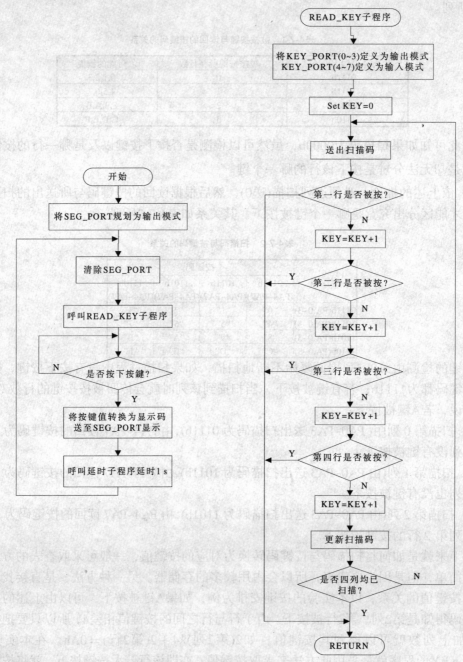

图4-7-3 流程图

2. 程序 4-7 4×4 键盘实验控制实验

```
1    ;PROGRAM : 4-7.ASM  (4-7.PRJ)                      2002.DEC.30.
2    ;FUNCTION: 4X4 MATRIX KEYPAD DEMO PROGRAM           BY STEVEN
3    #INCLUDE    HT46R23.INC
```

```
4              .CHIPHT46R23
5     ;--------------------------------------------------------------------
6     MY_DATA    .SECTION    'DATA'          ;== DATA SECTION ==
7     DEL1       DB     ?                    ;DELAY LOOP COUNT 1
8     DEL2       DB     ?                    ;DELAY LOOP COUNT 2
9     DEL3       DB     ?                    ;DELAY LOOP COUNT 3
10    COUNT      DB     ?
11    KEY        DB     ?                    ;KEY CODE REGISTER
12    ;--------------------------------------------------------------------
13    SEG_PORT   EQU    PB                   ;DEFINE 7-SEG PORT
14    SEG_PORTC  EQU    PBC                  ;DEFINE 7-SEG PORT CONTROL REG
15    KEY_PORT   EQU    PA                   ;DEFINE KEYPAD PORT
16    KEY_PORTC  EQU    PAC                  ;DEFINE KEY PORT VONTROL REG
17
18    MY_CODE    .SECTION   AT 0 'CODE'      ;== PROGRAM SECTION ==
19         ORG         00H                   ;HT-46RXX RESET VECTOR
20         CLR         SEG_PORTC             ;CONFIG PA AS OUTPUT MODE
21    MAIN:
22         CLR         SEG_PORT              ;CLEAR SEG_PORT
23         CALL        READ_KEY              ;READ KEYPAD
24         MOV         A,16
25         XOR         A,KEY
26         SZ          Z                     ;IS KEY NOT PRESSED(KEY=16)?
27         JMP         MAIN                  ;YES,RE-READ
28         MOV         A,KEY
29         CALL        TRANS                 ;GET 7 SEG DISPLAY CODE
30         MOV         SEG_PORT,A            ;DISPLAY KEY VALUE
31         CALL        DELAY                 ;DELAY 1 SEC
32         JMP         MAIN
33    ;*********************************************************************
34    ;    SCAN 4x4 MATRIX ON KEY PORT AND RETURN THE CODE IN KEY REGISTER
35    ;        IF NO KEY BEEN PRESSED, KEY=16.
36    ;*********************************************************************
37    READ_KEY       PROC
38         MOV         A,11110000B
39         MOV         KEY_PORTC,A           ;CONFIG PORT B
40         SET         KEY_PORT              ;INITIAL PORT B
41         CLR         KEY                   ;INITIAL KEY REGISTER
42         MOV         A,04
43         MOV         COUNT,A               ;SET ROW COUNTER
44         CLR         C                     ;CLEAR CARRY FLAG
45    SCAN_KEY:
46         RLC         KEY_PORT              ;ROTATE SCANNING BIT
47         SET         C                     ;MAKE SURE C=1
48         SNZ         KEY_PORT.4            ;COLUMN 0 PRESSED?
49         JMP         END_KEY               ;YES
50         INC         KEY                   ;NO, INCREASE KEY CODE
51         SNZ         KEY_PORT.5            ;COLUMN 1 PRESSED?
52         JMP         END_KEY               ;YES
53         INC         KEY                   ;NO, INCREASE KEY CODE
54         SNZ         KEY_PORT.6            ;COLUMN 2 PRESSED?
55         JMP         END_KEY               ;YES
56         INC         KEY                   ;NO, INCREASE KEY CODE
57         SNZ         KEY_PORT.7            ;COLUMN 3 PRESSED?
58         JMP         END_KEY               ;YES
59         INC         KEY                   ;NO, INCREASE KEY CODE
60         SDZ         COUNT                 ;HAVE ALL ROWs BEEN CHECKED?
61         JMP         SCAN_KEY              ;NO, NEXT ROW
62    END_KEY:
63         RET
64    READ_KEY       ENDP
65    ;*************************************************************
66    ;              Delay about DEL1(100)*10ms
67    ;*************************************************************
68    DELAY          PROC
69         MOV         A,100
70         MOV         DEL1,A
```

```
71  DEL_1:  MOV     A,30
72          MOV     DEL2,A          ;SET DEL2 COUNTER
73  DEL_2:  MOV     A,110
74          MOV     DEL3,A          ;SET DEL3 COUNTER
75  DEL_3:  SDZ     DEL3            ;DEL3 DOWN COUNT
76          JMP     DEL_3
77          SDZ     DEL2            ;DEL2 DOWN COUNT
78          JMP     DEL_2
79          SDZ     DEL1            ;DEL1 DOWN COUNT
80          JMP     DEL_1
81          RET
82  DELAY   ENDP
;****************************************************************
;           RETURN THE TABLE VALUE INDEX BY A
83  ;****************************************************************
84  TRANS   PROC
85          ADDM    A,PCL
86          RET     A,00111111B     ;0
87          RET     A,00000110B     ;1
88          RET     A,01011011B     ;2
89          RET     A,01001111B     ;3
90          RET     A,01100110B     ;4
91          RET     A,01101101B     ;5
92          RET     A,01111101B     ;6
93          RET     A,00000111B     ;7
94          RET     A,01111111B     ;8
95          RET     A,01100111B     ;9
96          RET     A,01110111B     ;a
97          RET     A,01111100B     ;b
98          RET     A,01011000B     ;c
99          RET     A,01011110B     ;d
100         RET     A,01111001B     ;E
101         RET     A,01110001B     ;F
102 TRANS   ENDP
103         END
```

程序说明

- 7~11 依序定义变量地址。
- 13~16 定义 KEY_PORT 为 PA，定义 SEG_PORT 为 PB。
- 19 声明存储器地址由 000h 开始(HT46xx Reset Vector)。
- 20 定义 SEG_PORT 为输出模式。
- 22 清除七段显示器的显示值。
- 23~27 调用扫描键盘子程序，完成后判断索引值(KEY)内容是否为 16，若成立表示键盘无输入，重新扫描键盘(MAIN)；反之，则键盘已有值，程序继续执行。
- 28~30 根据 KEY 调用查表子程序，取得相对应的七段显示器显示码并送到 SEG_PORT 显示。
- 31~32 延时 1 s 后，程序重新执行。
- 38~39 将 KEY_PORT 定义为半输入(Bit7~4)半输出(Bit3~0)模式。
- 40 将 KEY_PORT 内容设置为 0FFh。
- 41~43 将索引值(KEY)清除为 0，并将键盘扫描次数寄存器 COUNT 设置为 4(因为有 4 列要逐一扫描)。
- 44 将进位标志位(C)设为 0，目的在第 46 行处进行带进位位左移时将 0 由进位标志位移到 KEY_PORT.0，而 0 在哪个位位置即代表哪个列要检测是否有按键被按下。
- 46 将 0 带进位位左移到要检的位位置(只限 Bit0~Bit3，因为 Bit4~Bit7 已定义为输入模式)。
- 47 将进位标志位(C)设为 1，防止有多余的 0 干扰键盘扫描。
- 48~49 扫描 KEY_PORT.4 是否为 1，若成立表示该列无按键被按下；反之则表示按键有值输入。以第一次程序执行到此为例，首先让 KEY_PORT.0 设置为 0，再一一判断在 KEY_PORT.0 所对应的这个列上是否有键被按，由于需让芯片接受按键输入，KEY_PORT 需定义成一半输入(High Nibble：Bit7~4)、一半输出(Low Nibble：Bit3~0)的模式，因此只须检测由 KEY_PORT.7~4 对应到 KEY_PORT.0 的 4 个按键是否受到 KEY_PORT.0 的影响。若该列没有按键被按下，则 KEY_PORT.7~4 内容都为 1；若有按键按下，则 KEY_PORT.7~4 会受到 KEY_PORT.0 影响而变成 0(注：KEY_PORT.4~ KEY_PORT.7 只有一个位会受影响)。所以，这 2 行程序相当于是检测该列的第 0 行是否被按下(即 KEY_PORT.4 是否为 0)，若是则表示已检测到按键被按下，跳至到 END_KEY。

50~52　将按键值(KEY)加 1，其余如同 48~49 行动作，检测 KEY_PORT.5(第 1 行)。
53~55　将按键值(KEY)加 1，其余如同 48~49 行动作，检测 KEY_PORT.6(第 2 行)。
56~58　将按键值(KEY)加 1，其余如同 48~49 行动作，检测 KEY_PORT.7(第 3 行)。
59~63　将按键值(KEY)加 1，判断是否已经由第 0 列扫描至第 3 列(共 4 次，所以 COUNT 的初始值设置为 4，且每扫描一列 COUNT 就减 1)，若成立，则表示键盘已扫描完成，无任何值从键盘输入，回到主程序继续执行；反之，则表示还未扫描完毕，将回到 SCAN_KEY 继续扫描下一列。
68~82　DELAY 子程序，延时时间的计算请参考 4-1 节。
86~104　七段显示码数据建表区。

本程序执行时会将检测到的按键值直接显示在七段显示器上，而 READ_KEY 子程序则负责键盘检测工作，当其传回值(即 KEY 寄存器)为 16 时，表示键盘上并无键被按下；反之若 KEY ≠ 16，则其值即代表被按按键的值。当检测到有按键被按下时，即根据 KEY 寄存器调用查表子程序，取得相对应的七段显示器显示码并送到 SEG_PORT 显示，1 s 后将显示值清除并重新执行按键检测的工作。查表子程序的原理，其实是利用改变 PC 值的方式配合"RET A,x"指令来完成，这在 4-2 节中已有完整的说明，请读者参阅该实验的内容。

4-7-5　动动脑 + 动动手

  □ 在 PB 上加上步进电机(如图 4-7-4 所示的电路)，并改写程序，使其可以由 4×4 键盘上的按键值来控制电机的转速、转向(0-正转；1-反转；2-加速；3-减速；4-停止)。

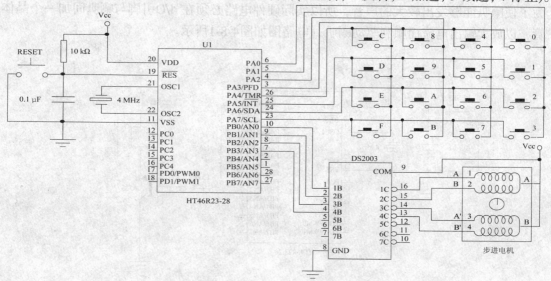

图4-7-4　4×4键盘与步进电机控制电路

4-8 喇叭发音控制实验

4-8-1 目的

利用HT46x23控制喇叭发出不同音调的声音。

4-8-2 学习重点

通过本实验,读者应了解如何控制喇叭发出不同音调的声音。本实验是以延时程序来控制喇叭发生的音调,因此对延时时间的计算还未彻底了解的读者,可通过4-1节所介绍的延时时间计算方式再复习一遍,以求融会贯通。其实,HT46xx系列都拥有PFD(Programmable Frequency Divider)的输出控制引脚,对于Buzzer或喇叭的控制极为方便,由于其牵涉定时器的搭配,因此留待定时器的相关实验再做应用。

4-8-3 电路图

在进行发音控制之前,必须先了解喇叭发音原理,如此才能针对喇叭的特性,再通过HT46xx的程序来加以控制,以发出不同音调的声音。喇叭发音的原理是利用喇叭上的线圈,将它通上断续的电流,就会造成薄膜的振动,以推动空气而产生声音。因此,若通上不同频率的信号就可以发出各种不同音调的声音。仔细分析使喇叭发出声音的信号,其实只不过是一连串不同频率的脉冲而已,所以若要让HT46xx来让喇叭发出声音,只要利用程序来控制一个I/O引脚,并使它能够不断输出0→1→0→1→0……循环变化的方波信号,再加上适当的延时时间,即可使喇叭发出不同频率的声音信号。但是由于HT46xx的I/O引脚所能提供的电流有限(V_{cc}=5 V时,拉电流为20 mA,灌电流为-10 mA),并无法完全驱动喇叭,为了使喇叭能够发出较大的声音,所以实际硬件电路必须在I/O引脚与喇叭间加一个晶体管,以放大输出电流而顺利驱动喇叭,电路图如图4-8-1所示。

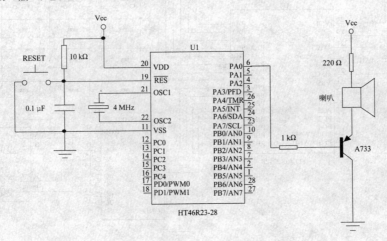

图4-8-1 喇叭发音控制电路

要控制I/O引脚输出音阶为Do、Re、Mi、……方波,首先必须要知道这些音阶的频率,如此驱动喇叭之后的音阶才会准确。要在 I/O 引脚上输出方波,则只要先计算出这个方波的周期,然后再以这个时间作为延时时间;每当到达延时时间之后,单片机就将I/O引脚的输出状态反相,然后再重复计时延时,等时间到了之后,再对I/O引脚做反相的动作,那么在I/O引脚上就会输出该频率的方波了。

计算各音阶频率的方法其实很简单,就是首先记住低音La的频率为440 Hz,然后每隔半度音程的频率就是前一个音的1.059倍($2^{1/12}$)。为方便读者记忆,可参考如图4-8-2所示的钢琴琴键位置与音阶间的关系。

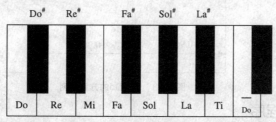

图4-8-2 钢琴的琴键位置与音阶

例如:Ti比La高一度音(两个半音就是一个全音),因此Ti和Do (Ti与Do差一个半音)的频率为:

$$f_{Ti}=440 \text{ Hz} \times 1.059 \times 1.059 = 493.9 \text{ Hz}$$

$$f_{Do}=440 \text{ Hz} \times 1.059 \times 1.059 \times 1.059 = 493.9 \text{ Hz} \times 1.059 = 523 \text{ Hz}$$

因此,依照以上的原则,可以算出从低音Do到高音Do音阶的频率,如表4-8-1所列。

表4-8-1 音阶—频率对照表

音 阶	频率/Hz	周期/ms	半周期/ms
Do	523	1.91	0.96
Do#	554	1.8	0.9
Re	587	1.7	0.85
Re#	622	1.6	0.8
Mi	659	1.52	0.76
Fa	698	1.43	0.72
Fa#	740	1.35	0.68
Sol	785	1.27	0.64
Sol#	831	1.2	0.6
La	880	1.14	0.57
La#	932	1.07	0.54
Ti	988	1.00	0.50
Do 高音	1047	0.96	0.48

因为方波的每一个周期有一半时间为High，另一半时间为Low，因此真正要定时的时间是音阶周期的一半(因为是当延时时间到了之后就将输出反相一次)。以音阶 Do为例，经查表4-8-1可知其频率为523 Hz，周期为1.91 ms，而半周期为0.96 ms；即送出High、Low的时间均为0.96 ms，如图4-8-3所示。

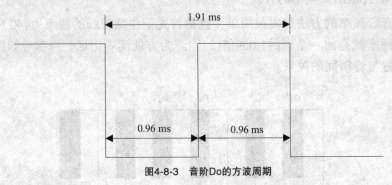

图4-8-3　音阶Do的方波周期

4-8-4　流程图及程序

1. 流程图(见图 4-8-4)

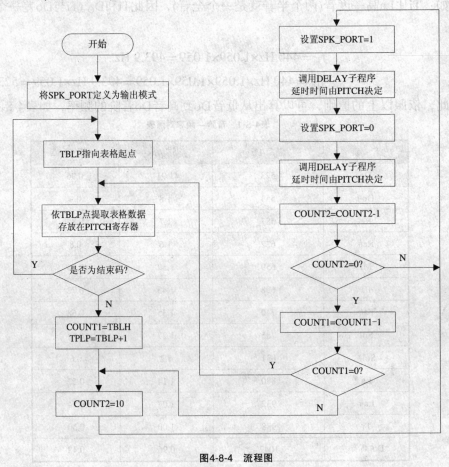

图4-8-4　流程图

2. 程序 4-8-1 喇叭发音实验

```
1   ;PROGRAM : 4-8.ASM  (4-8.PRJ)                      2002.DEC.30.
2   ;FUNCTION: GENERATE TONE DO~DO. BY USING DELAY SKILL      BY STEVEN
3   #INCLUDE    HT46R23.INC
4               .CHIPHT46R23
5   ;----------------------------------------------------------------
6   MY_DATA     .SECTION    'DATA'          ;== DATA SECTION ==
7   DEL1        DB      ?                   ;DELAY LOOP COUNT 1
8   DEL2        DB      ?                   ;DELAY LOOP COUNT 2
9   PITCH       DB      ?                   ;PITCH REGISTER
10  COUNT1      DB      ?                   ;DURATION COUNT 1
11  COUNT2      DB      ?                   ;DURATION COUNT 2
12  ;----------------------------------------------------------------
13  SPK_PORT    EQU     PA.0                ;DEFINE SPEAKER PORT
14  SPK_PORTC   EQU     PAC.0               ;DEFINE SPEAKER PORT CON. REG.
15
16  MY_CODE     .SECTION  AT 0 'CODE'       ;== PROGRAM SECTION ==
17          ORG     00H                     ;HT-46RXX RESET VECTOR
18          CLR     SPK_PORTC               ;CONFIG PA.0 AS OUTPUT MODE
19  MAIN:
20          MOV     A,TAB_PITCH_DURATION
21          MOV     TBLP,A                  ;INITIAL TABLE POINTER
22  NEXT_PITCH:
23          TABRDL  PITCH                   ;LOAD DATA TO PITCH AND DURATION TO TBLH
24          MOV     A,0
25          XOR     A,TBLH                  ;NOTE: TBLH IS READ ONLY!!!
26          SZ      Z                       ;IS THE LAST TONE?
27          JMP     MAIN                    ;YES. RESTART!!
28          MOV     A,TBLH                  ;NO, LOAD DURATION TO COUNT1
29          MOV     COUNT1,A
30          INC     TBLP                    ;UPDATE TABLE POINTER
31  LOOP:
32          MOV     A,10
33          MOV     COUNT2,A                ;INITIAL INNER LOOP COUNT
34  HI:
35          SET     SPK_PORT                ;SET SPK_PORT HIGH
36          MOV     A,PITCH
37          CALL    DELAY                   ;PITCH DELAY
38          CLR     SPK_PORT                ;RESET SPK_PORT
39          MOV     A,PITCH
40          CALL    DELAY                   ;PITCH DELAY
41          SDZ     COUNT2                  ;INNER LOOP OK?
42          JMP     HI                      ;NO.
43          SDZ     COUNT1                  ;OUTER LOOP OK?
44          JMP     LOOP                    ;NO.
45          JMP     NEXT_PITCH              ;YES, NEXT PITCH
46  ;****************************************************************
47  ;               Delay about DEL1*20us
48  ;****************************************************************
49  DELAY   PROC
50          MOV     DEL1,A
51  DEL_1:  MOV     A,5
52          MOV     DEL2,A                  ;SET DEL2 COUNTER
53  DEL_2:  SDZ     DEL2                    ;DEL2 DOWN COUNT
54          JMP     DEL_2
55          SDZ     DEL1                    ;DEL1 DOWN COUNT
56          JMP     DEL_1
57          RET
58  DELAY   ENDP
59          ORG     LASTPAGE
60  TAB_PITCH_DURATION:
61          DC      1000000/(523*2*20)+(523/(2*10)) SHL 8   ;DO TONE & DURATION
62          DC      1000000/(587*2*20)+(587/(2*10)) SHL 8   ;RE TONE & DURATION
63          DC      1000000/(659*2*20)+(659/(2*10)) SHL 8   ;MI TONE & DURATION
64          DC      1000000/(698*2*20)+(698/(2*10)) SHL 8   ;FA TONE & DURATION
65          DC      1000000/(785*2*20)+(785/(2*10)) SHL 8   ;SOL TONE & DURATION
```

```
66          DC      1000000/(880*2*20)+(880/(2*10)) SHL 8    ;LA TONE & DURATION
67          DC      1000000/(988*2*20)+(988/(2*10)) SHL 8    ;TI TONE & DURATION
68          DC      1000000/(1047*2*20)+(1047/(2*10)) SHL 8  ;DO. TONE & DURATION
69          DC      0                                        ;TABLE END CODE
70          END
```

程序说明

7~11   依序定义变量地址。

13~14   定义 SPK_PORT 为 PA0。

17   声明存储器地址由 000h 开始(HT46xx Reset Vector)。

18   定义 SPK_PORT 为输出。

20~21   将 TBLP 指向音符数据表的起点。

23   经查表取得欲发音的音阶频率参数存入 PITCH 寄存器，而控制音长的次数常量则存在 TBLH 寄存器中。

24~27   判断音长次数常量(TBLH 寄存器)内容是否为 0，成立则表示已提到最后一笔数值 0H(即音符数据表已经提完)，重新执行程序；反之，则继续执行发音的程序。

28~29   将控制音长的次数参数并存入 COUNT1 中，此值即为用来控制声音响的时间(即产生的脉冲数目)。

30   将 TBLP 加 1，指向下一笔音符数据。

32~33   设置计数器 COUNT2 为 10 与 COUNT 一起用来调整声音长度(注：调整脉冲产生的个数)。

35~37   喇叭开启(ON)，并调用延时子程序，其延时时间的长短由 PITCH 的内容来决定。

38~40   喇叭关闭(OFF)，并调用延时子程序，其延时时间的长短由 PITCH 的内容来决定。(注：此处喇叭开启与关闭的时间相同，只是为了产生 Duty Cycle＝50%的方波。)

41~42   判断 COUNT2－1 是否为 0，在此用来控制产生的脉冲个数。

43~45   判断 COUNT1－1 是否为 0，成立则表示此音阶已发音完毕，回到主程序进行下一个音阶的发音；反之，则继续产生脉冲(JMP LOOP)。

49~58   DELAY 子程序，延时时间的计算请参考 4-1 节。

60~69   音长数据建表区。

为了得到更精确的频率，DELAY子程序的延时时间为20~5 120 μs，由DEL1寄存器控制。至于每个音调(Pitch or Tone)的音长(Duration)则取决于输出的脉冲数目。如果不管输出的音调是什么，都输出相同的脉冲数，将会造成越高音发音越短的结果。由于本程序是希望各个音调都能有相同长度(0.5 s)的声音输出，所以就必须掌握不同音调所需输出的脉冲个数。

在程序中，再度使用查表法来提取这两项参数(音调及音长)。因此，每个音符在表中应占有两个值：音调参数及音长参数。假设所要产生的音调频率为F(Hz)，则所需的半周期时间为$(2\times F)^{-1}$ s。因为DELAY子程序的延时时间为DEL1×20 μs，因此产生F(Hz)所需的DEL1数值为：

$$(2\times F)^{-1} \div (20\times 10^{-6}) = 10^6 \div (2\times F\times 20)$$

此即为控制音调高低的参数值。而产生0.5 s、频率为F的音调所需的脉冲数目为$0.5/(F^{-1})=F/2$。因为HT46xx的寄存器只有8位，当$F\geqslant 512$之后就无法以一个寄存器来存放控制脉冲数目的

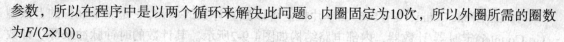

参数,所以在程序中是以两个循环来解决此问题。内圈固定为10次,所以外圈所需的圈数为$F/(2\times10)$。

通过此段说明,读者对于音符数据表中的数值应该十分清楚了吧?由于HT46x23的程序存储器大小为4 096×15位,在确定控制音调与音长的参数分别不会超过8位与7位的情况下,刻意将这两项参数集合在一个存储器位置中,以达到节省存储器空间的目的。此技巧在4-3节中已有详细的说明,还不了解的读者可以参阅该实验的内容。

4-8-5 动动脑+动动手

- 更改建表数值,试看看自己耳朵可以听见的频率范围为多少?
- 更改建表数值,使其唱一首自己所喜欢的短歌。
- 将4×4键盘程序与本实验结合(参考图4-8-5的电路),以按键控制喇叭发出的音调。

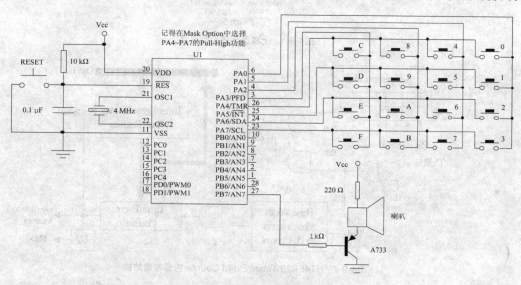

图4-8-5 4×4键盘与喇叭发音控制电路

4-9 Timer/Event Counter控制实验

4-9-1 目 的

利用HT46x23的Timer/Event Counter控制七段显示器的显示速度。

4-9-2 学习重点

通过本实验,读者应熟悉HT46x23的Timer/Event Counter控制方式。

4-9-3 电路图

七段显示器控制电路如图4-9-1所示。

Timer/Event Counter是HT46x23单片机上可以独立计数的单元，它是一个16位的上数型(Up-Counter)定时器/计数器，内部电路结构如图4-9-2所示。其计数的时钟脉冲来源、计数的时钟脉冲正负边缘都可以通过TMRC寄存器设置，请参考第2章中的相关说明。

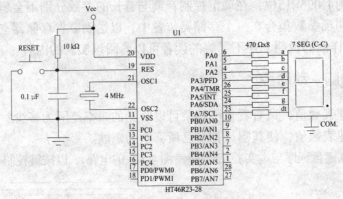

图4-9-1 七段显示器控制电路

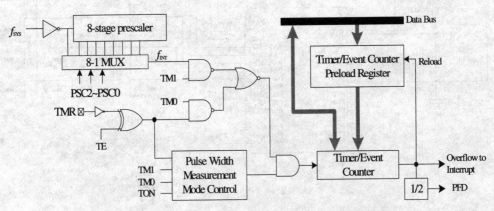

图4-9-2 HT46R23 Timer/Event Counter内部电路结构

定时器/计数器有3种工作模式，由TM1及TM0位选择。本实验利用定时模式(Timer Mode)来控制DELAY子程序的延时时间；所谓的定时模式是指芯片系统时钟脉冲(f_{SYS})通过8位的预除器后，作为计数的时钟脉冲来源(f_{INT})，并以TON位控制计数器是否开始计数。当计数器由FFFFh→0000h(产生溢位)时，除了会将TF(INT0.5)位设置为1外，并会同时由Preload Register载入起始数值继续计数；即其具有自动载入(Auto-reload)功能，除非是要更改计数的时间或次数，否则用户无须每次重新载入起始数值，以避免无谓的Over-head导致定时产生误差。如果定时器/计数器中断(Timer/Event Counter Interrupt)有效，在计数器溢位时也将产生中断请求的信号请求CPU服务，不过本实验暂不探讨中断的处理方式，待4-10节再做讨论。

HT46x23的TMRC控制器寄存器如图4-9-3所示。

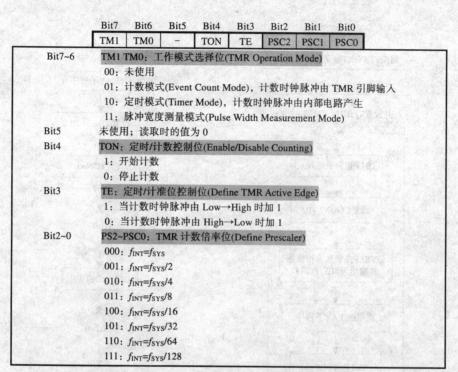

图4-9-3　HT46x23的TMRC控制寄存器

4-9-4　流程图及程序

1. 流程图(见图4-9-4)

2. 程序 4-9　定时模式控制实验

```
1    ;PROGRAM : 4-9.ASM  (4-9.PRJ)                         2002.DEC.30.
2    ;FUNCTION: DISPLAY 0~9 FOR 7 SEG USING TMR CONTROL    BY STEVEN
3    #INCLUDE    HT46R23.INC
4                .CHIPHT46R23
5    ;-----------------------------------------------------------------
6    MY_DATA     .SECTION    'DATA'              ;== DATA SECTION ==
7    DEL1        DB      ?
8    COUNT       DB      ?
9    ;-----------------------------------------------------------------
10   SEG_PORT    EQU     PA                      ;DEFINE 7-SEG PORT
11   SEG_PORTC   EQU     PAC                     ;DEFINE 7-SEG PORT CONTROL REG.
12
13   MY_CODE     .SECTION    AT 0 'CODE'         ;== PROGRAM SECTION ==
14           ORG     00H                         ;HT-46RXX RESET VECTOR
15           CLR     SEG_PORTC                   ;CONFIG SEG_PORT AS OUTPUT MODE
16           MOV     A,10000010B                 ;CONFIG TMR 0 IN MODE 2(TIMER MODE)

17           MOV     TMRC,A                      ;fINT=fSYS/4  (4MHz/4)
18           MOV     A,LOW (65536-50000)
19           MOV     TMRL,A                      ;SET TMR INITIAL VALUE LO-BYTE
20           MOV     A,HIGH (65536-50000)
21           MOV     TMRH,A                      ;SET TMR INITIAL VALUE HI-BYTE
22   MAIN:
23           MOV     A,TAB_7_SEG
24           MOV     TBLP,A                      ;INITIAL POINTER START ADDRESS
```

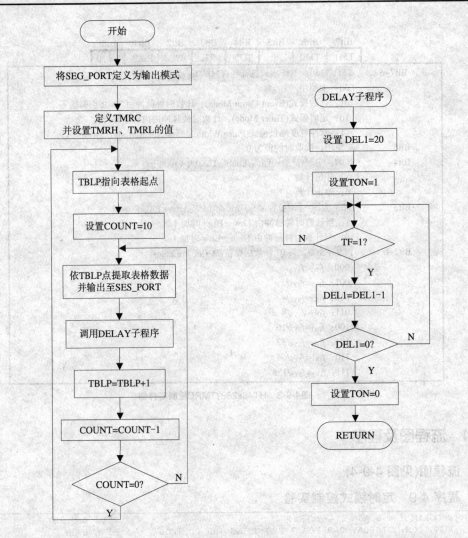

图4-9-4 流程图

```
25          MOV     A,10
26          MOV     COUNT,A              ;SET DATA COUNT OF TAB_7_SEG
27   LOOP:
28          TABRDL  SEG_PORT             ;READ TABLE AND SEND TO SEG_PORT
29          CALL    DELAY                ;DELAY TBLH*10mS
30          INC     TBLP                 ;UPDATE TABLE POINTER
31          SDZ     COUNT                ;COUNT DOWN BY 1
32          JMP     LOOP                 ;JUMP IF NOT TABLE END
33          JMP     MAIN                 ;RESTART
34   ;********************************************************************
35   ;           Delay about DEL1*(TMR)*1us   IF PSC=010
36   ;********************************************************************
37   DELAY  PROC
38          MOV     A,20
39          MOV     DEL1,A               ;SET OUTER COUNTER
40          SET     TON                  ;START 50mS TIMER COUNTING
41   DEL_1: SNZ     TF                   ;TIMER OVERFLOW?
42          JMP     DEL_1                ;NO, WAIT!
43          CLR     TF                   ;YES, 50mS IS OVER.
44          SDZ     DEL1                 ;IS 1 SEC OVER?
45          JMP     DEL_1                ;NO.
46          CLR     TON                  ;YES, STOP TMR.
```

```
47              RET
48      DELAY   ENDP
49              ORG     LASTPAGE
50      TAB_7_SEG:
51              DC      00111111B       ;7 SEG CODE FOR COMMOM CATHOD LED
52              DC      00000110B       ;CONNECT BIT 0~7 FOR SEGMENT A~H
53              DC      01011011B       ;LOW-BYTE FOR 7 SEG DISPLAY CODE
54              DC      01001111B
55              DC      01100110B
56              DC      01101101B
57              DC      01111101B
58              DC      00000111B
59              DC      01111111B
60              DC      01100111B
                END
61
```

程序说明

- **7~8**　依序定义变量地址。
- **10~11**　定义 SEG_PORT 为 PA。
- **14**　声明存储器地址由 000h 开始(HT46xx Reset Vector)。
- **15**　定义 SEG_PORT 为输出。
- **16~17**　设置 TMRC：
  TM1 TM0＝10：指定 Timer/Event Counter 为定时模式，即其计数时钟脉冲来源为 f_{SYS}；
  TON＝0：Timer 暂不计数；
  TE＝0：当计数时钟脉冲由 High→Low 时加 1；
  PSC2~PSC0＝010：计数倍率 1:4，即 $f_{INT}=f_{SYS}/4(1\mu s)$。
- **18~21**　载入计数常量数值到 TMRL、TMRH 寄存器,由于为上数型的 16 位计数器,因此必须以 65 536(2^{16})减去所要计数的数值再放入计数寄存器中。由于所要的计数时间为 50 ms,所以利用 HIGH、LOW 两个伪指令取出(65 536－50 000)的高、低位组，再分别存放在 TMRL、TMRH 寄存器。
- **23~24**　将 TBLP 指向七段显示码的数据起始地址。
- **25~26**　设置计数器 COUNT 为 10。
- **28**　根据 TBLP 的值查表，并将结果输出到 SEG_PORT 显示。
- **29**　调用 DELAY 子程序，延时 1 s。
- **30**　TBLP＝TBLP+1，指向下一笔显示数据。
- **31~33**　判断 COUNT 是否为 0，成立则表示七段显示器已显示到 9，需重新开始；反之，表示还未显示完毕，继续显示下一个数值。
- **38~39**　设置 DEL1＝20。
- **40**　启动计数器开始计数。
- **41~42**　在此 DELAY 子程序做了一些小改变,将原本的 DEL2 与 DEL3 所控制的循环改成利用定时器来计数,通过检查 TF 是否为 1 来判断是否已计数(50 ms)完毕,成立则继续执行；反之，则回到 DEL_1 重新检查 TF。
- **43**　清除 TF 标志位。
- **44~47**　判断 DEL1－1 是否为 0，若是则代表已延时 1 s(50 ms×20)，停止计数器的计数动作(CLR TON)并返回主程序；否则继续检查 TF 的动作(JMP DEL_1)。
- **50~60**　七段显示码数据建表区。

程序中的DELAY子程序配合Timer/Event Counter使用，其时间的计算公式如下：

$$延时时间 = f_{INT}^{-1} \times DEL1(20) \times 计数到溢位的次数$$

因为在程序中将Timer/Event Counter的定时时钟脉冲定义为内部f_{SYS}/4的信号，在$f_{SYS}=4$ MHz、TMRH_TMRL=65 536－50 000时，上式的延时时间＝DEL1×50 ms＝1 s。其次，由于HT46xx的Timer/Event Counter具备了自动载入(Auto-Reload)功能，所以每当计数溢位时，并不需要重新载入起始数值。

4-9-5 动动脑 + 动动手

  📖 试着通过改变预除比例的方式(PSC2~PSC0)，让显示的时间随着数字而改变。

4-10 Timer/Event Counter中断控制实验

4-10-1 目的

利用HT46x23 Timer/Event Counter的中断以及PFD(Programmable Frequency Devider)功能，控制喇叭发出不同音调的声音。

4-10-2 学习重点

通过本实验，读者应了解如何应用HT46x23 Timer/Event Counter的中断功能；同时对于PFD(Programmable Frequency Divider)的输出控制也应得心应手。

4-10-3 电路图(见图4-10-1)

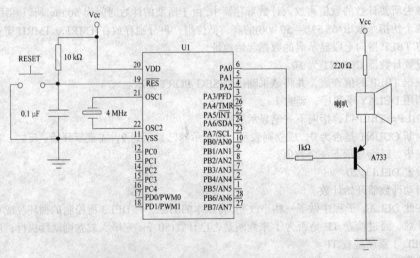

图4-10-1　Timer/Event Counter中断控制电路

在4-9节中，已经介绍了Timer/Event Counter的使用方式，相信读者对其工作方式也有了初步的了解。但是，4-9节是使用轮询(Polling)的方式来检查TF标志位，用以判断Timer/Event Counter是否已经计数溢位(Overflow)，CPU必须一直去检查"TF"标志位的状态，因此无法去处理其他的事务。虽然计数的工作是留给Timer/Event Counter单元负责，但轮询的方式对于整体的效率来讲，并无太大的帮助。因此，在本实验中将采用中断控制的方式，也就是说"当计数溢位时，由Timer/Event Counter以中断的方式主动告知CPU"。因此，在中断发生之前，CPU可以去执行其他的程序，待有中断发生时再去执行相关的中断程序(Interrupt Service Routine；ISR)。如此才能使CPU更有效率，而不是让CPU浪费在一些无谓的等待上。

既然要使用Timer/Event Counter的中断机制，在实验之前就必须将HT46x23中断相关特性及控制寄存器彻底弄清楚，在此仅将与Timer/Event Counter中断相关的寄存器列于图4-10-2，至于详细的说明请读者务必参考2-4节的说明。

HT46x23提供了4种不同的中断：外部中断、定时器/计数器中断、A/D转换中断以及I^2C BUS中断，分别以INTC0与INTC1寄存器来控制，其中INT0与Timer/Event Counter的中断有关，特将其整理如下。

	Bit7	Bit6	Bit5	Bit4	Bit3	Bit2	Bit1	Bit0
	—	ADF	TF	EIF	EADI	ETI	EEI	EMI

Bit7	未使用；读取时的值为0
Bit6	ADF：AD 转换器中断要求标志位(AD Converter Request Flag) 1：AD 转换已经完成，设置此位告知 CPU 0：无 AD 完成转换的中断要求
Bit5	TF：TMR 计数溢位中断要求标志位(Timer/Counter Request Flag) 1：定时器/计数器已经溢位，设置此位告知 CPU 0：无定时器/计数器溢位的中断要求
Bit4	EIF：外部中断要求标志位(External Interrupt Request Flag) 1：有外部中断产生，设置此位告知 CPU 0：无外部中断产生
Bit3	EADI：AD 转换中断控制位(AD Converter Interrupt control) 1：AD 转换器转换完成的中断有效 0：AD 转换器的中断禁止
Bit2	ETI：TMR 中断控制位(Timer/Counter Interrupt control) 1：定时器/计数器溢位中断有效 0：定时器/计数器溢位中断禁止
Bit1	EEI：外部中断控制位(External Interrupt control) 1：外部中断有效 0：外部中断禁止
Bit0	EMI：整体中断控制位(Master Interrupt control) 1：有效，此时 CPU 是否接受中断，需视个别的中断控制位而定 0：所有中断禁止

图4-10-2　HT46x23的INTC0控制寄存器

EMI位可视为HT46x23是否接受中断的控制总枢纽，当EMI位为0时，表示HT46x23将不接受任何中断(不论EHI、EADI、ETI、EEI的设置是什么)；反之当EMI位为1时，则需视EHI、EADI、ETI、EEI位的设置状况再决定是否接受中断。如果在定时器/计数器中断已有效且栈寄存器还有空间的情况下，若产生定时器/计数器溢位的情形，则CPU将跳到008h地址执行程序。

在4-8节中，已经介绍过以程序控制HT46x23，让喇叭产生不同音调输出的方式，其中是以产生脉冲的次数来控制音调的长短。这种方式在拥有Timer/Event Counter中断资源可供运用的HT46xx单片机上来说，感觉有点笨拙，所以本实验采用Timer/Event Counter中断方式予以改良。基本上，是以每次Timer/Event Counter计数溢位时，将I/O输出引脚状态反相的方式来产生所要的频率。所以，若要产生F(Hz)的方波，那么所需的Timer/Event Counter计数时间为$T=(2\times F)^{-1}$，换算成所需计数脉冲数为：

$$所需计数脉冲数 = \frac{T}{\frac{1}{f_{SYS}} \times 计数倍率} = \frac{f_{SYS}}{F \times 2 \times 计数倍率}$$

但是，Timer/Event Counter是一个上数型计数器，所以TMRH、TMRL的起始数值需设置为：

$$65536 - \frac{f_{SYS}}{F \times 2 \times 计数倍率}$$

可是这是一个16位的数值，而HT46x23程序存储器的每一个空间只有15位(HT46x22、HT46x47为14位)，怎么办呢？解决的方式，其一是将数据分为两个空间存放，不过这除了增加存储器空间的浪费之外，也会增加程序的复杂度。而笔者想到的解法是：如果能够找到一个预除比例，使得不管F的数值为何，都能满足下式：

$$65280 \leq \left(65536 - \frac{f_{SYS}}{F \times 2 \times 计数倍率}\right) \leq 65535$$

由于在符合上式的条件时，参数数据的高8字节都是FFh，因此只须建立参数的低8字节数据表并且在程序中再利用指令将TMH寄存器设为FFh就可以了，也就是说将16位的Timer/Event Counter当成8位来使用。在程序中所选用的分频比例为1:16，所以上式可以改写成：

$$256 - \frac{f_{SYS}}{F \times 2 \times 16}$$

这是程序中产生各音阶所需的时间常量表格数据的由来。

4-10-4 流程图及程序

1. 流程图(见图4-10-3)

第4章 基础实验篇

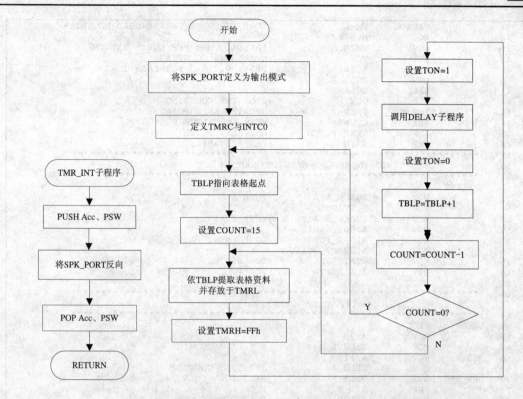

图4-10-3 流程图

2. 程序 4-10 Timer/EventCounter 中断控制实验

```
1       ;PROGRAM : 4-10.ASM  (4-10.PRJ)                     2002.DEC.30.
2       ;FUNCTION: GENERATE TONE DO~DO. BY TMR INTERRUPT SKILL   BY STEVEN
3       ;NOTE    : USING TMR AS A 8-BIT TIMER
4       #INCLUDE HT46R23.INC
5                      .CHIPHT46R23
6       ;---------------------------------------------------------------
7       MY_DATA  .SECTION   'DATA'         ;== DATA SECTION ==
8       DEL1          DB       ?           ;DELAY LOOP COUNT 1
9       DEL2          DB       ?           ;DELAY LOOP COUNT 2
10      DEL3          DB       ?           ;DELAY LOOP COUNT 3
11      COUNT         DB       ?           ;PITCH COUNT
12      STACK_A       DB       ?           ;STACK BUFFER FOR ACC
13      STACK_STATUS  DB       ?           ;STACK BUFFER FPR PSW
14      ;---------------------------------------------------------------
15      OSC           EQU      4000000     ;fSYS=4 MHz
16      ;---------------------------------------------------------------
17      SPK_PORT      EQU      PA          ;DEFINE SPEAKER PORT
18      SPK_PORTC     EQU      PAC.0       ;DEFINE SPEAKER PORT CON. REG.
19
20      MY_CODE  .SECTION  AT 0 'CODE'     ;== PROGRAM SECTION ==
21                ORG       00H            ;HT-46RXX RESET VECTOR
22                JMP       START
23                ORG       08H            ;HT-46RXX TMR INTERRUPT VECTOR
24                JMP       TMR_INT
25      START:
26                CLR       SPK_PORTC      ;CONFIG SPK_PORT AS OUTPUT MODE
27                MOV       A,00000101B    ;ENABLE GLOBAL AND TMR INTERRUPT
28                MOV       INTC0,A
29                MOV       A,10000100B    ;CONFIG TMR 0 IN MODE 2(TIMER MODE)
30                MOV       TMRC,A         ;fINT=fSYS/16(4MHz/16).
31      MAIN:
32                MOV       A,TAB_PITCH
33                MOV       TBLP,A         ;INITIAL TABLE POINTER
34                MOV       A,15
35                MOV       COUNT,A        ;SET COUNTER FOR PITCH TABLE
36      NEXT_PITCH:
```

145

```
37              TABRDL      TMRL                ;LOAD PITCH TO TMR PE-LOAD REGISTER
38              MOV         A,0FFH
39              MOV         TMRH,A              ;PRELOAD TMRH=FFH FOR 8-BIT TMR
40              SET         TON                 ;START TMR COUNTING
41              MOV         A,40                ;DELAY 0.4 SEC FOR EACH TONE
42              MOV         DEL1,A
43              CALL        DELAY
44              CLR         TON                 ;STOP TMR COUNTING
45              INC         TBLP                ;UPDATE TABLE POINTER
46              SDZ         COUNT               ;WERE ALL PITCH PLAYED?
47              JMP         NEXT_PITCH          ;NO, NEXT PITCH
48              JMP         MAIN                ;YES,RESTART.
49   ;*********************************************************************
50   ;                 TIMER INTERRUPT SERVICE ROUTINE
51   ;*********************************************************************
52   TMR_INT:
53              MOV         STACK_A,A           ;PUSH A
54              MOV         A,STATUS
55              MOV         STACK_STATUS,A      ;PUSH STATUS
56              MOV         A,00000001B
57              XORM        A,SPK_PORT          ;COMPLEMENT PA.0
58              MOV         A,STACK_STATUS
59              MOV         STATUS,A            ;POP STATUS
60              MOV         A,STACK_A           ;POP A
61              RETI
62   ;*********************************************************************
63   ;                 Delay about DEL1*10ms
64   ;*********************************************************************
65   DELAY      PROC
66              MOV         A,30
67              MOV         DEL2,A              ;SET DEL2 COUNTER
68   DEL_2:     MOV         A,110
69              MOV         DEL3,A              ;SET DEL3 COUNTER
70   DEL_3:     SDZ         DEL3                ;DEL3 DOWN COUNT
71              JMP         DEL_3
72              SDZ         DEL2                ;DEL2 DOWN COUNT
73              JMP         DEL_2
74              SDZ         DEL1                ;DEL1 DOWN COUNT
75              JMP         DELAY
76              RET
77   DELAY      ENDP
78              ORG         LASTPAGE
79   TAB_PITCH:                                 ;PITCH CONSTANT FOR fINT=fSYS/16
80              DC          256-OSC/(523*2*16)    ;DO  TONE TIME CONSTANT (LO-BYTE)
81              DC          256-OSC/(587*2*16)    ;RE  TONE TIME CONSTANT (LO-BYTE)
82              DC          256-OSC/(659*2*16)    ;MI  TONE TIME CONSTANT (LO-BYTE)
83              DC          256-OSC/(698*2*16)    ;FA  TONE TIME CONSTANT (LO-BYTE)
84              DC          256-OSC/(785*2*16)    ;SO  TONE TIME CONSTANT (LO-BYTE)
85              DC          256-OSC/(880*2*16)    ;LA  TONE TIME CONSTANT (LO-BYTE)
86              DC          256-OSC/(998*2*16)    ;TI  TONE TIME CONSTANT (LO-BYTE)
87              DC          256-OSC/(523*2*16*2)  ;DO. TONE TIME CONSTANT (LO-BYTE)
88              DC          256-OSC/(587*2*16*2)  ;RE. TONE TIME CONSTANT (LO-BYTE)
89              DC          256-OSC/(659*2*16*2)  ;ME. TONE TIME CONSTANT (LO-BYTE)
90              DC          256-OSC/(698*2*16*2)  ;FA. TONE TIME CONSTANT (LO-BYTE)
91              DC          256-OSC/(785*2*16*2)  ;SO. TONE TIME CONSTANT (LO-BYTE)
92              DC          256-OSC/(880*2*16*2)  ;LA. TONE TIME CONSTANT (LO-BYTE)
93              DC          256-OSC/(998*2*16*2)  ;TI. TONE TIME CONSTANT (LO-BYTE)
94              DC          256-OSC/(523*2*16*4)  ;DO. TONE TIME CONSTANT (LO-BYTE)
95              END
```

程序说明

8~13　依序定义变量地址。

15　定义 OSC(即 $f_{SYS}=4$ MHz)。

17~18　定义 SPK_PORT 为 PA。

21　声明存储器地址由 000h 开始(HT46xx Reset Vector)。

23　声明存储器地址由 008h 开始，此即 Timer/Event Counter 中断向量地址。

26　定义 SPK_PORT 为输出模式。

行号	说明
27~28	中断总开关(EMI)与 Timer/Event Counter 中断功能(ETI)有效。
29~30	定义 TMRC，设置：
	TM1TM0＝10：Timer Mode(Internal clock)；
	TON＝0：还未启动计数功能；
	PSC<2:0>＝100：计数倍率 1:16(即 $f_{INT}=f_{SYS}/16$)。
32~33	将 TBLP 指向表格数据的起始地址。
34~35	设置计数器 COUNT＝15，因为总共有 15 个不同的音阶。
37	通过查表取得欲发音的音阶频率参数并存入 TMRL 寄存器。
38~39	将 TMRH 设置为 FFh，至于其原因，请读者参考 4-10-3 小节的说明。
40	启动 Timer/Event Counter 开始计数，此后只要发生计数溢位的情况，SPK_PORT 就会反向一次。
41~43	延时 0.4 s，即即音长的控制。
44	停止 Timer/Event Counter 计数功能，即停止喇叭发出声音。
45	将 TBLP 加 1，指向下一笔音符数据的地址。
46~48	判断 COUNT－1 否为 0，成立则表示 15 个音阶已发音完毕，回到主程序重新控制发音；反之，则继续提取下一个时间常量(JMP LOOP)，产生下一个音阶。
52~61	Timer/Event Counter 中断服务子程序，首先将 Acc 与 PSW(即状态寄存器；Status)暂存起来，其次运用 "XORM A,SPK_PORT" 指令将 SPK_PORT 反向，在取回 Acc 与 PSW 的原值之后返回主程序。提醒读者：<u>中断服务子程序的最后一个指令可以是 RET 或 RETI 指令，不同的是：RETI 指令在返回主程序之前会先将 EMI 位设置为 1(中断使能)，而 RET 指令则不会</u>。
65~77	DELAY 子程序，延时时间的计算请参考 4-1 节。
79~94	音调数据建表区。

程序中有关产生音调的音长(即每个音调持续的时间)，由DELAY子程序控制。当CPU执行DELAY子程序时Timer/Event Counter仍持续计数，待其计数到溢位时(FFFFh→0000h)再以中断方式让CPU跳到中断地址去执行程序。由于HT46xx的Timer/Event Counter具备"自动重新载入"的功能，所以当计数溢位之后会自动将时间常量重新载入到Timer/Event Counter中继续计数，这是相当方便的。有些单片机的计数单元就不具备自动重新载入，此时用户就必须自行将时间常量重新载入到计数器中，否则下一次产生计数器溢位的时间将是"计数器最大计数数值×定时时钟脉冲周期"，而非原来的时间。这点也请读者在使用无自动重新载入功能的单片机时特别留意。

其实上述让喇叭发音的方式，主要是想让读者熟悉HT46xx系列的中断机制与Timer/Event Counter的控制方式，如果只是单纯想让喇叭发音，其实还有更简易的方法，这可得归功于HT46xx的贴心设计——PFD(可编程分频器；Programmable Frequency Divider)。HT46xx的PA3可当成一般的I/O使用，也可以当成PFD输出，需在Options中选择。在PFD选项有效的情况下，若PA3被定义为输入模式(PAC3＝1)，则PA3仍可当成一般的输入引脚使用。若PAC3＝0(输出模式)，此时如果设置PA3＝1即开始输出PFD信号，其所产生的频率为定时器/计数器溢位频率的一半；设置PA3＝0时将停止PFD信号的输出，并强迫使PA3的引脚呈现低电位的输出状态。这不是正符合我们的需求吗?为能让读者确实了解PFD的控制方式，接下来就以4×4键盘搭配喇叭发音控制程序为例做说明，其电路图如图4-10-4所示。

HT46xx 单片机原理与实践

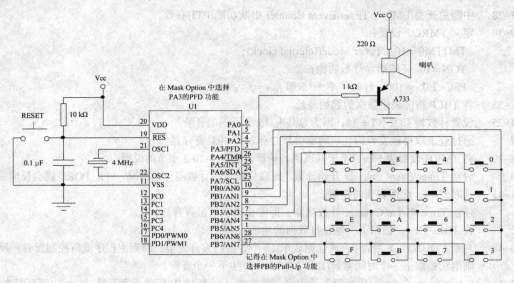

图4-10-4 4×4键盘搭配喇叭发音电路图

4-10-5 流程图及程序

1. 流程图(见图 4-10-5)

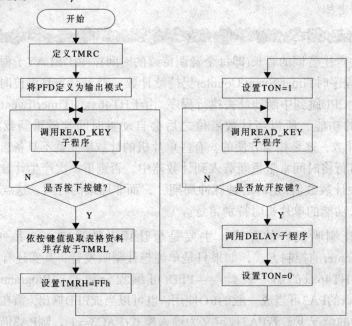

图4-10-5 PFD控制实验流程图

2. 程序 4-10-1 PFD 控制实验

```
1  ;PROGRAM : 4-10-1.ASM   (4-10-1.PRJ)              2002.DEC.30.
2  ;FUNCTION: GENERATE TONE DO~DO. BY PFD FUNCTION   BY STEVEN
3  ;NOTE    : USING TMR AS A 8-BIT TIMER
4  ;MASK OPTION:  ENABLE PA3 PROGRAMMABLE FREQUENCY DEVIDER
5  #INCLUDE   HT46R23.INC
6           .CHIPHT46R23
```

```
7       ;------------------------------------------------------------
8       MY_DATA     .SECTION    'DATA'          ;== DATA SECTION ==
9       DEL1        DB      ?                   ;DELAY LOOP COUNT 1
10      DEL2        DB      ?                   ;DELAY LOOP COUNT 2
11      DEL3        DB      ?                   ;DELAY LOOP COUNT 3
12      COUNT       DB      ?                   ;UNIVERSAL COUNTER
13      KEY         DB      ?                   ;KEYCODE REGISTER
14      ;------------------------------------------------------------
15      OSC         EQU     4000000             ;fSYS=4 MHz
16      KEY_PORT    EQU     PB                  ;DEFINE KEYPAD PORT
17      KEY_PORTC   EQU     PBC                 ;DEFINE KEY PORT VONTROL REG.
18      PFDC        EQU     PAC.3
19
20      MY_CODE     .SECTION AT 0 'CODE'        ;== PROGRAM SECTION ==
21              ORG         00H                 ;HT-46RXX RESET VECTOR
22              MOV         A,10000100B         ;CONFIG TMR 0 IN MODE 2(TIMER MODE)
23              MOV         TMRC,A              ;fINT=fSYS/16(4MHz/16).
24              CLR         PFDC                ;ENABLE PFD OUTPUT
25      MAIN:
26              CALL        READ_KEY            ;SCAN KEY_PAD
27              MOV         A,KEY
28              XOR         A,16
29              SZ          Z                   ;IS ANY KEY PRESSED?
30              JMP         MAIN                ;NO, RE-READ
31              MOV         A,KEY               ;YES!
32              CALL        TRANS               ;GET CORRESPONDING TIME CONSTAND
33              MOV         TMRL,A              ;LOAD PITCH TO TMR PE-LOAD REGISTER
34              MOV         A,0FFH
35              MOV         TMRH,A              ;PRELOAD TMRH=FFH FOR 8-BIT TMR
36              SET         TON                 ;START TMR COUNTING
37      WAIT_KEY_RELEASED:
38              CALL        READ_KEY            ;SCAN KEY_PAD FOR KEY RELEASED
39              MOV         A,KEY
40              XOR         A,16
41              SNZ         Z                   ;IS KEY STILL PRESSED?
42              JMP         WAIT_KEY_RELEASED   ;YES, WAIT KEY RELEASED
43              MOV         A,10                ;DELAY 0.1 SEC
44              MOV         DEL1,A
45              CALL        DELAY
46              CLR         TON                 ;STOP TMR COUNTING(DISABLE PFD)
47              JMP         MAIN                ;RESTART.
48      ;************************************************************************
49      ;   SCAN 4x4 MATRIX ON KEY PORT AND RETURN THE CODE IN KEY REGISTER
50      ;   IF NO KEY BEEN PRESSED, KEY=16.
51      ;************************************************************************
52      READ_KEY    PROC
53              MOV         A,11110000B
54              MOV         KEY_PORTC,A         ;CONFIG PORT B
55              SET         KEY_PORT            ;INITIAL PORT B
56              CLR         KEY                 ;INITIAL KEY REGISTER
57              MOV         A,04
58              MOV         COUNT,A             ;SET ROW COUNTER
59              CLR         C                   ;CLEAR CARRY FLAG
60      SCAN_KEY:
61              RLC         KEY_PORT            ;ROTATE SCANNING BIT
62              SET         C                   ;MAKE SURE C=1
63              SNZ         KEY_PORT.4          ;COLUMN 0 PRESSED?
64              JMP         END_KEY             ;YES.
65              INC         KEY                 ;NO, INCREASE KEY CODE.
66              SNZ         KEY_PORT.5          ;COLUMN 1 PRESSED?
67              JMP         END_KEY             ;YES.
68              INC         KEY                 ;NO, INCREASE KEY CODE.
69              SNZ         KEY_PORT.6          ;COLUMN 2 PRESSED?
70              JMP         END_KEY             ;YES.
71              INC         KEY                 ;NO, INCREASE KEY CODE.
72              SNZ         KEY_PORT.7          ;COLUMN 3 PRESSED?
73              JMP         END_KEY             ;YES.
```

```
74          INC     KEY                     ;NO, INCREASE KEY CODE.
75          SDZ     COUNT                   ;HAVE ALL ROWs BEEN CHECKED?
76          JMP     SCAN_KEY                ;NO, NEXT ROW.
77  END_KEY:
78          RET
79  READ_KEY    ENDP
80  ;**********************************************************************
81  ;              Delay about DEL1*10 ms
82  ;**********************************************************************
83  DELAY   PROC
84          MOV     A,30
85          MOV     DEL2,A                  ;SET DEL2 COUNTER
86  DEL_2:  MOV     A,110
87          MOV     DEL3,A                  ;SET DEL3 COUNTER
88  DEL_3:  SDZ     DEL3                    ;DEL3 DOWN COUNT
89          JMP     DEL_3
90          SDZ     DEL2                    ;DEL2 DOWN COUNT
91          JMP     DEL_2
92          SDZ     DEL1                    ;DEL1 DOWN COUNT
93          JMP     DELAY
94          RET
95  DELAY   ENDP
96  ;**********************************************************************
97  ;           RETURN THE TABLE VALUE INDEX BY A
98  ;**********************************************************************
99  TRANS   PROC
100         ADDM    A,PCL                   ;PITCH CONSTANT FOR fINT=fSYS/16
101         RET     A,256-OSC/(523*2*16)    ;DO  TONE TIME CONSTANT (LO-BYTE)
102         RET     A,256-OSC/(587*2*16)    ;RE  TONE TIME CONSTANT (LO-BYTE)
103         RET     A,256-OSC/(659*2*16)    ;MI  TONE TIME CONSTANT (LO-BYTE)
104         RET     A,256-OSC/(698*2*16)    ;FA  TONE TIME CONSTANT (LO-BYTE)
105         RET     A,256-OSC/(785*2*16)    ;SO  TONE TIME CONSTANT (LO-BYTE)
106         RET     A,256-OSC/(880*2*16)    ;LA  TONE TIME CONSTANT (LO-BYTE)
107         RET     A,256-OSC/(998*2*16)    ;TI  TONE TIME CONSTANT (LO-BYTE)
108         RET     A,256-OSC/(523*2*16*2)  ;DO. TONE TIME CONSTANT (LO-BYTE)
109         RET     A,256-OSC/(587*2*16*2)  ;RE. TONE TIME CONSTANT (LO-BYTE)
110         RET     A,256-OSC/(659*2*16*2)  ;ME. TONE TIME CONSTANT (LO-BYTE)
111         RET     A,256-OSC/(698*2*16*2)  ;FA. TONE TIME CONSTANT (LO-BYTE)
112         RET     A,256-OSC/(785*2*16*2)  ;SO. TONE TIME CONSTANT (LO-BYTE)
113         RET     A,256-OSC/(880*2*16*2)  ;LA. TONE TIME CONSTANT (LO-BYTE)
114         RET     A,256-OSC/(998*2*16*2)  ;TI. TONE TIME CONSTANT (LO-BYTE)
115         RET     A,256-OSC/(523*2*16*4)  ;DO. TONE TIME CONSTANT (LO-BYTE)
116         RET     A,256-OSC/(587*2*16*4)  ;RE. TONE TIME CONSTANT (LO-BYTE)
117         END
```

程序说明

9~13   依序定义变量地址。

15   定义 OSC(即 $f_{SYS}=4$ MHz)。

16~18   定义 SPK_PORT 为 PA。

21   声明存储器地址由 000h 开始(HT46xx Reset Vector)。

22~23   定义 TMRC,设置:

    TM1TM0=10:Timer Mode(Internal clock);

    TON=0:还未启动计数功能;

    PSC<2:0>=100:计数倍率 1:16(即 $f_{INT}=f_{SYS}/16$)。

24   定义 PFD 为输出模式。

26   调用 READ_KEY 子程序读取按键值。

27~31   判断是否按下按键,若没有则再次读取按键值。

32~36 有按键被按下时,就根据按键值查表取得欲发音的音阶频率参数并存入 TMRL 寄存器,设置 TMRH=FFh 后,即启动 Timer/Event Counter 开始计数,此后只要发生计数溢位的情况,PFD 输出就会反向一次。至于将 TMRH 设置为 FFh 的原因,请读者参考 4-10-3 小节的说明。

38~46 检查按键是否已放开,如果按键仍是按着,就持续 PFD 的输出;反之若按键已放开,则再延时 0.1 s 后将 Timer/Event Counte 计数功能关闭,并重新检查按键的输入(JMP MAIN)。

52~79 READ_KEY 子程序。

83~95 DELAY 子程序,延时时间的计算请参考 4-1 节。

99~116 查表子程序,此种查表方式的原理,请读者参考 4-2 节中的说明。

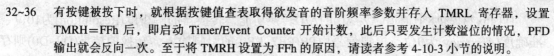

程序中运用了PFD输出功能,所以请读者一定要记得在Options中选择此项功能,要不然喇叭是发不出声音的。

4-11 A/D转换器控制实验

4-11-1 目 的

利用HT46x23的A/D转换器(Analog to Digital Converter),将模拟电压的变化直接以二进制方式显示在LED上。

4-11-2 学习重点

通过本实验,读者应熟悉HT46x23的A/D转换器与A/D中断的控制方式。

4-11-3 电路图(见图4-11-1)

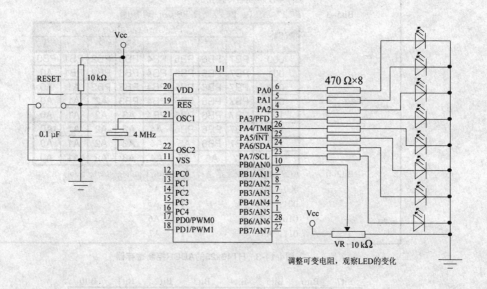

图4-11-1 A/D转换器控制电路

HT46x23提供8个通道(A7~A0)的A/D转换功能,转换器的解析度(Resolution)为10位、准确度(Accuracy)为9位,模拟信号由PB(PB7~PB0)的引脚输入,转换的结果(D9~D0)则存放在ADRH与ADRL寄存器中。本实验将以可变电阻控制AN0(即PB0)的模拟输入电压,然后将转换后的高8位结果直接以二进制的方式在LED上显示。有关HT46xxA/D转换的相关说明请读者参阅2-9节的内容,图4-11-2、图4-11-3和图4-11-4仅将相关的控制寄存器示出,方便读者参考。

Bit7	Bit6	Bit5	Bit4	Bit3	Bit2	Bit1	Bit0
TEST	—	—	—	—	—	ADCS1	ADCS0

Bit7	TEST:芯片测试位(Test Mode Only)
	此为IC芯片厂测试用的位,用户不可以使用
Bit6~2	未使用;读取时的值为0
Bit4	TON:定时/计数控制位(Enable/Disable Counting)
	1:开始计数
	0:停止计数
Bit1~0	ADCS1~ADCS0:A/D转换时钟脉冲选择位(ADC Clock Source)
	00: $f_{SYS}/2$
	01: $f_{SYS}/8$
	01: $f_{SYS}/32$
	11: 未定义

图4-11-2　HT46x23的ACSR控制寄存器

Bit7	Bit6	Bit5	Bit4	Bit3	Bit2	Bit1	Bit0
START	$\overline{EOC}$	PCR2	PCR1	PCR0	ACS0	ACS1	ACS0

Bit7	START:A/D转换起始控制位(Starts the A/D Conversion)
	0→1→0: A/D转换器开始转换
	0→1: A/D转换器回至复位状态(Reset A/D Converter)
Bit6	$\overline{EOC}$: 转换完成状态标志位(End of A/D Conversion)
	1: A/D转换器正进行转换中
	0: A/D转换器转换完成
Bit5~3	PCR2~PCR0:PB控制位(PB Control Bits)

PCR2~0	PB Configuration						
	7	6	5	4	3	2	1	0
000	PB7	PB6	PB5	PB4	PB3	PB2	PB1	PB0
001	PB7	PB6	PB5	PB4	PB3	PB2	PB1	A0
010	PB7	PB6	PB5	PB4	PB3	PB2	A1	A0
011	PB7	PB6	PB5	PB4	PB3	A2	A1	A0
100	PB7	PB6	PB5	PB4	A3	A2	A1	A0
101	PB7	PB6	PB5	A4	A3	A2	A1	A0
110	PB7	PB6	A5	A4	A3	A2	A1	A0
111	A7	A6	A5	A4	A3	A2	A1	A0

Bit2~0	ACS2~0:模拟通道选择位(Analog Channel Seclection)
	000: A0　　100: A4
	001: A1　　101: A5
	010: A2　　110: A6
	011: A3　　111: A7

图4-11-3　HT46x23的ADCR控制寄存器

Bit7	Bit6	Bit5	Bit4	Bit3	Bit2	Bit1	Bit0
—	ADF	TF	EIF	EADI	ETI	EEI	EMI

图4-11-4　HT46x23的INTC0控制寄存器

当A/D转换完成时,会将INTC0寄存器的ADF位设置为1,在应用时可以由ADF位来判断是否已经完成转换(转换完成时"EOC"位会被清除为0,也可以用来判断转换是否已经完成)。但是,也可以用中断的方式加以控制,当EMI位为1且EADI=1(A/D中断有效)时,若A/D转换完成,则CPU将跳至00Ch地址执行程序。

4-11-4 程序及流程图

1. 流程图(见图 4-11-5)

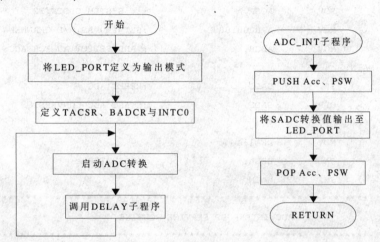

图4-11-5 ADC控制实验流程图

2. 程序 4-11-1 ADC 控制实验

```
1    ;PROGRAM : 4-11.ASM   (4-11.PRJ)          2002.DEC.30.
2    ;FUNCTION: ADC DEMO PROGRAM               BY STEVEN
3    #INCLUDE    HT46R23.INC
4                .CHIPHT46R23
5    ;------------------------------------------------------------
6    MY_DATA     .SECTION   'DATA'        ;== DATA SECTION ==
7    DEL1        DB    ?                  ;DELAY LOOP COUNT 1
8    DEL2        D     ?                  ;DELAY LOOP COUNT 2
9    DEL3        DB    ?                  ;DELAY LOOP COUNT 3
10   COUNT       DB    ?                  ;PITCH COUNT
11   STACK_A     DB    ?                  ;STACK BUFFER FOR A
12   STACK_PSW   DB    ?                  ;STACK BUFFER FOR PSW
13   ;------------------------------------------------------------
14   SADC        EQU   ADCR.7             ;DEFINE SADC AS ADC START BIT
15   LED_PORT    EQU   PA                 ;DEFINE LED_PORT
16   LED_PORTC   EQU   PAC                ;DEFINE LED_PORT CONTROL REG.
17   MY_CODE     .SECTION  AT 0 'CODE'    ;== PROGRAM SECTION ==
18       ORG     00H                      ;HT-46RXX RESET VECTOR
```

```
19          JMP       START
20          ORG       0CH                     ;HT-46RXX ADC INTERRUPT VECTOR
21          JMP       ADC_INT
22   START:
23          CLR       LED_PORTC               ;CONFIG PA AS OUTPUT
24          MOV       A,00001001B             ;ENABLE GLOBAL AND ADC INTERRUPT
25          MOV       INTC0,A
26          MOV       A,0000000B              ;SET A/D CONVERTER CLOCK SOURCE
27          MOV       ACSR,A                  ; = SYSTEM CLOCK/32
28          MOV       A,00001000B             ;SET PB0 AS A/D CHANNEL
29          MOV       ADCR,A                  ; AND SELECT A0 FOR ADC
30   MAIN:
31          SET       SADC                    ;RESET ADC
32          CLR       SADC                    ;START ADC CONVERSION
33          MOV       A,5
34          MOV       DEL1,A
35          CALL      DELAY                   ;DELAY 100ms
36          JMP       MAIN
37   ;**********************************************************************
38   ;               ADC INTERRUPT SERVICE ROUTINE
39   ;**********************************************************************
40   ADC_INT:
41          MOV       STACK_A,A               ;PUSH A
42          MOV       A,STATUS
43          MOV       STACK_PSW,A             ;PUSH STATUS
44          MOV       A,ADRH                  ;GET HIGH BYTE A/D RESULT
45          MOV       LED_PORT,A              ;SEND TO PA
46          MOV       A,STACK_PSW
47          MOV       STATUS,A                ;POP STATUS
48          MOV       A,STACK_A               ;POP A
49          RETI
50   ;**********************************************************************
51   ;               Delay about DEL1*10ms
52   ;**********************************************************************
53   DELAY PROC
54          MOV       A,30
55          MOV       DEL2,A                  ;SET DEL2 COUNTER
56   DEL_2: MOV       A,110
57          MOV       DEL3,A                  ;SET DEL3 COUNTER
58   DEL_3: SDZ       DEL3                    ;DEL3 DOWN COUNT
59          JMP       DEL_3
60          SDZ       DEL2                    ;DEL2 DOWN COUNT
61          JMP       DEL_2
```

```
62          SDZ     DEL1                ;DEL1 DOWN COUNT
63          JMP     DELAY
64          RET
65  DELAY   ENDP
66  END
```

程序说明

7~12   依序定义变量地址。

14   定义 SADC 为 ADCR.7。

15~16   定义 LED_PORT 为 PA。

18   声明存储器地址由 000h 开始(HT46xx Reset Vector)。

20   声明存储器地址由 00Ch 开始，此即 A/D 转换中断向量地址。

23   定义 LED_PORT 为输出模式。

24~25   中断总开关(EMI)与 A/D 转换中断功能(EADI)有效。

26~27   定义 ACSR，设置 A/D 转换的时钟脉冲来源为 $f_{SYS}/32$。

28~29   定义 ADCR，设置如下：
        ASC2~ASC0=001：选择 A0 通道；
        PCR2~PCR0=001：将 PB0 定义为模拟输入通道。

31~32   启动 A/D 转换。

33~36   延时 0.1 s 后重新启动 A/D 转换。

40~49   A/D 转换中断服务子程序，首先将 Acc 与 PSW(即状态寄存器；Status)暂存起来，其次将 A/D 转换的高 8 位数值输出到 LED_PORT，在取回 Acc 与 PSW 的原值之后返回主程序。提醒读者：中断服务子程序的最后一个指令可以是 RET 或 RETI，不同的是：RETI 指令在返回主程序之前会先将 EMI 位还原为 1(中断有效)，而 RET 指令则不会。

53~65   DELAY 子程序，延时时间的计算请参考 4-1 节。

程序本身相当容易，执行时是以每0.1 s更新一次，将转换值显示在LED上。ADCR寄存器中的START位，是控制A/D转换器停滞于复位状态或开始进行转换的控制开关，当START由0→1是令A/D转换器回到复位状态；当START由0→1→0，则是要求A/D转换器开始针对选择的模拟通道进行转换，当转换完成时A/D转换器会自动将$\overline{EOC}$位清除为0。用户可利用$\overline{EOC}$位判断A/D转换器是否已经完成转换，为了确保转换器的正常动作，在$\overline{EOC}$位还未被清除之前，应该让START位维持在0。

想提醒读者有关HT46xx转换时间的注意事项：请参考图4-11-6 A/D转换的时序图，依时序图所示A/D转换器完成一次转换约需花费76个T_{AD}的时间(即转换时间；Conversion Time)，而所指的就是转换的时钟脉冲周期。以f_{sys}＝4 MHz为例，若选择ADCS1 ADCS0＝01，则此时的T_{AD}＝2μs，转换时间＝152μs。不过请读者注意原厂数据手册的一项限制：T_{AD}≥1μs，也就是说HT46x23单片机的A/D转换器最短的转换时间为76μs，如果所选择的转换时钟脉冲(ADC Clock Source)<1μs，则并不保证转换结果的正确性。笔者曾经试着将转换时间缩短至0.5μs，发现A/D转换动作仍旧正常，不过在此还是不鼓励读者以超过原厂数据手册的规范来使用芯片。

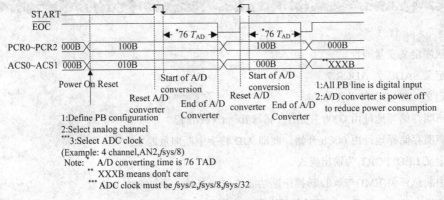

图4-11-6　HT46x23的A/D转换时序图

4-11-5　动动脑 + 动动手

  更改程序，使得 A/D 转换值与 LED 显示呈线性关系；也就是当模拟输入电压越大时，LED 就亮得越多，反之亦然。

4-12　外部中断控制实验

4-12-1　目　的

  本实验将利用中断控制步进电机的正转、反转。在未检测到中断之前，HT46x23控制七段显示器显示0~9，一旦检测到硬件中断信号，CPU立即执行中断服务子程序，让步进电机正转、反转一圈后再返回主程序继续执行。

4-12-2　学习重点

  通过本实验，读者应了解HT46xx的中断控制方式，同时也应学习主程序与中断服务子程序间的参数分配问题。

4-12-3　电路图

  本实验只是将4-3节的七段显示器控制与实验4-6的步进电机控制加以结合，因此在电路原理上不再赘述，请读者参阅这两个实验的内容。外部中断控制电路如图4-12-1所示。

所谓外部中断,是指当HT46xx的$\overline{\text{INT}}$(PA5)引脚降为低电位所产生的中断情况,请参考图4-12-2的中断时序。

在外部中断有效的情况下,HT46xx是在T2正边缘检查$\overline{\text{INT}}$引脚状态,如果在低电位连续维持两个T2状态,则CPU将跳至中断服务子程序(中断向量:004h),因此"外部中断信号应至少维持一个指令周期以上"。由于本实验的中断信号是由用户按下按键来触动,所以绝对满足此项要求。

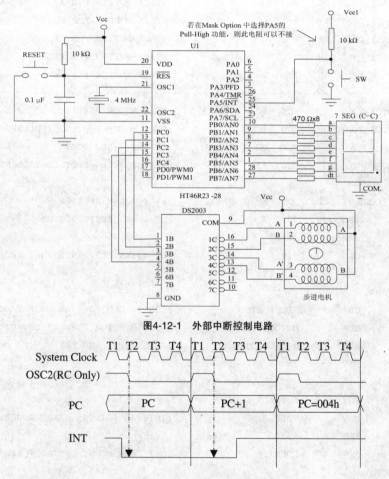

图4-12-1 外部中断控制电路

图4-12-2 HT46xx的中断时序

4-12-4 流程图及程序

1. 流程图(见图 4-12-3)

2. 程序 4-12 外部中断控制实验

```
1   ;PROGRAM : 4-12.ASM    (4-12.PRJ)              2002.DEC.30.
2   ;FUNCTION: EXTERNAL INTERRUPT DEMO PROGRAM     BY STEVEN
3   #INCLUDE  HT46R23.INC
4            .CHIPHT46R23
```

```
5       ;------------------------------------------------------------
6       MY_DATA     .SECTION    'DATA'          ;== DATA SECTION ==
7       DEL1        DB      ?                   ;DELAY LOOP COUNT 1
8       DEL2        DB      ?                   ;DELAY LOOP COUNT 2
9       DEL3        DB      ?                   ;DELAY LOOP COUNT 3
10      COUNT       DB      ?                   ;DISPLAY COUNT
11      COUNT2      DB      ?                   ;4-PHASE COUNT
12      STACK_A     DB      ?                   ;STACK BUFFER FOR A
13      STACK_STATUS DB     ?                   ;STACK BUFFER FOR PSW
14      STACK_DEL1  DB      ?                   ;STACK BUFFER FOR DEL1
15      STACK_DEL2  DB      ?                   ;STACK BUFFER FOR DEL2
16      STACK_DEL3  DB      ?                   ;STACK BUFFER FOR DEL3
17      ;------------------------------------------------------------
18      MOTOR_PORT      EQU     PC              ;DEF. MOTOR PORT
19      MOTOR_PORTC     EQU     PCC             ;DEF. MOTOR PORT CONTROL REG.
20      SEG_PORT        EQU     PB              ;DEFINE 7-SEG PORT
21      SEG_PORTC       EQU     PBC             ;DEFINE 7-SEG PORT CONTROL REG.
22
23      MY_CODE     .SECTION    AT 0 'CODE'     ;== PROGRAM SECTION ==
24          ORG         00H                     ;HT-46RXX RESET VECTOR
25          JMP         START
26          ORG         04H                     ;HT-46RXX EXT. INTERRUPT VECTOR
27          JMP         EXT_INT
28      START:
29          CLR         SEG_PORTC               ;CONFIG SEG_PORT AS OUTPUT MODE
30          CLR         MOTOR_PORTC             ;CONFIG MOTOR_PORT AS OUTPUT MODE
31          MOV         A,00000011B             ;ENABLE EMI AND EEI
32          MOV         INTC0,A
33      MAIN:
34          MOV         A,TAB_7_SEG
35          MOV         TBLP,A                  ;INITIAL POINTER START ADDRESS
36          MOV         A,10
37          MOV         COUNT,A                 ;SET DATA COUNT OF TAB_7_SEG
38      LOOP:
39          TABRDL      SEG_PORT                ;READ TABLE AND SEND TO SEG_PORT
40          MOV         A,250                   ;DELAY 0.25 SEC
41          MOV         DEL1,A                  ;SET DELAY FACTOR
42          CALL        DELAY                   ;DELAY TBLH*10mS
43          INC         TBLP                    ;UPDATE TABLE POINTER
44          SDZ         COUNT                   ;COUNT DOWN BY 1
45          JMP         LOOP                    ;JUMP IF NOT TABLE END
46          JMP         MAIN                    ;RESTART
47      ;****************************************************************
48      ;           EXTERNAL INTERRUPT SERVICE ROUTINE
```

```
49  ;*********************************************************************
50  EXT_INT:
51          MOV     STACK_A,A               ;PUSH A
52          MOV     A,STATUS
53          MOV     STACK_STATUS,A          ;PUSH STATUS
54          MOV     A,DEL1
55          MOV     STACK_DEL1,A            ;PUSH DEL1
56          MOV     A,DEL2
57          MOV     STACK_DEL2,A            ;PUSH DEL2
58          MOV     A,DEL3
59          MOV     STACK_DEL3,A            ;PUSH DEL3
60          MOV     A,50                    ;CW MOTION
61          MOV     COUNT2,A                ;SET STEP COUNT
62  LOOP_1:
63          MOV     A,10000B
64          MOV     MOTOR_PORT,A            ;SET INITIAL CODE
65  LOOP_2:
66          RR      MOTOR_PORT              ;GET EXCIATED CODE SEND TO MOTOR_PORT
67          MOV     A,10                    ;DELAY 10ms
68          MOV     DEL1,A
69          CALL    DELAY                   ;FOR DELAY
70          SNZ     MOTOR_PORT.0            ;4 PHASE OVER?
71          JMP     LOOP_2                  ;NO, NEXT PHASE
72          SDZ     COUNT2                  ;1 TURN?
73          JMP     LOOP_1                  ;NO, NEXT STEP PHASE
74          MOV     A,STACK_DEL3            ;POP DEL3
75          MOV     DEL3,A
76          MOV     A,STACK_DEL2            ;POP DEL2
77          MOV     DEL2,A
78          MOV     A,STACK_DEL1            ;POP DEL1
79          MOV     DEL2,A
80          MOV     A,STACK_STATUS
81          MOV     STATUS,A                ;POP STATUS
82          MOV     A,STACK_A               ;POP A
83          CLR     MOTOR_PORT              ;FOR POWER SAVING
84          CLR     EIF                     ;CLEAR EXT. INT CAUSED BY BOUNCING
85          RETI
86  ;*********************************************************************
87  ;                       Delay about DEL1*1ms
88  ;*********************************************************************
89  DELAY PROC
90          MOV     A,03
91          MOV     DEL2,A                  ;SET DEL2 COUNTER
92  DEL_2:  MOV     A,110
```

```
93              MOV     DEL3,A              ;SET DEL3 COUNTER
94    DEL_3:    SDZ     DEL3                ;DEL3 DOWN COUNT
95              JMP     DEL_3
96              SDZ     DEL2                ;DEL2 DOWN COUNT
97              JMP     DEL_2
98              SDZ     DEL1                ;DEL1 DOWN COUNT
99              JMP     DELAY
100             RET
101   DELAY ENDP
102             ORG     LASTPAGE
103   TAB_7_SEG:
104             DC      0001111100111111B   ;7 SEG CODE FOR COMMOM CATHOD LED
105             DC      0010100000000110B   ;CONNECT BIT 0~7 FOR SEGMENT A~H
106             DC      0011110010111011B   ;LOW-BYTE FOR 7 SEG DISPLAY CODE
107             DC      0101000010011111B   ;HIGH 7 BITs FOR DELAY FACTOR
108             DC      0110010011100110B
109             DC      0111100011101101B
110             DC      1000110011111101B
111             DC      1010000000000111B
112             DC      1011010011111111B
113             DC      1100100011100111B
114             END
```

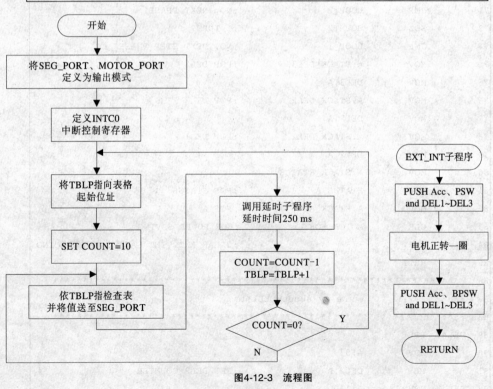

图 4-12-3 流程图

第4章 基础实验篇

程序说明

行号	说明
7~16	依序定义变量地址。
18~21	定义 MOTOR_PORT 为 PC，SEG_PORT 为 PB。
24	声明存储器地址由 000h 开始(HT46xx 复位向量)。
26	声明存储器地址由 004h 开始(HT46xx 外部中断向量)。
29~30	将 MOTOR_PORT、SEG_PORT 定义成输出模式。
31~32	中断总开关(EMI)与外部中断(EEI)有效，HT46xx 可以开始接受外部中断。
34~35	将 TBLP 指向七段显示器显示码的数据起始地址。
36~37	将 COUNT 计数器设置为 10。
39	通过查表将欲显示的七段显示器的数值取出，并输出到 SEG_PORT。
40~42	延时 250 ms(250×1 ms)。
43	TBLP 指针加 1，指向下一笔七段显示器显示码。
44~46	判断 COUNT−1 是否等于 0，成立则重新执行程序(JUMP MAIN)；反之，则显示下一个数值(JMP LOOP)。
51~59	中断服务子程序的开始，首先将 ACC 与 Status 数值保留。由于主程序与中断服务子程序都会调用 DELAY 子程序，因此也一并把 DELAY 子程序中使用到的变量(DEL1~DEL3)给存储起来。
60~61	将 COUNT2 设置为 50，至于 COUNT2 的作用是什么，下面将会有详细的说明。
63~64	MOTOR_PORT 的值设置为 10 000b，此时电机未被励磁。
66	将 MOTOR_PORT 带进位位右移，此时步进电机会前进一步。
67~69	延时 10 ms(10×1 ms)。
70~71	检查 MOTRO_PORT.0 是否等于 1，即判断是否已将 4 个相位全部激发完毕，若不成立表示需继续激发下个相位(JMP LOOP_2)。
72~73	判断步进电机是否已被连续激发 4 个相位 50 次(即 200 Steps)，成立则准备返回主程序；反之，则重新激发 4 相位使步进电机正转(JMP LOOP_1)。
74~82	取回所有进入中断服务子程序时所保留的变量值。
83	关闭步进电机的励磁信号，此举的主要目的，是希望在还未进入中断服务子程序之前，电机处于未励磁的状态，以免不必要的电流损耗。
84	清除外部中断标志位(EIF)，请参考以下的说明。
85	返回主程序。
89~101	DELAY 子程序，延时时间的计算请参考 4-1 节。
103~113	七段显示码数据与延时常量建表区。

主程序摘自4-3节七段显示器控制实验，中断服务子程序则请读者参阅实验4-6的步进电机控制实验，其基本的控制原理在此不再赘述。不同的是步进电机控制部分由原来的双向励磁改为单向励磁；而励磁信号原本为查表获得，此处以更简易的带进位位右移方式产生。由于主程序与中断服务子程序都会调用DELAY子程序，为了避免中断后将原来的寄存器内容改变，致使返回主程序时的延时时间不正确，所以把DELAY子程序中使用到的变量

(DEL1~DEL3)存储起来,如此可以使主程序与中断服务子程序共用一个子程序,达到缩减程序代码长度的目的。

请注意第84行的"CLR EIF"指令,细心的读者或许还记得2-4节中的叙述"当进入中断服务子程序执行后,相对应的状态标志位会由单片机自动清除为0,那么此行指令似乎有点多此一举的感觉。其实不然,请读者参考图4-12-4并回忆4-5节中对于弹跳现象的说明。读者可以试着将"CLR EIF"指令删除,然后看看电机的转动圈数有何不同。读者将会发现有时候明明只按下一次中断按钮,可是电机却转了两圈?这可不是程序写错了,而是弹跳造成的现象。如前所述,当进入外部中断服务子程序时,EIF标志位会由单片机自动清除为0,但是T_1以后的弹跳现象又使HT46xx的中断机制误以为又有中断要求产生,所以又再一次将EIF位设置为1,当第一次返回主程序后,CPU又再次检查到EIF=1,所以又再一次执行中断服务子程序。

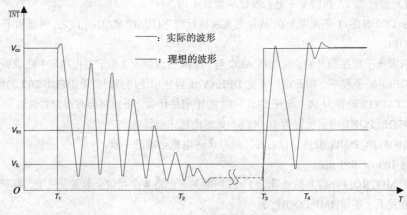

图4-12-4 弹跳现象

原程序执行时,如果读者按下按键不放,则依常理判断一旦电机转完一圈返回主程序之后,因为 $\overline{INT}$ (PA5)引脚仍维持低电位,所以中断服务子程序应该一直重复被执行(也就是说电机应该转个不停)。可是事实却不然,电机仍是只有转完一圈后就直接返回主程序。这说明了HT46xx的中断机制是属于负边缘触发(Negative-Edge Trigger)而非准位触发(Level Trigger),就是说 $\overline{INT}$ 引脚上一定要有1→0的状态变化,中断机制才会检测得到中断的发生。读者可以试试看按键一直按下不放的情形,来验证此特性。或许读者会发现,"偶而"在放开按键时电机又多转了一圈(即中断服务子程序又多执行一次),这种状况请读者参考图4-12-4,这是按键放开时的弹跳现象所造成的,读者应该明白了吧。而所谓的"偶而"是指电机已经转完一圈后,回到主程序继续执行显示动作的时候。如果是在电机转动期间就将按钮放开,弹跳现象所造成的中断事件虽会导致EIF位再度被设置为1,但是中断服务子程序中的"CLR EIF"指令又将其清除,所以电机不会发生连续转两圈的情形。

4-12-5 动动脑 + 动动手

📖 将第84行的指令删除,重新编译并执行程序,会有什么现象?为什么呢?

4-13 PWM接口控制实验

4-13-1 目的

本实验将利用脉宽调制(PWM)的技巧及人类视觉暂留的特性，让LED呈现不同的亮度。

4-13-2 学习重点

通过本实验，读者对于HT46xx脉冲宽度调变的技巧应透彻了解，并能利用视觉暂留的特性达到显示的效果。

4-13-3 电路图

如图4-13-1所示，当连接到LED输出引脚的输出信号为High时，则LED为顺偏(Forward Bias)，所以LED会亮；反之，当连接至LED输出引脚的输出信号为Low时，则并没有电流流过LED，所以LED不会亮。脉宽调制是利用工作周期(Duty Cycle)的改变来达到调制的目的，通常用来控制直流电机的转速、电机转向等，而本实验将利用它来控制LED的亮度。

参考图4-13-2的A、B及C的波形，如果将这些波形由HT46x23的I/O端口输出，会得到什么结果呢？毫无疑问，A会使得LED导通而始终发亮，但是B、C两组波形"可能"会造成LED忽亮忽灭。若能控制这两个波形的周期，使其小于人类视觉暂留的时间，这样就不会看到LED的闪烁。以功率的角度来探讨B、C两组波形，由于其频率相同而单位时间内B所提供的功率大于C，因此B波形会使LED较C波形要来得亮；而A波形使LED永远导通，所以LED最亮。

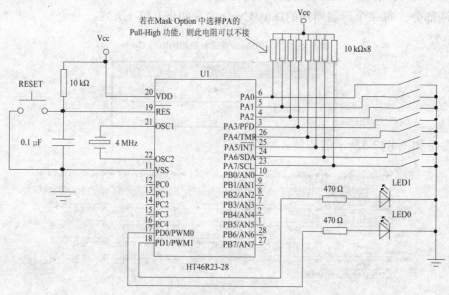

图4-13-1　PWM控制LED亮度电路

163

在未提供脉宽调制功能的单片机上，如果要实现PWM输出的话，通常必须以软件来完成。一般以软件来产生PWM输出的缺点是：PWM周期无法太高。而HT46x23提供了两个通道的PWM输出，必须在Options中选用此功能，此时PWM信号将由PD0、PD1输出，PWM计数器的时钟脉冲来源为f_{SYS}，而其Duty Cycle及调制周期(Modulation Cycle)分别由PWM0与PWM1寄存器控制。

当于Options选用PWM的功能后，若设置PD0(PD1)为1，PWM信号即开始由PD0(PD1)输出；若设置为0，则PWM信号停止输出，并迫使PD0(PD1)引脚维持在低电位状态。PWM的输出有两种模式：(6+2)Mode与(7+1)Mode，必须在Options中予以指定，以下说明这两种模式的不同。

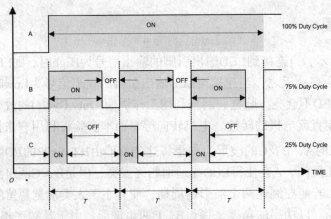

图4-13-2　PWM波形示意图

(6+2)Mode：此时PWM周期被分割成4个调制周期(Modulation Cycle 0~3)，每个调制周期为64个计数时钟脉冲，而PWM寄存器的控制位区分为DC(PWM7~PWM2)与AC(PWM1~PWM0)两部分，每一个调制周期的Duty Cycle关系如表4-13-1所列。

表4-13-1　(6+2)Mode各调制周期的Duty Cycle

参　数	AC(0~3)	Duty Cycle
Modulation Cycle i i=0~3	i<AC	$\frac{DC+1}{64}$
	I≥AC	$\frac{DC}{64}$

请参考如图4-13-3所示的图例。

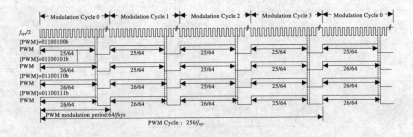

图4-13-3　(6+2)PWM Mode图例

(7+1)Mode：此时PWM周期被分割成两个调制周期(Modulation Cycle 0~1)，每个调制周期为128个计数时钟脉冲，而PWM寄存器的控制位区分为DC(PWM.7~ PWM.1)与AC(PWM.0)两部分，而每一个调制周期的Duty Cycle关系如表4-13-2所列。

表4-13-2 (7+1)Mode各调制周期的Duty Cycle

参　数	AC(0~1)	Duty Cycle
Modulation Cycle i i=0~1	i<AC	$\frac{DC+1}{128}$
	i≥AC	$\frac{DC}{128}$

请参考如图4-13-4所示的图例。

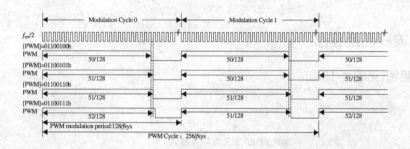

图4-13-4 (7+1)PWM Mode图例

本实验拟以(7+1) Mode的PWM输出来控制LED的亮度，而且可通过指拨开关的设置来改变其亮度。请读者务必记得在Options中选择相关的功能，如图4-13-5所示。

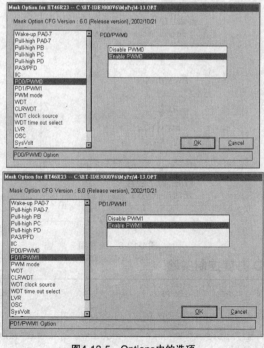

图4-13-5 Options中的选项

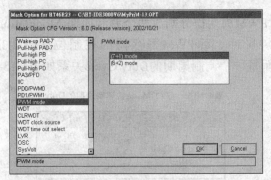

图4-13-5　Options中的选项（续）

4-13-4　流程图及程序

1. 流程图(见图 4-13-6)

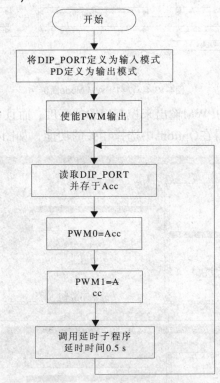

图4-13-6　流程图

2. 程序 4-13　LED 亮度控制实验

```
1    ;PROGRAM : 4-13.ASM (4-13.PRJ)                  2002.DEC.30.
2    ;FUNCTION: PWM LIGHT CONTROL DEMO PROGRAM         BY STEVEN
3    ;NOTE    : PWM1 AND PMW0 SHOULD BE ENABLED, SEL 7+1 MODE (MASK OPTION).
4    #INCLUDE    HT46R23.INC
```

```
 5                  .CHIPHT46R23
 6  ;------------------------------------------------------------------
 7  MY_DATA     .SECTION    'DATA'          ;== DATA SECTION ==
 8  DEL1        DB      ?                   ;DELAY LOOP COUNT 1
 9  DEL2        DB      ?                   ;DELAY LOOP COUNT 2
10  DEL3        DB      ?                   ;DELAY LOOP COUNT 3
11  ;------------------------------------------------------------------
12  DIP_PORT    EQU     PA                  ;DEFINE DIP PORT
13  DIP_PORTC   EQU     PAC                 ;DEFINE DIP PORT CONTROL REG.
14
15  MY_CODE     .SECTION  AT 0 'CODE'       ;== PROGRAM SECTION ==
16          ORG         00H                 ;HT-46RXX RESET VECTOR
17          SET         DIP_PORTC           ;CONFIG DIP_PORT AS INPUT MODE
18          CLR         PDC                 ;CONFIG PD AS OUTPUT PWM MODE
19          SET         PD                  ;ENABLE PWM OUTPUT
20  MAIN:
21          MOV         A,DIP_PORT          ;READ DIP SWITCH
22          MOV         PWM0,A              ;SET PWM0 DUTY AND MODE
23          CPLA        ACC                 ;COMPLEMENT A
24          MOV         PWM1,A              ;SET PWM1 DUTY AND MODE
25          CALL        DELAY               ;DELAY 0.5s
26          JMP         MAIN                ;RESTART
27  ***********************************************************************
28  ;              Delay about DEL1*10ms
29  ***********************************************************************
30  DELAY   PROC
31          MOV         A,50
32          MOV         DEL1,A              ;SET DEL1 COUNTER
33  DEL_1:  MOV         A,30
34          MOV         DEL2,A              ;SET DEL2 COUNTER
35  DEL_2:  MOV         A,110
36          MOV         DEL3,A              ;SET DEL3 COUNTER
37  DEL_3:  SDZ         DEL3                ;DEL3 DOWN COUNT
38          JMP         DEL_3
39          SDZ         DEL2                ;DEL2 DOWN COUNT
40          JMP         DEL_2
41          SDZ         DEL1                ;DEL1 DOWN COUNT
```

```
42          JMP     DEL_1
43          RET
44  DELAY   ENDP
45          END
```

程序说明

8~10   依序定义变量地址。

12~13   定义 DIP_PORT 为 PA。

16   声明存储器地址由 000h 开始(HT46xx Reset Vector)。

17~18   定义 DIP_PORT(PA)为输入模式、PD(PWM0、PWM1)为输出模式。

19   设置 PD=11，即开始输出 PWM 脉冲。

21   读取 DIP_PORT 并存于 Acc 寄存器。

22~24   将 Acc 与 Acc 的补数值分别放入 PWM0 与 PWM1 寄存器，用以控制输出脉冲的 Duty Cycle。

25   调用 DELAY 子程序，延时 0.5 s。

26   重新执行程序(JMP MAIN)。

30~44   DELAY 子程序，延时时间的计算请参考实验 4-1。

本实验是以HT46x23的PWM0与PWM1输出接口控制LED的亮度，请读者务必记得在Options中选择此项功能，由于两组Duty Cycle的控制寄存器(PWM0、PWM1)互成补数关系，因此读者将发现LED0与LED1的亮度是一个较暗、一个较亮；而在Duty Cycle=50%时，两个LED的亮度相同。

4-13-5 动动脑 + 动动手

- 试着改写本实验的程序，使 LED 亮度的变化更快速。
- 将 PWM 输出模式改为(6+2)Mode，重做本实验。

4-14 WDT控制实验

4-14-1 目　的

利用HT46x23的看门狗定时器(Watch Dog Timer)功能，控制七段显示器重复显示0~9的数值。

4-14-2 学习重点

通过本实验，读者应熟悉HT46x23的WDT的工作原理及其控制方式。

4-14-3 电路图（见图4-14-1）

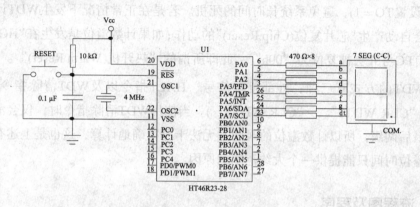

图4-14-1　WDT控制七段显示器电路

看门狗定时器(Watch Dog Timer；WDT)是目前绝大多数单片机上都会提供的设备，其主要功能是避免因不可预期的因素而造成系统长时间的瘫痪。HT46x23的WDT计数时钟脉冲可以是内置的自振式*RC*振荡器(Free-Running On-Chip RC Oscillator)，或是指令周期时钟脉冲($f_{SYS}/4$)，请参考图4-14-2。

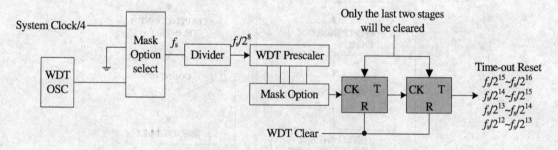

图4-14-2　HT46xx的WDT内部结构

只要在Options中使能及设置看门狗定时器的相关选项，一旦接上电源后WDT即开始计数。当WDT计数时钟脉冲选用内部*RC*自振式振荡器(WDT OSC)时，计数的时钟脉冲周期约为65μs(工作于5 V时)，其计数溢位的最短时间约为300~600 ms；最长时间约为2.3~4.7 s，计数溢位时间的控制必须由Options中的预除比例(WDT Prescaler)选项加以设置。当选用内部*RC*自振式振荡器作为WDT计数时钟脉冲时，即使单片机已进入"省电模式(HALT Mode)"，WDT仍会继续计数。由于计数时钟脉冲是由内部*RC*电路产生，其容易受温度及工作电压的影响，使用时须特别注意。

当WDT计数时钟脉冲选用指令周期时钟脉冲($f_{SYS}/4$)时,计数的动作与上述相同,计数溢位的最短时间约为$(f_S/2^{12})^{-1}$~$(f_S/2^{13})^{-1}$ s;最长时间约为$(f_S/2^{15})^{-1}$~$(f_S/2^{16})^{-1}$ s。当单片机进入"HOLD Mode"时会切断系统的时钟脉冲(f_{SYS}),因此WDT的计数动作也将随之停止。

如果WDT产生计数溢位,HT46xx会自动复位(Reset)回到初始状态,让程序从头开始执行(此时会设置TO=1),避免系统长时间的死机。若是在正常情况下发生WDT计数溢位,此时系统会自动产生"芯片复位(Chip Reset)"的动作;如果计数溢位是发生在"HOLD Mode"时,则只有PC与SP会被复位为"00h",此即所谓的"热开机(Warm Reset)"。

清除WDT的方式有三种:外部复位信号、HALT指令以及WDT清除指令(如"CLR WDT"、"CLR WDT1"、"CLR WDT2")。当使用WDT清除指令时,仅会清除看门狗计数器的最后两级,所以计数溢位的时间并无法十分精确地计算,这也是上述有关最短、最长计数溢位时间只能提供一个大约范围的原因。

4-14-4 流程图及程序

1. 流程图(见图 4-14-3)

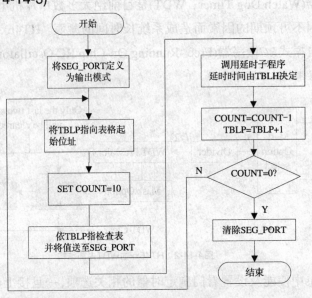

图4-14-3 流程图

执行本程序时,请注意Options中的选项需设置(见图4-14-4)如下:

```
WDT: Enable;
CLRWDT: One clear instruction;
WDT clock source: WDTOSC;
WDT time out select: WDT clock source/16384.
```

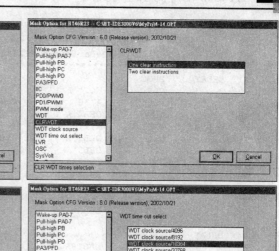
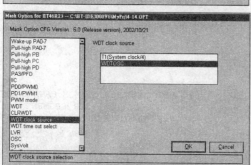
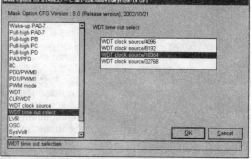

图4-14-4 Option中的选项设置

2. 程序 4-14-1 WDT 控制实验

```
1    ;PROGRAM : 4-14.ASM  (4-14.PTJ)                  2002.DEC.30.
2    ;FUNCTION: WDT DEMO PROGRAM                       BY STEVEN
3    ; NOTE  : MASK OPTION
4    ;              WDT                 : ENABLE
5    ;              CLRWDT              : ONE CLEAR INSTRUCTION
6    ;              WDT CLOCK SOURCE: WDT OSC
7    ;              WDT TIMEOUT SEL.: WDT CLOCK OURCE/16384
8    #INCLUDE   HT46R23.INC
9               .CHIP HT46R23
10   ;--------------------------------------------------------------
11   MY_DATA    .SECTION    'DATA'         ;== DATA SECTION ==
12   DEL1       DB      ?                  ;DELAY LOOP COUNT 1
13   DEL2       DB      ?                  ;DELAY LOOP COUNT 2
14   DEL3       DB      ?                  ;DELAY LOOP COUNT 3
15   COUNT      DB      ?
16   ;--------------------------------------------------------------
17   SEG_PORT       EQU     PA             ;DEFINE 7-SEG PORT
18   SEG_PORTC      EQU     PAC            ;DEFINE 7-SEG PORT CONTROL REG.
19
20   MY_CODE    .SECTION  AT 0 'CODE'      ;== PROGRAM SECTION ==
21              ORG       00H              ;HT-46RXX RESET VECTOR
22              CLR       SEG_PORTC        ;CONFIG SEG_PORT AS OUTPUT MODE
23   MAIN:
24              MOV       A,TAB_7_SEG
25              MOV       TBLP,A           ;INITIAL POINTER START ADDRESS
26              MOV       A,10
27              MOV       COUNT,A          ;SET DATA COUNT OF TAB_7_SEG
28   LOOP:
29              TABRDL    SEG_PORT         ;READ TABLE AND SEND TO SEG_PORT
30              MOV       A,50             ;GET DELAY FACTOR
31              MOV       DEL1,A           ;SET DELAY FACTOR
32              CALL      DELAY            ;DELAY TBLH*10mS
33              INC       TBLP             ;UPDATE TABLE POINTER
34              SDZ       COUNT            ;COUNT DOWN BY 1
35              JMP       LOOP             ;JUMP IF NOT TABLE END
36              CLR       SEG_PORT         ;TURN OFF 7-SEG LED
```

```
37      STOP:       JMP         STOP                    ;IDEL LOOP
38      ;***************************************************************
39      ;                       Delay about DEL1*10ms
40      ;***************************************************************
41      DELAY       PROC
42                  CLR         WDT                     ;MASK THIS LINE TO SEE DIFFERENCE
43                  MOV         A,30
44                  MOV         DEL2,A                  ;SET DEL2 COUNTER
45      DEL_2:      MOV         A,110
46                  MOV         DEL3,A                  ;SET DEL3 COUNTER
47      DEL_3:      SDZ         DEL3                    ;DEL3 DOWN COUNT
48                  JMP         DEL_3
49                  SDZ         DEL2                    ;DEL2 DOWN COUNT
50                  JMP         DEL_2
51                  SDZ         DEL1                    ;DEL1 DOWN COUNT
52                  JMP         DELAY
53                  RET
54      DELAY       ENDP
55                  ORG         LASTPAGE
56      TAB_7_SEG:
57                  DC          000000000111111B        ;7 SEG CODE FOR COMMOM CATHOD LED
58                  DC          000000000000110B        ;CONNECT BIT 0~7 FOR SEGMENT A-H
59                  DC          000000010011011B        ;LOW-BYTE FOR 7 SEG DISPLAY CODE
60                  DC          000000001001111B        ;HIGH 7 BITs FOR DELAY FACTOR
61                  DC          000000001100110B
62                  DC          000000001101101B
63                  DC          000000001111101B
64                  DC          000000000000111B
65                  DC          000000001111111B
66                  DC          000000001100111B
67                  END
```

程序说明

12~15   依序定义变量地址。

17~18   定义 SEG_PORT 为 PA。

21   声明存储器地址由 000h 开始(HT46xx Reset Vector)。

22   定义 SEG_PORT 为输出模式。

18~20   清除索引值 INDEX 与设置 COUNT 为 10。

24~25   将 TBLP 指向七段显示器显示码的起始地址。

26~27   设置 COUNT 为 10。

29   通过查表将欲显示在七段显示器上的数值取出,并送到 SEG_PORT 显示。

30~32   延时 0.5 s。

33   将 TBLP 加 1,指向下一个七段显示器显示码。

34~35   判断 COUNT−1 是否为 0,若不为 0 则显示下一个数值。

36~37   关闭七段显示器,进入无限循环状态(STOP:JMP STOP)。目的是为了观察 WDT 的超时复位功能,WDT 计数溢位后,会自动进行 Reset 的动作。

41~54   DELAY 子程序,请特别留意第 42 行的"CLR WDT"指令。

56~66   七段显示码数据建表区。

程序本身相当容易了解,执行时是以0.5 s一次的速度将0~9依序显示在七段显示器上,然后在清除七段显示器之后进入"STOP:JMP STOP"的无限循环。可是约2~3 s之后,HT46x23却自动重新开始执行程序。这主要是因为Options看门狗定时器的功能有效,且选

择WDT time out select＝WDT Clock source/16 384，所以当程序执行到无穷循环时，WDT因未能及时被清除而导致系统的复位。"CLR WDT"指令应该放在何处，其实只要确定在WDT定时溢位之前能够将其清除就可以了。在本实验将"CLR WDT"指令插在延时时间为0.5 s的DELAY子程序(第42行)，因为笔者确定在0~9的显示过程中，至少每隔0.5 s就会调用一次DELAY子程序。所以在正常情况下，是不会发生WDT超时复位的情形。读者可以试着将"CLR WDT"指令由DELAY子程序中删除，看看会有什么结果？此时七段显示器大约只能重复显示0~5而已。为什么呢？读者应该有办法回答了吧。

在Options中有一个选项用来选择看门狗的清除次数，当选用一次清除时，只要执行"CLR WDT"即可达到清除WDT的目的；但当选择两次清除时，必须"CLR WDT1"与"CLR WDT2"两个指令都被执行后，方可达到将WDT清除的效果，如此可以再降低系统跳至无限循环导致死机的机会。

4-14-5 动动脑＋动动手

- 更改"WDT time out select"选项，看看程序执行起来会有什么不一样的结果？并试着去分析其原因。
- 一般在进入"HALT Mode"(省电模式)后，为了能够真正达到省电的效果，都不希望因为WDT定时溢位的关系一直将系统唤醒，所以Options中"WDT clock source"的选项会选用T_1(System clock/4)。试着更改"WDT clock source"＝T_1(System clock/4)，看看程序执行起来会有什么不一样的结果？并试着去分析其原因。

4-15 "HALT Mode"省电模式控制实验

4-15-1 目　的

本实验以按键控制七段显示器的显示；每按一次按键(SW1)，七段显示器的显示就加1，否则显示器上的数值维持不变。但若超过预定的时间未按按键，则HT46x23即进入"HALT Mode"省电模式；待用户按下RESET、INT或SW1按键之后，则重新恢复按键检测及计数的动作。

4-15-2 学习重点

通过本实验，读者了解HT46xx省电模式的控制及运用，以及各种不同的唤醒方式(Wake-Up)。

4-15-3 电路图（见图4-15-1）

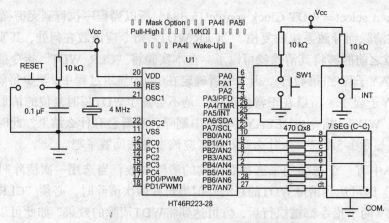

图4-15-1 "HALT Mode"省电模式控制电路

在许多单片机的应用场合，为了方便携带，电池通常是唯一的电源供应来源，如电视遥控器、汽车遥控器、随身听、手机及互动性的玩具等。如果所设计的产品无法达到省电的要求，而必须经常更换电池，必然无法受到消费大众的认同。为了达到省电的目的，HT46xx提供了"HALT Mode(省电模式)"功能，一旦系统进入省电模式，HT46xx单片机大约只消耗数μA的电流。本实验将深入探讨HT46x23的省电模式及唤醒方式。

HT46x23执行HALT指令后，系统随即进入省电模式，单片机并将执行以下的工作：
◇ 关闭系统时钟脉冲(但若WDT计数时钟脉冲选用内部RC自振式振荡器(WDT OSC)时，WDT仍会继续计数)；
◇ 所有内部数据存储器(RAM)的内容维持不变；
◇ 清除WDT，并重新开始计数(若WDT计数时钟脉冲选用WDT OSC)；
◇ 清除TO，并设置PD＝1。

进入省电模式后，I/O端口会保持在执行HALT指令之前的状态。所以若要达到更佳的省电效果，最好在进入省电模式之前也一并将外围的负载元件一起关闭，以减少电流的损耗。

让HT46x23由省电模式重新恢复工作的方式有4种，分别是：
① 外部硬件复位(External Reset)：即在 $\overline{RES}$ 引脚输入 Low 电位，至于复位后的寄存器状态，请参考 2-11 节中的说明。
② WDT 定时溢位复位：注意要使用此种方式唤醒时，必须在 Options 选择 WDT OSC 作为 WDT 计数时钟脉冲。

上述的两种方式都是以复位的方式来唤醒单片机，不同的是WDT定时溢位所产生的复位，是所谓的"热复位(Warm Reset)"，此时仅有SP及PC寄存器被重新设置为00h，单片机的内部电路均维持原来的状态。

③ 中断唤醒：如果在省电模式中，有中断请求发生，致使状态标志位(EIF、TF、ADF、HIF)由 0 变为 1，将使单片机脱离省电模式。此时，若相对的中断使能且栈寄存器还有空间存放返回地址，CPU 将跳至对应的中断向量去提取指令执行；否则 CPU 将执行 HALT 的下一行指令。
④ PA 有 1→0 的准位变化发生：使用此种方式唤醒的前提是在 Options 中必须选择 Wake-Up 的功能，唤醒后的 CPU 将执行 HALT 的下一行指令。本程序利用 PA4

第4章 基础实验篇

的唤醒功能，请读者务必于 Options 中勾选，如图 4-15-2 所示。

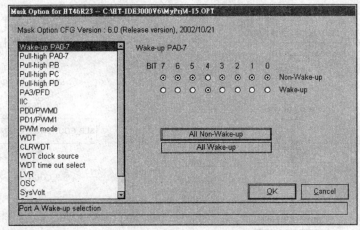

图4-15-2　勾选PA4的唤醒功能

如果需要区分省电模式下的单片机究竟是由哪一种方式唤醒的话，用户可以依据TO与PDF位来加以分辨，请参考表4-15-1和表4-15-2。

表4-15-1　影响TO与PDF位的事件

事　件	TO	PDF
Power-Up	0	0
WDT计数溢位	0	u[(1)]
HALT指令	0	1
"CLR WDT"指令[(2)]	0	0
"CLR WDT1"与"CLR WDT2"指令[(3)]	0	0

表4-15-2　Reset后，TO与PDF位的状态

TO	PDF	RESET起因
0	0	$\overline{\text{RES}}$ during Power-up
0	0	正常运作下的 $\overline{\text{RES}}$ 复位
0	1	省电模式下的 $\overline{\text{RES}}$ 复位
1	0	正常运作下的WDT计数溢位复位
1	1	省电模式下的WDT计数溢位复位

注：(1)u 表示维持原来的状态。
　　(2)当选择 WDT 为一次清除模式时。
　　(3)当选择 WDT 为两次清除模式，且两个指令都执行过后。

4-15-4　流程图及程序

1. 流程图(见图 4-15-3)

2. 程序 4-15　"HALT Mode"省电模式实验

```
1       ;PROGRAM : 4-15.ASM (4-15.PRJ)                  2002.DEC.30.
2       ;FUNCTION: HALT FUNCTION DEMO PROGRAM            BY STEVEN
3       ; NOTE   : MASK OPTION
4       ;          ENABLE PA.0 WAKE-UP FUNCTION
5       #INCLUDE   HT46R23.INC
6                  .CHIPHT46R23
7       ;--------------------------------------------------------------
8       MY_DATA    .SECTION   'DATA'        ;== DATA SECTION ==
9       DEL1       DB         ?             ;DELAY LOOP COUNT 1
10      DEL2       DB         ?             ;DELAY LOOP COUNT 2
11      DEL3       DB         ?             ;DELAY LOOP COUNT 3
12      COUNT      DB         ?             ;COUNTER
13      SLP_COUNT  DB         ?             ;SLEEP COUNT
14      ;--------------------------------------------------------------
15      SEG_PORT   EQU        PB            ;DEFINE 7-SEG PORT
```

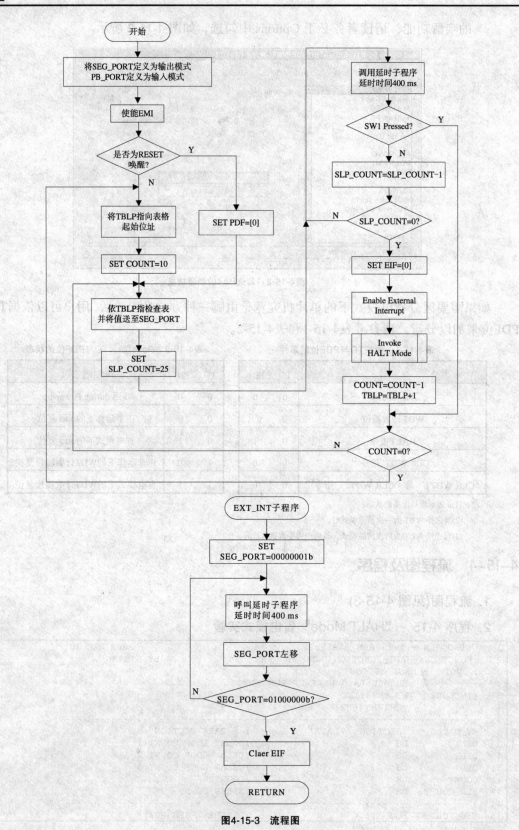

图4-15-3 流程图

```
16      SEG_PORTC       EQU     PBC                     ;DEFINE 7-SEG PORT CONTROL REG.
17      PB_PORT         EQU     PA.4                    ;DEFINE PUSH BUTTON PORT
18      PB_PORTC        EQU     PAC.4                   ;DEFINE PUSH BUTTON PORT CONTROL REG.
19
20      MY_CODE         .SECTION  AT 0 'CODE'           ;== PROGRAM SECTION ==
21              ORG     00H                             ;HT-46RXX RESET VECTOR
22              JMP     START
23              ORG     04H                             ;HT-46RXX EXT. INTERRUPT VECTOR
24              JMP     EXT_INT
25      START:
26              CLR     SEG_PORTC                       ;CONFIG   SEG_PORT AS OUTPUT MODE
27              SET     PB_PORTC                        ;CONFIG PB_PORT AS INPUT MODE
28              SET     EMI                             ;ENABLE GLOBAL INTERRUPT
29              SNZ     PDF                             ;RESET FROM POWER DOWN?
30              JMP     MAIN                            ;NO.
31              CLR     WDT                             ;YES, CLEAR PDF.
32              JMP     LOOP                            ;JUMP TO DISPLAY LAST DIGIT.
33      MAIN:
34              MOV     A,TAB_7_SEG
35              MOV     TBLP,A                          ;INITIAL POINTER START ADDRESS
36              MOV     A,10
37              MOV     COUNT,A                         ;SET DATA COUNT OF TAB_7_SEG
38      LOOP:
39              TABRDL  SEG_PORT                        ;READ TABLE AND SEND TO SEG_PORT
40              MOV     A,25
41              MOV     SLP_COUNT,A                     ;INITIAL SLEEP COUNT
42      WAIT_KEY:
43              CALL    DELAY                           ;DELAY 0.4 SEC
44              SNZ     PB_PORT                         ;KEY PRESSED?
45              JMP     LOOP_1                          ;YES, JUMP TO LOOP_1
46              SDZ     SLP_COUNT                       ;SLP_COUNT OVER? (25*0.4 SEC=10 SEC.)
47              JMP     WAIT_KEY                        ;NO.
48              CLR     SEG_PORT                        ;YES, CLEAR DISPLAY
49              MOV     A,00000000B
50              CLR     EIF                             ;DISABLE EXT INT DURING NORMAL OPERATION
51              SET     EEI                             ;ENABLE EXTERNAL INTERRUPT
52              HALT                                    ;GOTO SLEEP.
53              CLR     EEI                             ;DISABLE EXTERNAL INTERRUPT
54      LOOP_1:
55              INC     TBLP                            ;UPDATE TABLE POINTER
56              SDZ     COUNT                           ;COUNT DOWN BY 1
57              JMP     LOOP                            ;JUMP IF NOT TABLE END
58              JMP     MAIN                            ;RESTART
59      ;****************************************************************************
60      ;               EXTERNAL INTERRUPT SERVICE ROUTINE
61      ;****************************************************************************
62      EXT_INT:
63              SET     SEG_PORT.0                      ;TURN SEG_PORT.0 LED ON
64      SHIFT:
65              CALL    DELAY                           ;DELAY 0.2 SEC
66              RL      SEG_PORT                        ;SHIFT THE LED
67              SNZ     SEG_PORT.6                      ;SEG_PORT.6=1?
68              JMP     SHIFT                           ;NO, SHIFT AGAIN.
69              CLR     EIF                             ;TO AVOID THE INT. CAUSED BY BOUNCING
70              RETI                                    ;YES, RETURN
71      ;****************************************************************************
72      ;                       Delay about DEL1(40)*10ms
73      ;****************************************************************************
74      DELAY   PROC
75              MOV     A,40
76              MOV     DEL1,A                          ;SET DEL1 COUNTER
77      DEL_1:  MOV     A,30
78              MOV     DEL2,A                          ;SET DEL2 COUNTER
79      DEL_2:  MOV     A,110
80              MOV     DEL3,A                          ;SET DEL3 COUNTER
81      DEL_3:  SDZ     DEL3                            ;DEL3 DOWN COUNT
82              JMP     DEL_3
```

```
83              SDZ        DEL2              ;DEL2 DOWN COUNT
84              JMP        DEL_2
85              SDZ        DEL1              ;DEL1 DOWN COUNT
86              JMP        DEL_1
87              RET
88   DELAY      ENDP
89              ORG        LASTPAGE
90   TAB_7_SEG:
91              DC         000000000111111B  ;7 SEG CODE FOR COMMOM CATHOD LED
92              DC         000000000000110B  ;CONNECT BIT 0~7 FOR SEGMENT A~H
93              DC         000000001011011B  ;LOW-BYTE FOR 7 SEG DISPLAY CODE
94              DC         000000001001111B
95              DC         000000001100110B
96              DC         000000001101101B
97              DC         000000001111101B
98              DC         000000000000111B
99              DC         000000001111111B
100             DC         000000001100111B
101             END
```

程序说明

9~13   依序定义变量地址。

15~16   定义 SEG_PORT 为 PB。

17~18   定义 PB_PORT 为 PA.4。

21   声明存储器地址由 000h 开始(HT46x23 Reset Vector)。

23   声明存储器地址由 004h 开始(HT46x23 External Interrupt Vector)。

26~27   分别定义 SEG_PORT 为输出模式、PB_PORT 为输入模式。

28   中断总开关(EMI)有效。虽然在此将 EMI 位设置为 1，但是因为其余的中断控制位(EEI、ETI、EADI、EHI)在系统复位过程中被全部清除为 0(禁止)，因此仍不会产生中断的情况。

29~32   判断状态寄存器中的 "PDF(Power Down Flag)" 位是否为 1。若成立，则表示目前 CPU 的状态是由省电模式中以硬件 RESET 方式将其唤醒，回到一般状态，显示进入省电模式前的数值。至于第 31 行的 "CLR WDT" 指令的主要目的，并非在于清除 WDT 的内容(因为本实验在 Options 选项中并未将 WDT 使能)，而是要将 PDF 位清除为 0，并继续睡眠模式前的工作("JMP LOOP")。反之，表示 CPU 为一般状态下的复位，程序重新执行("JMP MAIN")。

34~35   将 TBLP 指向七段显示器显示码的起始地址。

36~37   设置 COUNT=10。

39   通过查表将欲显示的数值取出，并送到 SEG_PORT 显示。

40~41   设置计数器 SLP_COUNT 为 25，目的是在经过 1 s(25×400 ms)后，若仍没有按键(SW1)输入则进入省电模式。

43   调用 DELAY 子程序，延时 0.4 s。

44~45   检查是否有按键(SW1)输入。

46~47   判断 SLP_COUNT 是否为 0，若成立，表示已经 10 s 都没有按键(SW1)输入，准备进入省电模式；反之，则表示还未到达 10 s，继续计数 SLP_COUNT。

48~51   首先将七段显示器关闭，以节省电源损耗；接着清除 EIF 标志位，以消除还未进入省电模式前，INT 按键被按的记录；在使能外部中断控制位后随即进入省电模式。

52   被唤醒时先使能外部中断。如果是以 Wake-Up 功能唤醒时(按下 SW1)，即执行此行指令；如果是以中断唤醒(按下 INT)，会先执行外部中断服务子程序后才执行此行指令。

54   TPLP 加 1，指向下一笔七段显示器显示码的地址。

55~57   判断 COUNT−1 是否为 0，成立则表示已经显示至最后一个数值(9)，重新由 0 开始显示；反之，则表示还未显示完毕，继续显示下个数值。

行号	说明
61~69	外部中断服务子程序。当 HT46x23 进入进入省电模式后,若用户按下 INT 按键(即以中断方式唤醒),才会执行此子程序。执行时是将七段显示器的 a~f 段循序点亮,每段约点亮 400 ms 的时间。请注意第 68 行"CLR EIF"指令,其主要目的,是防止 INT 按键放开时的弹跳现象导致中断服务子程序被重复执行的情形。
73~87	DELAY 子程序,延时时间的计算请参考 4-1 节。
86~104	七段显示码数据建表区。

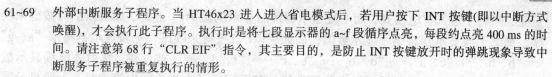

本程序是以RESET硬件复位、外部中断以及PA4的Wake-Up功能来唤醒CPU,所以请读者务必记得在Options中把看门狗定时器关闭。

程序中以PDF位来区分是否为开机复位(Power-On)RESET,如果是开机复位则PDF=0;而如果是由省电模式中被唤醒的复位,则PDF=1(这主要是因为执行HALT指令时,系统会将WDT清除为0、PDF设置为1)。SLP_COUNT主要是用来控制何时进入省电模式,范例程序中每隔400 ms检查一次按键(SW1),如果25次之后仍未按下按键(大约是10 s),则进入省电模式。为了真正达到省电的目的,在系统进入省电模式之前先将七段显示器关闭,以减少电流的损耗。

本实验并未启动看门狗定时器,读者或许会质疑:"为什么程序中会用到'CLR WDT'指令呢?",这主要是因为程序中以PDF位的值来决定是否要将显示值归0。如果是第一次开机执行(Power-On Reset)PDF位=0,则将显示值归0;如果是由省电模式中被唤醒的复位(PDF=1),则显示进入省电模式前的值(显示值不归0)。

但是,若显示值已经增加到9时,仍旧是要回到0开始显示,所以如果是由省电模式中被唤醒,必须将PDF位清除为0,程序才得以正常地运行。而"CLR PDF"指令对PDF位并无作用,所以程序中"CLR WDT"指令的主要目的是用来将PDF位重设为0,并非是清除WDT。

程序执行时,七段显示器显示值会随着SW1按下的次数而递增,如果一直按着SW1不放,显示值大约每0.4 s递增一次;如果在10 s内没有按下SW1按键,系统就会进入省电模式,此时七段显示器不再显示任何数值。请读者细心体验一下不同的唤醒方式(假设进入省电模式的最后显示值为8):

◇ 按下 RESET 按键:此时七段显示器将显示进入省电模式前所显示的数值(8),如果是在还未进入省电模式时按下 RESET 按键,则七段显示器将显示 0。
◇ 按下 SW1 按键:此时七段显示器将显示进入省电模式前所显示的下一笔数值(9)。
◇ 按下 INT 按键:此时先循序点亮七段显示器的 a~f 段,然后接着显示进入省电模式前所显示的下一笔数值(9),在还未进入省电模式时,INT 按键是不具任何效果的。

请读者观察第49行"CLR EIF"指令,要注意的是在第50行外部中断的功能才有效,为什么需要在第49行执行清除外部中断标志位的动作呢?首先请读者注意,不管中断是否已经有效,HT46xx的中断机制都会将是否有中断请求发生的事件记录在中断标志位(EIF、TF、ADF或HIF)。

所以第49行将外部中断标志位清除的目的,是为了排除在还未进入省电模式前,用户按下INT按键的事件所造成的非预期中断。请读者将第49行指令删除,然后故意在还未进入省电模式前按一次INT按键,会发现10s内若未按SW1按键,七段显示器的a~f段会先循序点亮，然后才进入省电模式，而七段显示器最后显示的段为g。此时，如果以SW1按键来唤醒，那么七段显示器将继续显示数值；而如果以INT按键来唤醒，七段显示器将依"a、g"→"b、h"→"c、a"→"d、b"→"e、c"→"f、d"顺序被点亮，然后才继续显示数值。至于为什么七段显示器会同时亮两个段，其原因就是在进入省电模式时，七段显示器的g段仍是点亮的状态(PB6=1)。

再者，在没有启动中断的情形下，所有的中断事件都可用来唤醒省电模式中的单片机，读者可以将程序第50行"SET EEI"指令删除，此时仍可以按下INT按键来唤醒HT46x23，但是唤醒后的执行动作却与用SW1来唤醒是完全相同。所以若中断未被有效，以中断方式唤醒时CPU并不会跳到中断向量地址去执行程序，而是直接执行HALT的下一行指令。不过提醒读者，如果在进入省电模式前，中断标志位为0已经被设置为1，其对应的中断事件即丧失唤醒功能。请读者将第49及50行指令一并删除，然后故意在还未进入省电模式前按一次INT按键，会发现在HT46x23进入省电模式后，已无法再用INT按键将其唤醒。所以在未有效外部中断的情况下，第49行"CLR EIF"指令是为了确保在进入省电模式EIF=0，这样方能保证INT按键唤醒功能不致失效。

4-15-5 动动脑 + 动动手

- 省电模式的运用相当重要，像家里电视、音响的遥控器都是很好的实例。当用户选好某家电台之后，可能几分钟甚至几小时都不会再去动遥控器了，因此务必进入省电模式节省电池的电力损耗。但本实验中只检查 10 s 后随即进入省电模式的状态，似乎也太短了些。可否延长其进入的时间呢？
- 延长了进入省电模式的时间之后，读者可以通过数字电表来测量芯片的电流损耗，比较进入省电模式前、后的电流损耗究竟差了多少。
- 在 Options 中选择 WDT 功能，如果用户超过 T 时间仍未唤醒单片机，则单片机也可以自动唤醒。此时 WDT Clock Source 应该如何选择？T 的最大值为多少？

4-16 I²C串行接口控制实验

4-16-1 目　的

本实验利用HT46x23的I²C串行传输接口，完成4×4键盘按键值的显示动作，并辨别由Master端所送出的Device ID与Slave Address是否相符，以决定是否显示按键值。

4-16-2 学习重点

通过本实验，读者对于HT46xx 的I²C串行传输接口控制方式应透彻地了解，尤其对于I²C接口的相关控制寄存器的意义更需了如指掌。

4-16-3 电路图

第4章 基础实验篇

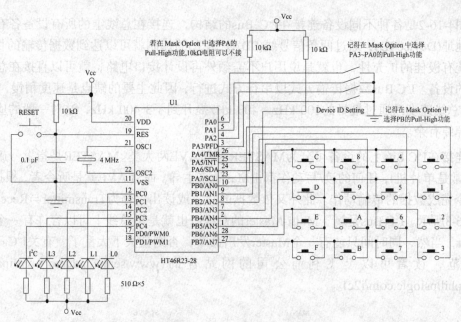

图4-16-1 I^2C串行接口控制电路

I^2C(Inter IC)又可称为I^2C Bus，它是由飞利浦公司研发出来的串行存取方式，目前有许多制造商(请参考表4-16-1)提供此标准接口的相关产品，产品种类高达150余种，足见I^2C Bus确有其独到的魅力。如图4-16-2所示为I^2C Bus连接图例。

表4-16-1 I^2C标准接口的相关产品制造商

制造商	产品
Philips	Audio/Video, Memory、I/O、PLL、E^2PROM、Microprocessors
Xicor	E^2PROM、E2pot (digital controllable potmeter)
Maxim	A/D、D/A、E2pot
Analog Devices	A/D、D/A
Arizona	E^2PROM
Exel	E^2PROM
Catalyst	E^2PROM
Plessey	PLLsynthesizers
National	AD/DA、Audio
Siemens	E^2PROM、PLL Synthesisers Audio/Video Circuits
Atmel	E^2PROM
ISSI	E^2PROM
Holtek	E^2PROM、Microprocessors

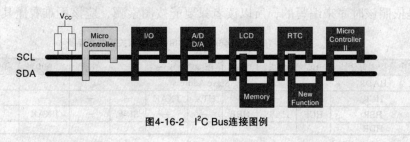

图4-16-2 I^2C Bus连接图例

图4-16-2是各种不同设备连接在I^2C Bus的结构，连接在总线上的所有设备各有自己的设备地址(Device ID)，通过两条信号线(SDA、SCL)的控制就可以达到数据传输的目的，而且系统有极佳的扩充性，也就是说用户不需额外再设计接口电路，就可以直接在总线上加入新的设备。I^2C Bus数据传输方式以串行方式进行，因此主要的缺点是速度稍慢，但是I^2C Bus的总线速度已通过最早的100 kHz、400 kHz提升到了3 400 kHz，这对一般的应用来说已是绰绰有余。

连接在I^2C Bus上的设备可分为MASTER与SLAVE两大类。MASTER是指发号施令的设备(通常是单片机)，任何的读/写动作都由Master来主导。而SLAVE就是听令者，根据Master的命令完成数据的传输动作。此外又依数据的写入或读出，分为Transmitter与Receiver，如图4-16-3所示。I^2C Bus是属于Multi-Master的总线，也就是说总线上可以容许有一个以上的Master，当然，同时间只能有一个Master发出命令，否则就天下大乱了。有关I^2C Bus的详细规范，读者可以上飞利浦公司的网站查询(www.semiconductors.philips.com 或 www.philipslogic.com/i2c)。

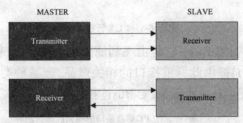

图4-16-3　I^2C Bus上的设备分类

HT46xx也提供Slave I^2C串行传输功能，此时数据传输是通过SCL(Serial Clock；PA6)与SDA(Serial Data；PA7)两条信号线控制完成。HT46x23提供两种数据传输模式：被动式的发送模式(Slave Transmit Mode)与被动式的接收模式(Slave Receive Mode)。所谓的被动(Slave)是指HT46x23的I^2C串行接口无法主动对其他设备提出数据传输的要求，而须由I^2C Bus的控制者(Master)主动存取其数据。因此HT46x23的SCL是输入信号，由Master提供存取所需的参考时钟脉冲；而SDA则须视其传输模式，可能为输入或输出的状态。本实验是以单个HT46x23同时当成Master与Slave的方式，完成数据传输的动作。

HT46x23的I^2C串行接口相关控制寄存器如表4-16-2所列，请读者查阅相关章节参考。这么多的控制位看似复杂，但其实只要根据I^2C Bus时序图(图4-16-4)一一了解，其实也不算太难。整个I^2C串行接口的传输过程，其实就是依照此时序图来完成，而本实验中的许多子程序也是依照该时序来编写的，所以读者只要能按图索骥，应该不难看懂其控制方式。

表4-16-2　HT46x23的I^2C串行接口相关控制寄存器

特殊功能寄存器	Bit7	Bit6	Bit5	Bit4	Bit3	Bit2	Bit1	Bit0	参考章节
HADR	Slave Address								2-8
HCR	HEN	—	—	HTX	TXAK	—	—	—	2-8
HSR	HCF	HASS	HBB	—	—	SRW	—	RXAK	2-8
HDR									2-8

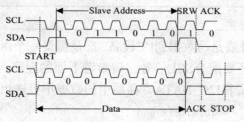

图4-16-4　I²C Bus时序图

首先是START Condition，由Bus Master送出，告诉所有连接在I²C Bus的设备准备监视由Bus Master所送出的Slave Address。START Condition的条件是：当SCL＝1时，如果SDA由1→0，这用HT46xx的位指令就可轻易地达成。请参考以下的程序，其中"CALL DELAY10"的目的是为符合t_{IIC}(I²C Bus Clock Period)最少必须为$64×(f_{SYS})^{-1}$的要求，请参阅原厂的数据手册。

```
1     ;****************************************************************
2     ;              GENERATE I2C START CONDITION
3     ;****************************************************************
4     I2C_START    PROC
5          CLR       SCLC              ;CONFIG SLC AS OUTPUT MODE
6          CLR       SDAC              ;CONFIG SDA AS OUTPUT MODE
7          CLR       SCL               ;SET SCL=0
8          CLR       SDA               ;SET SDA=0
9          CALL      DELAY_10          ;DELAY
10         SET       SCL               ;SET SCL=1
11         SET       SDA               ;SET SDA=1
12         CALL      DELAY_10          ;DELAY
13         CLR       SDA               ;SET SDA=0
14         CALL      DELAY_10          ;DELAY
15         CLR       SCL               ;SET SCL=0
16         CALL      DELAY_10          ;DELAY
17         RET
18    ;I2C_START    ENDP
```

接下来是继续由Bus Master送出7位设备地址(Slave Address)以及所要执行的动作(SRW位)。SRW＝1表示Bus Master要读取该设备的数据，所以被选择到的Bus Slave需进入Transmit Mode准备送数据到总线上。若SRW＝0表示Bus Master要写入数据到该设备，所以被选择到的Bus Slave需进入Receive Mode准备接收总线上的数据。不过要有设备对此Slave Address产生回应，Bus Master才能进行读/写的过程。Bus Maste发送完SRW位之后，会在第9个位时间读取SDA的状态，若SDA＝1，表示该设备存在，Bus Master可以存取该设备的数据；若SDA＝0，表示并没有设备地址与Slave Address相符，Bus Master必须以STOP Condition结束此次传输过程。STOP Condition的条件是：当SCL＝1时，如果SDA由0→1，请参考以下的程序。

```
1     ;****************************************************************
2     ;              GENERATE I2C STOP CONDITION
3     ;****************************************************************
4     I2C_STOP     PROC
5          CLR       SDAC              ;CONFIG SDA AS OUTPUT MODE
6          CLR       SCLC
7          CLR       SCL               ;SET SCL=0
8          CLR       SDA               ;SET SDA=0
9          CALL      DELAY_10          ;DELAY
10         CLR       SDA               ;SET SDA=0
11         CALL      DELAY_10          ;DELAY
12         SET       SCL               ;SET SCL=1
13         CALL      DELAY_10          ;DELAY
14         SET       SDA               ;SET SDA=1
```

15		RET
16	I2C_STOP	ENDP

不管是送出设备地址或是写数据到I^2C设备,都是需要一个由SDA送出所要送出串行数据的过程,这其中当然还要配合SCL信号,请参考下面的程序,流程图如图4-16-5所示,WRITE_BYTE子程序负责将I2C_DATA寄存器中的数据串行送出。

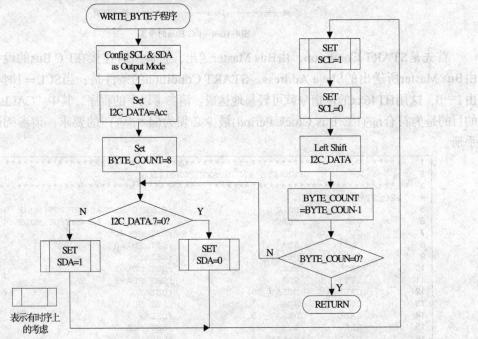

图4-16-5 WRITE_BYTE子程序流程图

```
1    ;************************************************************
2    ;           SERIAL OUT DATA IN Acc VIA SDA & SCL
3    ;************************************************************
4    WRITE_BYTE  PROC
5         CLR      SDAC                ;CONFIG SDA AS OUTPUT MODE
6         CLR      SCLC
7         CLR      SCL                 ;SET SCL=0
8         CLR      SDA                 ;SET SDA=0
9         MOV      I2C_DATA,A          ;RESERVED DATA IN TX BUFFER
10        MOV      A,8                 ;SET 8 BIT COUNTER
11        MOV      BYTE_COUNT,A
12   WNB_0:
13        SZ       I2C_DATA.7          ;IS MSB=0?
14        JMP      WRITE_1             ;NO,JUMP TO WRITE_1
15   WRITE_0:
16        CLR      SDA                 ;SET SDA=0
17        JMP      WNB_1               ;JUMP TO WNB_1
18   WRITE_1:
19        SET      SDA                 ;SET SDA=1
20   WNB_1:
21        CALL     DELAY_10            ;DELAY
22        SET      SCL                 ;SET SCL=1
23        CALL     DELAY_10            ;DELAY
24        CLR      SCL                 ;SET SCL=0
25        CALL     DELAY_10            ;DELAY
26        RL       I2C_DATA            ;SHIFT TX BUFFER
27        SDZ      BYTE_COUNT          ;BYTE_COUNT-1=0?
28        JMP      WNB_0               ;NO, WRITE NEXT BIT
29        RET                          ;YES, RETURN.
30   WRITE_BYTE  ENDP
```

而Bus Master由I^{2}C设备读回数据时，必须在第9位送出ACK或NO ACK信号，以表示是否要继续读取数据。由于本实验一次只读取一个字节，因此第25~30行设置ACK＝1(即送出NO_ACK信号，所以HT46x23单片机的RXAK会为0)。如图4-16-6所示为READ_BYTE子程序流程图。

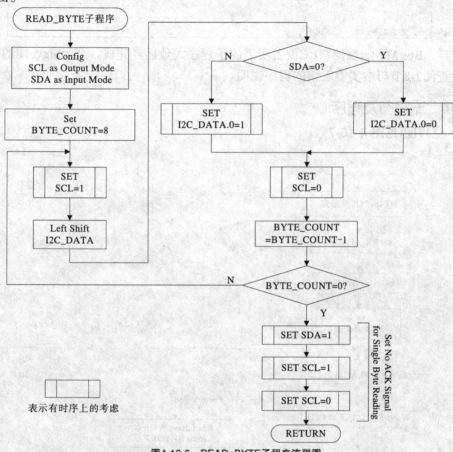

图4-16-6 READ_BYTE子程序流程图

```
1    ;****************************************************************
2    ;              SERIAL IN DATA TO Acc VIA SDA & SCL
3    ;****************************************************************
4    READ_BYTE    PROC
5         SET        SDAC                ;CONFIG SDA AS INTPUT MODE
6         MOV        A,8                 ;SET 8 BIT COUNTER
7         MOV        BYTE_COUNT,A
8    RNB_0:
9         SET        SCL                 ;SET SCL=0
10        CALL       DELAY_10            ;DELAY
11        RLC        I2C_DATA            ;SHIFT RX BUFFER
12        SZ         SDA                 ;SDA = 0?
13        JMP        READ_1              ;NO, JUMP TO READ_1
14   READ_0:
15        CLR        I2C_DATA.0          ;YES, SET LSB=0
16        JMP        RNB_1               ;JUMP TO RNB_1
17   READ_1:
18        SET        I2C_DATA.0          ;SET LSB=1
19   RNB_1:
20        CALL       DELAY_10            ;DELAY
21        CLR        SCL                 ;SET SCL=0
22        CALL       DELAY_10            ;DELAY
```

```
23      SDZ     BYTE_COUNT              ;BYTE_COUNT-1 = 0?
24      JMP     RNB_0                   ;NO, READ NEXT BIT
25      MOV     A,I2C_DATA              ;RELOAD RX DATA TO Acc
26      CLR     SDAC                    ;YES, SEND NO_ACK_SIGNAL
27      CLR     SDA
28      SET     SCL
29      CALL    DELAY_10
30      SET     SDA
31      CLR     SCL
32      RET                             ;YES, RETURN
33 READ_BYTE    ENDP
```

这样，Bus Master端控制I^2C设备所需的子程序大致已经完成，至于Slave端的相关设置请读者查阅2-8节与本实验后续的程序说明。

4-16-4 流程图及程序

1. 流程图(见图 4-16-7)

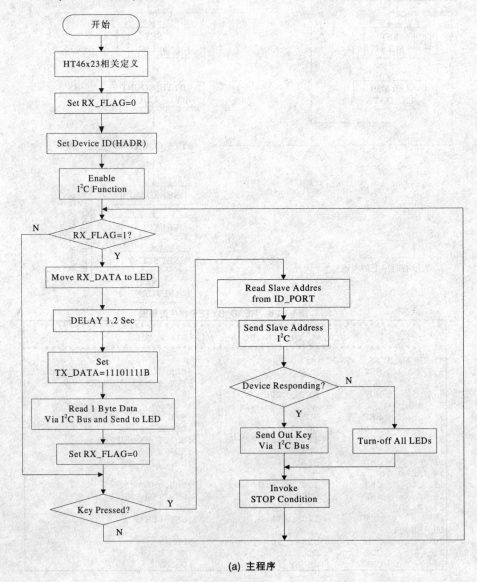

(a) 主程序

第 4 章 基础实验篇

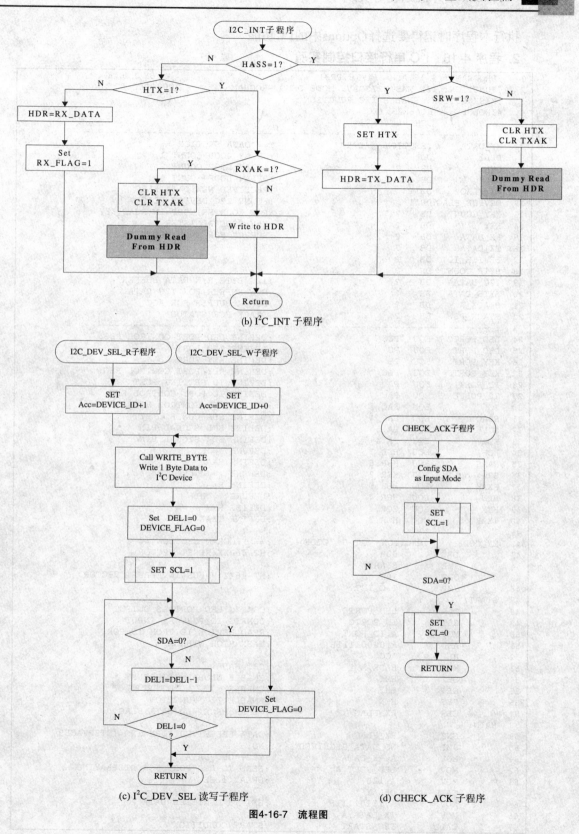

(b) I²C_INT 子程序

(c) I²C_DEV_SEL 读写子程序

(d) CHECK_ACK 子程序

图4-16-7 流程图

执行本程序时记得要选择Options中的I^2C功能。

2. 程序 4-16 I^2C 串行接口控制实验

```
1       ;PROGRAM : 4-16.ASM (4-16.PRJ)              BY STEVEN
2       ;FUNCTION: I2C MASTER/SLAVE MODE DEMO PROGRAM  2002.DEC.07.
3       ;MASK OPTION: ENABLE I2C FUNCTION
4       #INCLUDE   HT46R23.INC
5                  .CHIPHT46R23
6       ;-----------------------------------------------------------------
7       MY_DATA     .SECTION    'DATA'          ;== DATA SECTION ==
8       DEL1        DB    ?                     ;DELAY LOOP COUNT 1
9       DEL2        DB    ?                     ;DELAY LOOP COUNT 2
10      DEL3        DB    ?                     ;DELAY LOOP COUNT 3
11      RX_FLAG     DBIT                        ;RECEIVED NEW DATA FLAG
12      DEVICE_FLAG DBIT                        ;FOUND I2C DEVICE FLAG
13      KEY_COUNT   DB    ?                     ;ROW COUNTER USED IN READ_KEY
14      KEY         DB    ?                     ;KEY CODE REGISTER
15      RX_DATA     DB    ?                     ;BUFFER FOR RECEIVED DATA
16      TX_DATA     DB    ?
17      DEVICE_ID   DB    ?
18      BYTE_COUNT  DB    ?
19      I2C_DATA    DB    ?                     ;I2C BYTE R/W DATA BUFFER
20      BYTE_DATA   DB    ?                     ;R/W BUFFER FOR I2C DATA
21      STACK_A     DB    ?                     ;STACK BUFFER FOR ACC
22      STACK_PSW   DB    ?                     ;STACK FOR STATUS
23      ;-----------------------------------------------------------------
24      LED_PORTC   EQU   PCC                   ;DEFINE LED PORT CONTROL REG.
25      LED_PORT    EQU   PC                    ;DEFINE LED PORT
26      KEY_PORT    EQU   PB                    ;DEFINE KEY PORT
27      KEY_PORTC   EQU   PBC                   ;DEFINE KEY PORT CONTROL REG.
28      ID_PORT     EQU   PA                    ;DEFINE ID SETTING PORT
29      ID_PORTC    EQU   PAC                   ;DEFINE ID PORT CONTROL REG.
30      SCLC        EQU   PAC.5                 ;DEFINE SCL CONTROL BIT
31      SDAC        EQU   PAC.4                 ;DEFINE SDA CONTROL BIT
32      SCL         EQU   PA.5                  ;DEFINE SCL SIGNAL PIN
33      SDA         EQU   PA.4                  ;DEFINE SDA SIGNAL PIN
34      HCF         EQU   HSR.7                 ;DEFUNE   HCF
35      HAAS        EQU   HSR.6                 ;DEFINE HAAS
36      SRW         EQU   HSR.2                 ;DEFUNE   SRW
37      RXAK        EQU   HSR.0                 ;DEFINE RXAK
38      HEN         EQU   HCR.7                 ;DEFUNE   HEN
39      HTX         EQU   HCR.4                 ;DEFINE HTX
40      TXAK        EQU   HCR.3                 ;DEFINE TXAK
41
42      MY_CODE     .SECTION  AT 0 'CODE'       ;== PROGRAM SECTION ==
43              ORG       00H                   ;HT-46RXX RESET VECTOR
44              JMP       START
45              ORG       10H                   ;HT-46XX H-BUS INTERRUPT VECTOR
46              JMP       I2C_INT
47      START:
48              CLR       LED_PORTC             ;CONFIG LED PORT AS OUTPUT
49              SET       ID_PORTC              ;CONFIG ID_PORT AS INPUT
50              MOV       A,ID_PORT             ;READ DEVICE ID FROM ID_PORT
51              AND       A,00001111B           ;MASK HIGH NIBBLE
52              RL        ACC
53              MOV       HADR,A                ;SET HADR
54              SET       EMI                   ;ENABLE GLOBAL INTERRUPT
55              SET       EHI                   ;ENABLE  I2C INTERRUPT
56              SET       HEN                   ;ENABLE I2C FUNCTION
57              CLR       RX_FLAG               ;CLEAR RECEIVED DATA FLAG
58      MAIN:
59              SNZ       RX_FLAG               ;DATA ERECEIVED FORM I2C INTERFACE?
60              JMP       NO_DATA_RECEIVED      ;NO.
61              MOV       A,RX_DATA             ;YES, GET DATA
62              MOV       LED_PORT,A            ;SEND TO LED PORT FOR DISPLAYING
63              MOV       A,120                 ;DELAY 1 SEC
64              CALL      DELAY
65              MOV       A,11101111B           ;SET TRANSMITTED DATA
66              MOV       TX_DATA,A
67              CALL      I2C_START             ;START CONDITION
```

```
 68            CALL       I2C_DEV_SEL_R         ;I2C DATA READ COMMAND
 69            CALL       READ_BYTE             ;START TO READ 1 BYTE DATA
 70            CALL       I2C_STOP              ;STOP CONDITION FOR I2C
 71            MOV        A,I2C_DATA            ;GET DATA
 72            MOV        LED_PORT,A            ;SEND TO LED PORT
 73            CLR        RX_FLAG
 74 NO_DATA_RECEIVED:
 75            CALL       READ_KEY              ;READ KEYPAD
 76            MOV        A,16
 77            XOR        A,KEY
 78            SZ         Z                     ;KEY PRESSED ?
 79            JMP        MAIN                  ;NO, RE-READ KEYPAD
 80            CLR        DEVICE_FLAG           ;CLEAR DIVICE FOUND FLAG
 81            MOV        A,ID_PORT             ;READ DEVICE ID FROM ID_PORT
 82            AND        A,00001111B           ;MASK HIGH NIBBLE
 83            RL         ACC
 84            MOV        DEVICE_ID,A           ;SET I2C ADDRESS
 85            CALL       I2C_START             ;START CONDITION
 86            CALL       I2C_DEV_SEL_W         ;I2C DATA WRITE COMMAND
 87            MOV        A,00001111B
 88            XOR        A,KEY                 ;INVERSE KEY CODE FOR DISPLAYNG
 89            SNZ        DEVICE_FLAG           ;DEVICE RESPOUNDING ?
 90            SET        LED_PORT              ;NO, TURN-OFF ALL LEDS.
 91            SZ         DEVICE_FLAG           ;DEVICE RESPOUNDING ?
 92            CALL       WRITE_BYTE            ;YES, WRITE DATA
 93            CALL       I2C_STOP
 94            JMP        MAIN                  ;READ NEXT KEY
 95 ;**********************************************************************
 96 ;              I2C INTERRUPT SERVICE ROUTINE
 97 ;**********************************************************************
 98 I2C_INT    PROC
 99            MOV        STACK_A,A             ;PUSH ACC
100            MOV        A,STATUS
101            MOV        STACK_PSW,A           ;PUSH STATUS
102            SNZ        HAAS                  ;HAAS = 1?
103            JMP        DATA_RT               ;NO, JUMP DATA RX/TX PROCESSING
104            SNZ        SRW                   ;ADRESS MATCH! TX MODE?
105            JMP        RX_MODE               ;NO, JUMP TO RX MODE
106 TX_MODE:
107            SET        HTX                   ;SET TX MODE
108            MOV        A,TX_DATA
109            MOV        HDR,A                 ;WRITE DATA TO HDR
110            JMP        END_I2C_INT
111 RX_MODE:
112            CLR        HTX                   ;SET RX MODE
113            CLR        TXAK                  ;ACKNOWLEDGE
114            MOV        A,HDR                 ;DUMMY READ FROM HDR
115            JMP        END_I2C_INT
116 DATA_RT:
117            SNZ        HTX                   ;TX MODE?
118            JMP        DATA_RX               ;NO, JUMP TO DATA RX MODE
119 DATA_TX:
120            SZ         RXAK                  ;RXAK = 0? CONTINUE DATA READ?
121            JMP        WITH_ACK              ;YES!
122 NO_ACK:
123            CLR        HTX
124            CLR        TXAK
125            MOV        A,HDR                 ;DUMMY READ FORM HDR
126            JMP        END_I2C_INT
127 WITH_ACK:
128            MOV        A,TX_DATA
129            MOV        HDR,A                 ;WRTIE TO HDR
130            JMP        END_I2C_INT
131 DATA_RX:
132            MOV        A,HDR                 ;SET CURSOR ADDRESS
133            MOV        RX_DATA,A             ;COPY TO RX_DATA BUFFER
134            SET        RX_FLAG
135 END_I2C_INT:
136            MOV        A,STACK_PSW
137            MOV        STATUS,A              ;POP PSW
138            MOV        A,STACK_A             ;POP ACC
```

```
139         RETI
140 I2C_INT  ENDP
141 ;**********************************************************************
142 ;             GENERATE I2C START CONDITION
143 ;**********************************************************************
144 I2C_START  PROC
145         CLR      SCLC
146         CLR      SDAC            ;CONFIG SDA AS OUTPUT MODE
147         CLR      SCL             ;SET SCL=0
148         CLR      SDA             ;SET SDA=0
149         CALL     DELAY_10        ;DELAY
150         SET      SDA             ;SET SDA=1
151         SET      SCL             ;SET SCL=1
152         CALL     DELAY_10        ;DELAY
153         CLR      SDA             ;SET SDA=0
154         CALL     DELAY_10        ;DELAY
155         CLR      SCL             ;SET SCL=0
156         CALL     DELAY_10        ;DELAY
157         RET
158 I2C_START  ENDP
159 ;**********************************************************************
160 ;       SEND I2C DEVICE SELECT CODE (R/W) TO DEVICE
161 ;**********************************************************************
162 I2C_DEV_SEL_W   PROC             ;WRITE MODE
163         MOV      A,DEVICE_ID     ;LOAD A WITH DEVICE SELECT CODE
164         CLR      ACC.0           ;SET SRW T0 0 FOR WRITING TO I2C
165         JMP      DEV_SEL
166 I2C_DEV_SEL_R:                   ;READ MODE
167         MOV      A,DEVICE_ID     ;LOAD A WITH DEVICE SELECT CODE
168         SET      ACC.0           ;SET SRW T0 1 FOR READING FROM I2C
169 DEV_SEL:
170         CALL     WRITE_BYTE      ;SEND OUT 1 BYTE DATA IN Acc
171         CALL     CHECK_ACK       ;WAIT FOR DEVICE ACK SIGNAL
172         RET
173 I2C_DEV_SEL_W   ENDP
174 ;**********************************************************************
175 ;             GENERATE I2C STOP CONDITION
176 ;**********************************************************************
177 I2C_STOP   PROC
178         CLR      SDAC            ;CONFIG SDA AS OUTPUT MODE
179         CLR      SCLC
180         CLR      SCL             ;SET SCL=0
181         CLR      SDA             ;SET SDA=0
182         CALL     DELAY_10        ;DELAY
183         CLR      SDA             ;SET SDA=0
184         CALL     DELAY_10        ;DELAY
185         SET      SCL             ;SET SCL=1
186         CALL     DELAY_10        ;DELAY
187         SET      SDA             ;SET SDA=1
188         RET
189 I2C_STOP   ENDP
190 ;**********************************************************************
191 ;         SERIAL OUT DATA IN Acc VIA SDA & SCL
192 ;**********************************************************************
193 WRITE_BYTE  PROC
194         CLR      SDAC            ;CONFIG SDA AS OUTPUT MODE
195         CLR      SCLC
196         CLR      SCL             ;SET SCL=0
197         CLR      SDA             ;SET SDA=0
198         MOV      I2C_DATA,A      ;RESERVED DATA IN TX BUFFER
199         MOV      A,8             ;SET 8 BIT COUNTER
200         MOV      BYTE_COUNT,A
201 WNB_0:
202         SZ       I2C_DATA.7      ;IS MSB = 0?
203         JMP      WRITE_1         ;NO, JUMP TO WRITE_1
204 WRITE_0:
205         CLR      SDA             ;SET SDA=0
206         JMP      WNB_1           ;JUMP TO WNB_1
207 WRITE_1:
208         SET      SDA             ;SET SDA=1
209 WNB_1:
```

```
210           CALL        DELAY_10                ;DELAY
211           SET         SCL                     ;SET SCL=1
212           CALL        DELAY_10                ;DELAY
213           CLR         SCL                     ;SET SCL=0
214           CALL        DELAY_10                ;DELAY
215           RL          I2C_DATA                ;SHIFT TX BUFFER
216           SDZ         BYTE_COUNT              ;BYTE_COUNT-1 = 0?
217           JMP         WNB_0                   ;NO, WRITE NEXT BIT
218           RET                                 ;YES, RETURN.
219 WRITE_BYTE  ENDP
220 ;**********************************************************************
221 ;             SERIAL IN DATA TO Acc VIA SDA & SCL
222 ;**********************************************************************
223 READ_BYTE   PROC
224           SET         SDAC                    ;CONFIG SDA AS INTPUT MODE
225           MOV         A,8                     ;SET 8 BIT COUNTER
226           MOV         BYTE_COUNT,A
227 RNB_0:
228           SET         SCL                     ;SET SCL=0
229           CALL        DELAY_10                ;DELAY
230           RLC         I2C_DATA                ;SHIFT RX BUFFER
231           SZ          SDA                     ;SDA = 0?
232           JMP         READ_1                  ;NO, JUMP TO READ_1
233 READ_0:
234           CLR         I2C_DATA.0              ;YES, SET LSB=0
235           JMP         RNB_1                   ;JUMP TO RNB_1
236 READ_1:
237           SET         I2C_DATA.0              ;SET LSB=1
238 RNB_1:
239           CALL        DELAY_10                ;DELAY
240           CLR         SCL                     ;SET SCL=0
241           CALL        DELAY_10                ;DELAY
242           SDZ         BYTE_COUNT              ;BYTE_COUNT-1 = 0?
243           JMP         RNB_0                   ;NO, READ NEXT BIT
244           MOV         A,I2C_DATA              ;RELOAD RX DATA TO Acc
245           CLR         SDAC                    ;YES, SEND NO_ACK_SIGNAL
246           CLR         SDA
247           SET         SCL
248           CALL        DELAY_10
249           SET         SDA
250           CLR         SCL
251           RET                                 ;YES, RETURN
252 READ_BYTE   ENDP
253 ;**********************************************************************
254 ;             WAIT FOR ACK SIGNAL FROM I2C DEVICE
255 ;**********************************************************************
256 CHECK_ACK   PROC
257           SET         SDAC                    ;CONFIG    SDA AS INPUT MODE
258           CLR         SCLC
259           CALL        DELAY_10                ;DELAY
260           SET         SCL                     ;SET SCL=1
261           CALL        DELAY_10                ;DELAY
262           CLR         DEL1
263           SET         DEVICE_FLAG             ;ASSUME DEVICE FOUND
264 WAIT_ACK:
265           SNZ         SDA                     ;ACK SIGNAL SET ?
266           JMP         GET_ACK                 ;YES.
267           SDZ         DEL1                    ;NO. OUT OF WAIT ACK TIME
268           JMP         WAIT_ACK                ;NO. WAIT AGAIN
269           CLR         DEVICE_FLAG             ;NO DEVICE FOUND!
270 GET_ACK:
271           CALL        DELAY_10                ;DELAY
272           CLR         SCL                     ;SET SCL=0
273           CALL        DELAY_10                ;DELAY
274           RET
275 CHECK_ACK   ENDP
276 ;**********************************************************************
277 ;             DELAY 10us FOR TIMING CONSIDERATION
278 ;**********************************************************************
279 DELAY_10:
280           TABRDL      DEL1                    ;NULL READ
```

```
281         TABRDL      DEL1                    ;DELAY 10 INS. CYCLES
282         TABRDL      DEL1
283         TABRDL      DEL1
284         RET
285 ;************************************************************************
286 ;SCAN 4x4 MATRIX ON KEY PORT AND RETURN THE CODE IN KEY REGISTER
287 ; IF NO KEY BEEN PRESSED, KEY=16.
288 ;************************************************************************
289 READ_KEY    PROC
290         MOV         A,11110000B
291         MOV         KEY_PORTC,A             ;CONFIG KEY_PORT
292         SET         KEY_PORT                ;INITIAL KEY PORT
293         CLR         KEY                     ;INITIAL KEY REGISTER
294         MOV         A,04
295         MOV         KEY_COUNT,A             ;SET ROW COUNTER
296         CLR         C                       ;CLEAR CARRY FLAG
297 SCAN_KEY:
298         RLC         KEY_PORT                ;ROTATE SCANNING BIT
299         SET         C                       ;MAKE SURE C=1
300         SNZ         KEY_PORT.4              ;COLUMN 0 PRESSED?
301         JMP         END_KEY                 ;YES.
302         INC         KEY                     ;NO, INCREASE KEY CODE.
303         SNZ         KEY_PORT.5              ;COLUMN 1 PRESSED?
304         JMP         END_KEY                 ;YES.
305         INC         KEY                     ;NO, INCREASE KEY CODE.
306         SNZ         KEY_PORT.6              ;COLUMN 2 PRESSED?
307         JMP         END_KEY                 ;YES.
308         INC         KEY                     ;NO, INCREASE KEY CODE.
309         SNZ         KEY_PORT.7              ;COLUMN 3 PRESSED?
310         JMP         END_KEY                 ;YES.
311         INC         KEY                     ;NO, INCREASE KEY CODE.
312         SDZ         KEY_COUNT               ;HAVE ALL ROWs BEEN CHECKED?
313         JMP         SCAN_KEY                ;NO, NEXT ROW.
314 END_KEY:
315         RET
316 READ_KEY    ENDP
317 ;************************************************************************
318 ;                   Delay about DEL1*10ms
319 ;************************************************************************
320 DELAY   PROC
321         MOV         DEL1,A                  ;SET DEL1 COUNTER
322 DEL_1:  MOV         A,30
323         MOV         DEL2,A                  ;SET DEL2 COUNTER
324 DEL_2:  MOV         A,110
325         MOV         DEL3,A                  ;SET DEL3 COUNTER
326 DEL_3:  SDZ         DEL3                    ;DEL3 DOWN COUNT
327         JMP         DEL_3
328         SDZ         DEL2                    ;DEL2 DOWN COUNT
329         JMP         DEL_2
330         SDZ         DEL1                    ;DEL1 DOWN COUNT
331         JMP         DEL_1
332         RET
333 DELAY   ENDP
334         END
```

程序说明

8~22   依序定义变量地址。

24~25   定义 LED_PORT 为 PC。

26~27   定义 KEY_PORT 为 PB。

28~29   定义 ID_PORT 为 PA，实际上 ID 地址只由 PA3~PA0 4 位输入，PA7~PA4 另有用途。

30~33   定义 SCL 为 PA5、SDA 为 PA4，作为控制 I^2C 串行接口传输的引脚。

34~40   依序定义 HT46x23 中与 I^2C 串行传输接口的相关控制位的名称，以增加程序的可读性与方便性。

43   声明存储器地址由 000h 开始(HT46x23 Reset Vector)。

行号	说明
45	声明存储器地址由 010h 开始(HT46x23 I²C Interrupt Vector)。
48~49	定义 LED_PORT 为输出模式、ID_PORT 为输入模式。
50~53	由 ID_PORT 读回指拨开关设置数值，并取其低 4 位当成 Slave Mode I²C 的设备 ID。
54~56	中断总开关(EMI)与 I²C Interrupt(EHI)有效，并启动 I²C 串行接口的传输功能。
57	清除 RX_FLAG，程序中将以此位作为是否由 I²C 串行接口接收到数据的判断(1 表示接收到数据)，因此先将其清除为 0。
59~60	判断 I²C 串行传输接口是否接收到数据，若有(RX_FLAG＝1)则进行数据显示的动作；若无(RX_FLAG＝0)就继续扫描 4×4 键盘(JMP NO_DATA_RECEIVED)。
61~64	将 I²C 串行传输接口所接收到数据送至 LED_PORT 显示，并调用 DELAY 子程序延时 1.2 s。
65~66	将 TX_DATA 寄存器设为 1110111b，作为下次由 I²C 串行接口所送出的数据；此数据稍后由 HT46x23 的 SDA 读回之后将送到 LED_PORT，此时只有代表 DEVICE_ID 相同的 LED(I²C)会亮之外，原来显示的按键值将消失。
67~68	送出 I²C START 条件与 Slave Address(DEVICE_ID)。注意，此时 R/W＝1，表示 Master 要由 I²C Device 读取数据，所以 HT46x23 的 I²C 接口必须准备送出数据。
69	调用 READ_BYTE 子程序，通过 SDA 与 SCL 的控制从 I²C 接口读回 8 位的数据。
70	调用 I2C_STOP 子程序产生 STOP 条件，结束此次 I²C 接口的数据传输。
71~73	将 I²C 串行传输接口所接收到数据送到 LED_PORT 显示，并清除 RX_FLAG 表示此按键值已经完成显示。
75~79	调用 READ_KEY 子程序扫描键盘，完成后判断索引值(KEY)内容是否为 16，若成立表示键盘无输入，重新扫描键盘(MAIN)；反之，则键盘已被按下，开始将按键值通过 I²C 接口传送出去的过程。
80	清除 DEVICE_FLAG，程序中将以此位判断是否有与 DEVICE_ID 相符的设备回应(1 表示与 DEVICE_ID 相符)，因此先将其清除为 0。
81~84	由 ID_PORT 读回指拨开关设置数值，并取其低 4 位当成要存取的 I²C 设备的设备 ID。
85~86	送出 I²C START 条件与设备选择码(DEVICE_ID)。注意，此时 R/W＝0，表示 Master 要写数据到 I²C 设备，所以 HT46x23 的 I²C 接口必须准备接收数据。
87~88	先将按键值取反相；由于 PC 驱动 LED 的方式是 0 才会亮，所以先将按键值反相，以便看到对应的二进制数值显示。
89~93	当调用 I²C_DEV_SEL_W 子程序时，该程序会检查是否有 I²C 设备对所送出的 DEVICE_ID 产生回应(ACK＝0)，若无设备回应则传回 DEVICE_FLAG＝0；若有则传回 DEVICE_FLAG＝1。此段程序是依 DEVICE_FLAG 为 0 或 1 分别完成以下动作： DEVICE_FLAG＝1：表示有设备回应，调用 WRITE_BYTE 子程序，通过 SDA 与 SCL 的控制将按键值送至 I²C 接口，然后调用 I2C_STOP 子程序产生 STOP 条件，结束此次 I²C 接口的数据传输。 DEVICE_FLAG＝0：表示没有设备回应，将 LED 全部熄灭后直接调用 I2C_STOP 子程序产生 STOP 条件，结束此次 I²C 接口传输。
94	重新执行程序(JMP MAIN)。
98~140	I2C_INT 中断服务子程序，此段程序主要是根据原厂数据手册中的"I²C Bus 中断服务子程序"的流程所编写，流程图如图 4-16-8 所示。

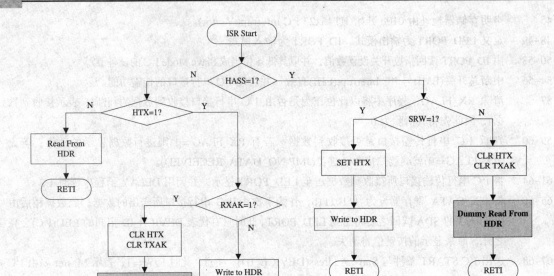

图4-16-8 I²C_INT中断服务子程序流程图

102~103 依据 HAAS 位分析是 Device ID 相符所产生的中断(HASS=1)，或是 8 位数据发送/接收完毕所产生的中断(HASS=0，若是则 JMP DATA_RT)。

104~115 Device ID 相符所产生的中断：再依据 SRW 位分析为 Receive Mode(SRW=0)或 Transmit Mode(SRW=1)。若为 Transmit Mode，就将 HTX 设置为 1，并将欲通过 I²C 接口发送的数据(TX_DATA)写入 HDR 寄存器。如果是 Receive Mode，就将 HTX 与 TXAK 位设置为 0，并依数据手册中的要求，对 HDR 寄存器执行一次"Dummy Read"(MOV A,HDR)。

116~135 8 位数据发送/接收完毕所产生的中断：再依据 HTX 位分析为 Receive Mode(HTX=0)或 Transmit Mode(HTX=1)。

如果是 Transmit Mode，表示 8 位数据已经传输完毕，再根据 RXAK 位分析 Master 是否送出 ACK 信号，以决定是否要继续送出数据。如果 RXAK=1，表示 Master 端要继续读数据，所以再次将 TX_DATA 写至 HDR 寄存器；如果 RXAK=0，则清除 HTX 与 TXAK 位，并依数据手册中的要求，对 HDR 寄存器执行一次"Dummy Read"(MOV A,HDR)。本实验一次只读取一个字节，因此 RXAK=0。

如果是 Receive Mode，表示 8 位数据已经接收完毕，接收的数据复制到 RX_DATA 寄存器，并设置 RX_FLAG=1。

144~158 I2C_START 子程序，产生 START 条件。

162~173 I2C_DEV_SEL_W 与 I2C_DEV_SEL_R 子程序，将设备选择码(DEVICE_ID)输出到 I²C Bus。I2C_DEV_SEL_W 子程序会设置 R/W 位=0，表示是对 I²C 设备做写入动作；I2C_DEV_SEL_R 子程序会设置 R/W 位=1，表示是对 I²C 设备进行读取动作。程序中会调用 WRITE_BYTE 子程序，将设备选择码以串行方式一一送出。此外，调用 CHECK_ACK 子程序的目的，是看看是否有 I²C 设备对此 DEVICE_ID 产生回应。

177~189 I2C_STOP 子程序，产生 STOP 条件。

193~219 WRITE_BYTE 子程序，将 Acc 寄存器的数据由 SDA 引脚循序移出(配合 SCL 信号)。

223~251 READ_BYTE 子程序，将 SDA 引脚上的信号循序移入(配合 SCL 信号)Acc 寄存器。由于本实验一次只读取一个字节，因此第 244~249 行设置 ACK=1(即送出 NO_ACK 信号，所以 HT46x23 的 RXAK 会为 0)。

255~274 CHECK_ACK 子程序，等待 I^2C 设备回应 ACK 信号，第 261~268 行将检查的次数设为 256 次。如果经过 256 次的检查 ACK 信号仍未送出，就代表 I^2C Bus 上没有与 DEVICE_ID 相符的设备，此时设置 DEVICE_FLAG=0。

278~283 DELAY_10 子程序，延时 10 个指令周期的时间，以确保 I^2C 接口来得及反应。

288~315 READ_KEY 子程序，请参考 4-7 节中的说明。

319~332 DELAY 子程序，延时时间的计算请参考 4-1 节。

本节实验以HT46x23同时以Master与Slave的方式来完成按键值显示的动作，虽然在同一个单片机上将数据通过I^2C接口传输，感觉上有些"多此一举"，但通过此实验，在只有一个ICE可用的情况下，读者还是可以熟悉HT46x23 I^2C接口串行传输的控制方式，这就是本节实验的主要目的。但心急的读者可以参考5-13节，该节实验是两个单片机通过I^2C接口串行传输数据的实验，在实用上较有意义。

在程序一开始时，会先由ID_PORT读回开关设置值，并存于HADR寄存器作为I^2C接口的Slave Address，此后HADR寄存器便不再改变。然而主程序每次与I^2C接口传输之前，都会再重新由ID_PORT读回开关设置值当成要存取的Device ID，所以在正常情况下，如果用户不去变动Port C的指拨开关，那么按键值将正常显示在LED3~LED0上，而代表Slave Address与Device ID相符的I^2C LED也将点亮。程序一旦执行之后，如果读者再去改变Port C的指拨开关(即更改Device ID)，那么LED将不再显示键值，而I2C LED也将熄灭，表示未发现与Device ID相符的I^2C设备。

4-16-5 动动脑 + 动动手

- 试说明 I2C_INT 中断服务子程序中各行指令的意义。
- 是否可以改采 Polling 方式完成与本实验相同的功能呢？

第 5 章

进阶实验篇

经过基础实验篇的洗礼，相信读者对于 HT46xx 的硬件、程序写作技巧以及开发环境的使用，都已经有初步的认识。程序其实要多写、多错、多调试，然后方能由错误中去累积经验，增强本身的逻辑观念及程序编写功力。延续上一章的学习成果，本章将继续介绍几个更深入的实验：直流电机控制、LCD、多颗七段显示器与点矩阵 LED 的扫描等,通过这些实验,相信必定能够让读者对于 HT46xx 单片机的运用能有更深一层的了解。

5-1　直流电机控制实验

5-2　马表——多颗七段显示器控制实验

5-3　静态点矩阵LED控制实验

5-4　动态点矩阵LED控制实验

5-5　LCD字形显示实验

5-6　LCD自建字形实验

5-7　LCD与4×4键盘控制实验

5-8　LCD的DD/CG RAM读取控制实验

5-9　LCD的4位控制模式实验

5-10　比大小游戏实验

5-11　中文显示型LCD控制实验

5-12　半矩阵式(Half-Matrix)键盘与LCD控制实验

5-13　HT46xx I^2C Mater-Slave传输实验

第5章 进阶实验篇

5-1 直流电机控制实验

5-1-1 目 的

本实验将利用脉宽调制(PWM)的技巧,让直流电机呈现不同的转速,并可以使用按键控制直流电机速度的加、减。

5-1-2 学习重点

通过本实验,读者应了解如何以PWM方法控制直流电机转速的快慢。

5-1-3 电路图（见图5-1-1）

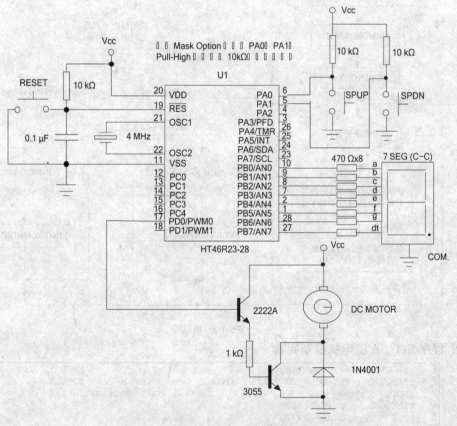

图5-1-1 直流电机控制电路

直流电机转动速度的快慢由电机的电势大小来决定,其正反转则由电压极性来决定,一般转速大小和电机两端电压成线性,其电压越大转速越快,反之则越小。所谓的PWM即脉宽

HT46xx 单片机原理与实践

调制,利用固定频率来改变工作周期,使电机产生ON-OFF的时间长短来决定转速快慢。HT46xx的PWM输出接口已经在实验4-13中详细地探讨过了,请读者自行参考。此实验利用3055晶体管来做微控制直流电机的开关;当晶体管导通时,直流电机开始转动。图中的二极管俗称"飞轮二极管(Free-Whelling Diode)",其主要的功能是保护晶体管。因为当晶体管瞬间截止时,电机内部的线圈所产生的感应电势会对晶体管造成损害。所以电路中采用二极管将此感应电势予以旁路,以保护晶体管不致受损。再次提醒读者,执行本实验的程序时务必使能Options中的PWM0,并选定(7+1)Mode。

5-1-4 流程图及程序

1. 流程图(见图5-1-2)

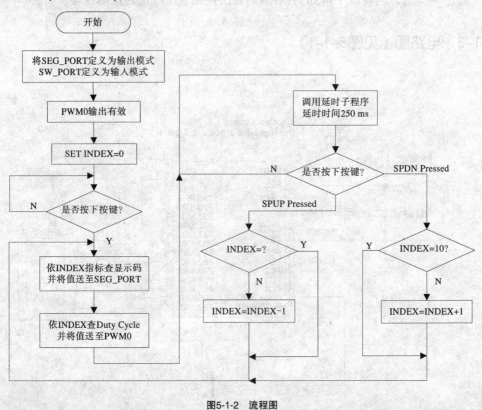

图5-1-2 流程图

2. 程序 5-1 直流电机控制实验

```
1    ;PROGRAM : 5-1 PROGRAM (5-1.PRJ)                         2002.DEC.20.
2    ;FUNCTION: PWM DC MOTOR SPEED CONTROL DEMO PROGRAM    BY STEVEN
3    ;NOTE    : PWM0 SHOULD BE ENABLED, SEL 7+1 MODE (MASK OPTION).
4    #INCLUDE     HT46R23.INC
5                 .CHIP HT46R23
6    ;--------------------------------------------------------------
7    MY_DATA      .SECTION    'DATA'         ;== DATA SECTION ==
8    DEL1         DB     ?                   ;DELAY LOOP COUNT 1
9    DEL2         DB     ?                   ;DELAY LOOP COUNT 2
10   DEL3         DB     ?                   ;DELAY LOOP COUNT 3
11   INDEX        DB     ?                   ;TABLE INDEX
```

```
12      ;-------------------------------------------------------------------
13      SW_PORT         EQU     PA                      ;DEFINE SWITCH PORT
14      SW_PORTC        EQU     PAC                     ;DEFINE SWITCH PORT CONTROL REG.
15      SEG_PORT        EQU     PB                      ;DEFINE 7 SEG PORT
16      SEG_PORTC       EQU     PBC                     ;DEFINE 7 SEG PORT CONTROL REG.
17      SPUP_SW         EQU     SW_PORT.0               ;DEFINE SPEED UP SW
18      SPDN_SW         EQU     SW_PORT.1               ;DEFINE SPEED DOWN SW
19
20      MY_CODE         .SECTION    AT 0 'CODE'         ;== PROGRAM SECTION ==
21              ORG     00H                             ;HT-46RXX RESET VECTOR
22              SET     SW_PORTC                        ;CONFIG SW_PORT AS INPUT MODE
23              CLR     SEG_PORTC                       ;CONFIG SEG_PORT AS OUTPUT PWM MODE
24              CLR     PDC                             ;CONFIG PD AS OUTPUT PWM MODE
25              CLR     SEG_PORT                        ;CLEAR DISPLAY
26              CLR     PWM0
27              SET     PD.0                            ;ENABLE PWM OUTPUT
28              CLR     INDEX                           ;SET INDEX=0
29      WAIT_KEY:
30              SNZ     SPUP_SW                         ;SPUP_SW PRESSED?
31              JMP     MAIN                            ;YES!
32              SZ      SPDN_SW                         ;SPDN_SW PRESSED?
33              JMP     WAIT_KEY                        ;NO, WAIT KEY PRESSED.
34      MAIN:
35              MOV     A,INDEX
36              CALL    TRANS_SEG                       ;GET 7 SEG CODE
37              MOV     SEG_PORT,A                      ;DISPLAY
38              MOV     A,INDEX
39              CALL    TRANS_PWM                       ;GET DUTY CYCLE
40              MOV     PWM0,A                          ;SET PWM0
41      LOOP:
42              CALL    DELAY                           ;DELAY 0.25 SEC
43              SNZ     SPUP_SW                         ;SPEED UP SW PRESSED?
44              JMP     SPEED_UP                        ;YES, JUMP TO SPEED_UP
45              SNZ     SPDN_SW                         ;SPEED DOWN PRESSED?
46              JMP     SPEED_DOWN                      ;YES, JUMP TO SPEED_DOWN
47              JMP     LOOP                            ;NO, KEEP CURRENT SPEED.
48      SPEED_UP:
49              SZA     INDEX                           ;IS INDEX = 0?
50              DEC     INDEX                           ;NO, INDEX=INDEX-1
51              JMP     MAIN                            ;JUMP TO MAIN
52      SPEED_DOWN:
53              INCA    INDEX                           ;Acc=INDEX+1
54              SUB     A,10                            ;Acc >10 ?
55              SNZ     C                               ;YES.
56              INC     INDEX                           ;NO, INCREASE INDEX BY 1
57              JMP     MAIN                            ;JUMP TO MAIN
58      ;*******************************************************************
59      ;              GET DUTY CLCLE CONSTANT INDEXED BY Acc
60      ;*******************************************************************
61      TRANS_PWM       PROC
62              ADDM    A,PCL
63              RET     A,127*100/100 SHL 1     ;DC MOTOR PWM DUTY CYCLE TABLE
64              RET     A,127*98/100 SHL 1      ;DC MOTOR PWM DUTY CYCLE TABLE
65              RET     A,127*97/100 SHL 1      ;DC MOTOR PWM DUTY CYCLE TABLE
66              RET     A,127*96/100 SHL 1      ;DC MOTOR PWM DUTY CYCLE TABLE
67              RET     A,127*95/100 SHL 1      ;DC MOTOR PWM DUTY CYCLE TABLE
68              RET     A,127*93/100 SHL 1      ;DC MOTOR PWM DUTY CYCLE TABLE
69              RET     A,127*90/100 SHL 1      ;DC MOTOR PWM DUTY CYCLE TABLE
70              RET     A,127*86/100 SHL 1      ;DC MOTOR PWM DUTY CYCLE TABLE
71              RET     A,127*83/100 SHL 1      ;DC MOTOR PWM DUTY CYCLE TABLE
72              RET     A,127*80/100 SHL 1      ;DC MOTOR PWM DUTY CYCLE TABLE
73      TRANS_PWM       ENDP
74      ;*******************************************************************
75      ;              RETURN THE TABLE VALUE INDEX BY A
76      ;*******************************************************************
77      TRANS_SEG       PROC
78              ADDM    A,PCL
```

```
79          RET       A,01100111B         ;9
80          RET       A,01111111B         ;8
81          RET       A,00000111B         ;7
82          RET       A,01111101B         ;6
83          RET       A,01101101B         ;5
84          RET       A,01100110B         ;4
85          RET       A,01001111B         ;3
86          RET       A,01011011B         ;2
87          RET       A,00000110B         ;1
88          RET       A,00111111B         ;0
89  TRANS_SEG ENDP
90  ;****************************************************************
91  ;                 Delay about DEL1(10)*10ms
92  ;****************************************************************
93  DELAY     PROC
94          MOV       A,25
95          MOV       DEL1,A              ;SET DEL1 COUNTER
96  DEL_1:  MOV       A,30
97          MOV       DEL2,A              ;SET DEL2 COUNTER
98  DEL_2:  MOV       A,110
99          MOV       DEL3,A              ;SET DEL3 COUNTER
100 DEL_3:  SDZ       DEL3                ;DEL3 DOWN COUNT
101         JMP       DEL_3
102         SDZ       DEL2                ;DEL2 DOWN COUNT
103         JMP       DEL_2
104         SDZ       DEL1                ;DEL1 DOWN COUNT
105         JMP       DEL_1
106         RET
107 DELAY   ENDP
108         END
```

程序说明

8~11   依序定义变量地址。

13~18   定义 SW_PORT 为 PA，SEG_PORT 为 PB。

21   声明存储器地址由 000h 开始(HT46xx Reset Vector)。

22~23   定义 SEG_PORT_PORT、SW_PORT 为输出模式与输入模式。

26~27   清除 PWM0 寄存器，在第 27 行 PWM0 的输出有效之后，如果 PWM0 寄存器不为 0，直流电机就会开始转动，所以第 26 行的目的是设定 Duty Cycle=0%，虽然在第 27 行 PWM0 的输出有效，但在用户按下按键之前电机仍不会转动。

28   清除 INDEX 指针值。

29~33   等待用户按下按键。

35~37   显示目前的转速值(9：最快~0：最慢)。

38~40   依据 INDEX 值进行查表，并将结果存入 PWM0 寄存器；此举即改变输出波形的 Duty Cycle，而达到改变电机转速的目的。

42~47   每隔 250 ms 检查加速(SPUP：SPeed UP)与减速(SPDN：SPeed DowN)按键，用以判断是否要改变电机的转速。

49~51   判断 INDEX 是否为 0，若是，代表电机转速已经达到最高指数，维持 INDEX 的值；反之，表示电机还有加速空间，将 INDEX 减 1 以提高转速。

53~57   判断 INDEX 是否为 10，若是，代表电机转速已经达到最低指数，维持 INDEX 的值；反之，表示电机还有减速空间，将 INDEX 加 1 以降低转速。

61~73   Duty Cycle 查表子程序，请注意本实验使用(7+1)Mode 的 PWM 输出模式，此时 PWM 周期被分割成两个调制周期(Modulation Cycle 0~1)，每个调制周期为 128 个计数时钟，而 PWM 寄存器的控制位区分为 DC(PWM.7~ PWM.1)与 AC(PWM.0)两部分,每一个调制周期的 Duty Cycle 关系如表 5-1-1 所列。

表5-1-1 (7+1)Mode各调制周期的Duty Cycle

参 数	AC(0~1)	Duty Cycle
Modulation Cycle i i＝0~1	i＜AC	$\frac{DC+1}{128}$
	i≥AC	$\frac{DC}{128}$

因此,在查表数据中利用"SHL 1"伪指令将 Duty Cycle 左移一位,分别让 PWM0 寄存器的 AC Part＝0、DC Part＝Duty Cycle。

77~89 七段显示器显示码查表子程序。
93~107 DELAY 子程序,延迟 250 ms。

关于直流电机的PWM控制方式,其实与实验4-13中控制LED亮度的原理差不多,只是针对不同的设备特性要稍微有些不同。关于电机的控制,就必须了解其特性,电机毕竟是个机械设备,无法在瞬间达到其应有的转速。本实验只是很简单地利用按键来控制其转速,读者可以发现当按下按键时,电机需要一段时间方能达到定速,这就如同惯性一般;同理,由转动而静止时,也是需要一段时间。而电机由静止启动时,也需要耗费较大的电力。一般精确的电机控制,必须要考虑加速、减速的课题,有兴趣的读者可以多参考这方面的书籍。此外在进行本实验时,将控制芯片电路与电机的电源分开,可避免电源互相干扰。甚至可以用光隔离(如PC817、PC847)方式使正电源、地线与控制线全部分开,互不相连,以避免出现不正常动作。

5-1-5 动动脑＋动动手

- 试着改写本实验的程序,使电机的转速能有更多种变化。
- 试着让 Duty Cycle 再降低,看看电机在何种 Duty Cycle 下就无法转动了。

5-2 马表——多颗七段显示器控制实验

5-2-1 目 的

本实验利用人类视觉暂留的特性,让依序逐一点亮的七段显示器呈现同时点亮的错觉,以解决I/O端口不足的情况。本实验承接4-3节单颗七段显示器控制实验,但增加为4位数的计数器,配合HT46x23的Timer/Event Counter与外部中断的控制,制作一个定时分辨率为0.1 s的马表。

5-2-2 学习重点

通过本实验,读者能了解如何以扫描的方式来控制多颗七段显示器的显示。

5-2-3 电路图（见图5-2-1）

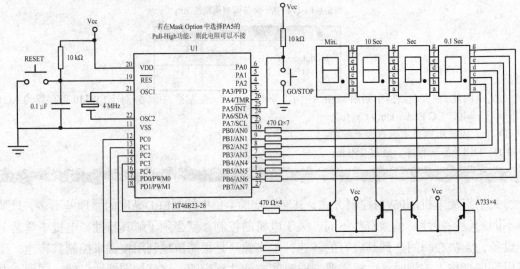

图5-2-1 多颗七段显示器控制电路

在一个系统中为了能够达到显示更多消息的目的，通常会有多颗七段显示器并排使用的情况。若以4-3节的方法来驱动七段显示器，就需要很多个I/O端口。例如，一个显示"时"和"分"的定时器就需要4颗七段显示器，这表示驱动用的单片机就需要有4组I/O端口，这对I/O资源极为珍贵的小型控制器而言，无疑是一种浪费。解决之道，就是采用"动态扫描"的驱动方式。利用此种驱动方式，就可节省许多I/O引脚。此法利用人类视觉暂留的现象(约30~45 ms)，利用程序控制晶体管，以分时(Time Devision)的方式来依序点亮每一颗七段显示器，只要轮流显示的速度控制得当，那么虽然是一次只点亮一颗七段显示器，但是在视觉效果上却是所有显示器都同时亮着，而且也达到了节省I/O引脚的目的。本实验使用Timer/Event Counter中断与外部中断来达到定时马表的功能，其控制寄存器仅列于表5-2-1，其控制方式请读者参考相关章节。

表5-2-1 本实验相关控制寄存器

特殊功能寄存器	Bit7	Bit6	Bit5	Bit4	Bit3	Bit2	Bit1	Bit0	参考章节
INTC0	—	ADF	TF	EIF	EADI	ETI	EEI	EMI	2-4
TMRC	TM1	TM0	—	TON	TE	PSC2	PSC1	PSC0	2-5
TMRH									2-5
TMRL									2-5

5-2-4 流程图及程序

1. 流程图(见图 5-2-2)

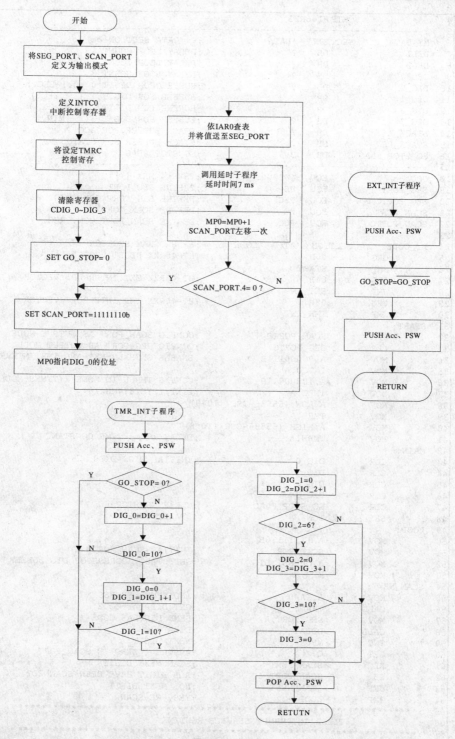

图5-2-2 流程图

2. 程序5-2 多颗七段显示器控制实验

```
1   ;PROGRAM : 5-2.ASM  (5-2.PRJ)            BY STEVEN
2   ;FUNCTION: MULTI 7-SEG LED DEMO PROGRAM  2002.DEC.07.
3   #INCLUDE    HT46R23.INC
```

```
4                   .CHIP HT46R23
5       ;-----------------------------------------------------------------
6       MY_DATA     .SECTION    'DATA'          ;== DATA SECTION ==
7       DEL1        DB     ?                    ;DELAY LOOP COUNT 1
8       DEL2        DB     ?                    ;DELAY LOOP COUNT 2
9       DEL3        DB     ?                    ;DELAY LOOP COUNT 3
10      DIG_0       DB     ?                    ;BUFFER FOR DIGIT 0 (0.1 SEC.)
11      DIG_1       DB     ?                    ;BUFFER FOR DIGIT 1 (    SEC.)
12      DIG_2       DB     ?                    ;BUFFER FOR DIGIT 2 (10 SEC.)
13      DIG_3       DB     ?                    ;BUFFER FOR DIGIT 3(   MIN.)
14      STACK_A     DB     ?                    ;STACK BUFFER FOR ACC
15      STACK_PSW   DB     ?                    ;STACK BUFFER FOR STATUS
16      GO_STOP_FLAG DB    ?                    ;GO_STOP_FLAG
17      ;-----------------------------------------------------------------
18      OSC         EQU    4000000              ;fSYS=4 MHz
19      SEG_PORT    EQU    PB                   ;DEFINE SEG_PORT
20      SEG_PORTC   EQU    PBC                  ;DEFINE SEG_PORT CON. REG.
21      SCAN_PORT   EQU    PC                   ;DEFINE SCAN_PORT
22      SCAN_PORTC  EQU    PCC                  ;DEFINE SCAN_PORT CON. REG.
23
24      MY_CODE     .SECTION AT 0 'CODE'        ;== PROGRAM SECTION ==
25              ORG     00H                     ;HT-46RXX RESET VECTOR
26              JMP     START
27              ORG     04H                     ;HT-46RXX EXT INTERRUPT VECTOR
28              JMP     EXT_INT
29              ORG     08H                     ;HT-46RXX TMR INTERRUPT VECTOR
30              JMP     TMR_INT
31      START:
32              CLR     SCAN_PORTC              ;CONFIG SCAN_PORT AS OUTPUT MODE
33              CLR     SEG_PORTC               ;CONFIG SEG_PORT AS OUTPUT MODE
34              MOV     A,00000111B             ;ENABLE GLOBAL, EXT AND TMR INTERRUPT
35              MOV     INTC0,A
36              MOV     A,10000011B             ;CONFIG TMR 0 IN MODE 2(TIMER MODE)
37              MOV     TMRC,A                  ;fINT=fSYS/8(4MHz/8).
38              MOV     A,LOW (65536-OSC/(10*8))
39              MOV     TMRL,A
40              MOV     A,HIGH (65536-OSC/(10*8))
41              MOV     TMRH,A                  ;DELAY 100ms TIME CONSTANT
42      MAIN:
43              CLR     DIG_0                   ;INITIAL DIG3~DIG0
44              CLR     DIG_1
45              CLR     DIG_2
46              CLR     DIG_3
47              CLR     GO_STOP_FLAG            ;CLEAR GO_STOP_FLAG
48              SET     TON                     ;START TIMER
49      LOOP:
50              MOV     A,11111110B
51              MOV     SCAN_PORT,A             ;INITIAL SCAN_PORT
52              MOV     A,OFFSET DIG_0          ;GET START ADDRESS OF DIG BUFFER
53              MOV     MP0,A
54      SCAN_NEXT:
55              MOV     A,IAR0                  ;INDIRECT ADDRESSING
56              CALL    TRANS                   ;GET DISPLAY CODE
57              MOV     SEG_PORT,A              ;SEND DISPLAY CODE
58              CALL    DELAY                   ;DELAY 7mS
59              INC     MP0                     ;NEXT DIGIT
60              SET     C                       ;ASSURE 1 ZERO IN SCAN_PORT
61              RLC     SCAN_PORT               ;SCAN NEXT DIGIT
62              SZ      SCAN_PORT.4             ;ALL DIGIT HAVE BEEN SCANNED?
63              JMP     SCAN_NEXT               ;NO, NEXT DIFIT
64              JMP     LOOP                    ;YES, RE-SCAN
65      ;*****************************************************************
66      ;           TIMER INTERRUPT SERVICE ROUTINE
67      ;*****************************************************************
68      TMR_INT:
69              MOV     STACK_A,A               ;PUSH A
70              MOV     A,STATUS
71              MOV     STACK_PSW,A             ;PUSH STATUS
72              SNZ     GO_STOP_FLAG.0          ;GO_STOP_FLAG = 1?
73              JMP     END_TMR_INT             ;NO, STOP TIME
74              INC     DIG_0                   ;INCREASE DIG0
```

```
75        MOV      A,DIG_0
76        SUB      A,10
77        SNZ      Z                       ;DIG0 = 10?
78        JMP      END_TMR_INT             ;NO., RETURN
79        CLR      DIG_0                   ;YES, RESET DIG0
80        INC      DIG_1                   ;INCREASE DIG1 BY 1
81  LOOP_1:
82        MOV      A,DIG_1
83        SUB      A,10
84        SNZ      Z                       ;DIG1 = 10?
85        JMP      END_TMR_INT             ;NO., RETURN
86        CLR      DIG_1                   ;YES, RESET DIG1
87        INC      DIG_2                   ;INCREASE DIG2 BY 1
88  LOOP_2:
89        MOV      A,DIG_2
90        SUB      A,6
91        SNZ      Z                       ;DIG2 = 5?
92        JMP      END_TMR_INT             ;NO., RETURN
93        CLR      DIG_2                   ;YES, RESET DIG2
94        INC      DIG_3                   ;INCREASE DIG2 BY 1
95  LOOP_3:
96        MOV      A,DIG_3
97        SUB      A,10
98        SZ       Z                       ;DIG3 = 10?
99        CLR      DIG_3                   ;YES, RESET DIG0
100 END_TMR_INT:
101       MOV      A,STACK_PSW
102       MOV      STATUS,A                ;POP STATUS
103       MOV      A,STACK_A
104       RETI                              ;POP A
105 ;*********************************************************************
106 ;         EXTERNAL INTERRUPT SERVICE ROUTINE
107 ;*********************************************************************
108 EXT_INT   :
109       MOV      STACK_A,A               ;PUSH A
110       MOV      A,STATUS
111       MOV      STACK_PSW,A             ;PUSH STATUS
112       MOV      A,00000001B
113       XORM     A,GO_STOP_FLAG          ;COMPLEMENT GO_STOP_FLAG
114       MOV      A,STACK_PSW
115       MOV      STATUS,A                ;POP STATUS
116       MOV      A,STACK_A               ;POP A
117       RETI
118 ;*********************************************************************
119 ;         RETURN THE TABLE VALUE INDEX BY A
120 ;*********************************************************************
121 TRANS PROC
122       ADDM     A,PCL                   ;COMMON ANODE 7-SEG LED CODE
123       RET      A,11000000B             ;'0'
124       RET      A,11111001B             ;'1'
125       RET      A,10100100B             ;'2'
126       RET      A,10110000B             ;'3'
127       RET      A,10011001B             ;'4'
128       RET      A,10010010B             ;'5'
129       RET      A,10000010B             ;'6'
130       RET      A,11111000B             ;'7'
131       RET      A,10000000B             ;'8'
132       RET      A,10011000B             ;'9'
133 TRANS ENDP
134 ;*********************************************************************
135 ;              Delay about DEL1*1ms
136 ;*********************************************************************
137 DELAY PROC
138       MOV      A,07
139       MOV      DEL1,A                  ;SET DEL1 COUNTER
140 DEL_1: MOV     A,03
141       MOV      DEL2,A                  ;SET DEL2 COUNTER
142 DEL_2: MOV     A,110
143       MOV      DEL3,A                  ;SET DEL3 COUNTER
144 DEL_3: SDZ     DEL3                    ;DEL3 DOWN COUNT
145       JMP      DEL_3
```

```
146         SDZ     DEL2                    ;DEL2 DOWN COUNT
147         JMP     DEL_2
148         SDZ     DEL1                    ;DEL1 DOWN COUNT
149         JMP     DEL_1
150         RET
151 DELAY   ENDP
152         END
```

程序说明

行号	说明
7~16	依序定义变量地址。请读者注意，由于在程序中采用间接寻址法(Indirect Addressing)依序显示 DIG0~DIG3 的数值，因此这 4 个寄存器的地址必须是连续的，不可以分开。
18	定义 $f_{SYS}=4$ MHz。
19~22	定义 SCAN_PORT 为 PC，SEG_PORT 为 PB。
25	声明存储器地址由 000h 开始(HT46x23 Reset Vector)。
27	声明存储器地址由 004h 开始(HT46x23 Externsl Interrupt Vector)。
29	声明存储器地址由 008h 开始(HT46x23 Time/Event Counter Interrupt Vector)。
32~33	将控制显示七段显示器各线段的 SEG_PORT，以及控制扫描的 SCAN_PORT 全设置为输出模式。
34~35	EMI、ETI 及 EEI 等中断控制位有效。
36~37	将 Time/Event Counter 定义为 Timer 模式，并设置计数时脉 $f_{INT}=f_{SYS}/8$。
38~41	设置 TMRH 与 TMRL 的值，使 Time/Event Counter 计数溢位的时间为 100 ms。
43~46	清除寄存器 DIG_0、DIG_1、DIG_2 与 DIG_3，这 4 个寄存器在程序中分别存储秒的十分位数、秒的个位数、秒的十位数及分钟的计数值。
47	清除 GO/STOP 标志位，此标志位在 Time/Event Counter 中断服务子程序中是用来判断是否计时的依据(0：停止计时，1：继续计时)。
48	启动 Time/Event Counter 开始计数。
50~61	此段程序是逐一点亮(扫描)DIG_0~DIG_3；46 其中是以间接寻址法(MP0 配合 IAR0)，循序将 DIG_0~DIG_3 的数值通过 TRANS 子程序转换为七段显示码后输出到 SEG_PORT，并配合 SCAN_PORT 的左移动作点亮对应的显示器。同时，每个显示器被点亮的时间由 DELAY 子程序所控制，大约为 7 ms。或许读者会质疑在第 50 行已经将 SCAN_PORT 设为 11111110b，只要以 "RL SCAN_PORT" 不就可以达到一次点亮一个显示器的目的了吗？为何要化简为繁用 60、61 两行指令来完成呢？如果是以 PA 来控制的话，上述的作法绝对正确可行。但问题是 SCAN_PORT 为 PC，而 HT46x23 的 PC 只有 5 位(PC4~PC0)，虽然第 50 行已经将 SCAN_PORT 设为 11111110b，但第一次执行 "RL SCAN_PORT" 指令的结果，会使 SCAN_PORT=11111100b(一次点亮两个显示器)。这主要是因为对于实际上不存在的位(PC6、PC7)，若读取其值的话将得到 "0" 所造成的。笔者特别以 PC 来控制扫描，其目的就是想特别提醒读者注意此特性。
62~64	判断是否已扫描到最后一个显示器，若是，则重新开始扫描；否则继续扫描下一个显示器。
68~104	Time/Event Counter 中断服务子程序，程序会先判断 GO/STOP 标志位，决定是否要定时，若 GO/STOP 标志位=0，表示为停止定时模式，直接返回主程序；反之，若 GO/STOP 标志位=1，表示为定时模式，此时先将记录 "1/10 秒" 的寄存器(DIG_0)加 1，然后再依序判断 "秒" 寄存器(DIG_1)、"十秒" 寄存器(DIG_2)与 "分" 寄存器(DIG_3)是否需要更新。
108~107	外部中断服务子程序，此段程序的目的是将控制是否定时的 GO/STOP 标志位予以反向。
121~133	七段显示器查表子程序。
137~151	DELAY 子程序，延时 7 ms。

关于共阳极(Common Anode)七段显示器的字形码，请读者参阅表4-3-1。本实验的重点就在于延时时间的掌握，如果延时时间不够长，显示的数字就不清楚；反之，若延时时间太

长,就会看到数字闪烁,甚至看到数字是逐一点亮的情形。由于视觉暂留的时间约30~45 ms,而在此让每一个数字点亮后延时约7 ms(4×7＝28 ms),不至超过视觉暂留时间,所以可以清楚看见七段显示器上的数字。如果读者执行时发现七段显示器仍有些闪动的话,可以试着缩短DELAY子程序的延时时间加以改善。在程序执行时,用户可以通过GO/STOP按键来控制定时器开始或暂停计数,若想让时间归零,可以按下RESET按键达到此目的。

5-2-5 动动脑＋动动手

- 试着将DELAY子程序的时间延长,观察七段显示器的显示情形。
- 试着将DELAY子程序的时间缩短,观察七段显示器的显示情形。

5-3 静态点矩阵LED控制实验

5-3-1 目 的

本实验将利用视觉暂留的特性,在8×8点矩阵LED上显示字形。

5-3-2 学习重点

通过本实验,读者能了解如何以扫描的方式来控制8×8点矩阵LED的显示。

5-3-3 电路图(见图5-3-1)

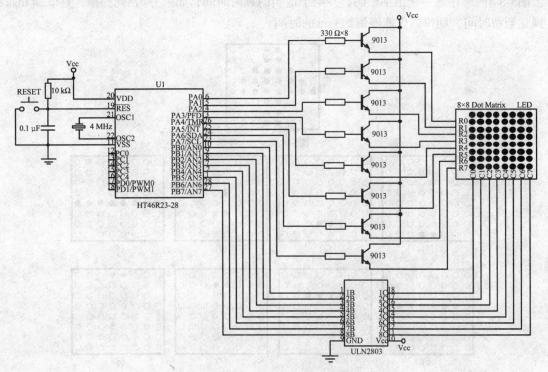

图5-3-1 8×8点矩阵LED控制电路

点矩阵(Dot Matrix)LED在各式广告牌上都看得到,其控制方式其实与5-2节七段显示器的扫描控制差不多,也是利用人类视觉暂留的特性,让依序一排一排点亮点矩阵的LED,呈现同时发亮的错觉。只要适当地控制各排所需点亮的字形码与点亮的时间,就可以看到各种字形了。

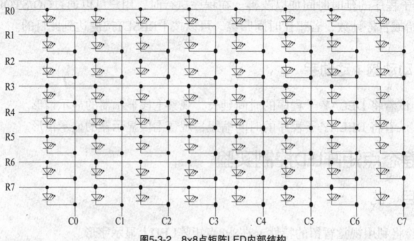

图5-3-2 8×8点矩阵LED内部结构

图5-3-2为8×8点矩阵LED的内部电路,一般应用上是将其分成行(Column;C0~C7)、列(Row;R0~R7)加以控制。例如本实验的电路中(见图5-3-1所示),是将HT46x23的PB作为列扫描的控制;而PA则作为字形输出的行控制。如果要显示如图5-3-3所示的"↑"符号,则就需依图5-3-4的顺序逐一送出字形码,并给予适当的延时时间;如果分为8列扫描,且取24 ms的视觉暂留时间,则每列仅能停留约3 ms的时间。

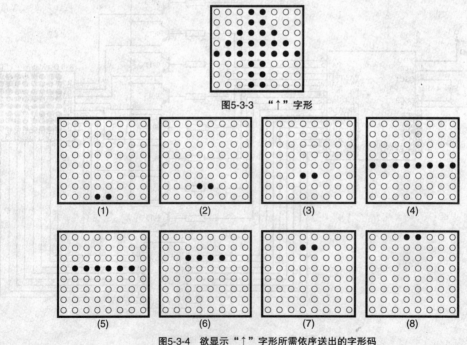

图5-3-3 "↑"字形

图5-3-4 欲显示"↑"字形所需依序送出的字形码

5-3-4 流程图及程序

1. 流程图(见图 5-3-5)

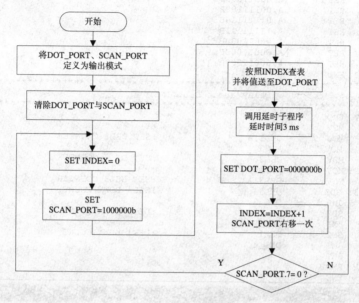

图5-3-5 流程图

2. 程序 5-3 静态点矩阵 LED

```
1    ;PROGRAM : 5-3.ASM  (5-3.PRJ)                      BY STEVEN
2    ;FUNCTION: STATIC 8X8 DOT-MATRIX LED DEMO PROGRAM  2002.DEC.07.
3    #INCLUDE    HT46R23.INC
4                .CHIP HT46R23
5    ;--------------------------------------------------------------
6    MY_DATA     .SECTION    'DATA'        ;== DATA SECTION ==
7    DEL1        DB      ?                 ;DELAY LOOP COUNT 1
8    DEL2        DB      ?                 ;DELAY LOOP COUNT 2
9    DEL3        DB      ?                 ;DELAY LOOP COUNT 3
10   INDEX       DB      ?                 ;TABLE INDEX
11   ;--------------------------------------------------------------
12   DOT_PORT    EQU     PA                ;DEFINE DOT_PORT
13   DOT_PORTC   EQU     PAC               ;DEFINE DOT_PORT CON. REG.
14   SCAN_PORT   EQU     PB                ;DEFINE SCAN_PORT
15   SCAN_PORTC  EQU     PBC               ;DEFINE SCAN_PORT CON. REG.
16
17   MY_CODE     .SECTION  AT 0 'CODE'     ;== PROGRAM SECTION ==
18          ORG     00H                    ;HT-46RXX RESET VECTOR
19          CLR     SCAN_PORT              ;TURN OFF ALL SCAN LINE
20          CLR     DOT_PORT               ;CLEAR DISPLAY DATA
21          CLR     SCAN_PORTC             ;CONFIG SCAN_PORT AS OUTPUT MODE
22          CLR     DOT_PORTC              ;CONFIG DOT_PORT AS OUTPUT MODE
23          SET     SCAN_PORT.7
24   MAIN:
25          CLR     INDEX                  ;INITIAL TABLE INDEX
26   NEXT_ROW:
27          MOV     A,INDEX
28          CALL    TRANS                  ;GET DISPLAY CODE
29          MOV     DOT_PORT,A             ;SEND TO DOT_PORT
30          CALL    DELAY                  ;DELAY 3mS
31          INC     INDEX                  ;INCREASE INDEX BY 1
32          CLR     DOT_PORT               ;TURN OFF ALL LED (GHOST EFFECT)
33          RR      SCAN_PORT              ;SCAN NEXT ROW
34          SNZ     SCAN_PORT.7            ;ALL LINES HAVE BEEN SCANNED?
35          JMP     NEXT_ROW               ;NO, SCAN NEXT LINE
36          JMP     MAIN                   ;YES, RE-START
37   ;**************************************************************
```

```
38  ;                   RETURN THE TABLE VALUE INDEX BY A
39  ;*********************************************************************
40  TRANS   PROC
41          ADDM        A,PCL
42          RET         A,00011000B
43          RET         A,00011000B
44          RET         A,00111100B
45          RET         A,01111110B
46          RET         A,11111111B
47          RET         A,00011000B
48          RET         A,00011000B
49          RET         A,00011000B
50  TRANS   ENDP
51  ;*********************************************************************
52  ;                       Delay about DEL1*1ms
53  ;*********************************************************************
54  DELAY   PROC
55          MOV         A,02
56          MOV         DEL1,A              ;SET DEL1 COUNTER
57  DEL_1:  MOV         A,03
58          MOV         DEL2,A              ;SET DEL2 COUNTER
59  DEL_2:  MOV         A,110
60          MOV         DEL3,A              ;SET DEL3 COUNTER
61  DEL_3:  SDZ         DEL3                ;DEL3 DOWN COUNT
62          JMP         DEL_3
63          SDZ         DEL2                ;DEL2 DOWN COUNT
64          JMP         DEL_2
65          SDZ         DEL1                ;DEL1 DOWN COUNT
66          JMP         DEL_1
67          RET
68  DELAY   ENDP
69          END
```

程序说明

7~10   依序定义变量地址。

12~15   定义 DOT_PORT 与 SCAN_PORT 分别为 PA 与 PB。

18   声明存储器地址由 000h 开始(HT46xx Reset Vector)。

19~22   定义 DOT_PORT 与 SCAN_PORT 为输出模式，并将两个 PORT 均清除为 0，其目的在关闭点矩阵的所有 LED。

23   将 SCAN_PORT.7 设置为 1，扫描点矩阵的第一排。

25   清除索引值 INDEX 为 0。

27~29   调用 TRANS 子程序，根据 INDEX 的值提出欲显示的 LED 码并输出至 DOT_PORT。

30   调用 DELEY 子程序，延时 3 ms。

31   将索引值 INDEX 加 1。

32   将 DOT_PORT 清除为 0，在扫描下一排之前先将 DOT_PORT 清除，以避免产生"鬼影"现象。

33~36   判断是否已扫描完毕，成立，则重新扫描第一排；反之继续扫描下一排。

程序中是以 PB 作为字形码的输出，而以 PA 作为扫描列的控制。有时为了获得更高分辨率的字形或显示多个字形，可能会增加 LED 的列数及行数，此时每列扫描的时间势必将缩短。太短的停留时间会使 LED 无法在短时间内完全导通，而使 LED 的亮度大打折扣。解决的方式是：增加通过 LED 的电流，使 LED 在导通的瞬间达到应有的亮度。不过此刻要特别注意，因为这时候的大电流是针对瞬间导通时所设计的，切记勿让此电流长时间流经 LED，否则将会损害电路。

5-3-5 动动脑 + 动动手

📖 改写本实验的程序，使显示的字形由"↑"改为"↓"。

5-4 动态点矩阵LED控制实验

5-4-1 目的

承续实验5-3静态点矩阵LED字形显示，本实验在8×8点矩阵LED上显示向上移动的字形。

5-4-2 学习重点

通过本实验，读者能运用编程技巧，在8×8点矩阵LED上做动态显示，并应更熟悉HT46xx"间接寻址"的运用。

5-4-3 电路图（见图5-4-1）

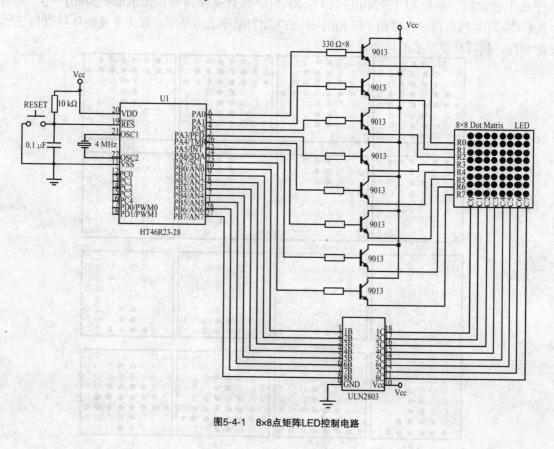

图5-4-1　8×8点矩阵LED控制电路

5-3节中在点矩阵LED上显示静态的字形，相信读者已经十分清楚该控制方式，那么只需要再使用一点编程上的技巧，就可以让字形达到移动的效果。

参考图5-4-2，如果在程序存储器(PROM)中建立图中所需的16种图案的字形码，然后再让每个字形重复显示几次，也就是说依序由(1)到(16)让字形在点矩阵上停留120 ms(5次×8列×3 ms/列)，应该就可以看到"↑"符号移出点矩阵后再移回原位置的效果。至于移动速度的快慢，则取决于每个字形重复显示的次数。问题是这种做法，一个字形码需要约8字节的程序存储器，光是一个字形的移动效果就必须耗费掉128字节(8×16)，万一要多几个字形的移动特效，耗费掉大半的程序存储器不说，光是建表的工程就可能浪费掉许多宝贵的时间。所以，上述控制字形移动的做法虽然正确，但是否有其他的替代方案可行，以避免耗费存储器并节省建表工程所需时间呢?答案当然是肯定的。读者可以再仔细观察图5-4-2，如果将原来的字形先发送到数据存储器(RAM)中，然后再将数据存储器中的字形依序移位、显示，是不是就达到一样的效果呢?简而言之，就是以编程技巧来换取程序所需的空间。

首先，在数据存储器中保留16字节作为字形移动缓冲区(BUFFER[0]~BUFFER[15])，要显示字形时，将8字节的字形码由程序存储器先发送到BUFFER[0]~BUFFER[7]的数据存储器中，并将BUFFER[8]~BUFFER[15]清除为0(见图5-4-3的(1))。紧接着将BUFFER[0]~BUFFER[7]循序送出显示后，再将BUFFER[0]~BUFFER[15]中的数据循环移位(Circular Shift)一次(见图5-4-3的(2))，然后再一次将BUFFER[0]~BUFFER[7]循序送出显示。如此重复循环移位与显示的动作，就可获得字形的移动效果了。

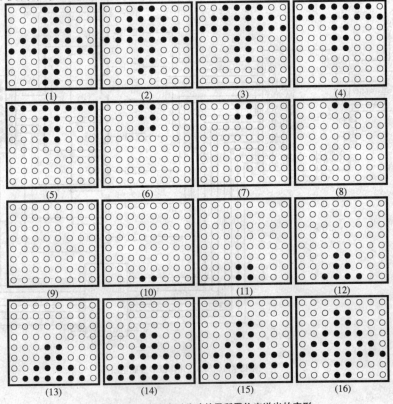

图5-4-2　产生字形移动效果所需依序送出的字形

第 5 章　进阶实验篇

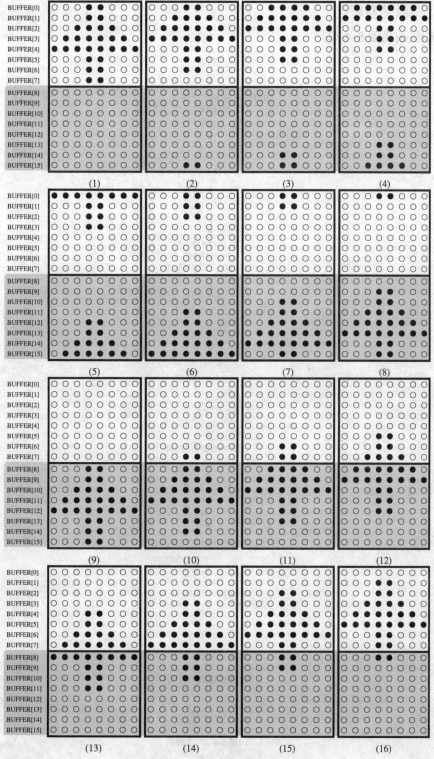

图5-4-3　以BUFFER产生字形移动效果

5-4-4 流程图及程序

1. 流程图(见图5-4-4)

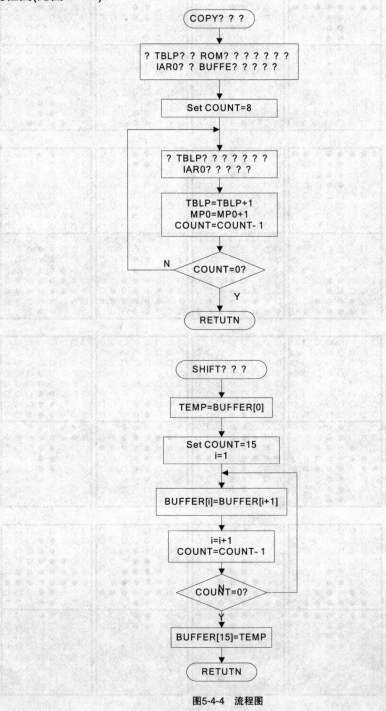

图5-4-4 流程图

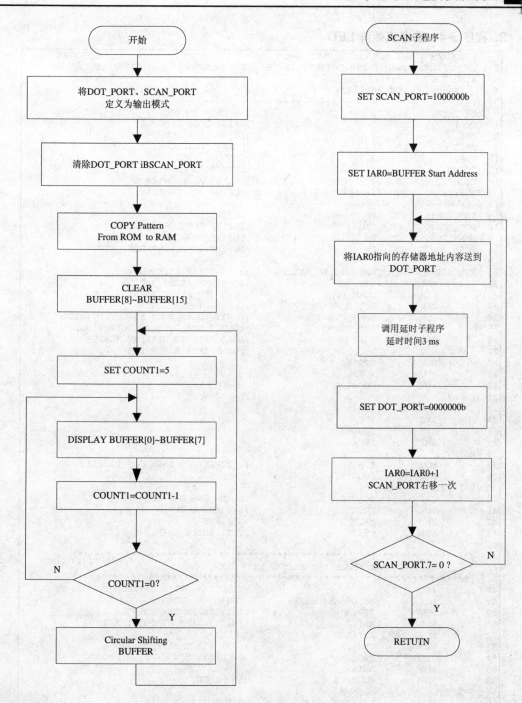

图5-4-4 流程图(续)

2. 程序 5-4 动态点矩阵 LED

```
1    ;PROGRAM : 5-4.ASM  (5-4.PRJ)                        BY STEVEN
2    ;FUNCTION: DYNAMIC 8X8 DOT-MATRIX LED DEMO PROGRAM    2002.DEC.07.
3    #INCLUDE   HT46R23.INC
4               .CHIP HT46R23
5    ;------------------------------------------------------------
6    MY_DATA     .SECTION    'DATA'       ;== DATA SECTION ==
7    DEL1        DB    ?                  ;DELAY LOOP COUNT 1
8    DEL2        DB    ?                  ;DELAY LOOP COUNT 2
9    DEL3        DB    ?                  ;DELAY LOOP COUNT 3
10   TEMP        DB    ?                  ;TEMPORY REGISTER
11   COUNT       DB    ?                  ;UNIVERSAL COUNTER
12   COUNT1      DB    ?                  ;UNIVERSAL COUNTER 1
13   BUFFER      DB    16 DUP(?)          ;BUFFER FOR DISPLAY DATA
14   ;------------------------------------------------------------
15   DOT_PORT    EQU   PA                 ;DEFINE DOT_PORT
16   DOT_PORTC   EQU   PAC                ;DEFINE DOT_PORT CON. REG.
17   SCAN_PORT   EQU   PB                 ;DEFINE SCAN_PORT
18   SCAN_PORTC  EQU   PBC                ;DEFINE SCAN_PORT CON. REG.
19
20   MY_CODE     .SECTION   AT 0 'CODE'   ;== PROGRAM SECTION ==
21         ORG        00H                 ;HT-46RXX RESET VECTOR
22         CLR        SCAN_PORT           ;TURN OFF ALL SCAN LINE
23         CLR        DOT_PORT            ;CLEAR DISPLAY DATA
24         CLR        SCAN_PORTC          ;CONFIG SCAN_PORT AS OUTPUT MODE
25         CLR        DOT_PORTC           ;CONFIG DOT_PORT AS OUTPUT MODE
26         CALL       COPY                ;LOAD DATA TO DISPLAY BUFFER
27         MOV        A,OFFSET BUFFER+8   ;CLEAR LAST 8 BUFFERS
28         MOV        MP0,A
29         MOV        A,8
30         MOV        COUNT,A
31   CLEAR:
32         CLR        IAR0
33         INC        MP0
34         SDZ        COUNT
35         JMP        CLEAR
36   MAIN_1:
37         MOV        A,5                 ;5 TIMES FOR EACH DOT_DISPLAY
38         MOV        COUNT1,A
39   MAIN_2:
40         CALL       SCAN                ;SCAN LINE 0~7
41         SDZ        COUNT1              ;5 TIMES OVER?
42         JMP        MAIN_2              ;NO, RESCAN
43         CALL       SHIFT               ;YES, SHIFT BUFFER 1 BYTE
44         JMP        MAIN_1              ;RE DO
45   ;****************************************************************
46   ;           SHIFT DISPLAY BUFFER BY 1 LOCATION
47   ;****************************************************************
48   SHIFT   PROC                         ;CIRCULAR SHIFT DATA IN BUFFER BY ONE
49         MOV        A, OFFSET BUFFER    ;TOTAL 16 BYTES
50         MOV        MP0,A               ;INITIAL MEMORY POINTER 0
51         MOV        MP1,A               ;INITIAL MEMORY POINTER 1
52         MOV        A,IAR0
53         MOV        TEMP,A              ;SAVE FIRST DATA
54         INC        MP0                 ;INCREASE MP0 BY 1
55         MOV        A,16-1
56         MOV        COUNT,A             ;SET COUNT
57   SHIFT_1:
58         MOV        A,IAR0              ;GET BUFFER+1 DATA
59         MOV        IAR1,A              ;COPY TO BUFFER
60         INC        MP0                 ;INCREASE MP0 BY 1
61         INC        MP1                 ;INCREASE MP1 BY 1
62         SDZ        COUNT               ;COUNT-1 = 0?
63         JMP        SHIFT_1             ;NO, NEXT BYTE
64         MOV        A,TEMP              ;YES.
65         MOV        IAR1,A              ;RE-LOAD FIRST DATA TO BUFFER END
66         RET
67   SHIFT   ENDP
68   ;****************************************************************
```

```
69  ;           COPY DISPLAY CODE TO DISPLAY BUFFER
70  ;********************************************************************
71  COPY    PROC                                ;COPY 5 BYTES PATTEN TO BUFFER
72          MOV         A,OFFSET BUFFER
73          MOV         MP0,A                   ;INITIAL MEMORY POINTER 0
74          MOV         A,8
75          MOV         COUNT,A                 ;SET BYTE COUNTER
76          MOV         A,OFFSET TABLE          ;INITIAL TABLE POINTER
77          MOV         TBLP,A
78  COPY_1:
79          TABRDL      IAR0                    ;READ TABLE DATA TO BUFFER
80          INC         TBLP                    ;INCREASE TABLE POINTER
81          INC         MP0                     ;POINT TO NEXT BUFFER ADDRESS
82          SDZ         COUNT                   ;COUNT-1 = 0?
83          JMP         COPY_1                  ;NO, COPY NEXT BYTE
84          RET                                 ;YES, RETURN
85  COPY    ENDP
86  ;********************************************************************
87  ;               SCAN EACH LINE ONE TIME
88  ;********************************************************************
89  SCAN    PROC                                ;DISPLAY DATA IN BUFFER 1 TIMES
90          MOV         A,OFFSET BUFFER
91          MOV         MP0,A                   ;MEMORY POINTER 0
92  MAIN:
93          CLR         SCAN_PORT               ;TURN OFF ALL SCAN LINE
94          SET         SCAN_PORT.7             ;TURN ON TOP LINE
95  NEXT_ROW:
96          MOV         A,IAR0
97          MOV         DOT_PORT,A              ;SEND TO DOT_PORT
98          CALL        DELAY                   ;DELAY 3ms
99          INC         MP0
100         CLR         DOT_PORT                ;TURN OFF ALL LED (GHOST EFFECT)
101         RR          SCAN_PORT               ;SCAN NEXT ROW
102         SNZ         SCAN_PORT.7             ;ALL LINES HAVE BEEN SCANNED?
103         JMP         NEXT_ROW                ;NO, SCAN NEXT LINE
104         RET                                 ;YES, RETURN
105 SCAN    ENDP
106 ;********************************************************************
107 ;                   Delay about DEL1*1ms
108 ;********************************************************************
109 DELAY   PROC
110         MOV         A,03
111         MOV         DEL1,A                  ;SET DEL1 COUNTER
112 DEL_1:  MOV         A,03
113         MOV         DEL2,A                  ;SET DEL2 COUNTER
114 DEL_2:  MOV         A,110
115         MOV         DEL3,A                  ;SET DEL3 COUNTER
116 DEL_3:  SDZ         DEL3                    ;DEL3 DOWN COUNT
117         JMP         DEL_3
118         SDZ         DEL2                    ;DEL2 DOWN COUNT
119         JMP         DEL_2
120         SDZ         DEL1                    ;DEL1 DOWN COUNT
121         JMP         DEL_1
122         RET
123 DELAY   ENDP
124         ORG         LASTPAGE
125 TABLE:  DC          0000000000011000B
126         DC          0000000000111000B
127         DC          0000000001111110B
128         DC          0000000011111111B
129         DC          0000000000011000B
130         DC          0000000000011000B
131         DC          0000000000011000B
132         DC          0000000000011000B
133         END
```

程序说明

7~13   依序定义变量地址。

15~18   定义 DOT_PORT 与 SCAN_PORT 分别为 PA 与 PB。

21   声明存储器地址由 000h 开始。

行号	说明
22~25	定义 DOT_PORT 与 SCAN_PORT 为输出模式,并将其清除为 0。
26	调用 COPY 子程序,将 8 字节的字形码由程序存储器发送到数据存储器 BUFFER[0]~[7]中。
27~35	运用间接寻址模式,将 BUFFER[8]~[15]的内容予以清除。
37~38	设置计数器 COUNT 为 5,此数值决定 BUFFER[0]~[7]的重复扫描次数,其值越大则点矩阵显示一个字形的时间越久,即图形移动速度越慢。
40	调用 SCAN 子程序,将 BUFFER[0]~[7]的字形码显示一次。
41~42	判断是否已重复显示 5 次,若成立,则进行发送数据的子程序;反之则继续显示。
43	调用 SHIFT 子程序,将 BUFFER[0]~[15]的数据循环移位一次。
44	重新执行程序。
48~67	SHIFT 子程序,运用间接寻址模式将寄存器 BUFFER[0]~[15]内的数据循环移位一次,如下所示:

71~85	COPY 子程序,利用间接寻址模式将 8 字节的字形码由程序存储器发送到的数据存储器 BUFFER[0]~[7]中。
89~105	SCAN 子程序,利用间接寻址模式将 BUFFER[0]~[7]的字形码显示一次。其扫描方法与 5-3 节静态点矩阵 LED 的扫描方法相同,只不过是改成提取 BUFFER[0] ~ [7]中的数据予以显示。
109~123	DELAY 子程序,延时 3 ms。

COPY子程序的功能,是运用间接寻址模式将8字节的字形码由程序存储器复制到数据存储器BUFFER[0]~BUFFER[7]。SCAN子程序,负责将BUFFER[0]~ BUFFER[7]的字形码配合不同的扫描码一一送出显示,每送出一笔字形码即调用DELAY子程序一次,延时3 ms(与实验5-3的做法相同,只不过是将显示的存储器由程序存储器改成数据存储器)。SHIFT子程序则担负起将BUFFER中的数据循环移位一次的工作,以制造字形移动的显示效果。程序中每次都将BUFFER[0]~BUFFER[7]的字形码重复显示5次,所以字形是以每列120 ms(5×8×3 ms)的速度向上移动,读者可以试着更改次数来调整字形移动的快慢。

5-4-5 动动脑 + 动动手

- 试着改写本实验的程序,使字形的移动由上移变成往下移。
- 是否可以试着写出向左、向右移动的控制程序呢?
- 请分析若将程序中第 15~18 行改成如下的定义,点矩阵显示的图案会有什么变化?请读者先在分析中预测其结果,然后与实际的执行结果对比。

```
15   DOT_PORT         EQU  PB      ;DEFINE DOT_PORT
16   DOT_PORTC        EQU  PBC     ;DEFINE DOT_PORT CON. REG.
17   SCAN_PORT        EQU  PA      ;DEFINE SCAN_PORT
18   SCAN_PORTC       EQU  PAC     ;DEFINE SCAN_PORT CON. REG.
```

5-5　LCD字形显示实验

液晶显示器(Liquid Crystal Display;LCD),可分为文字(Text)型以及绘图(Graphic)型两大类,本实验将以介绍文字形液晶显示器为主,由于它的显示方式是电压驱动,因此电路本身所需的工作电流非常低,再加上又有许多的内置字形,所以被广泛应用于讲究人机接口的各种场合,例如传真机、影印机、计算机及各式数字仪器上。目前我们所买到的文字型LCD 模块,不管是可以显示几行几个字,通常它内部都使用由日本HITACHI公司制造的HD44780A模块来控制,所以即使购买的LCD是不同厂牌或型号的文字型LCD,只要控制芯片相同,其控制方法都是一样的。HD44780A控制器内部共有80个字节,可供存储单片机送过来的显示数据。本节将针对其控制方式与特性加以介绍。

1. LCD 模块引脚说明

市售 LCD 模块一般都为14个引脚,且功能及引脚位置大致相同,但实际应用时仍需先确定其引脚规格。另外,具背光功能的LCD会有额外的2个引脚,作为背光设备的电源输入之用。各引脚的功能说明,请参考表5-5-1。

表5-5-1　LCD模块引脚说明

引脚号	引脚符号	方　向	名称及功能
1	Vss	—	电源接地端(Ground)
2	VDD	—	电源正端:接+5 V
3	Vo	—	亮度调整电压输入端(Contrast Adjustment Voltage),输入+0 V时字符最清晰
4	RS	I	寄存器选择:Low 为指令寄存器(IR); High 为数据寄存器(DR)
5	R/W	I	读/写控制:Low 为写入 LCD; High 为读取 LCD 数据
6	E	I	有效信号:Low 为 LCD 有效;Low 为 LCD 有效
7~14	DB0~DB7	I/O	数据总线:以 8 位控制方式时,DB0 ~ DB7 都有效。若以 4 位控制方式时,则仅 DB4~ DB7 有效,DB0~DB3 不必连接

2. 如何让 LCD 显示数据

HD44780内部存储显示数据的存储器(共80字节),称为Display Data RAM(DD RAM)。以20(字)×2(行)LCD为例,其DD RAM的地址分配,第一行为00h ~ 13h;第二行为40h ~ 53H。只要先设置LCD功能,再将欲显示数据的字符码(Character Code;恰好为ASCII Code)写到LCD内部的DD RAM, LCD就会将这个字在其对应的位置上显示出来。例如:要在LCD的第一行显示一个A,那么只要先设置显示地址为00h,再将41h(A的ASCII码)写入DD RAM即可。若欲显示在第二行时,则需设置显示地址为40h,至于显示地址与DD RAM的对应关系请参考图5-5-1。

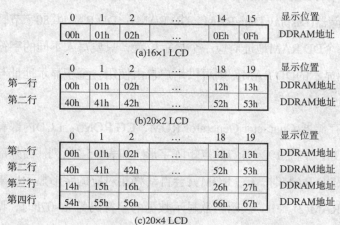

图5-5-1　显示地址与DD RAM的对应关系

3. HD44780 的结构与功能

HD44780的主要特性如下：

- ◆ 80字节的DD RAM(Display Data Memory)。
- ◆ 有一内置192个5×7文字型的CG ROM(Character Generator ROM)。
- ◆ 64字节的CG RAM(Character Generator RAM)，可提供用户自建8个字形(每个字形由8字节组成)。
- ◆ 内部有两个寄存器：指令寄存器(Instruction Register；IR)与数据寄存器(Data Register；DR)。

IR寄存器仅可写入，其功能是接收命令用以设置LCD，例如清除显示、LCD 功能设置、DD RAM或CG RAM地址设置等命令，稍后将会针对LCD的指令加以详细的说明。

DR寄存器则有以下功能：

- ◆ 作为外部写数据到LCD DD RAM或CG RAM时的数据缓冲区。
- ◆ 作为外部读取LCD DD RAM或CG RAM数据时的数据缓冲区。在读出数据后，DD RAM或CG RAM的下一个地址的内容会被自动放入DR寄存器，以备外界读取。

4. LCD 内部的标志位与寄存器

① 忙碌标志位(Busy Flag，BF)：因为 LCD在处理内部工作时，通常要花上40μs～1.64 ms的时间，但外部控制的芯片(如HT46xx)执行一条指令，通常只要几个μs，如果单片机将命令或数据持续不断地送给LCD，那么LCD会因为反应不及而导致大部分数据都无法正常写入。所以单片机在存取LCD数据之前，一定要先检查BF标志位。若BF＝1，表示LCD正在处理内部的工作，此时外部用来控制LCD的芯片无法对LCD做任何写入的动作；反之，当BF＝0时，则表示外部控制芯片可以存取LCD的数据。

② 数据显示存储器(Display Data RAM，DD RAM)：DD RAM共80字节，用以存放欲显示数据的字符码。一个DD RAM地址对应一个显示位置，只要写入不同的字符码即可显示不同的字形，未对应到显示的DD RAM，则可供用户自由使用。至于显示地址与DD RAM的对应关系，请参考图5-5-1。

③ 字符产生器存储器(Character Generator ROM，CG ROM)：LCD内部有存放内置字形的ROM，称为CG ROM，内置192个5×7的点矩阵字形，如表5-5-2所列。当外部控制芯片将这些字形的字符码写入DD RAM时，CG ROM就会自动将相对应的字形显示出来。而字形栏中的CG RAM(1)、CG RAM(2)、…，表示当欲显示数据的字符码为0h~07h时，调出CG RAM里由用户自建的造型码加以显示，而不是CG ROM内置的字形。

第5章 进阶实验篇

表5-5-2 字符码与字形对应表

④ 自建字符产生器存储器(Character Generator RAM，CG RAM)：CG RAM提供用户存放8个自己所设计的5×7点矩阵造型数据码(若不需要显示光标时，则可设计5×8点矩阵造型数据码)，以便显示所需的特殊字符造型。字符码与CG RAM地址的对应关系，请参考表5-5-3。当显示数据的字符码为00h~07h，即可调出相对存放于CG RAM位置内的字形并加以显示。

表5-5-3　字符码与CG RAM地址及自造字形对应表

Character Codes (DDRAM data) 7 6 5 4 3 2 1 0 High　　　　Low	CGRAM Address 5 4 3 2 1 0 High　Low	Character Patterns (CGRAM data) 7 6 5 4 3 2 1 0 High　　　　Low
0 0 0 0 * 0 0 *	0 0 0 0 0 0 0 0 0 1 0 0 1 0 0 0 1 1 0 1 0 0 0 1 0 1 0 1 1 0 0 1 1 1 1 0 0 0 1 0 0 1 1 0 1 0	* * * 0 0 0 0 0 ↕ 0 0 0 0 0 1 0 1 1 0 1 1 0 0 1 1 0 0 0 1 1 0 0 0 1 1 1 1 1 0 1 0 0 0 0 1 0 0 0 0 1 0 0 0 0 * * * 0 0 0 0 0
	1 0 1 1 1 1 0 0 1 1 0 1 1 1 1 0 1 1 1 1	* * * * * * * * ↕　　　↕ * * * * * * * *
	0 0 0 0 0 0 0 1	* * * ↕
0 0 0 0 * 1 1 *	1 1 0 1 0 1 0 1 0	* * *
	1 0 1 1 1 1 0 0 1 1 0 1 1 1 1 0 1 1 1 1	* * * * * * * * ↕　　　↕ * * * * * * * *

⑤ 地址计数器(Address Counter，AC)：AC用来指示写数据到DD RAM或CG RAM的地址。使用地址设置命令将地址写入IR寄存器后，地址信号就会由IR寄存器传到AC，当数据写入或读出后，AC的内容值会自动加1或减1(由进入模式命令中的I/D值决定)。

5. LCD 模块指令说明

表5-5-4为LCD模块的控制指令，LCD执行这些命令要花40μs~1.64 ms的时间，用户只要将这些命令写入IR寄存器，即可设置及控制LCD。

表5-5-4 LCD模块指令表

指令	指令码									指令说明		执行时间/μs
	RS	R/W	DB7	DB6	DB5	DB4	DB3	DB2	DB1	DB0		
清除显示器	0	0	0	0	0	0	0	0	0	1	DD RAM里的所有地址填入空白码 20h，AC设置为00h，I/D设置为1	1640
光标归位	0	0	0	0	0	0	0	0	1	X	DD RAM的AC设为0H，光标回到左上角第一行的第一个位置，DD RAM内容不变	1640
进入模式	0	0	0	0	0	0	0	1	I/D	S	I/D=0 CPU写数据到 DD RAM 或读取数据之后AC减1，光标会向左移动 I/D=1 CPU写数据到 DD RAM 或读取数据之后AC加1，光标会向右移动 S=0 显示器画面不因读写数据而移动 S=1 CPU写数据到 DD RAM 后，整个显示器会向左移(若 I/D=0)或向右移(若 I/D=1)，但从DD RAM读取数据时则显示器不会移动	40
显示器 ON/OFF 控制	0	0	0	0	0	0	1	D	C	B	显示器控制： D=0 所有数据不显示 D=1 显示所有数据 光标控制： C=0 不显示光标 C=1 显示光标 光标闪数控制： B=0 不闪烁 B=1 光标所在位置的字会闪烁	40
光标或显示器移动	0	0	0	0	0	1	S/C	R/L	X	X	S/C=0 R/L=0 光标位置向左移(AC值减1) S/C=0 R/L=1 光标位置向右移(AC值加1) S/C=1 R/L=0 显示器与光标一起向左移 S/C=1 R/L=1 显示器与光标一起向右移	40
功能设置	0	0	0	0	1	DL	N	F	X	X	设置数据位长度： DL=0 使用4位(DB7~DB4)控制模式 DL=1 使用8位(DB7~DB0)控制模式 设置显示器的行数： N=0 单行显示 N=1 双行显示两行 设置字形： F=0 5×7 点矩阵字形 F=1 5×10 点矩阵字形	40
CG RAM 地址设置	0	0	0	1	CGRAM Address						将CG RAM的地址(DB5~DB0)写入AC	40
DDRAM 地址设置	0	0	1	DDRAM Address							将DD RAM的地址(DB6~DB0)写入AC	40
读取忙碌标志位和地址	0	1	BF	Address Counter							BF=1，表示目前LCD正忙着内部的工作，因此无法接受外部的命令，必须等到BF=0之后，才可以接受外部的命令。在(DB6~DB0)可读出AC值	0
写数据到CG RAM 或 DD RAM	1	0	Write Data								将数据写入DD RAM或CG RAM	40
从CG RAM 或 DD RAM 读取数据	1	1	Read Data								读取CG RAM或DD RAM数据	40

(1) 清除显示器(Clear Display)

指令码	RS	R/W	DB7	DB6	DB5	DB4	DB3	DB2	DB1	DB0
	0	0	0	0	0	0	0	0	0	1

此指令会将DD RAM里的所有地址填入空白码(20h)，且将DDRAM的地址计数器(即AC)设置为00h，并将I/D设置为1(即当写数据到 DD RAM或读取数据之后AC会加1，光标会向右移动)。

(2) 光标归位(Cursor Home)

指令码	RS	R/W	DB7	DB6	DB5	DB4	DB3	DB2	DB1	DB0
	0	0	0	0	0	0	0	0	1	X

X：Don't Care

此指令会将DD RAM的地址计数器(AC)设置为00h，但不会改变DD RAM内部的值；它与第一个指令的差异就是不会清除DD RAM内原有的数据。

(3) 进入模式(Entry Mode)

指令码	RS	R/W	DB7	DB6	DB5	DB4	DB3	DB2	DB1	DB0
	0	0	0	0	0	0	0	1	I/D	S

此指令会设置光标移动的方向，以及显示器是否要移动，其中：

I/D=0 　当外部写数据到DD RAM或从DD RAM读取数据之后，地址计数器(AC)将会被减1，因此光标会向左移动。

I/D=1 　当外部写数据到DD RAM或从DD RAM读取数据之后，地址计数器(AC)将会被加1，因此光标会向右移动。

S=1 　当外部写数据到DD RAM后，整个显示器会向左移(若I/D=0)或向右移(若I/D=1)，但从DD RAM读取数据时则显示器不会移动。

S=0 　显示器画面不会因为外部读/写数据而移动。

(4) 显示器ON/OFF控制(Display ON/OFF)

指令码	RS	R/W	DB7	DB6	DB5	DB4	DB3	DB2	DB1	DB0
	0	0	0	0	0	0	1	D	C	B

此指令控制了整个显示器和光标的ON(显示)或OFF(不显示)，以及光标是否闪烁。

　　　　D=0　所有数据都不显示；　　　D=1　显示所有数据；

　　　　C=0　不显示光标；　　　　　　C=1　显示光标；

　　　　B=0　光标不闪烁；　　　　　　B=1　光标闪烁。

(5) 光标或显示器移动(Cursor or Display Shift)

指令码	RS	R/W	DB7	DB6	DB5	DB4	DB3	DB2	DB1	DB0
	0	0	0	0	0	1	S/C	R/L	X	X

此指令可以在不改变显示数据的情况下，移动光标位置或控制显示器向左或向右移动。S/C和R/L值表示的动作如表5-5-5所列。

表5-5-5　S/C和R/L值表示的动作列表

S/C	R/L	动　作
0	0	光标位置向左移（AC值减1）
0	1	光标位置向右移（AC值加1）
1	0	显示的数据连同光标一起向左移
1	1	显示的数据连同光标一起向右移

注：如果只移动显示数据，则AC的内容不会改变。

以20×2的LCD为例，虽然可以看见的DD RAM为40个，但是LCD屏幕上却不一定只有对应到00h~13h与40h~53h，用户可以控制LCD屏幕左、右移动以观看不同的地址数据，其关系如下所示：

26h	27h	00h	01h	12h	...	11h	12h	13h	14h	15h	...
56h	67h	40h	41h	42h	...	51h	52h	53h	54h	55h	...

屏幕不移动时，LCD显示与DD RAM的地址关系

26h	27h	00h	01h	12h	...	11h	12h	13h	14h	15h	...
56h	67h	40h	41h	42h	...	51h	52h	53h	54h	55h	...

屏幕左移一次时，LCD显示与DD RAM的地址关系

26h	27h	00h	01h	12h	...	11h	12h	13h	14h	15h	...
56h	67h	40h	41h	42h	...	51h	52hh	53h	54h	55h	...

屏幕右移一次时，LCD显示与DD RAM的地址关系

(6) 功能设置(FunctionSet)

指令码	RS	R/W	DB7	DB6	DB5	DB4	DB3	DB2	DB1	DB0
	0	0	0	0	1	DL	N	F	X	X

在设置LCD时，功能设置(FunctionSet)指令必须最先执行。其指令功能说明如下：

① 设置数据位长度

　　DL＝0　使用4位(DB7~DB4)控制模式，数据的读/写必须分成两次完成(先读/写高4位，然后再读/写低4位)。

　　DL＝1　使用8位(DB7~DB0)控制模式。

② 设置显示器的行数

　　N＝0　单行显示。

　　N＝1　双行显示。

③ 设置字形

　　F＝0　5×7点矩阵字形。

　　F＝1　5×10点矩阵字形。注意：设置为5×10点矩阵字形时，LCD将只显示单行。

图5-5-2与图5-5-3分别是LCD与外部单片机以4位及8位总线连接方式的接口。如图5-5-2所示，使用4位总线模式时，必须接到LCD的DB7~DB4这4条总线上，这样LCD方能正常工作，

此时必须将一个字节的数据分成两次读/写，并且是高4位先读/写之后才是低4位的读/写，如此才算真正完成一个字节数据的存取动作。

而以8位总线连接方式时，则只需读/写一次即完成存取一个字节数据的动作，这是4位与8位总线连接接口最大的不同。虽然4位控制模式较为繁琐，但在I/O资源有限的情况之下，有时又不得不采用。

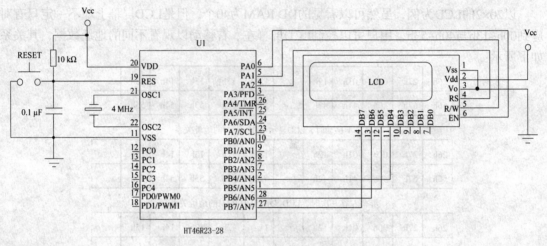

图5-5-2　LCD 4位总线连接接口

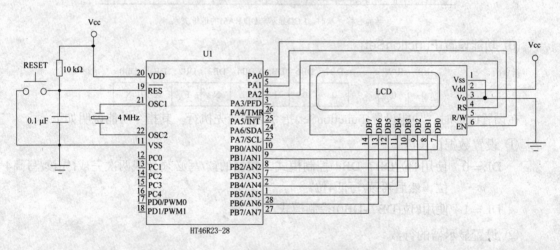

图5-5-3　LCD 8位总线连接接口

(7) CG RAM地址设置(Set CG RAM Address)

指令码	RS	R/W	DB7	DB6	DB5	DB4	DB3	DB2	DB1	DB0
	0	0	0	1	A	A	A	A	A	A

此指令将CG RAM的地址(DB5~DB0)写入地址计数器(AC)。用户在建立自己的字形时，必须先利用此指令设置所欲输入字形的地址。

(8) DD RAM地址设置(Set DD RAM Address)

指令码	RS	R/W	DB7	DB6	DB5	DB4	DB3	DB2	DB1	DB0
	0	0	1	A	A	A	A	A	A	A

此指令将DD RAM的地址(DB6~DB0)写入地址计数器(AC)内。用户可以利用此指令，改变字形在LCD上的显示位置。

(9) 读取忙碌标志位(BF)和地址(AC)(Read Busy Flag and Address)

指令码	RS	R/W	DB7	DB6	DB5	DB4	DB3	DB2	DB1	DB0
	0	1	BF	A	A	A	A	A	A	A

当BF＝1时，表示目前HD44780正忙着内部的工作，因此LCD无法接受外部输入的命令或数据，必须等到BF＝0之后，才可以接受命令或数据。在读取BF的同时，也会读到地址定数器的值(DB6~DB0)，读出的地址可能是CG RAM的地址，或是DD RAM的地址，须视先前的地址指令是设置CG RAM或DD RAM的地址而定。

(10) 写数据到CG RAM或DD RAM(Write Data to CG or DD RAM)

指令码	RS	R/W	DB7	DB6	DB5	DB4	DB3	DB2	DB1	DB0
	1	0	D	D	D	D	D	D	D	D

此指令功能为，依前一次地址指令所设置的RAM地址(CG RAM或DD RAM)，将数据(DB7~DB0)写入DD RAM或CG RAM中。

(11) 自CG RAM或DD RAM读取数据(Read Data from CG or DD RAM)

指令码	RS	R/W	DB7	DB6	DB5	DB4	DB3	DB2	DB1	DB0
	1	1	D	D	D	D	D	D	D	D

此指令功能为读取CG RAM或DD RAM的数据。至于究竟是读CG RAM或DD RAM，则须依前一次地址设置指令所设置的RAM(CG RAM或DD RAM)而定。

6. LCD模块读取/写入时序与常用子程序(WLCMD and WLCMC)

图5-5-4是从LCD读取一个字节的时序图。表5-5-6为其时序列表。首先，需设置RS(RS＝0选择命令寄存器IR，RS＝1选择数据寄存器DR)，用以选择要读出LCD中的哪一个寄存器(IR或DR)，同时将R/W引脚设为High(表示要做读取的动作)，延时t_{AS}时间(最少需140 ns)后，再将E升为High，延时t_{DDR}时间(最大320 ns)后，此时CPU便可以读取DB7~DB0上的数据。CPU在读取数据之后，必须将E的信号设置为Low，如此便完成了一个字节的读取动作。

观察图5-5-5写入时序。要将一个字节数据写入LCD时，首先需设置RS(RS＝0选择命令寄存器IR，RS＝1选择数据寄存器DR)，以选择要将数据写入LCD中的哪一个寄存器(IR或DR)，同时将R/W引脚设为Low(表示要做写的动作)，延时t_{AS}时间(最少需140 ns)之后再将E引脚设置为High(写入动作有效)，其次将所想要写入LCD的数据放到DB7~DB0。最少延时t_{DSW}的时间之后，才可以将E设置为Low，从而完成数据写入动作。E由Low转态到High再转态为Low的时间，称为PW_{EN}，至少必须维持450 ns。

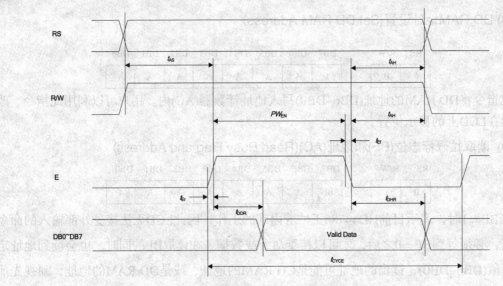

图5-5-4 LCD的读取时序图

表5-5-6 LCD的读取时序列表

符号	项目	最小值	最大值	单 位
t_{CYCE}	Enable Cycle Time	1000	—	ns
PW_{EN}	Enable Pulse Width	450	—	ns
t_{Er}、t_{Ef}	Enable Rise and Fall Time	—	25	ns
t_{AS}	Setup Time	140	—	ns
t_{AH}	Address Hold Time	10	—	ns
T_{DDR}	Data Delay Time	—	320	ns
t_H	Data Hold Time	20	—	ns

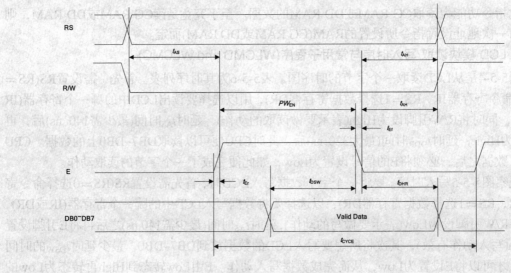

图5-5-5 LCD的写入时序

第 5 章 进阶实验篇

表5-5-7 LCD的写入时序列表

符号	项目	最小值	最大值	单位
t_{CYCE}	Enable Cycle Time	1000	—	ns
PW_{EN}	Enable Pulse Width	450	—	ns
t_{Er}、t_{Ef}	Enable Rise and Fall Time	—	25	ns
t_{AS}	Setup Time	140	—	ns
t_{AH}	Address Hold Time	10	—	ns
t_{DSW}	Data Setup Time	195	—	ns
t_H	Data Hold Time	10	—	ns

在了解LCD的读/写时序关系之后，要来控制LCD就不是难事了。

以图5-5-3 8位总线连接接口为例，只要写程序让HT46xx的输入/输出引脚按照时序图上的要求变化就行了。

现在，就介绍一个以后实验中经常会使用到的子程序：WLCMD及WLCMC，分别负责将数据写到LCD的DR及IR寄存器。虽然是两个子程序，但是从图5-5-5可以看出，其实写入DR或IR寄存器只有"一线之隔"，即RS＝1或0的差异而已。因此，可将这两个子程序加以集成，以达到节省程序存储器的目的。当然，如前面的说明"必须确定LCD有空的时候(BF＝0)，才能再次写数据或送命令给LCD"，所以在子程序中也包含了检查LCD是否忙碌的部分。请参考以下的程序，其流程图如图5-5-6所示。

```
1       ;*********************************************************************
2       ;             LCD DATA/COMMAND WRITE PROCEDURE
3       ;*********************************************************************
4       WLCMD   PROC
5               SET     DC_FLAG                 ;SET DC_FLAG=1 FOR DATA WRITE
6               JMP     WLCM
7       WLCMC:
8               CLR     DC_FLAG                 ;SET DC_FLAG=0 FOR COMMAND WRITE
9       WLCM:
10              SET     LCD_DATAC               ;CONFIG LCD_DATA AS INPUT MODE
11              CLR     LCD_CONTR               ;CLEAR ALL LCD CONTROL SIGNAL
12              SET     LCD_RW                  ;SET RW SIGNAL (READ)
13              NOP                             ;FOR TAS
14              SET     LCD_EN                  ;SET EN HIGH
15              NOP                             ;FOR TDDR
16      WF:
17              SZ      LCD_READY               ;IS LCD BUSY?
18              JMP     WF                      ;YES, JUMP TO WAIT
19              CLR     LCD_DATAC               ;NO, CONFIG LCD_DATA AS OUTPUT MODE
20              MOV     LCD_DATA,A              ;LATCH DATA/COMMAND ON LCD DATA PORT
21              CLR     LCD_CONTR               ;CLEAR ALL LCD CONTROL SIGNAL
22              SZ      DC_FLAG                 ;IS COMMAND WRITE?
23              SET     LCD_RS                  ;NO, SET RS HIGH
24              SET     LCD_EN                  ;SET EN HIGH
25              NOP
26              CLR     LCD_EN                  ;SET EN LOW
27              RET
28      WLCMD   ENDP
```

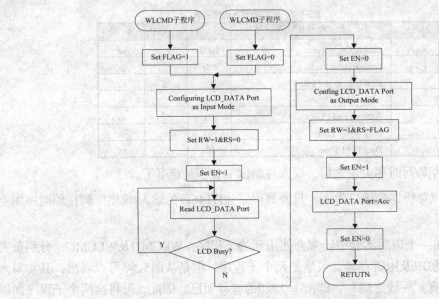

图5-5-6 流程图

读者只要先把HT46x23用来控制E、RS与R/W的Port定义为输出模式,然后将要写入LCD的数据(LCD的指令或字形码)放在Acc寄存器,再调用此子程序就可以。程序中在设置RS与R/W后,在设置E之前插入了一个NOP指令,这是为了要再次提醒读者t_{AS}时间(最少需140 ns)的重要性。如果HT46x23工作在4 MHz频率的情形下,此行指令并不需要;但若使用较快速的芯片(或提高工作频率)来控制LCD时,就必须对LCD时序上的时间仔细考虑,否则,就算是控制信号变化的程序没有错误,LCD也不一定会正常动作。

5-5-1 目 的

在LCD上显示两行字形。

5-5-2 学习重点

通过本实验,读者能熟悉LCD的基本控制方式,而且硬件上的相关电路及程序的控制都能得心应手,操控自如。

5-5-3 电路图

本实验采用8位LCD控制方式。关于LCD的相关命令以及特性请读者先参阅本实验之前的相关说明。

如图5-5-7所示为LCD(8位)控制电路。

第 5 章 进阶实验篇

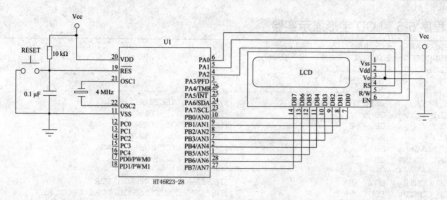

图5-5-7 LCD(8位)控制电路

5-5-4 流程图及程序

1. 流程图(见图 5-5-8)

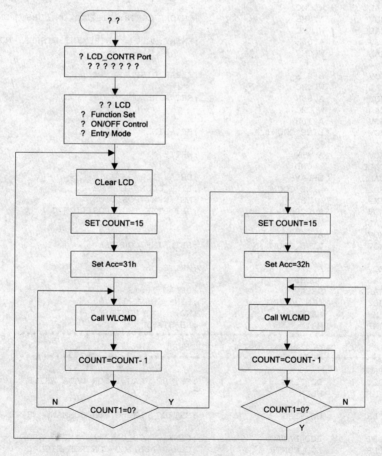

图5-5-8 流程图

2. 程序 5-5　LCD 字形显示实验

```
1       ;PROGRAM : 5-5.ASM  (5-5.PRJ)              BY STEVEN
2       ;FUNCTION: 16X2 LCD DEMO PROGRAM           2002.DEC.07.
3       #INCLUDE    HT46R23.INC
4                   .CHIP HT46R23
5       ;-----------------------------------------------------------------
6       MY_DATA     .SECTION    'DATA'          ;== DATA SECTION ==
7       DEL1        DB      ?                   ;DELAY LOOP COUNT 1
8       DEL2        DB      ?                   ;DELAY LOOP COUNT 2
9       DEL3        DB      ?                   ;DELAY LOOP COUNT 3
10      DC_FLAG     DBIT                        ;LCD DATA/COMMAND FLAG
11      LINE_COUNT  DB      ?                   ;DISPLAY LINE COUNT
12      ;-----------------------------------------------------------------
13      LCD_CONTR   EQU     PA                  ;DEFINE LCD CONTROL PORT
14      LCD_CONTRC  EQU     PAC                 ;DEFINE LCD CONTROL PORT CON. REG.
15      LCD_DATA    EQU     PB                  ;DEFINE LCD DATA PORT
16      LCD_DATAC   EQU     PBC                 ;DEFINE LCD DATA PORT CON. REG.
17      LCD_EN      EQU     LCD_CONTR.0         ;DEFINE EN CONTROL PIN
18      LCD_RW      EQU     LCD_CONTR.1         ;DEFINE RW CONTROL PIN
19      LCD_RS      EQU     LCD_CONTR.2         ;DEFINE RS CONTROL PIN
20      LCD_READY   EQU     LCD_DATA.7          ;DEFINE READY BIT OF LCD
21      MY_CODE     .SECTION  AT 0 'CODE'       ;== PROGRAM SECTION ==
22              ORG         00H                 ;HT-46RXX RESET VECTOR
23              CLR         LCD_CONTRC          ;CONFIG LCD_CONTR PORT AS OUTPUT MODE
24              MOV         A,38H               ;FUNCTION SET: 8-BIT,2-LINE,5X10 DOTS
25              CALL        WLCMC
26              MOV         A,0FH               ;ON/OFF CONTR: DISPLAY ON,CURSOR ON,BLINKING ON
27              CALL        WLCMC
28              MOV         A,06H               ;ENTRY MODE : INCREMENT,DISPLAY NOT SHIFT
29              CALL        WLCMC
30      MAIN:
31              MOV         A,01H               ;CLEAR DISPLAY
32              CALL        WLCMC
33              MOV         A,80H               ;SET LINE ONE, POSITION 0
34              CALL        WLCMC
35              MOV         A,10H
36              MOV         LINE_COUNT,A        ;SET LINE COUNTER
37      MAIN_1:
38              MOV         A,31H               ;WRITE '1'
39              CALL        WLCMD
40              CALL        DELAY               ;DELAY 100mS
41              SDZ         LINE_COUNT          ;LINE_COUNT-1 = 0?
42              JMP         MAIN_1              ;NO, NEXT POSITION
43              MOV         A,0C0H              ;SET LINE TWO, POSITION 0
44              CALL        WLCMC
45              MOV         A,16
46              MOV         LINE_COUNT,A        ;SET LINE COUNTER
47      MAIN_2:
48              MOV         A,32H               ;WRITE '2'
49              CALL        WLCMD
50              CALL        DELAY               ;DELAU 100mS
51              SDZ         LINE_COUNT          ;LINE_COUNT-1 = 0?
52              JMP         MAIN_2              ;NO, NEXT POSITION
53              JMP         MAIN                ;RE-START
54      ;*****************************************************************
55      ;           LCD DATA/COMMAND WRITE PROCEDURE
56      ;*****************************************************************
57      WLCMD   PROC
58              SET         DC_FLAG             ;SET DC_FLAG=1 FOR DATA WRITE
59              JMP         WLCM
60      WLCMC:
61              CLR         DC_FLAG             ;SET DC_FLAG=0 FOR COMMAND WRITE
62      WLCM:
63              SET         LCD_DATAC           ;CONFIG LCD_DATA AS INPUT MODE
64              CLR         LCD_CONTR           ;CLEAR ALL LCD CONTROL SIGNAL
65              SET         LCD_RW              ;SET RW SIGNAL (READ)
66              NOP                             ;FOR TAS
67              SET         LCD_EN              ;SET EN HIGH
68              NOP                             ;FOR TDDR
69      WF:
```

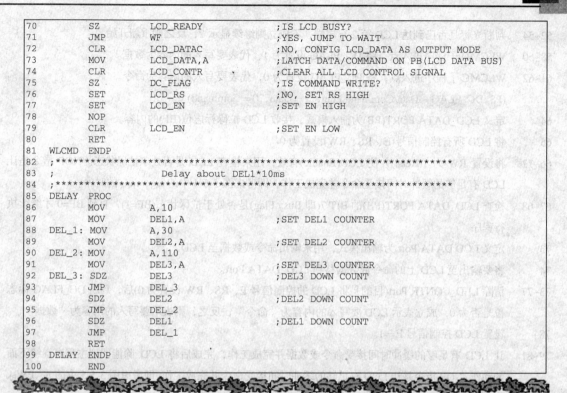

程序说明

7~11　依序定义变量地址。

13~16　定义 LCD_DATA Port 与 LCD_CONTR Port 分别为 PB 与 PA。

17~20　定义 LCD 的控制信号引脚。

23　声明存储器地址由 000h 开始。

24　将 LCD_CONTR Port(PA)定义为输出模式。

25~26　将 LCD 设置为双行显示(N = 1)、使用 8 位(DB7~DB0)控制模式(DL = 1)、5×7 点矩阵字形(F = 0)。

27~28　将 LCD 设置为显示所有数据(D = 1)、显示光标(C = 1)、光标所在位置的字会闪烁(B = 1)。

29~30　将 LCD 的地址标志位(AC)设为递加(I/D = 1)、显示器画面不因读/写数据而移动(S = 0)。

32~33　将 LCD 整个显示器清空。

34~35　将光标移至第 1 行的第 1 个位置。

36~37　在此 LCD 一行共有 16 个位置，因此 COUNT_LINE 设为 10h。

39~41　LCD 的字形表与 ASCII 的对应关系，请参阅表 5-5-2 LCD 字符码与字形对应表。ASCII 中 31h 为 1，在此将 1 写入 LCD。加入延时子程序目的是为了能看到 LCD 上显示 1，不会因为速度过快而无法看见逐字显示的结果。

42~43　判断光标是否已到达 LCD 第一行尾端，不成立则继续显示 1。

44~47　光标已到达 LCD 第一行尾端，所以将光标由第 1 行移至第 2 行的第 1 个位置，并将 COUNT_LINE 复位为 10h。

49~51　如同 36~38 行的动作，差别在于显示的字形为 2(32h)。

52~54	判断光标是否已到达 LCD 第 2 行尾端，不成立则继续显示 2，反之则重新开始。
58~60	WLCMD 子程序进入点，将 DC_FLAG 设置为 1，代表要写入的内容为数据。
61~62	WLCMC 子程序进入点，将 DC_FLAG 设置为 0，代表要写入的内容为命令。 (PS. DC_FLAG：Data/Command Flag； 1＝Data, 0＝Command)
64	定义 LCD_DATA PORT(PB)为输入模式，接收 LCD 忙碌标志位(BF)的内容。
65	将 LCD 所有控制信号（E、RS、RW）设置为 0。
66~72	将设置 RW 为 1(Read)并使 LCD(Enable)有效，用来检查 LCD 是否处于忙碌状态。NOP 指令是让 LCD 有足够的缓冲时间接受命令并完成工作，即考虑图 5-5-4 中的 t_{AS}、t_{DDR} 的时序。
67~68	检查 LCD_DATA PORT(PB)的 BIT 7(即 Busy Flag)是否处于忙碌状态(BF=1)，直到 BF=0 才继续执行程序。
73	定义 LCD DATA Port 为输出模式，用来输出命令或数据至 LCD。
74	将要输出至 LCD 上的命令或数据移到 LCD_DATA Port。
73~77	清除 LED_CONTR Port(目的是将 LCD 的控制信号 E、RS、RW 设置为 0)后，判断 DC_FLAG 标志位是否为 0，成立表示 LCD 要写入的内容为"命令"；反之，则表示要写入的内容为"数据"。
78	设置 LCD 控制信号 E=1。
79~81	让 LCD 有足够的缓冲时间接受命令或数据并完成工作，完成后将 LCD 除能回到主程序继续下面的动作。NOP 指令是让 LCD 有足够的缓冲时间接受命令并完成工作，即考虑图 5-5-5 中的 t_{Er}、PW_{EN} 的时序。
86~99	DELAY 子程序，延时 0.1 s。

本程序主要在 16×2 LCD 的第 1 行及第 2 行上分别显示 1、2 的字形。程序一开始先对 LCD 做初始化的设置，如设置 LCD 为 8 位控制模式、2 行显示、5×7 的点矩阵字形、显示光标、AC 值加 1、清除等，接下来就是分别把 AC 设为 DD RAM 的第 2 行(00h)与第 2 行并显示 1、2 的字形。所有的命令与字形码的写入，都是通过 WLCMC 与 WLCMD 两个子程序来完成，对于这些子程序不了解的读者，烦请再次参阅之前的说明。

5-5-5 动动脑＋动动手

  试着改写程序，让 1、2 字形由原来第 1 行显示完后再换第 2 行显示的方式，改成是上下交替显示的方式。

5-6 LCD 自建字形实验

5-6-1 目 的

在 LCD 上自建"王"、"田"、"|"、"/"、"—"、"\"等字形，并利用程序的技巧产生转动的效果。

5-6-2 学习重点

通过本实验,读者能熟悉如何在LCD的CG RAM(Character Generator RAM)中建立自己的字形。

5-6-3 电路图

如图5-6-1所示为LCD(8位)控制电路。

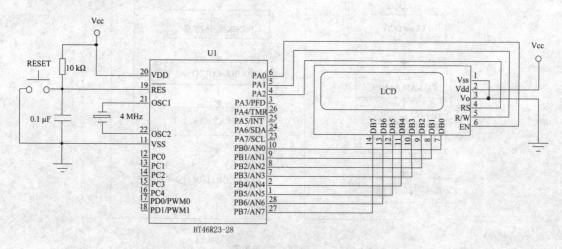

图5-6-1 LCD(8位)控制电路

LCD内部的CG RAM提供用户存放8个自己所设计的5×7点矩阵造型数据码,以便显示用户所需的特殊字符造型。字符码与CG RAM地址的对应关系,请参考表5-5-3。当显示数据的字符码为00h~07h,即可调出相对存放于CG RAM位置内的字形并加以显示。图5-6-2是本实验中要自行建立的字形码。

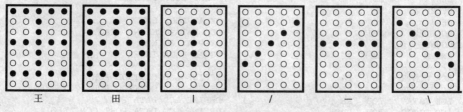

图5-6-2 要建立的字形码

5-6-4 流程图及程序

1. 流程图(见图 5-6-3)

HT46xx 单片机原理与实践

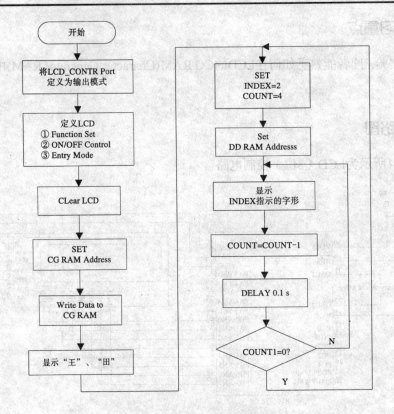

图5-6-3 流程图

2. 程序 5-6 LCD 自建字形实验

```
1   ;PROGRAM : 5-6.ASM (5-6.PRJ)              BY STEVEN
2   ;FUNCTION: 16X2 LCD CG-RAM DEMO PROGRAM   2002.DEC.07.
3   #INCLUDE    HT46R23.INC
4               .CHIP HT46R23
5   ;--------------------------------------------------------
6   MY_DATA     .SECTION   'DATA'       ;== DATA SECTION ==
7   DEL1        DB    ?                 ;DELAY LOOP COUNT 1
8   DEL2        DB    ?                 ;DELAY LOOP COUNT 2
9   DEL3        DB    ?                 ;DELAY LOOP COUNT 3
10  DC_FLAG     DBIT                    ;LCD DATA/COMMAND FLAG
11  COUNT       DB    ?                 ;DISPLAY COUNT
12  INDEX       DB    ?                 ;CGRAM DATA INDEX
13  ;--------------------------------------------------------
14  LCD_CONTR   EQU   PA                ;DEFINE LCD CONTROL PORT
15  LCD_CONTRC  EQU   PAC               ;DEFINE LCD CONTROL PORT CON. REG.
16  LCD_DATA    EQU   PB                ;DEFINE LCD DATA PORT
17  LCD_DATAC   EQU   PBC               ;DEFINE LCD DATA PORT CON. REG.
18  LCD_EN      EQU   LCD_CONTR.0       ;DEFINE EN CONTROL PIN
19  LCD_RW      EQU   LCD_CONTR.1       ;DEFINE RW CONTROL PIN
20  LCD_RS      EQU   LCD_CONTR.2       ;DEFINE RS CONTROL PIN
21  LCD_READY   EQU   LCD_DATA.7        ;DEFINE READY BIT OF LCD
22
23  MY_CODE     .SECTION   AT 0 'CODE'  ;== PROGRAM SECTION ==
24       ORG   00H                      ;HT-46RXX RESET VECTOR
25       CLR   LCD_CONTRC               ;CONFIG LCD_CONTR AS OUTPUT MODE
26       MOV   A,38H                    ;FUNCTION SET: 8-BIT,2-LINE,5X10 DOTS
27       CALL  WLCMC
28       MOV   A,0CH                    ;ON/OFF CONTR: DISPLAY ON,CURSOR BLINKING OFF
29       CALL  WLCMC
30       MOV   A,06H                    ;ENTRY MODE : INCREMENT,DISPLAY NOT SHIFT
31       CALL  WLCMC
32       MOV   A,01H                    ;CLEAR DISPLAY
```

```
 33            CALL       WLCMC
 34            MOV        A,01000000B           ;SET CGRAM START ADDRESS
 35            CALL       WLCMC
 36            MOV        A,6*8
 37            MOV        COUNT,A               ;SET CGRAM WRITE COUNTER
 38            MOV        A,OFFSET TAB_CGRAM    ;GET TABLE START ADDRESS
 39            MOV        TBLP,A
 40    NEXT:
 41            TABRDL     ACC                   ;READ DATA
 42            CALL       WLCMD                 ;WRITE TO CG_RAM
 43            INC        TBLP                  ;INCREASE POINTER BY 1
 44            SDZ        COUNT                 ;COUNT-1 = 0?
 45            JMP        NEXT                  ;NO, NEXT DATA
 46    MAIN:
 47            MOV        A,86H                 ;SET LINE 1 POSITION 6
 48            CALL       WLCMC
 49            CLR        ACC                   ;WRITE CGRAM DATA 0
 50            CALL       WLCMD
 51            MOV        A,01H                 ;WRITE CGRAM DATA 1
 52            CALL       WLCMD
 53    MAIN_1:
 54            MOV        A,02H
 55            MOV        INDEX,A               ;SET CGRAM DATA INDEX
 56            MOV        A,04H
 57            MOV        COUNT,A               ;SET CGRAM DATA COUNTER
 58    MAIN_2:
 59            MOV        A,0C5H                ;SET LINE 2 POSITION 5
 60            CALL       WLCMC
 61            MOV        A,INDEX
 62            CALL       WLCMD                 ;WRITE CGRAM DATA TO DDRAM
 63            CALL       WLCMD                 ;WRITE CGRAM DATA TO DDRAM
 64            CALL       WLCMD                 ;WRITE CGRAM DATA TO DDRAM
 65            CALL       WLCMD                 ;WRITE CGRAM DATA TO DDRAM
 66            CALL       DELAY                 ;JUST FOR SEEING THE DISPLAY
 67            INC        INDEX                 ;INCREASE INDEX BY 1
 68            SDZ        COUNT                 ;COUNT-1 = 0?
 69            JMP        MAIN_2                ;NO, NEXT DATA
 70            JMP        MAIN_1                ;YES, JUMP TO MAIN_2
 71    ;***************************************************************
 72    ;           LCD DATA/COMMAND WRITE PROCEDURE
 73    ;***************************************************************
 74    WLCMD  PROC
 75            SET        DC_FLAG               ;SET DC_FLAG=1 FOR DATA WRITE
 76            JMP        WLCM
 77    WLCMC:
 78            CLR        DC_FLAG               ;SET DC_FLAG=0 FOR COMMAND WRITE
 79    WLCM:
 80            SET        LCD_DATAC             ;CONFIG LCD_DATA AS INPUT MODE
 81            CLR        LCD_CONTR             ;CLEAR ALL LCD CONTROL SIGNAL
 82            SET        LCD_RW                ;SET RW SIGNAL (READ)
 83            NOP                              ;FOR TAS
 84            SET        LCD_EN                ;SET EN HIGH
 85            NOP                              ;FOR TDDR
 86    WF:
 87            SZ         LCD_READY             ;IS LCD BUSY?
 88            JMP        WF                    ;YES, JUMP TO WAIT
 89            CLR        LCD_DATAC             ;NO, CONFIG LCD_DATA AS OUTPUT MODE
 90            MOV        LCD_DATA,A            ;LATCH DATA/COMMAND ON PB(LCD DATA BUS)
 91            CLR        LCD_CONTR             ;CLEAR ALL LCD CONTROL SIGNAL
 92            SZ         DC_FLAG               ;IS COMMAND WRITE?
 93            SET        LCD_RS                ;NO, SET RS HIGH
 94            SET        LCD_EN                ;SET EN HIGH
 95            NOP
 96            CLR        LCD_EN                ;SET EN LOW
 97            RET
 98    WLCMD  ENDP
 99    ;***************************************************************
100    ;                Delay about DEL1*10ms
101    ;***************************************************************
102    DELAY  PROC
103            MOV        A,10
```

```
104            MOV       DEL1,A              ;SET DEL1 COUNTER
105  DEL_1:    MOV       A,30
106            MOV       DEL2,A              ;SET DEL2 COUNTER
107  DEL_2:    MOV       A,110
108            MOV       DEL3,A              ;SET DEL3 COUNTER
109  DEL_3:    SDZ       DEL3                ;DEL3 DOWN COUNT
110            JMP       DEL_3
111            SDZ       DEL2                ;DEL2 DOWN COUNT
112            JMP       DEL_2
113            SDZ       DEL1                ;DEL1 DOWN COUNT
114            JMP       DEL_1
115            RET
116  DELAY     ENDP
117            ORG       LASTPAGE
118  TAB_CGRAM:
119            DC        00011111B           ;CHARACTER '王'
120            DC        00000100B
121            DC        00000100B
122            DC        00011111B
123            DC        00000100B
124            DC        00000100B
125            DC        00011111B
126            DC        00000000B
127
128            DC        00011111B           ;CHARACTER '田'
129            DC        00010101B
130            DC        00010101B
131            DC        00011111B
132            DC        00010101B
133            DC        00010101B
134            DC        00011111B
135            DC        00000000B
136
137            DC        00000000B           ;CHARACTER '|'
138            DC        00000100B
139            DC        00000100B
140            DC        00000100B
141            DC        00000100B
142            DC        00000100B
143            DC        00000000B
144            DC        00000000B
145
146            DC        00000000B           ;CHARACTER '/'
147            DC        00000001B
148            DC        00000010B
149            DC        00000100B
150            DC        00001000B
151            DC        00010000B
152            DC        00000000B
153            DC        00000000B
154
155            DC        00000000B           ;CHARACTER '-'
156            DC        00000000B
157            DC        00000000B
158            DC        00011111B
159            DC        00000000B
160            DC        00000000B
161            DC        00000000B
162            DC        00000000B
163
164            DC        00000000B           ;CHARACTER '\'
165            DC        00010000B
166            DC        00001000B
167            DC        00000100B
168            DC        00000010B
169            DC        00000001B
170            DC        00000000B
171            DC        00000000B
172            END
```

程序说明

行号	说明
7~12	依序定义变量地址。
14~17	定义 LCD_DATA Port 与 LCD_CONTR Port 分别为 PB 与 PA。
18~21	定义 LCD 的控制信号引脚。
24	声明存储器地址由 000h 开始(HT46xx Reset Vector)。
25	将 LCD_CONTR Port(PA)定义为输出模式。
26~27	将 LCD 设置为双行显示(N = 1)、使用 8 位(DB7 ~ DB0)控制模式(DL = 1)、5×7 点矩阵字形(F = 0)。
28~29	将 LCD 设置为显示所有数据(D = 1)、显示光标(C = 1)、光标所在位置的字会闪烁(B = 1)。
30~31	将 LCD 的地址标志位(AC)设置为递加(I/D = 1)、显示器画面不因读/写数据而移动(S = 0)。
32~33	将 LCD 整个显示器清空。
34~35	设置 CG RAM 的地址为 00h。
36~39	因为需要写入 6 个自建字形至 CG RAM 中,而建立一个字形需写入 8 次,所以计数器 COUNT 设置为 48,并将 TBLP 指向字形数据的起始地址。
41~42	通过查表将自建字形的数据取出并写入 CG RAM 中。
43~45	将 TBLP 加 1 后,判断 COUNT−1 是否为 0。成立,表示所有自建字形已输入完毕;反之,表示自建字形还未完全写入 CG RAM 中,继续在 CG RAM 中填入字形数据。
47~48	设置 LCD 的 AC 值(Address Counter)在第 1 行的中间。
49~52	分别将自建字形"王"、"田"显示于 LCD 上。
54~57	分别设置索引值 INDEX 为 2 以及计数器 COUNT 为 4。
57~58	设置 LCD 的 AC 值(Address Counter)在第 2 行第 5 个位置。
59~65	运用索引的方式将自建字形写入(执行程序时 LCD 上将可以看到自建字形"丨"、"/"、"—"、"\"交互地显示)。
66	调用 DELAY 子程序,延时 0.1 s。
67	将索引值 INDEX 加 1。
68~70	判断 COUNT−1 是否为 0。成立则表示自建字形都显示过了,重新显示(不断重复结果就如同一个小棒子不断地在旋转);反之,则显示下一个字形。
74~98	WLCMD 与 WLCMC 子程序。
102~116	DELAY 子程序,延时 0.1 s。
118~172	CG RAM 自建字形数据定义区。

程序一开始先对LCD做t初始化的设置,接下来就是连续填入"田"、"中"、"丨"、"/"、"—"、"\"等6个5×7字形的字形码,请参考表5-5-3。然后设置DD RAM地址,让"田"、"中"两个字形显示在第1行中央的地址。其次于第2行中央地址显示"丨"、"/"、"—"、"\"字形,由于每次显示字形都将DD RAM的地址重新设置,所以"丨"、"/"、"—"、"\"字形都显示在相同的位置上,再加上DELAY时间的控制,读者可以在第2行上看到旋转的图形。

5-6-5 动动脑 + 动动手

- 试着在 CG RAM 中建立一些简单的中文字形,如 "甲"、"上"、"下" 等。
- 改变 DELAY 子程序的延时时间,观察旋转图形的变化。

5-7 LCD与4×4键盘控制实验

5-7-1 目 的

将4×4键盘上的按键值(0~F)在LCD上显示。

5-7-2 学习重点

通过本实验,了解I/O Port共用时该留意的事项。

5-7-3 电路图(见图5-7-1)

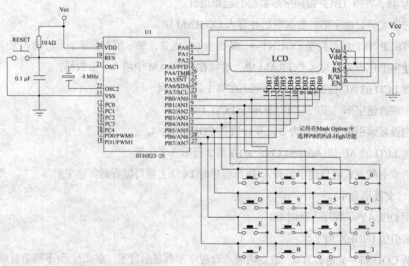

图5-7-1 LCD与4×4键盘控制电路

为了节省I/O Port,我们把LCD的数据总线与4×4键盘共用。因为LCD的所有数据与命令都必须通过E(Enable)信号加以控制,只要确定在执行按键扫描时不要对LCD做任何读/写,就不会影响LCD的动作。有关4×4键盘与LCD的数据,请参考4-7节与5-5节中的说明。

5-7-4 流程图及程序

1. 流程图(见图 5-7-2)

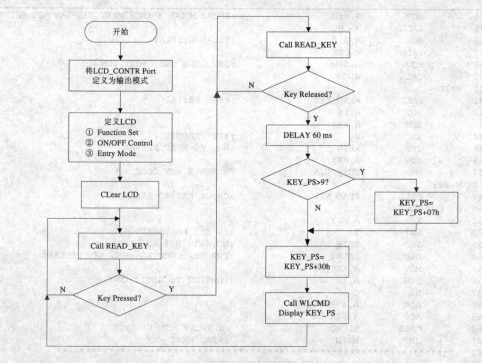

图5-7-2 流程图

2. 程序 5-7　LCD 与 4×4 键盘控制实验

```
1   ;PROGRAM : 5-7.ASM (5-7.PRJ)                                BY STEVEN
2   ;FUNCTION: 16X2 LCD vs. 4x4 KEYPAD DEMO PROGRAM             2002.DEC.07.
3   #INCLUDE     HT46R23.INC
4                .CHIP HT46R23
5   ;---------------------------------------------------------------------
6   MY_DATA      .SECTION  'DATA'         ;== DATA SECTION ==
7   DEL1         DB    ?                  ;DELAY LOOP COUNT 1
8   DEL2         DB    ?                  ;DELAY LOOP COUNT 2
9   DEL3         DB    ?                  ;DELAY LOOP COUNT 3
10  DC_FLAG      DBIT                     ;LCD DATA/COMMAND FLAG
11  COUNT        DB    ?                  ;UNIVERSAL COUNTER
12  KEY_COUNT    DB    ?                  ;ROW COUNTER USED IN READ_KEY
13  KEY          DB    ?                  ;KEY CODE REGISTER
14  KEY_PS       DB    ?                  ;KEY CODE REGISTER
15  ;---------------------------------------------------------------------
16  LCD_CONTR    EQU   PA                 ;DEFINE LCD CONTROL PORT
17  LCD_CONTRC   EQU   PAC                ;DEFINE LCD CONTROL PORT CON. REG.
18  LCD_DATA     EQU   PB                 ;DEFINE LCD DATA PORT
19  LCD_DATAC    EQU   PBC                ;DEFINE LCD DATA PORT CON. REG.
20  LCD_EN       EQU   LCD_CONTR.0        ;DEFINE EN CONTROL PIN
21  LCD_RW       EQU   LCD_CONTR.1        ;DEFINE RW CONTROL PIN
22  LCD_RS       EQU   LCD_CONTR.2        ;DEFINE RS CONTROL PIN
23  LCD_READY    EQU   LCD_DATA.7         ;DEFINE READY BIT OF LCD
24  KEY_PORT     EQU   PB                 ;DEFINE KEY PORT
25  KEY_PORTC    EQU   PBC                ;DEFINE KEY PORT CONTROL REG.
26
27  MY_CODE      .SECTION  AT 0 'CODE'    ;== PROGRAM SECTION ==
28               ORG   00H                ;HT-46RXX RESET VECTOR
29               CLR   LCD_CONTRC         ;CONFIG LCD_CONTR AS OUTPUT MODE
30               MOV   A,38H              ;FUNCTION SET: 8-BIT,2-LINE,5X10 DOTS
31               CALL  WLCMC
32               MOV   A,0CH              ;ON/OFF CONTR: DISPLAY ON,CURSOR BLINKING OFF
33               CALL  WLCMC
```

```
34              MOV     A,06H                   ;ENTRY MODE : INCREMENT,DISPLAY NOT SHIFT
35              CALL    WLCMC
36              MOV     A,01H                   ;CLEAR DISPLAY
37              CALL    WLCMC
38              MOV     A,80H                   ;SET LINE ONE, POSITION 0
39              CALL    WLCMC
40      MAIN:
41              CALL    READ_KEY                ;READ KEYPAD
42              MOV     A,16
43              XOR     A,KEY
44              SZ      Z                       ;KEY PRESSED ?
45              JMP     MAIN                    ;NO, RE-READ KEYPAD
46              MOV     A,KEY
47              MOV     KEY_PS,A                ;RESERVE KEY CODE
48      WAIT:
49              CALL    READ_KEY                ;CHECK KEYPAD RELEASED
50              MOV     A,16
51              XOR     A,KEY
52              SNZ     Z                       ;IS KEY RELEASED?
53              JMP     WAIT                    ;NO,WAIT KEY RELEASE
54              CALL    DELAY                   ;FOR DE-BOUNCING WHEN KEY RELEASE
55              MOV     A,KEY_PS
56              SUB     A,10                    ;CONVERT TO ASCII CODE
57              SZ      C
58              ADD     A,40H-30H-9
59              ADD     A,30H+10
60              CALL    WLCMD                   ;DISPLAY KEY IN DATA
61              JMP     MAIN                    ;READ NEXT KEY
62      ;******************************************************************
63      ;           LCD DATA/COMMAND WRITE PROCEDURE
64      ;******************************************************************
65      WLCMD   PROC
66              SET     DC_FLAG                 ;SET DC_FLAG=1 FOR DATA WRITE
67              JMP     WLCM
68      WLCMC:
69              CLR     DC_FLAG                 ;SET DC_FLAG=0 FOR COMMAND WRITE
70      WLCM:
71              SET     LCD_DATAC               ;CONFIG LCD_DATA AS INPUT MODE
72              CLR     LCD_CONTR               ;CLEAR ALL LCD CONTROL SIGNAL
73              SET     LCD_RW                  ;SET RW SIGNAL (READ)
74              NOP                             ;FOR TAS
75              SET     LCD_EN                  ;SET EN HIGH
76              NOP                             ;FOR TDDR
77      WF:
78              SZ      LCD_READY               ;IS LCD BUSY?
79              JMP     WF                      ;YES, JUMP TO WAIT
80              CLR     LCD_DATAC               ;NO, CONFIG LCD_DATA AS OUTPUT MODE
81              MOV     LCD_DATA,A              ;LATCH DATA/COMMAND ON PB(LCD DATA BUS)
82              CLR     LCD_CONTR               ;CLEAR ALL LCD CONTROL SIGNAL
83              SZ      DC_FLAG                 ;IS COMMAND WRITE?
84              SET     LCD_RS                  ;NO, SET RS HIGH
85              SET     LCD_EN                  ;SET EN HIGH
86              NOP
87              CLR     LCD_EN                  ;SET EN LOW
88              RET
89      WLCMD   ENDP
90      ;******************************************************************
91      ;   SCAN 4x4 MATRIX ON KEY PORT AND RETURN THE CODE IN KEY REGISTER
92      ; IF NO KEY BEEN PRESSED, KEY=16.
93      ;******************************************************************
94      READ_KEY    PROC
95              MOV     A,11110000B
96              MOV     KEY_PORTC,A             ;CONFIG KEY_PORT
97              SET     KEY_PORT                ;INITIAL KEY PORT
98              CLR     KEY                     ;INITIAL KEY REGISTER
99              MOV     A,04
100             MOV     KEY_COUNT,A             ;SET ROW COUNTER
```

第5章 进阶实验篇

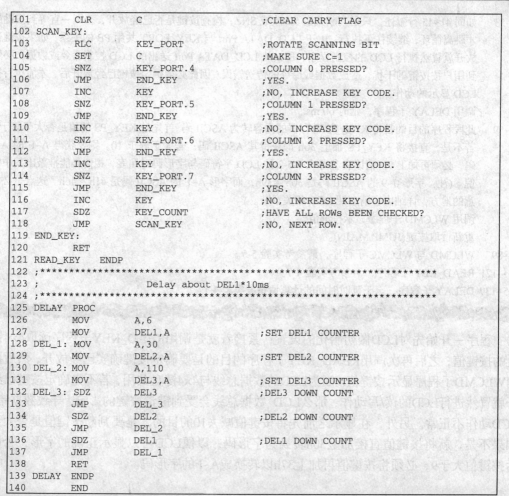

```
101         CLR     C                       ;CLEAR CARRY FLAG
102 SCAN_KEY:
103         RLC     KEY_PORT                ;ROTATE SCANNING BIT
104         SET     C                       ;MAKE SURE C=1
105         SNZ     KEY_PORT.4              ;COLUMN 0 PRESSED?
106         JMP     END_KEY                 ;YES.
107         INC     KEY                     ;NO, INCREASE KEY CODE.
108         SNZ     KEY_PORT.5              ;COLUMN 1 PRESSED?
109         JMP     END_KEY                 ;YES.
110         INC     KEY                     ;NO, INCREASE KEY CODE.
111         SNZ     KEY_PORT.6              ;COLUMN 2 PRESSED?
112         JMP     END_KEY                 ;YES.
113         INC     KEY                     ;NO, INCREASE KEY CODE.
114         SNZ     KEY_PORT.7              ;COLUMN 3 PRESSED?
115         JMP     END_KEY                 ;YES.
116         INC     KEY                     ;NO, INCREASE KEY CODE.
117         SDZ     KEY_COUNT               ;HAVE ALL ROWs BEEN CHECKED?
118         JMP     SCAN_KEY                ;NO, NEXT ROW.
119 END_KEY:
120         RET
121 READ_KEY    ENDP
122 ;************************************************************
123 ;               Delay about DEL1*10ms
124 ;************************************************************
125 DELAY   PROC
126         MOV     A,6
127         MOV     DEL1,A                  ;SET DEL1 COUNTER
128 DEL_1:  MOV     A,30
129         MOV     DEL2,A                  ;SET DEL2 COUNTER
130 DEL_2:  MOV     A,110
131         MOV     DEL3,A                  ;SET DEL3 COUNTER
132 DEL_3:  SDZ     DEL3                    ;DEL3 DOWN COUNT
133         JMP     DEL_3
134         SDZ     DEL2                    ;DEL2 DOWN COUNT
135         JMP     DEL_2
136         SDZ     DEL1                    ;DEL1 DOWN COUNT
137         JMP     DEL_1
138         RET
139 DELAY   ENDP
140         END
```

程序说明

- 7~14 依序定义变量地址。
- 16~19 定义 LCD_DATA Port 与 LCD_CONTR Port 分别为 PB 与 PA。
- 20~23 定义 LCD 的控制信号引脚。
- 24~25 定义 KEY_PORT 为 PB。
- 28 声明存储器地址由 000h 开始(HT46xx Reset Vector)。
- 29 将 LCD_CONTR Port(PA)定义为输出模式。
- 30~31 将 LCD 设置为双行显示(N = 1)、使用 8 位(DB7~DB0)控制模式(DL = 1)、5×7 点矩阵字形(F = 0)。
- 32~33 将 LCD 设置为显示所有数据(D = 1)、显示光标(C = 1)、光标所在位置的字会闪烁(B = 1)。
- 34~35 将 LCD 的地址标志位(AC)设置为递加(I/D = 1)、显示器画面不因读/写数据而移动(S = 0)。
- 36~37 将 LCD 整个显示器清空。
- 38~39 将光标移至第 1 行的第 1 个位置。
- 41~45 调用 READ_KEY 扫描键盘子程序，完成后判断 KEY 值是否为 16，若成立，表示键盘还未被按下，重新扫描键盘(JMP MAIN)；反之，则代表键盘已有输入，则程序继续执行。
- 46~47 将由键盘扫描所得到的结果存储到 KEY_PS 寄存器。

243

行号	说明
48~53	如同 41~45 行叙述，只是判断式由 SZ 改成 SNZ，检查按键是否已经放开，否则一直等到按键放开才跳离循环，继续往下执行。由于 LCD_DATA Port 与 KEY_PORT 共用 PB 的关系，如果按键还未放开就冒然执行 LCD 的控制程序，那么从 LCD_DATA Port 送出的 LCD 控制命令或数据，必然受到用户按按键的干扰，造成数据或命令的传送错误。因此必须确定按键已经放开后，才能执行控制 LCD 显示的动作。
54	调用 DELAY 子程序，延时 60 ms。
55~59	此段程序的目的是将 KEY_PS 寄存器的内容转为 ASCII 码。首先是 KEY_PS 的值是否大于等于 10，若不是，直接将 KEY_PS 加上 30h 即获得其 ASCII 码；若大于等于 10，则要获得 A~F 的 ASCII 码，必须再加上 7。请读者参考表 5-5-2 的 LCD 字符码与字形码对照表，相信必能推敲出其中的端倪。(注：字形 0~9 的 ASCII 码是 30h ~ 39h，而字形 A~F 的 ASCII 码是 41h ~ 46h。这点是必须注意的地方，详细内容请参照 LCD 相关章节。)
60	调用 WLCMD 子程序显示按键值。
61	重新读取按键(JUMP MAIN)。
65~89	WLCMD 与 WLCMC 子程序，请参考实验 5-5。
94 ~ 121	READ_KEY 子程序，请参考实验 4-7。
125~139	DELAY 子程序，至于延时时间的计算请参考实验 4-1。

程序一开始先对LCD做初始化的设置，紧接着就是调用READ_KEY子程序读取4×4键盘上的按键值；之后再次调用READ_KEY子程序的目的是要确定按键确实已经放开，然后再调用WLCMD子程序显示按键值。因为LCD的数据总线与4×4键盘共用，若不先确定按键已经放开就冒然进行LCD的读/写动作，那么LCD 数据总线会受到键盘按键的影响，导致数据错误、LCD动作不正常。另外，在显示之前先将按键值减去10的目的，是要判断按键值是否大于9，如果不是，就将按键值直接加上30h转换为字形码，以便LCD可以显示正确的字形；否则表示按键值大于9，必须将按键值再加上37h以转换成A~F的字形码。

5-7-5 动动脑 + 动动手

- 本范例程序中当按下 16 次按键后 LCD 会将字形存储在 10h 以后的 DD RAM 地址，而不是立即换至第 2 行开始显示，所以中间会有一段期间无法将按键值显示。请改写程序，让按键值无间断地在第 1 行与第 2 行显示。
- 修改程序，让第 1 个按键值显示在第 1 行，让第 2 个按键值显示在第 2 行。
- 写一段程序，当按键值为奇数时，则在第 1 行显示按键值；反之，如果按键值为偶数，则将按键值显示在第 2 行。

5-8 LCD的DD/CG RAM读取控制实验

5-8-1 目　的

以Timer/Event Counter产生一个0~ F的随机数显示在LCD的第1行上，在LCD的第2行以跑马灯的方式配合按键的控制，判断用户按下按键时，光标位置上的数值与之前所产生的随机数值是否相同。

5-8-2 学习重点

通过本实验，了解如何由LCD的DD RAM或CG RAM读回数据。此外，也应对利用Timer/Event Counter产生随机数的方式有所认识。

5-8-3 电路图（见图5-8-1）

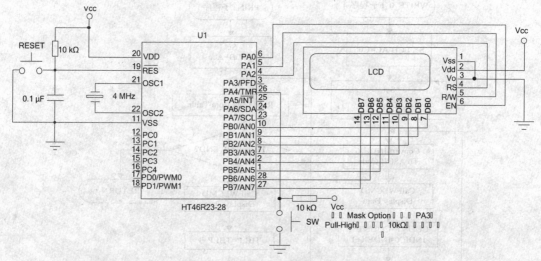

图5-8-1 LCD与按键控制电路

本实验希望呈现如下的执行结果：首先，在LCD第1行显示由Timer/Event Counter产生的随机数(0~F)。其次，在第2行显示0~F的字形，并制造光标在字形上移动的效果，如下所示(此图例中产生的随机数值为8)：

							8								
0	1	2	3	4		6	7	8	9	A	B	C	D	E	F

当用户按下按键之后，再由程序判断此时光标停留位置上的数值与随机数是否相同，分别显示以下的画面：

M	I	S	S	!			8								
0	1	2	3	4		6	7	8	9	A	B	C	D	E	F

							8		B	I	N	G	O	!	
0	1	2	3	4	5	6	7	8	9	A	B	C	D	E	F

随机数的产生有许多方法，由于本实验使用按键有无被按下的判断，用户很难(几乎不可能)控制每一次按下按键的时间都相同，因此利用Timer/Event Counter来产生随机数应是最容易、最方便的选择。

在5-5节中，已经介绍了如何依照LCD读/写时序的要求，而写出相关的子程序(WLCMC、WLCMD)，如果要由LCD的DD RAM或CG RAM读回数据，其实也只是"依样画葫芦"——"根据LCD DD/CG RAM的读/写命令及时序，依序设置控制引脚的电位变化"就可以了。请读者参考图5-5-4、图5-5-5及表5-5-3、表5-5-6及表5-5-7，并对照RLCM子程序及其流程图来理解。

5-8-4 流程图及程序

1. 流程图(见图 5-8-2)

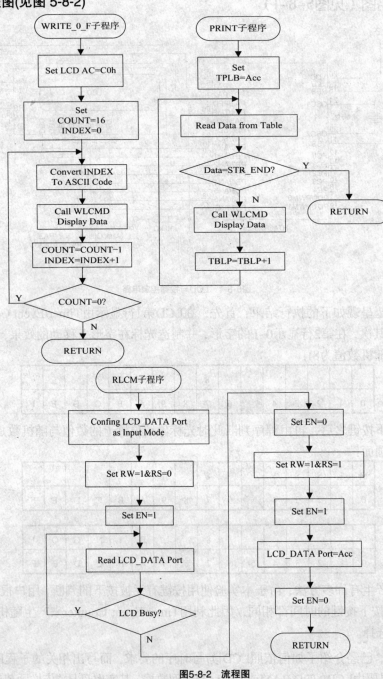

图5-8-2 流程图

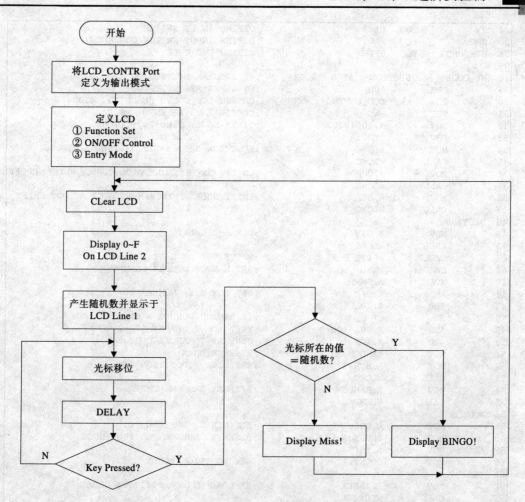

图5-8-2 流程图（续）

2. 程序 5-8　LCD DD/CG RAM 读取控制实验

```
1    ;PROGRAM : 5-8.ASM  (5-8.PRJ)                          BY STEVEN
2    ;FUNCTION: 16X2 LCD DD/CG RAM READ/WRITE DEMO PROGRAM  2002.DEC.07.
3    #INCLUDE    HT46R23.INC
4                .CHIP HT46R23
5    ;--------------------------------------------------------------------
6    MY_DATA     .SECTION    'DATA'         ;== DATA SECTION ==
7    DEL1        DB     ?                   ;DELAY LOOP COUNT 1
8    DEL2        DB     ?                   ;DELAY LOOP COUNT 2
9    DEL3        DB     ?                   ;DELAY LOOP COUNT 3
10   DC_FLAG     DBIT                       ;LCD DATA/COMMAND FLAG
11   TEMP        DB     ?                   ;TEMPORY BUFFER
12   COUNT       DB     ?                   ;UNIVERSAL COUNTER
13   RANDOM DB   ?                          ;RANDOM BUFFER
14   INDEX       DB     ?                   ;INDEX REGISTER
15   ;--------------------------------------------------------------------
16   LCD_CONTR   EQU    PA                  ;DEFINE LCD CONTROL PORT
17   LCD_CONTRC  EQU    PAC                 ;DEFINE LCD CONTROL PORT CON. REG.
18   LCD_DATA    EQU    PB                  ;DEFINE LCD DATA PORT
19   LCD_DATAC   EQU    PBC                 ;DEFINE LCD DATA PORT CON. REG.
20   LCD_EN      EQU    LCD_CONTR.0         ;DEFINE EN CONTROL PIN
21   LCD_RW      EQU    LCD_CONTR.1         ;DEFINE RW CONTROL PIN
22   LCD_RS      EQU    LCD_CONTR.2         ;DEFINE RS CONTROL PIN
23   LCD_READY   EQU    LCD_DATA.7          ;DEFINE READY BIT OF LCD
```

```
24  SW          EQU     PA.3                    ;DEFINE INPUT SWITCH
25  SWC         EQU     PAC.3                   ;DEFINE INPUT SWITCH CONTR REG.
26  STR_END     EQU     0FEH                    ;STRING END CODE
27
28  MY_CODE     .SECTION  AT 0 'CODE'           ;== PROGRAM SECTION ==
29              ORG     00H                     ;HT-46RXX RESET VECTOR
30              CLR     LCD_CONTRC              ;CONFIG LCD_CONTR AS OUTPUT MODE
31              SET     SWC                     ;CONFIG PA.3 AS INPUT MODE
32              MOV     A,10010000B             ;TIMER MODE, Fint=Fsys, TON=1
33              MOV     TMRC,A
34              MOV     A,38H                   ;FUNCTION SET: 8-BIT,2-LINE,5X10 DOTS
35              CALL    WLCMC
36              MOV     A,0CH                   ;ON/OFF CONTR: DISPLAY ON,CURSOR BLINKING OFF
37              CALL    WLCMC
38              MOV     A,06H                   ;ENTRY MODE  : INCREMENT,DISPLAY NOT SHIFT
39              CALL    WLCMC
40  RESTART:
41              MOV     A,01H                   ;CLEAR DISPLAY
42              CALL    WLCMC
43              CALL    WRITE_0_F               ;WRITE 0~F ON LINE 2
44              SWAPA   TMRL                    ;GET RANDOM NUMBER FORM TMR
45              XOR     A,TMRH
46              AND     A,0FH                   ;GET LOW Byte NUMBER (0~F)
47              SUB     A,10
48              SZ      C                       ;IS RANDOM NUMBER > 9?
49              ADD     A,40H-30H-9             ;YES, ADD 7 OFFSET FOR 9 TO A
50              ADD     A,30H+10                ;CONVERT TO ASCII CODE
51              MOV     RANDOM,A                ;RESERVE DATA
52              MOV     A,80H+7                 ;SET LINE 1,POSITION 7
53              CALL    WLCMC
54              MOV     A,RANDOM                ;DISPLAY RANDOM NUMBER
55              CALL    WLCMD
56  MAIN_1:
57              MOV     A,0C0H                  ;SET LIN 2, POSITION 0
58              MOV     INDEX,A                 ;RESERVED CURSOR POSITION CODE
59              MOV     A,16
60              MOV     COUNT,A                 ;SET POSITION COUNT
61  MAIN_2:
62              MOV     A,INDEX                 ;RE-LOAD CURSOR POSITION CODE
63              CALL    WLCMC
64              CALL    RLCM                    ;READ DDRAM DATA
65              MOV     TEMP,A                  ;RESERVED DATA
66              MOV     A,INDEX                 ;RESTORE CURSOR POSITION
67              CALL    WLCMC
68              MOV     A,0FFH                  ;MASK WITH "■"
69              CALL    WLCMD
70              MOV     A,08H                   ;DELAY 80ms
71              CALL    DELAY
72              SNZ     SW                      ;IS KEY PRESSED?
73              JMP     MAIN_3                  ;YES, JUMP TO MAIN_3
74              MOV     A,INDEX                 ;NO,RESTORE CURSOR POSITION
75              CALL    WLCMC
76              MOV     A,TEMP                  ;RESTORE DDRAM DATA
77              CALL    WLCMD
78              INC     INDEX                   ;INCREASE INDEX BY 1
79              SDZ     COUNT                   ;COUNT-1 = 0?
80              JMP     MAIN_2                  ;NO,JUMP TO MAIN_2
81              JMP     MAIN_1                  ;YES, RESTART
82  MAIN_3:
83              MOV     A,RANDOM
84              XOR     A,TEMP
85              SZ      Z                       ;IS TEMP = RANDOM?
86              JMP     BINGO                   ;YES,JUMP TO BINGO
87  MISS:
88              MOV     A,80H                   ;NO, DISPLAY "MISS!" MESSAGE
89              CALL    WLCMC                   ;SET LINE 1 POSITION 0
90              MOV     A,OFFSET STR1           ;LOAD STRING START ADDRESS
```

```
91              CALL        PRINT                   ;PRINT ON LCD
92              MOV         A,100                   ;DELAY 1 SEC.
93              CALL        DELAY
94              JMP         RESTART                 ;RESTART
95      BINGO:
96              MOV         A,80H+10                ;DISPLAY "BINGO!"
97              CALL        WLCMC                   ;SET LINE 1 POSITION 10
98              MOV         A,OFFSET STR2           ;LOAD STRING START ADDRESS
99              CALL        PRINT                   ;PRINT ON LCD
100             MOV         A,250                   ;DELAY 2.5 SEC.
101             CALL        DELAY
102             JMP         RESTART                 ;RESTART
103     ;**********************************************************************
104     ;           WRITE "0"~"F" ON LCD PROCEDURE
105     ;**********************************************************************
106     WRITE_0_F   PROC
107             MOV         A,0C0H                  ;SET LINE 2
108             CALL        WLCMC
109             MOV         A,16
110             MOV         COUNT,A                 ;SET COUNTER=16
111             CLR         INDEX
112     WRITE:
113             MOV         A,INDEX
114             SUB         A,10                    ;CONVERT TO ASCII CODE
115             SZ          C
116             ADD         A,40H-30H-9
117             ADD         A,30H+10
118             CALL        WLCMD                   ;WRITE DATA
119             INC         INDEX                   ;INCREASE INDEX BY 1
120             SDZ         COUNT                   ;COUNT-1 = 0?
121             JMP         WRITE                   ;NO,NEXT DIGIT
122             RET                                 ;YES, RETURN
123     WRITE_0_F   ENDP
124     ;**********************************************************************
125     ;           LCD PRINT PROCEDURE (START ADDRS=TBLP)
126     ;**********************************************************************
127     PRINT   PROC
128             MOV         TBLP,A                  ;LOAD TABLE POINTER
129     PRINT_1:
130             TABRDL      ACC                     ;READ STRING
131             XOR         A,STR_END
132             SZ          Z                       ;END OF STRING CHARACTER?
133             JMP         END_PRINT               ;YES, STOP PRINT
134             TABRDL      ACC                     ;NO, RE-READ
135             CALL        WLCMD                   ;SEND TO LCD
136             INC         TBLP                    ;INCREASE DATA POINTER
137             JMP         PRINT_1                 ;NEXT CHARACTER
138     END_PRINT:
139             RET                                 ;YES, RETURN
140     PRINT   ENDP
141     ;**********************************************************************
142     ;           LCD DATA READ PROCEDURE
143     ;**********************************************************************
144     RLCM    PROC
145             SET         LCD_DATAC               ;CONFIG LCD_DATA AS INPUT MODE
146             CLR         LCD_CONTR               ;CLEAR ALL LCD CONTROL SIGNAL
147             SET         LCD_RW                  ;SET RW=1
148             NOP                                 ;FOR TAS
149             SET         LCD_EN                  ;SET EN=1
150             NOP                                 ;FOR TDDR
151     RWF:
152             SZ          LCD_READY               ;IS LCD BUSY?
153             JMP         RWF                     ;YES,WAIT
154             CLR         LCD_CONTR               ;CLEAR ALL LCD CONTROL SIGNAL
155             SET         LCD_RW                  ;SET LCD_RW
156             SET         LCD_RS                  ;SET RS=1 (DATA READ)
157             NOP
```

```
158            SET       LCD_EN                    ;SET EN
159            NOP
160            MOV       A,LCD_DATA                ;READ LCD DATA
161            CLR       LCD_EN                    ;CLEAR EN
162            RET
163   RLCM     ENDP
164   ;***************************************************************
165   ;           LCD DATA/COMMAND WRITE PROCEDURE
166   ;***************************************************************
167   WLCMD    PROC
168            SET       DC_FLAG                   ;SET DC_FLAG=1 FOR DATA WRITE
169            JMP       WLCM
170   WLCMC:
171            CLR       DC_FLAG                   ;SET DC_FLAG=0 FOR COMMAND WRITE
172   WLCM:
173            SET       LCD_DATAC                 ;CONFIG LCD_DATA AS INPUT MODE
174            CLR       LCD_CONTR                 ;CLEAR ALL LCD CONTROL SIGNAL
175            SET       LCD_RW                    ;SET RW SIGNAL (READ)
176            NOP                                 ;FOR TAS
177            SET       LCD_EN                    ;SET EN HIGH
178            NOP                                 ;FOR TDDR
179   WF:
180            SZ        LCD_READY                 ;IS LCD BUSY?
181            JMP       WF                        ;YES, JUMP TO WAIT
182            CLR       LCD_DATAC                 ;NO, CONFIG LCD_DATA AS OUTPUT MODE
183            MOV       LCD_DATA,A                ;LATCH DATA/COMMAND ON PB(LCD DATA BUS)
184            CLR       LCD_CONTR                 ;CLEAR ALL LCD CONTROL SIGNAL
185            SZ        DC_FLAG                   ;IS COMMAND WRITE?
186            SET       LCD_RS                    ;NO, SET RS HIGH
187            SET       LCD_EN                    ;SET EN HIGH
188            NOP
189            CLR       LCD_EN                    ;SET EN LOW
190            RET
191   WLCMD    ENDP
192   ;***************************************************************
193   ;                    Delay about ACC*10ms
194   ;***************************************************************
195   DELAY    PROC
196            MOV       DEL1,A
197   DEL_1:   MOV       A,30
198            MOV       DEL2,A                    ;SET DEL2 COUNTER
199   DEL_2:   MOV       A,110
200            MOV       DEL3,A                    ;SET DEL3 COUNTER
201   DEL_3:   SDZ       DEL3                      ;DEL3 DOWN COUNT
202            JMP       DEL_3
203            SDZ       DEL2                      ;DEL2 DOWN COUNT
204            JMP       DEL_2
205            SDZ       DEL1                      ;DEL1 DOWN COUNT
206            JMP       DEL_1
207            RET
208   DELAY    ENDP
209            ORG       LASTPAGE
210   STR1:    DC        'MISS!',STR_END           ;DEFINE STRING DATA 1
211   STR2:    DC        'BINGO!',STR_END          ;DEFINE STRING DATA 2
212            END
```

程序说明

7~14   依序定义变量地址。

16~19   定义 LCD_DATA Port 与 LCD_CONTR Port 分别为 PB 与 PA。

20~23   定义 LCD 的控制信号引脚。

24~25   定义 SW 为 PA.3。

行	说明
26	定义字符串结束码 STR_END=0FEh，在 PRINT(打印字符串)子程序中，当读取的字符等于 STR_END 时，即表示该字符串已经全部显示完毕。
29	声明存储器地址由 000h 开始(HT46xx Reset Vector)。
30~31	将 LCD_CONTR Port(PA)定义为输出模式、SW 定义为输入模式。
32~33	将 Time/Event Counter 定义为 Timer Mode、分频比例 1:1($f_{INT}=f_{SYS}$)、TON=1(启动 TIMER 开始计数)。
34~35	将 LCD 设置为双行显示(N = 1)、使用 8 位(DB7 ~ DB0)控制模式(DL = 1)、5×7 点矩阵字形(F = 0)。
36~37	将 LCD 设置为显示所有数据(D = 1)、显示光标(C = 1)、光标不闪烁(B = 0)。
38~39	将 LCD 的地址标志位(AC)设置为递加(I/D = 1)、显示器画面不因读/写数据而移动(S = 0)。
41~42	将 LCD 整个显示器清空。
43	调用 WRITE_0_F 子程序，将 0~F 依序显示于 LCD 的第 2 行。
44~46	将 TMRL 寄存器的高 4 位与低 4 位交换、与 TMRH 寄存器执行 XOR 的动作，然后与常数 0Fh 做 AND，主要目的是取得一个小于或等于 F 的随机数。
47~51	将随机数转换为对应的 ASCII 码，并存放于 RANDOM 寄存器。
53~53	设置 DD RAM 的地址到 LCD 第 1 行的正中间。
54~55	调用 WLCMD 子程序，将随机数显示在 LCD 上。
57~60	设置索引值 INDEX 内容为 C0h(INDEX 是用来控制移动中的光标地址)，并设置计数器 COUNT 内容为 16(在此 LCD 可见范围为 16×2 个字节，由于游戏数字的范围为 0~F，因此只须扫描 16 次，下面将有详细的说明)。
62~63	将 DD RAM 的地址设置在与索引值 INDEX 所对应的地址(以程序第 1 次执行至此，索引值 INDEX 为 C0h 其所对应的地址即为显示器第 2 行的第 1 个字节)。
64~69	调用子程序 RLCM 读取 LCD_DATA 的内容值存入 TEMP 寄存器中，并执行第 66~67 行的动作。完成后写入 "■" 于目前 INDEX 与 LCD 所对应的地址，表示游戏中的光标正停留在此位置。
70~71	调用 DELAY 子程序，延时 80 ms。此延时时间的长短决定了光标移动的速度，读者可以改变延时的时间试试自己反应的快慢。
72~73	判断是否有按键输入，成立则至 MAIN_3 显示结果；反之，则继续扫描下一个 LCD 地址。
74~77	将刚刚设为 "■" 的地址恢复成正常的数字(以程序第一次执行至此，即将 "■" 改为 0)，原本地址所存放的内容在 TEMP 寄存器。
78	将索引值 INDEX 加 1，指向下一个 DD RAM 位置。
79~81	判断 COUNT−1 是否为 0，成立，则表示游戏的光标已将 0~F 扫描一次，依然无任何按键输入，重新由 0 开始扫描；反之，则进行下个地址的扫描工作。(注：设置 LCD 时已将 LCD 的光标显示关闭，在此设置了一个光标,利用 LCD 字形码中的 "■" 覆盖原来索引值 INDEX 对应的地址内容，来表示光标正停留在此位置。)
83~86	将答案(RANDOM)与结果(TEMP)互相对比，若结果等于 0(Z = 1)即表示答对；反之则答错了。
87~94	调用 PRINT 子程序，在 LCD 左上角显示 "MISS!" 并延时 100 ms，完成后程序重新开始。
96~102	调用 PRINT 子程序，在 LCD 右上角显示 "BINGO!" 并延时 250 ms，完成后程序重新开始。
106~123	WRITE_0_F 子程序，在 LCD 的第 2 行依序显示 0~F。首先，设置 DD RAM 地址位于第 2 行的起始位置(C0h)，然后以 INDEX 与 COUNT 寄存器控制，一一显示 0~F。
127~140	PRINT 子程序，此子程序负责将定义好的字符串依序显示在 LCD 上。在调用此子程序之前，除了必须先设置好 LCD 的位置之外，还需先在 Acc 寄存器中指定字符串的起始地址(字符串必须存放于最末程序页)，并请在字符串的最后一个字符插入 STR_END，代表字符串退出。
144~163	RLCM 子程序，负责读取 LCD DDRAM 的内容，请参考实验中有关 LCD 读取子程序的说明。
145	定义 LCD_DATA(PB)为输入模式，接收 LCD 忙碌标志位(BF)的内容。
146	将 LCD 所有控制信号(E、RS、RW)设置为 0。
147~150	将设置 RW 为 1(Read)并使 LCD(Enable)有效，用来检查 LCD 是否处于忙碌状态，NOP 指令让 LCD 有足够的缓冲时间接受命令并完成工作，即考虑图 5-5-3 中的 T_{AS}、T_{DDR} 的时序。

152~153　检查 LCD_DATA Port(PB)的 BIT 7(即 Busy Flag)是否处于忙碌状态(BF=1)，直到 BF=0 才继续执行程序。
154　　　清除 LED_CONTR Port(目的是将 LCD 的控制信号 E、RS、RW 设置为 0，结束此次的读取动作)。
155~157　设置 RW=1、RS=1，表示准备进行 LCD DR(Data Register)的读取动作。
158　　　设置 LCD 控制信号 E=1
160　　　读取 LCD 的数据。
161　　　将 LCD 的控制信号 E 设置为 0，结束此次的读取动作。
167~191　WLCMD 与 WLCMC 子程序，请参考 5-5 节中的说明。
195~208　DELAY 子程序，延时时间的计算请参考 4-1 节中的说明。
210~211　字符串定义区，请注意每一个字符串必须以 STR_END 作为结束。

　　随机数的产生方式其实有各种不同的做法，而本实验因为使用按键检测，所以程序一开始就先启动HT46x23的Timer/Event Counter，由于每次按按键的时间都不相同，所以采用Timer/Event Counter来产生随机数是最方便的方法。DELAY子程序延时时间的长短决定了光标移动的速度，读者可以改变延时的时间试试自己反应的快慢。

5-8-5　动动脑+动动手

- 执行本实验的范例程序时，并未限定用户输入的时间。请读者修改程序，若光标已经重复扫描每个数值 10 次，而用户还未按下按键则及显示"MISS!"的错误画面。
- 修改程序，一开始提供用户 50 分的分数，答对一次加 5 分、答错一次扣 5 分，并将分数显示在 LCD 上。
- 改写程序，使得所产生的随机数重复率再降低。例如，在 10 次之内所产生的随机数均不得重复(有点类似 CD Player 的随机播放模式，在整张 CD 播放完毕之前，随机播放的曲目不得重复，在此假设 CD 的曲目为 10 首)。

5-9　LCD的4位控制模式实验

5-9-1　目　的

　　采用LCD的4位控制模式，在LCD的CG RAM建立"大"、"小"两个字形，并分别显示在LCD的第1、2排。

5-9-2　学习重点

　　通过本实验，了解LCD的4位控制模式。

5-9-3　电路图

　　如图5-9-1所示为4位总线连接接口电路图。

第 5 章　进阶实验篇

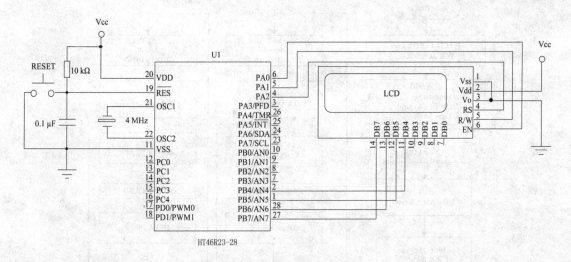

图5-9-1　4位总线连接接口

其实LCD的8位及4位控制模式,唯一的差别就是将要写入或读出的数据分成2次来处理,而且是先存取高4位(DB7~DB4)后再存取低4位(DB3~DB0)。图5-9-2的时序图是将指令写到LCD、由LCD读回忙碌标志位(BF)以及读回数据寄存器(DR)的时序范例。根据此时序范例,只要将实验5-5的WLCMC与WLCMD程序稍加修改即可。请读者参考后面的WLCMC_4与WLCMD_4的程序及流程图。

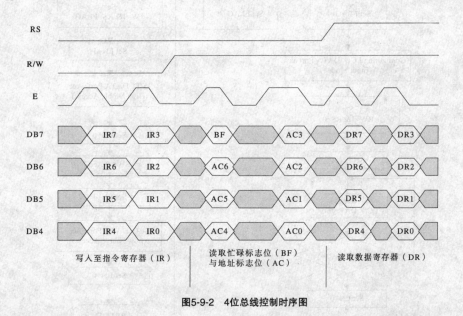

图5-9-2　4位总线控制时序图

5-9-4　流程图及程序

1. 流程图(见图 5-9-3)

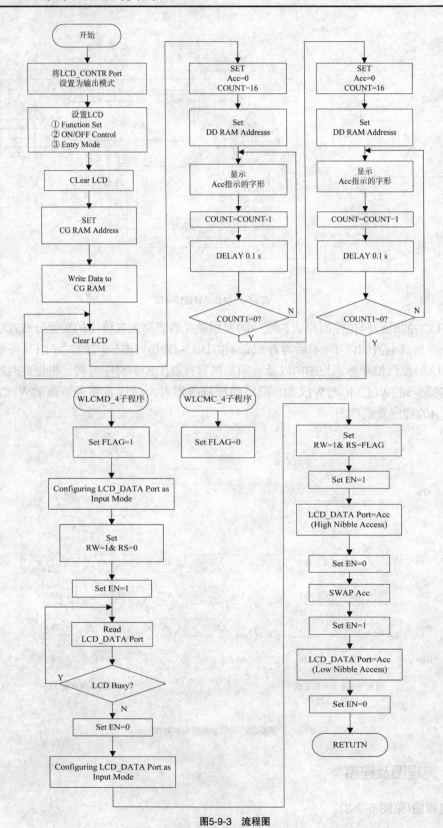

图5-9-3 流程图

2. 程序 5-9 LCD 4 位控制模式

```
1    ;PROGRAM : 5-9.ASM  (5-9.PRJ)                    BY STEVEN
2    ;FUNCTION: 16X2 LCD 4-BIT MODE DEMO PROGRAM      2002.DEC.07.
3    #INCLUDE    HT46R23.INC
4                .CHIP HT46R23
5    ;----------------------------------------------------------------
6    MY_DATA     .SECTION    'DATA'      ;== DATA SECTION ==
7    DEL1        DB      ?               ;DELAY LOOP COUNT 1
8    DEL2        DB      ?               ;DELAY LOOP COUNT 2
9    DEL3        DB      ?               ;DELAY LOOP COUNT 3
10   DC_FLAG     DBIT                    ;LCD DATA/COMMAND FLAG
11   LINE_COUNT  DB      ?               ;DISPLAY COUNT
12   ;----------------------------------------------------------------
13   LCD_CONTR   EQU     PA              ;DEFINE LCD CONTROL PORT
14   LCD_CONTRC  EQU     PAC             ;DEFINE LCD CONTROL PORT CON. REG.
15   LCD_DATA    EQU     PB              ;DEFINE LCD DATA PORT
16   LCD_DATAC   EQU     PBC             ;DEFINE LCD DATA PORT CON. REG.
17   LCD_EN      EQU     LCD_CONTR.0     ;DEFINE EN CONTROL PIN
18   LCD_RW      EQU     LCD_CONTR.1     ;DEFINE RW CONTROL PIN
19   LCD_RS      EQU     LCD_CONTR.2     ;DEFINE RS CONTROL PIN
20   LCD_READY   EQU     LCD_DATA.7      ;DEFINE READY BIT OF LCD
21
22   MY_CODE     .SECTION  AT 0 'CODE'   ;== PROGRAM SECTION ==
23           ORG     00H                 ;HT-46RXX RESET VECTOR
24           CLR     LCD_CONTRC          ;CONFIG LCD_CONTR AS OUTPUT MODE
25           MOV     A,28H               ;FUNCTION SET: 4-BIT,2-LINE,5X10 DOTS
26           CALL    WLCMC_4
27           MOV     A,0CH               ;ON/OFF CONTR: DISPLAY ON,CURSOR BLINKING OFF
28           CALL    WLCMC_4
29           MOV     A,06H               ;ENTRY MODE : INCREMENT,DISPLAY NOT SHIFT
30           CALL    WLCMC_4
31           MOV     A,01H               ;CLEAR DISPLAY
32           CALL    WLCMC_4
33           MOV     A,01000000B         ;SET CGRAM START ADDRESS
34           CALL    WLCMC_4
35           MOV     A,2*8
36           MOV     LINE_COUNT,A        ;SET CGRAM WRITE COUNTER
37           MOV     A,OFFSET TAB_CGRAM  ;GET TABLE START ADDRESS
38           MOV     TBLP,A
39   NEXT:
40           TABRDL  ACC                 ;READ DATA
41           CALL    WLCMD_4             ;WRITE TO CG_RAM
42           INC     TBLP                ;INCREASE POINTER BY 1
43           SDZ     LINE_COUNT          ;COUNT-1 = 0?
44           JMP     NEXT                ;NO, NEXT DATA
45   MAIN:
46           MOV     A,01                ;CLEAR DISPLAY
47           CALL    WLCMC_4
48           MOV     A,80H               ;SET LINE ONE
49           CALL    WLCMC_4
50           MOV     A,16
51           MOV     LINE_COUNT,A
52   MAIN_1:
53           MOV     A,0                 ;WRITE '小'
54           CALL    WLCMD_4
55           CALL    DELAY               ;JUST FOR SEEING THE DISPLAY
56           SDZ     LINE_COUNT
57           JMP     MAIN_1
58           MOV     A,0C0H              ;SET LINE TWO
59           CALL    WLCMC_4
60           MOV     A,16
61           MOV     LINE_COUNT,A
62   MAIN_2:
63           MOV     A,1                 ;WRITE '大'
64           CALL    WLCMD_4
```

```
65          CALL    DELAY                   ;JUST FOR SEEING THE DISPLAY
66          SDZ     LINE_COUNT
67          JMP     MAIN_2
68          JMP     MAIN
69  ;**********************************************************************
70  ;    4-BIT LCD DATA/COMMAND WRITE PROCEDURE
71  ;**********************************************************************
72  WLCMD_4     PROC
73          SET     DC_FLAG                 ;SET DC_FLAG=1 FOR DATA WRITE
74          JMP     WLCM_4
75  WLCMC_4:
76          CLR     DC_FLAG                 ;SET DC_FLAG=0 FOR COMMAND WRITE
77  WLCM_4:
78          SET     LCD_DATAC               ;CONFIG LCD_DATA AS INPUT MODE
79          CLR     LCD_CONTR               ;CLEAR ALL LCD CONTROL SIGNAL
80          SET     LCD_RW                  ;SET RW SIGNAL (READ)
81          NOP                             ;FOR TAS
82          SET     LCD_EN                  ;SET EN HIGH
83          NOP                             ;FOR TDDR
84  WF_4:
85          SZ      LCD_READY               ;IS LCD BUSY?
86          JMP     WF_4                    ;YES, JUMP TO WAIT
87          CLR     LCD_DATAC               ;NO, CONFIG LCD_DATA AS OUTPUT MODE
88          MOV     LCD_DATA,A              ;LATCH DATA/COMMAND ON PB(LCD DATA BUS)
89          CLR     LCD_CONTR               ;CLEAR ALL LCD CONTROL SIGNAL
90          SZ      DC_FLAG                 ;IS COMMAND WRITE?
91          SET     LCD_RS                  ;NO, SET RS HIGH
92          SET     LCD_EN                  ;SET EN HIGH
93          NOP
94          CLR     LCD_EN                  ;SET EN LOW
95          SWAPA   ACC
96          SET     LCD_EN                  ;SET EN HIGH
97          MOV     LCD_DATA,A
98          CLR     LCD_EN                  ;SET EN LOW
99          RET
100 WLCMD_4     ENDP
101 ;**********************************************************************
102 ;                    Delay about DEL1*10ms
103 ;**********************************************************************
104 DELAY   PROC
105         MOV     A,10
106         MOV     DEL1,A                  ;SET DEL1 COUNTER
107 DEL_1:  MOV     A,30
108         MOV     DEL2,A                  ;SET DEL2 COUNTER
109 DEL_2:  MOV     A,110
110         MOV     DEL3,A                  ;SET DEL3 COUNTER
111 DEL_3:  SDZ     DEL3                    ;DEL3 DOWN COUNT
112         JMP     DEL_3
113         SDZ     DEL2                    ;DEL2 DOWN COUNT
114         JMP     DEL_2
115         SDZ     DEL1                    ;DEL1 DOWN COUNT
116         JMP     DEL_1
117         RET
118 DELAY   ENDP
119         ORG     LASTPAGE
120 TAB_CGRAM:
121         DC      00000000B               ;CHARACTER '小'
122         DC      00000100B
123         DC      00000100B
124         DC      00010101B
125         DC      00010101B
126         DC      00000100B
127         DC      00001100B
128         DC      00000000B
129
130         DC      00000000B               ;CHARACTER '大'
131         DC      00000100B
```

132	DC	00000100B
133	DC	00011111B
134	DC	00000100B
135	DC	00001010B
136	DC	00010001B
137	DC	00000000B
138	END	

程序说明

7~11	依序定义变量地址。
13~16	定义 LCD_DATA Port 与 LCD_CONTR Port 分别为 PB 与 PA。
17~20	定义 LCD 的控制信号引脚。
23	声明存储器地址由 000h 开始(HT46xx Reset Vector)。
24	将 LCD_CONTR Port(PA)定义为输出模式。
25~26	将 LCD 设置为双行显示(N = 1)、使用 4 位(DB7 ~ DB4)控制模式(DL = 0)、5×7 点矩阵字形(F = 0)。
27~28	将 LCD 设置为显示所有数据(D = 1)、显示光标(C = 1)、光标不闪烁(B = 0)。
29~30	将 LCD 的地址标志位(AC)设置为递加(I/D = 1)、显示器画面不因读/写数据而移动(S = 0)。
31~32	将 LCD 整个显示器清空。
33~34	设置 CG RAM 的地址为 00H，准备写入自建字形。
35~38	设置计数器 LINE_COUNT 为 16 并将 TBLP 指向自建字形数据的起始地址。
40~41	通过查表将自建字形的数据取出写入 CG RAM 中。
42	将 TBLP 加 1。
43~44	判断 COUNT−1 是否为 0。成立，表示所有自建字形已输入完毕；反之，则表示自建字形还未建立完成，继续填入字形码。
46~47	将 LCD 整个显示器清空。
48~49	将光标移至第 1 行的第 0 个位置。
50~51	设置计数器 LINE_COUNT 为 16。
53~54	将自建字形"小"显示在 LCD 上。
55	调用 DELAY 子程序，延时 100 ms。
56~57	判断 LCD 光标是否已到第 1 行最后一个位置，成立则换到第 2 行；反之则继续显示。
58~59	将光标移到第 2 行的第 0 个位置。
60~61	设置计数器 LINE_COUNT 为 16。
63~64	将自建字形"大"显示于 LCD 上。
65	调用 DELAY 子程序，延时 100 ms。
66~68	判断 LCD 光标是否已到第 2 行最后一个位置，成立则重新从第 1 行显示，反之则继续显示。
72~74	WLCMD_4 子程序进入点，将 DC_FLAG 设置为 1，代表欲写入的内容为数据。
75~76	WLCMC_4 子程序进入点，将 DC_FLAG 设置为 0，代表欲写入的内容为命令。
	(PS. DC_FLAG：Data/Command Flag；1=Data、0=Command)
78	定义 LCD_DATA PORT(PB)为输入模式，接收 LCD 忙碌标志位(BF)的内容。
79	将 LCD 所有控制信号(E、RS、RW)设置为 0。
80~83	将设置 RW 为 1(Read)并 LCD(Enable)有效，用来检查 LCD 是否处于忙碌状态，NOP 指令让 LCD 有足够的缓冲时间接受命令并完成工作，即考虑图 5-5-3 中的 T_{AS}、T_{DDR} 的时序。
84~86	检查 LCD_DATA Port(PB)的 BIT 7(即忙碌标志)是否处于忙碌状态(BF=1)，直到 BF=0 才继续执行程序。
87	定义 LCD DATA Port 为输出模式，用来输出命令或数据到 LCD。
88	将要输出到 LCD 上的命令或数据移到 LCD_DATA Port，此阶段乃处理高 4 位(High Nibble)部分。

89~91 清除 LED_CONTR Port(目的是将 LCD 的控制信号 E、RS、RW 设置为 0)后，判断 DC_FLAG 标志位是否为 0，成立表示 LCD 要写入的内容为"命令"；反之，则表示要写入的内容为"数据"。

92 设置 LCD 控制信号 E=1。

93~94 让 LCD 有足够的缓冲时间接受命令或数据并完成工作，完成后将禁止 LCD 工作并回到主程序继续下面的动作。NOP 指令让 LCD 有足够的缓冲时间接受命令并完成工作，即考虑图 5-5-4 中的 t_{Er}、PW_{EN} 的时序。

95 将 Acc 的高、低 4 个位数据互换；准备将低 4 位(Low Nibble)的命令或数据移到 LCD_DATA Port。

96 设置 LCD 控制信号 E=1。

97 将低 4 位(Low Nibble)的命令或数据移到 LCD_DATA Port(PB7~PB4)。

98 将禁止 LCD 工作并回到主程序继续下面的动作。

104~118 DELAY 子程序，延时 0.1 s。关于延时时间的计算请读者参考 4-1 节。

120~137 自建字形数据定义区。

程序一开始先对LCD做初始化的设置，紧接着在设置CG RAM的地址之后开始建立"小"及"大"两个字形，最后就是重复在第1及第2行显示这两个自建的字形。除了LCD的控制子程序(WLCMC_4与WLCMD_4)改以4位处理之外，本实验与前两个实验实际上并没有什么差异。要提醒读者的是在图5-9-2的4位控制时序范例中，在读取忙碌标志位(BF)时，应将BF与光标地址(AC6~AC0)分两次读取，但因为本范例并不需要光标位置的信息，所以在控制子程序中，只把包含忙碌标志位的高4位读回做判断而已。

5-9-5 动动手 + 动动脑

📖 改写程序，让"大"、"小"两个字形变为上下交替显示。

5-10 比大小游戏实验

5-10-1 目 的

利用HT46x23的Timer/Event Counter产生两个随机数，让用户猜猜此两组数字的大小关系，由程序判断其正确性后再分别显示出"MISS!"或"BINGO!"的字符串，分别代表用户猜得"不正确"与"正确"。本实验采用LCD的4位控制模式。

5-10-2 学习重点

通过本实验，读者除了应更熟悉LCD的4位控制模式之外，也应了解十六进制与BCD码的转换关系。

5-10-3 电路图（见图5-10-1）

十六进制与BCD的转换，在许多单片机的应用场合都会使用到。基本的原理很简单，假设DIG0代表要转换的8位数值，DIG2、DIG1及DIG0分别代表转换为十进制之后的百位数、

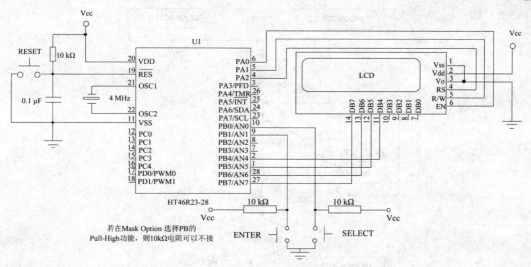

图5-10-1 比大小游戏实验电路

十位数及个位数,请参考HEX2BCD(Hex To BCD)子程序的流程图。因为本实验要将Timer/Event Counter所产生的8位随机数显示在LCD上,所以光是转成BCD码是不够的,必须再将其进一步转换成LCD的字符码(Chararcter Code),请参考流程图中的虚线路径,只要将转换好的BCD(0~9)再加上30h就可以了。此程序称之为HEX2ASCII(Hex To ASCII)子程序。本实验想在LCD上呈现如下的画面:

随	机	数	1		小	_									
					大										

首先将Timer/Event Counter产生的8位随机数显示在LCD上,同时在第1、2行上分别显示"大"、"小"。

接着用户可通过SELECT按钮控制光标(_)位置来选择第2个随机数是大于或小于随机数1。在用户按下ENTER按键之后,再由程序判断此时用户所选择的随机数大小关系是否正确,分别显示出以下的画面:

随	机	数	1		小										
随	机	数	2		大		B	I	N	G	O	!			

随	机	数	1		小		M	I	S	S	!				
随	机	数	2		大										

5-10-4 流程图及程序

1. 流程图(见图 5-10-2)

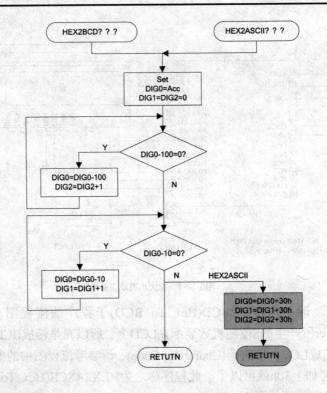
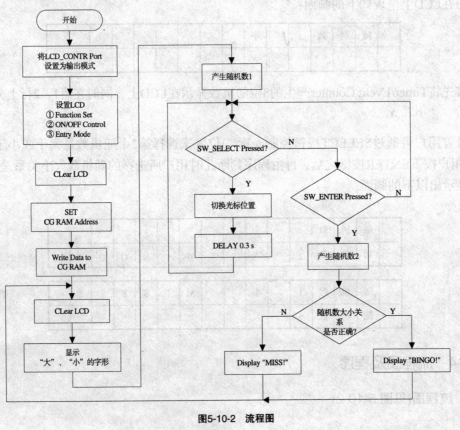

图5-10-2 流程图

2. 程序 5-10 比大小游戏实验

```
1   ;PROGRAM : 5-10.ASM  (5-10.PRJ)              BY STEVEN
2   ;FUNCTION: GREAT OR LESS LCD GAME DEMO PROGRAM   2002.DEC.07.
3   #INCLUDE    HT46R23.INC
4               .CHIP HT46R23
5   ;-----------------------------------------------------------------
6   MY_DATA     .SECTION    'DATA'          ;== DATA SECTION ==
7   DEL1        DB      ?                   ;DELAY LOOP COUNT 1
8   DEL2        DB      ?                   ;DELAY LOOP COUNT 2
9   DEL3        DB      ?                   ;DELAY LOOP COUNT 3
10  DC_FLAG     DBIT                        ;LCD DATA/COMMAND FLAG
11  LINE_COUNT  DB      ?                   ;DISPLAY COUNT
12  DIG0        DB      ?                   ;ASCII BUFFER FOR DIGIT 0
13  DIG1        DB      ?                   ;ASCII BUFFER FOR DIGIT 1
14  DIG2        DB      ?                   ;ASCII BUFFER FOR DIGIT 2
15  TEMP        DB      ?                   ;TEMPORY BUFFER
16  COUNT       DB      ?                   ;UNIVERSAL COUNTER
17  RANDOM DB   ?                           ;RANDOM BUFFER
18  INDEX       DB      ?                   ;INDEX REGISTER
19  CLICK       DB      ?
20  ;-----------------------------------------------------------------
21  LCD_CONTR   EQU     PA                  ;DEFINE LCD CONTROL PORT
22  LCD_CONTRC  EQU     PAC                 ;DEFINE LCD CONTROL PORT CON. REG.
23  LCD_DATA    EQU     PB                  ;DEFINE LCD DATA PORT
24  LCD_DATAC   EQU     PBC                 ;DEFINE LCD DATA PORT CON. REG.
25  LCD_EN      EQU     LCD_CONTR.0         ;DEFINE EN CONTROL PIN
26  LCD_RW      EQU     LCD_CONTR.1         ;DEFINE RW CONTROL PIN
27  LCD_RS      EQU     LCD_CONTR.2         ;DEFINE RS CONTROL PIN
28  LCD_READY   EQU     LCD_DATA.7          ;DEFINE READY BIT OF LCD
29  SW_SELECT   EQU     PB.0                ;DEFINE SELECT SWITCH PIN
30  SW_ENTER    EQU     PB.1                ;DEFINE ENTER SWITCH PIN
31  STR_END     EQU     0FEH                ;STRING END CODE
32
33  MY_CODE     .SECTION    AT 0 'CODE'     ;== PROGRAM SECTION ==
34          ORG     00H                     ;HT-46RXX RESET VECTOR
35          CLR     LCD_CONTRC              ;CONFIG LCD_CONTR AS OUTPUT MODE
36          MOV     A,10010000B             ;TIMER MODE, Fint=Fsys, TON=1
37          MOV     TMRC,A
38          MOV     A,28H                   ;FUNCTION SET: 4-BIT,2-LINE,5X10 DOTS
39          CALL    WLCMC_4
40          MOV     A,0FH                   ;ON/OFF CONTR: DISPLAY ON,CURSOR BLINKING ON
41          CALL    WLCMC_4
42          MOV     A,06H                   ;ENTRY MODE : INCREMENT,DISPLAY NOT SHIFT
43          CALL    WLCMC_4
44          MOV     A,01H                   ;CLEAR DISPLAY
45          CALL    WLCMC_4
46          MOV     A,01000000B             ;SET CGRAM START ADDRESS
47          CALL    WLCMC_4
48          MOV     A,2*8
49          MOV     LINE_COUNT,A            ;SET CGRAM WRITE COUNTER
50          MOV     A,OFFSET TAB_CGRAM      ;GET TABLE START ADDRESS
51          MOV     TBLP,A
52  NEXT:
53          TABRDL  ACC                     ;READ DATA
54          CALL    WLCMD_4                 ;WRITE TO CG_RAM
55          INC     TBLP                    ;INCREASE POINTER BY 1
56          SDZ     LINE_COUNT              ;COUNT-1 = 0?
57          JMP     NEXT                    ;NO, NEXT DATA
58  MAIN:
59          MOV     A,01                    ;CLEAR DISPLAY
60          CALL    WLCMC_4
61          MOV     A,TMRL                  ;GET TMRL
62          XOR     A,TMRH                  ;XOR WITH TMRH AS RANDOM NUMBER 1
63          MOV     RANDOM,A                ;RESERVED RANDOM NUMBER 1
64          CALL    HEX2ASCII               ;CONVERT RANDOM 1 TO ASCII CODE
65          MOV     A,82H                   ;SET LINE 1 POSITION 2
66          CALL    WLCMC_4
67          CALL    WRITE_DIG               ;DISPLAY RANDOM 1 ON LCD
68          MOV     A,86H                   ;SET LINE 1 POSITION 6
69          CALL    WLCMC_4
```

```
70          MOV     A,00                    ;DISPLAY "小"
71          CALL    WLCMD_4
72          MOV     A,0C6H                  ;SET LINE 2 POSITION 6
73          CALL    WLCMC_4
74          MOV     A,01                    ;DISPLAY "大"
75          CALL    WLCMD_4
76          MOV     A,87H                   ;SET LINE 1 POSITION 7 CODE ON CLICK
77          MOV     CLICK,A
78  LOOP_1:
79          MOV     A,01000000B             ;INVERSE BIT 6 FOR TOGGLE LINE 1 & 2
80          XOR     A,CLICK
81          MOV     CLICK,A
82          CALL    WLCMC_4                 ;SET POSITION AS CODE IN CLICK
83  WAIT:
84          SET     PBC                     ;CONFIG PB AS INPUT MODE
85          SNZ     SW_SELECT               ;SELECT SWITCH PRESSED?
86          JMP     LOOP_2                  ;YES, JUMP TO LOOP_2
87          SNZ     SW_ENTER                ;ENTER SWITCH PRESSED?
88          JMP     LOOP_3                  ;YES, JUMP TO LOOP_3
89          JMP     WAIT                    ;NO KEY PRESSED,WAIT KEY!
90  LOOP_2:
91          MOV     A,30                    ;DELAY 0.3 SEC
92          CALL    DELAY
93          JMP     LOOP_1
94  LOOP_3:
95          MOV     A,TMRL                  ;GET TMRL
96          XOR     A,TMRH                  ;XOR WITH TMRH AS RANDOM NUMBER 2
97          MOV     TEMP,A                  ;RESERVED RANDOM NUMBER 2
98          CALL    HEX2ASCII               ;CONVERT RANDOM 2 TO ASCII CODE
99          MOV     A,0C2H                  ;SET LINE 2 POSITION 2
100         CALL    WLCMC_4
101         CALL    WRITE_DIG               ;DISPLAY RANDOM 2 ON LCD
102         MOV     A,TEMP
103         SUB     A,RANDOM
104         SZ      C                       ;IS TEMP > RANDOM ?
105         JMP     R_G_T                   ;NO.
106         SNZ     CLICK.6                 ;IS USER SELECT RANDOM > TEMP?
107         JMP     BINGO                   ;YES, DISPLAY BINGO!
108         JMP     MISS                    ;NO, DISPLAY MISS!
109 R_G_T:
110         SNZ     CLICK.6                 ;IS USER SELECT TEMP > RANDOM?
111         JMP     MISS                    ;NO, DISPLAY MISS!
112         JMP     BINGO                   ;YES, DISPLAY BINGO!
113 MISS:
114         MOV     A,80H+9                 ;NO, DISPLAY "MISS!" MESSAGE
115         CALL    WLCMC_4                 ;SET LINE 1 POSITION 9
116         MOV     A,OFFSET STR1           ;LOAD STRING START ADDRESS
117         CALL    PRINT                   ;PRINT ON LCD
118         MOV     A,200                   ;DELAY 2 SEC.
119         CALL    DELAY
120         JMP     MAIN                    ;RESTART
121 BINGO:
122         MOV     A,0C0H+9                ;DISPLAY "BINGO!"
123         CALL    WLCMC_4                 ;SET LINE 2 POSITION 9
124         MOV     A,OFFSET STR2           ;LOAD STRING START ADDRESS
125         CALL    PRINT                   ;PRINT ON LCD
126         MOV     A,250                   ;DELAY 2.5 SEC.
127         CALL    DELAY
128         JMP     MAIN                    ;RESTART
129 ;****************************************************************
130 ;                   WRITE_DIG SUBROUTINE
131 ;****************************************************************
132 WRITE_DIG   PROC
133         MOV     A,DIG2
134         CALL    WLCMD_4
135         MOV     A,DIG1
136         CALL    WLCMD_4
137         MOV     A,DIG0
138         CALL    WLCMD_4
139         RET
```

```
140 WRITE_DIG   ENDP
141 ;**********************************************************************
142 ;      HEX2ASCII SUBROUTINE (CONVERT HEX DATA IN Acc TO DIG2~DIG0)
143 ;**********************************************************************
144 HEX2ASCII   PROC
145         MOV         DIG0,A
146         CLR         DIG1
147         CLR         DIG2
148 DIG_2:
149         SUB         A,100
150         SNZ         C
151         JMP         DIG_1
152         INC         DIG2
153         JMP         DIG_2
154 DIG_1:
155         ADD         A,100
156 DIG_1_1:
157         SUB         A,10
158         SNZ         C
159         JMP         DIG_0
160         INC         DIG1
161         JMP         DIG_1_1
162 DIG_0:
163         ADD         A,10
164         MOV         DIG0,A
165         MOV         A,30H                   ;BEFORE THIS LINE HEX TO UN-PACK BCD
166         ADDM        A,DIG2
167         ADDM        A,DIG1
168         ADDM        A,DIG0
169         RET
170 HEX2ASCII   ENDP
171 ;**********************************************************************
172 ;          LCD PRINT PROCEDURE (START ADDRS=TBLP)
173 ;**********************************************************************
174 PRINT   PROC
175         MOV         TBLP,A                  ;LOAD TABLE POINTER
176 PRINT_1:
177         TABRDL      ACC                     ;READ STRING
178         XOR         A,STR_END
179         SZ          Z                       ;END OF STRING CHARACTER?
180         JMP         END_PRINT               ;YES, STOP PRINT
181         TABRDL      ACC                     ;NO, RE-READ
182         CALL        WLCMD_4                 ;SEND TO LCD
183         INC         TBLP                    ;INCREASE DATA POINTER
184         JMP         PRINT_1                 ;NEXT CHARACTER
185 END_PRINT:
186         RET                                 ;YES, RETURN
187 PRINT   ENDP
188 ;**********************************************************************
189 ;      4-BIT LCD DATA/COMMAND WRITE PROCEDURE
190 ;**********************************************************************
191 WLCMD_4     PROC
192         SET         DC_FLAG                 ;SET DC_FLAG=1 FOR DATA WRITE
193         JMP         WLCM_4
194 WLCMC_4:
195         CLR         DC_FLAG                 ;SET DC_FLAG=0 FOR COMMAND WRITE
196 WLCM_4:
197         SET         LCD_DATAC               ;CONFIG LCD_DATA AS INPUT MODE
198         CLR         LCD_CONTR               ;CLEAR ALL LCD CONTROL SIGNAL
199         SET         LCD_RW                  ;SET RW SIGNAL (READ)
200         NOP                                 ;FOR TAS
201         SET         LCD_EN                  ;SET EN HIGH
202         NOP                                 ;FOR TDDR
203 WF_4:
204         SZ          LCD_READY               ;IS LCD BUSY?
205         JMP         WF_4                    ;YES, JUMP TO WAIT
206         CLR         LCD_DATAC               ;NO, CONFIG LCD_DATA AS OUTPUT MODE
207         MOV         LCD_DATA,A              ;LATCH DATA/COMMAND ON PB(LCD DATA BUS)
208         CLR         LCD_CONTR               ;CLEAR ALL LCD CONTROL SIGNAL
209         SZ          DC_FLAG                 ;IS COMMAND WRITE?
210         SET         LCD_RS                  ;NO, SET RS HIGH
```

```
211          SET      LCD_EN              ;SET EN HIGH
212          NOP
213          CLR      LCD_EN              ;SET EN LOW
214          SWAPA ACC
215          SET      LCD_EN              ;SET EN HIGH
216          MOV      LCD_DATA,A
217          CLR      LCD_EN              ;SET EN LOW
218          RET
219 WLCMD_4  ENDP
220 ;********************************************************************
221 ;                 Delay about ACC*10ms
222 ;********************************************************************
223 DELAY    PROC
224          MOV      DEL1,A
225 DEL_1:   MOV      A,30
226          MOV      DEL2,A              ;SET DEL2 COUNTER
227 DEL_2:   MOV      A,110
228          MOV      DEL3,A              ;SET DEL3 COUNTER
229 DEL_3:   SDZ      DEL3                ;DEL3 DOWN COUNT
230          JMP      DEL_3
231          SDZ      DEL2                ;DEL2 DOWN COUNT
232          JMP      DEL_2
233          SDZ      DEL1                ;DEL1 DOWN COUNT
234          JMP      DEL_1
235          RET
236 DELAY    ENDP
237          ORG      LASTPAGE
238 STR1:    DC       'MISS!',STR_END     ;DEFINE STRING DATA 1
239 STR2:    DC       'BINGO!',STR_END    ;DEFINE STRING DATA 2
240 TAB_CGRAM:
241          DC       00000000B           ;CHARACTER '小'
242          DC       00000100B
243          DC       00000100B
244          DC       00010101B
245          DC       00010101B
246          DC       00000100B
247          DC       00001100B
248          DC       00000000B
249
250          DC       00000000B           ;CHARACTER '大'
251          DC       00000100B
252          DC       00000100B
253          DC       00011111B
254          DC       00000100B
255          DC       00001010B
256          DC       00010001B
257          DC       00000000B
258          END
```

程序说明

7~19   依序定义变量地址。

21~24   定义 LCD_DATA Port 与 LCD_CONTR Port 分别为 PB 与 PA。

25~28   定义 LCD 的控制信号引脚。

29~30   定义 Push Button 的控制信号引脚。

31   定义字符串结束码 STR_END=FEh。

34   声明存储器地址由 000h 开始(HT46xx Reset Vector)。

35   将 LCD_CONTR Port(PA)定义为输出模式。

36~37   定义 Timer/Event Counter 的工作模式：Timer Mode、计数比例 1:1($f_{INT}=f_{SYS}$)、TON=1(启动 Timer/Event Counter 开始计数)。

38~39   将 LCD 设置为双行显示(N=1)、使用 4 位(DB7~DB4)控制模式(DL=0)、5×7 点矩阵字形(F=0)。

40~41   将 LCD 设置为显示所有数据(D=1)、显示光标(C=1)、光标不闪烁(B=0)。

行号	说明
42~43	将 LCD 的地址标志位(AC)设置为递加(I/D = 1)、显示器画面不因读/写数据而移动(S = 0)。
44~45	将 LCD 整个显示器清空。
46~47	设置 CG RAM 的地址为 00h，准备写入自建字形。
48~51	设置计数器 LINE_COUNT 为 16 并将 TBLP 指向自建字形数据的起始地址。
53~54	通过查表将自建字形的数据取出写入 CG RAM 中。
55	将 TBLP 加 1。
56~57	判断 COUNT−1 是否为 0。成立，表示所有自建字形已输入完毕；反之，则表示自建字形还未建立完成，继续填入字形码。
59~60	将 LCD 整个显示器清空。
61~64	利用 Timer/Event Counter 产生随机数存放在 RANDOM 寄存器，并调用 HEX2ASCII 子程序转换成可写入 LCD 的 ASCII 码，在此产生的随机数为"题目"。(注：详细说明请参考 HEX2ASCII 子程序的说明。)
65~67	将 LCD 的 AC 值设置在第 1 行第 2 个位置，并调用 WRITE_DIG 子程序将刚刚取得的随机数值显示在 LCD 上(即将 DIG0、DIG1、DIG2 内容显示在 LCD 上)。
68~71	在第 1 行第 6 个位置显示自建字形"小"。
72~75	在第 2 行第 6 个位置显示自建字形"大"。
76~77	设置 CLICK 寄存器为 87h，当成光标的位置。
79~82	通过 XOR 来做切换行数(用户每按一次 SELECT 按键，光标位置就换行)的功能。XOR 所改变的只有 Bit6 的内容，其余内容会保留住。
84~86	若有按键输入(SW_SELECT)，则将目前光标停留的行数切换到另一行，表现出可以选择"大"或"小"选项功能(例如：目前光标停留在第 1 行，若有键输入则跳到第 2 行)。
87~89	若有按键输入(SW_ENTER)表示用户已确定选择的内容，则显示结果(LOOP_3)；反之，则不断地等到确定为止(WAIT)。
91~93	调用 DELAY 子程序，延时 0.3 s 后进行切换行数的动作。
95~98	利用 Timer/Event Counter 产生第 2 笔随机数存放在 TEMP 寄存器，并调用 HEX2ASCII 子程序转换成可写入 LCD 的 ASCII 码，在此产生的随机数为"答案"。
99~101	将 LCD 的 AC 值设置在第 2 行第 2 个位置，并调用 WRITE_DIG 子程序将刚取得的随机数值显示在 LCD 上(即将 DIG0、DIG1、DIG2 内容显示在 LCD 上)。
102~112	将答案与结果相减(TEMP−RANDOM)判断是否有借位(C=1)。若成立，则表示题目提供的数值比结果还要小(C=1)，所以 CLICK 的判断式也修改成选择"大"(BIT6=0)表示错误，选择"小"(BIT6=1)表示正确。若题目比结果还要大(C=0)则 CLICK 判断的式子也跟着相反。
113~120	显示"MISS！"在第 1 行后重新开始执行程序。
121~128	显示"BINGO！"在第 2 行后重新开始执行程序。
132~140	WRITE_DIG 子程序，此子程序负责将 DIG0、DIG1、DIG2 三个寄存器的内容显示在 LCD 上。
144~170	HEX2ASCII 子程序，此子程序的功能是将 Acc 寄存器中的十六进制数据(00h~FFh)转换成 ASCII 码。转换后的结果分别存放在 DIG0(个位数)、DIG1(十位数)、DIG2(百位数)三个寄存器当中。
174~187	PRINT 子程序，此子程序负责将定义好的字符串依序显示在 LCD 上。在调用此子程序之前，除了必须先设置好 LCD 的位置之外，还需先在 Acc 寄存器中指定字符串的起始地址(字符串必须存放在最末程序页)，并在字符串的最后一个字符插入 STR_END，代表字符串结束。
182~214	WLCMD_4 与 WLCMC_4 子程序，请参考 5-9 节的说明。
223~236	DELAY 子程序，至于延时时间的计算请参考 4-1 节。
238~239	字符串定义区，每一个字符串务必以 STR_END 作为结束字符。
240~257	自建字形的字形码定义区。

5-10-5 动动脑+动动手

- 修改程序，一开始提供用户 100 分的分数，答对一次加 5 分、答错一次扣 5 分，并将分数显示在 LCD 上；而当用户分数小于 0 或大于 255 时，程序就不再产生随机数。
- 承上题，如果允许用户的分数超过 255，势必要一个 16 位的十六进制转 BCD 的转换程序，请尝试着写写看。
- 请修改程序，使产生的变量值局限在 0~99 之间。

5-11 中文显示型LCD控制实验

5-11-1 目 的

在中文显示型LCD上显示字形，并利用程序的技巧产生字符串移动的效果。

5-11-2 学习重点

通过本实验，读者能熟悉中文显示型LCD的控制方式，此外亦需能运用LCD的控制指令让液晶显示器展现不同的显示效果。

5-11-3 电路图（见图5-11-1）

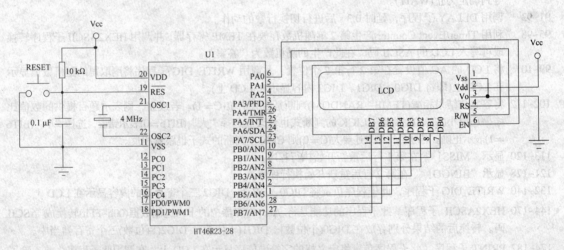

图5-11-1 LCD(8位)控制电路

在前几个实验中，利用LCD内部的CG RAM建立了几个简单的中文字形，由于分辨率不变(5×7点)，若要显示笔画较多的中文字形，可就必须以内置中文字形的中文显示型LCD来实践较为恰当。目前市面上也可买到此类型的液晶显示器，虽然价格不斐，但是其所显示的字形十分细致(见图5-11-2)。中文显示型LCD的控制方式其实与一般文字形LCD相当类似，所以关于其控制部分在此不再赘述，仅提醒读者相当重要的注意事项。本实验采用9×2的中文显示型LCD(编号：P-14B32CT)，此液晶显示器共有2行，每行可以显示9个中文字形(或18个英文字形)，其显示位置与DD RAM地址的对应关系如图5-11-3所示。

图5-11-2　中文LCD显示字形

	0	1	2	…	7	8	显示位置
第1行	00h	01h	02h	…	07h	08h	DD RAM地址
第2行	10h	11h	12h	…	17h	15h	DD RAM地址

图5-11-3　9×2中文显示型LCD 显示位置与DD RAM地址的对应关系

请注意第1行的最后一个位置(08h)与第2行的第0个位置(10h)的DD RAM地址并不连续。而每一个位置可以显示1个中文字形或是2个英文字形，这表示在中文显示型LCD上，一个DD RAM地址是可以存放2个字节的数据，当显示中文时，只须将中文的BIG-5码(2 字节)分成2次(LCD 8位控制模式)写至LCD的DD RAM即可；显示英文时，则与一般LCD的控制方式相同。不过要注意在写入中文时，要确定写入的BIG-5码一定要在同一个DD RAM地址内，如果是被拆成2个字节分别存放在不同DD RAM地址(例如位置00h的High Byte与01h的Low Byte)，将会显示乱码(请参考本实验的"动动手＋动动脑"部分)。

5-11-4　流程图及程序

1. 流程图(见图 5-11-4)

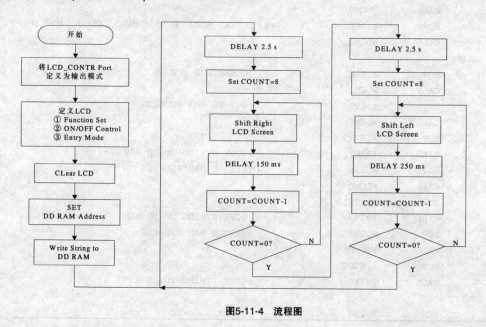

图5-11-4　流程图

2. 程序 5-11 中文显示型 LCD 控制实验

```
1   ;PROGRAM : 5-11.ASM  (5-11.PRJ)                    BY STEVEN
2   ;FUNCTION: 9X2 CHINESE LCD CONTROL DEMO PROGRAM    2002.DEC.07.
3       #INCLUDE   HT46R23.INC
4                  .CHIP HT46R23
5   ;------------------------------------------------------------
6   MY_DATA       .SECTION   'DATA'         ;== DATA SECTION ==
7   DEL1          DB         ?              ;DELAY LOOP COUNT 1
8   DEL2          DB         ?              ;DELAY LOOP COUNT 2
9   DEL3          DB         ?              ;DELAY LOOP COUNT 3
10  DC_FLAG       DBIT                      ;LCD DATA/COMMAND FLAG
11  COUNT         DB         ?              ;UNIVERSAL COUNTER
12  ;------------------------------------------------------------
13  LCD_CONTR     EQU        PA             ;DEFINE LCD CONTROL PORT
14  LCD_CONTRC    EQU        PAC            ;DEFINE LCD CONTROL PORT CON. REG.
15  LCD_DATA      EQU        PB             ;DEFINE LCD DATA PORT
16  LCD_DATAC     EQU        PBC            ;DEFINE LCD DATA PORT CON. REG.
17  LCD_EN        EQU        LCD_CONTR.0    ;DEFINE EN CONTROL PIN
18  LCD_RW        EQU        LCD_CONTR.1    ;DEFINE RW CONTROL PIN
19  LCD_RS        EQU        LCD_CONTR.2    ;DEFINE RS CONTROL PIN
20  LCD_READY     EQU        LCD_DATA.7     ;DEFINE READY BIT OF LCD
21  STR_END       EQU        0FEH           ;STRING END CODE
22
23  MY_CODE       .SECTION   AT 0 'CODE'    ;== PROGRAM SECTION ==
24                ORG        00H            ;HT-46RXX RESET VECTOR
25                CLR        LCD_CONTRC     ;CONFIG LCD_CONTR AS OUTPUT MODE
26                MOV        A,38H          ;FUNCTION SET: 8-BIT,2-LINE,5X10 DOTS
27                CALL       WLCMC
28                MOV        A,0CH          ;ON/OFF CONTR: DISPLAY ON,CURSOR BLINKING OFF
29                CALL       WLCMC
30                MOV        A,06H          ;ENTRY MODE : INCREMENT,DISPLAY NOT SHIFT
31                CALL       WLCMC
32  MAIN:
33                MOV        A,01H          ;CLEAR LCD SCREEN
34                CALL       WLCMC
35                MOV        A,89H          ;SET DD RAM ADDRESS
36                CALL       WLCMC
37                MOV        A,STR3
38                CALL       PRINT          ;COPY STRING TO DD RAM
39                MOV        A,83H          ;SET DD RAM ADDRESS
40                CALL       WLCMC
41                MOV        A,STR1
42                CALL       PRINT          ;COPY STRING TO DD RAM
43                MOV        A,091H         ;SET DD RAM ADDRESS
44                CALL       WLCMC
45                MOV        A,STR2
46                CALL       PRINT          ;COPY STRING TO DD RAM
47  SHIFT:
48                MOV        A,250          ;DELAY 2.5 SEC
```

```
49          CALL        DELAY
50          MOV         A,8                     ;SET COUNT=8
51          MOV         COUNT,A
52  SCREEN_SHIFT_RIGHT:
53          MOV         A,15                    ;DELAY 150ms
54          CALL        DELAY
55          MOV         A,1CH
56          CALL        WLCMC                   ;SET SCREEN SHIFT RIGHT
57          SDZ         COUNT                   ;COUNT-1 = 0?
58          JMP         SCREEN_SHIFT_RIGHT      ;NO, SHIFT RIGHT AGAIN
59          MOV         A,250                   ;DELAY 2.5 SEC
60          CALL        DELAY
61          MOV         A,8                     ;SET COUNT=8
62          MOV         COUNT,A
63  SCREEN_SHIFT_LEFT:
64          MOV         A,25                    ;DELAY 250ms
65          CALL        DELAY
66          MOV         A,18H                   ;SET SCREEN SHIFT LEFT
67          CALL        WLCMC
68          SDZ         COUNT                   ;COUNT-1 = 0?
69          JMP         SCREEN_SHIFT_LEFT       ;NO, SHIFT RIGHT AGAIN
70          JMP         SHIFT                   ;RESTART
71  ;*************************************************************************
72  ;           LCD PRINT PROCEDURE (START ADDRS=TBLP)
73  ;*************************************************************************
74  PRINT   PROC
75          MOV         TBLP,A                  ;LOAD TABLE POINTER
76  PRINT_1:
77          TABRDL      ACC                     ;READ STRING
78          XOR         A,STR_END
79          SZ          Z                       ;END OF STRING CHARACTER?
80          JMP         END_PRINT               ;YES, STOP PRINT
81          TABRDL      ACC                     ;NO, RE-READ
82          CALL        WLCMD                   ;SEND TO LCD
83          INC         TBLP                    ;INCREASE DATA POINTER
84          JMP         PRINT_1                 ;NEXT CHARACTER
85  END_PRINT:
86          RET                                 ;YES, RETURN
87  PRINT   ENDP
88  ;*************************************************************************
89  ;           LCD DATA/COMMAND WRITE PROCEDURE
90  ;*************************************************************************
91  WLCMD   PROC
92          SET         DC_FLAG                 ;SET DC_FLAG=1 FOR DATA WRITE
93          JMP         WLCM
94  WLCMC:
95          CLR         DC_FLAG                 ;SET DC_FLAG=0 FOR COMMAND WRITE
96  WLCM:
97          SET         LCD_DATAC               ;CONFIG LCD_DATA AS INPUT MODE
98          CLR         LCD_CONTR               ;CLEAR ALL LCD CONTROL SIGNAL
```

```
99            SET       LCD_RW                    ;SET RW SIGNAL (READ)
100           NOP                                 ;FOR TAS
101           SET       LCD_EN                    ;SET EN HIGH
102           NOP                                 ;FOR TDDR
103  WF:
104           SZ        LCD_READY                 ;IS LCD BUSY?
105           JMP       WF                        ;YES, JUMP TO WAIT
106           CLR       LCD_DATAC                 ;NO, CONFIG LCD_DATA AS OUTPUT MODE
107           MOV       LCD_DATA,A                ;LATCH DATA/COMMAND ON PB(LCD DATA BUS)
108           CLR       LCD_CONTR                 ;CLEAR ALL LCD CONTROL SIGNAL
109           SZ        DC_FLAG                   ;IS COMMAND WRITE?
110           SET       LCD_RS                    ;NO, SET RS HIGH
111           SET       LCD_EN                    ;SET EN HIGH
112           NOP
113           CLR       LCD_EN                    ;SET EN LOW
114           RET
115  WLCMD    ENDP
116  ;**********************************************************************
117  ;                   Delay about ACC*10ms
118  ;**********************************************************************
119  DELAY    PROC
120           MOV       DEL1,A
121  DEL_1:   MOV       A,30
122           MOV       DEL2,A                    ;SET DEL2 COUNTER
123  DEL_2:   MOV       A,110
124           MOV       DEL3,A                    ;SET DEL3 COUNTER
125  DEL_3:   SDZ       DEL3                      ;DEL3 DOWN COUNT
126           JMP       DEL_3
127           SDZ       DEL2                      ;DEL2 DOWN COUNT
128           JMP       DEL_2
129           SDZ       DEL1                      ;DEL1 DOWN COUNT
130           JMP       DEL_1
131           RET
132  DELAY    ENDP
133           ORG       LASTPAGE
134  STR1:    DC        'HT46xx',STR_END          ;DEFINE STRING DATA 1
135  STR2:    DC        '理论与实务宝典',STR_END ;DEFINE STRING DATA 2
136  STR3:    DC        '钟启仁 编著',STR_END    ;DEFINE STRING DATA 3
137           END
```

程序说明

7~11   依序定义变量地址。

13~16   定义 LCD_DATA Port 与 LCD_CONTR Port 分别为 PB 与 PA。

17~20   定义 LCD 的控制信号引脚。

21   定义字符串结束码 STR_END=FEh。

24   声明存储器地址由 000h 开始(HT46xx Reset Vector)。

25   将 LCD_CONTR Port(PA)定义为输出模式。

26~27   将 LCD 设置为双行显示($N=1$)、使用 8 位(DB7~DB0)控制模式($DL=1$)、5×7 点矩阵字形($F=0$)。

行号	说明
28~29	将 LCD 设置为显示所有数据(D = 1)、显示光标(C = 1)、光标所在位置的字不会闪烁(B = 0)。
30~31	将 LCD 的地址标志位(AC)设置为递加(I/D = 1)、显示器画面不因读/写数据而移动(S = 0)。
33~34	将 LCD 整个显示器清空。
36~36	设置 DD RAM 的地址为 09h。
37~38	调用 PRINT 子程序,通过查表将字符串(STR3)写入 DD RAM 中。
39~40	设置 DD RAM 的地址为 03h。
41~42	调用 PRINT 子程序,通过查表将字符串(STR1)写入 DD RAM 中。
43~44	设置 DD RAM 的地址为 11h。
45~46	调用 PRINT 子程序,通过查表将字符串(STR2)写入 DD RAM 中。
48~49	调用 DELAY 子程序,延时 2.5 s。
50~51	设置 COUNT=8。
53~54	调用 DELAY 子程序,延时 150 ms。
55~56	利用 LCD 命令,控制 LCD 屏幕右移一个字符位置,请参考表 5-5-5 的 LCD 指令表。
57~58	判断 COUNT−1 是否为 0。成立,则表示屏幕已经右移 8 次;反之,则继续右移。
59~60	调用 DELAY 子程序,延时 2.5 s。
61~62	设置 COUNT=8。
64~65	调用 DELAY 子程序,延时 250 ms。
66~67	利用 LCD 命令,控制 LCD 屏幕左移一个字符位置,请参考表 5-5-4 的 LCD 指令表。
68~70	判断 COUNT−1 是否为 0。成立,则表示屏幕已经左移 8 次,重新开始右移的动作(JMP SHIFT);反之,则继续左移。
74~87	PRINT 子程序,此子程序负责将定义好的字符串依序显示在 LCD 上。在调用此子程序之前,除了必须先设置好 LCD 的位置之外,还需先在 Acc 寄存器中指定字符串的起始地址(字符串必须存放在最末程序页 Last Page),并请在字符串的最后一个字符插入 STR_END,代表字符串结束。
91~115	WLCMD 与 WLCMC 子程序,请参考 5-5 节的说明。
119~132	DELAY 子程序,至于延时时间的计算请参考 4-1 节。
134~136	字符串定义区,每一个字符串务必以 STR_END 作为结束字符。

程序一开始先对LCD做初始化的设置,接下来就是连续在不同的DD RAM填入字符串STR3、STR2与STR1。在填入STR3之前,将AC(LCD Address Counter)设置为09h(即第9个显示位置),此时LCD的可见窗口范围为位置0~8,因此看不到STR3的字符串。而后利用LCD屏幕的左、右移指令,逐一改变LCD的可见窗口范围,再配合不同的延时时间而达到字符串左、右移动的效果。

5-11-5 动动脑 + 动动手

  将 STR3 的定义改变如下(在字符间插入空白字符),观察 LCD 显示的字形有何改变。
  STR3: DC '钟启仁　编著', STR_END ; DEFINE STRING DATA 3
  由于在中文字间插入了空白码(20h),致使第 2 个中文字的 BIG-5 码被拆成两个字节分别存放在 DD RAM 地址 8Ah(High Byte)与 8Bh(Low Byte),因此第 2 个中文字形无法正常显示。而第 2 个空白码恰巧被插入 8Bh(High Byte)地址,使得其后的中文 BIG-5 码存放在同一个 DD RAM 地址,所以除了第 2 个中文字形无法显示之外,其他字形显示均正常无误。
- 试着利用 LCD 控制指令,让 LCD 产生闪、灭的效果。
- 改变 DELAY 子程序的延时时间,使字形移动的速度变化。

5-12 半矩阵式(Half-Matrix)键盘与LCD控制实验

5-12-1 目 的

以半矩阵式键盘(Half-Matrix Key Pad)为输入设备,让用户可以输入大、小写的英文字母,并显示在LCD上。

5-12-2 学习重点

通过本实验,读者应熟悉半矩式键盘的控制方式,以便在多按键输入的应用场合加以运用;此外对LCD的控制应更加得心应手。

5-12-3 电路图(见图5-12-1)

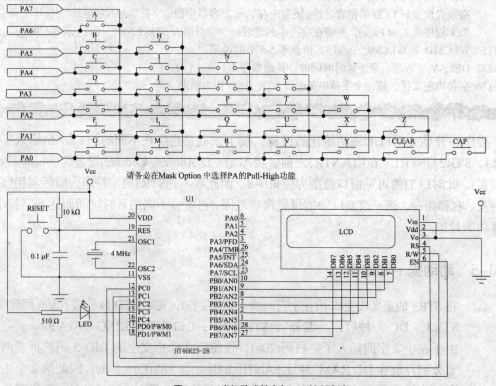

图5-12-1 半矩阵式键盘与LCD控制电路

第 5 章　进阶实验篇

在4-7节中，相信读者应该已经了解4×4矩阵式键盘(Matrix Key Pad)的工作原理，它是以8位的I/O端口搭配程序的运行，就可以区分出16个不同的按键。然而，在实践运用上(尤其在I/O资源弥足珍贵的单片机应用场合)，如果想要增加按键的个数，势必又得多腾出几根I/O引脚供按键扫描使用。例如本实验要求能够输入26个英文字母，万一单片机已经没有多余的I/O引脚可供使用的话，该怎么办呢？以下所介绍的半矩阵式键盘(Half-Matrix Key Pad)，是在一般实践设计上经常使用的技巧，只要一个8位的I/O端口并配合程序的运行，就能区分出28个不同的按键，足足比4×4矩阵式键盘多出了12个，正可满足本实验的需求。图5-12-2是半矩阵式键盘的硬件连接方式，在此笔者以HT46x23的PA做控制来做说明。

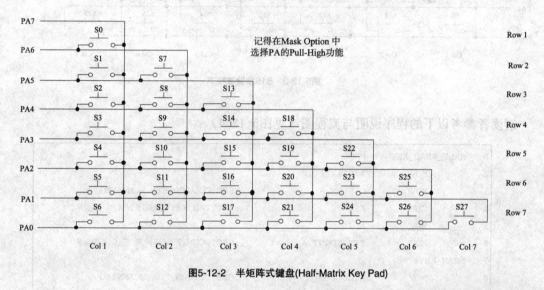

图5-12-2　半矩阵式键盘(Half-Matrix Key Pad)

半矩阵式键盘的扫描原理是依序让每一个I/O引脚送出(输出模式)0准位，每当送出0之后，就读回(输入模式)其他I/O引脚的状态，如果读回的状态出现0准位，就表示该行有按键被按下；至于是哪一列被按下，则可由读回的数值加以分析。因此每一根I/O引脚都可能被定义为输入或输出模式，不像4×4矩阵式键盘是4个I/O引脚固定作为输出，另4个则固定当成输入。

以图5-12-2为例，半矩阵式键盘的扫描是由PA7~PA0依序送出0(即逐行检查)，然后再读回PA的状态判断是哪一列被按下。倘若是S15按键被按下(见图5-12-3)，首先由PA7(输出模式)送出0，然后由PA6~PA0(输入模式)读回的状态为"?1111111B"，表示第1行(Col 1)没有按键被按下；接着由PA6(输出模式)送出0检查Col 2，然后由PA5~PA0读回的状态为"??111111B"，表示该行也没有按键被按下；当由PA5(输出模式)送出0检查Col 3时，从PA4~PA0读回的状态为"???11011B"，表示该行第5列的按键(即S15)被按下。(注："?"表示该引脚为输出模式或者是未接上按键，其状态不需理会。)

273

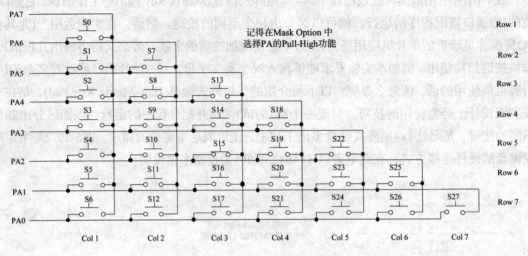

图5-12-3　S15按键被按下

请读者参考以下的程序说明与流程图（见图5-12-4）：

```
1   HALF_READ_KEY  PROC
2       MOV     A,01111111B
3       MOV     COLUMN,A            ;SET COLUMN FOR SCANNING
4       CLR     KEY                 ;INITIAL KEY REGISTER
5       MOV     A,7
6       MOV     KEY_COUNT,A         ;SET COUNTER=7 FOR 7 COLUMNS
7   SCAN_KEY:
8       MOV     A,COLUMN            ;GET COLUMN FOR SCANNING
9       MOV     KEY_PORTC,A         ;CONFIG KEY_PORT
10      MOV     KEY_PORT,A          ;SCANNING KEY PAD
11      MOV     A,KEY_COUNT         ;GET COLUMN COUNTER
12      ADDM    A,PCL               ;COMPUTATIONAL JUMP
13      NOP                         ;OFFSET FOR KEY_COUNT=0
14      JMP     $+24                ;KEY_COUNT=1
15      JMP     $+20                ;KEY_COUNT=2
16      JMP     $+16                ;KEY_COUNT=3
17      JMP     $+12                ;KEY_COUNT=4
18      JMP     $+8                 ;KEY_COUNT=5
19      JMP     $+4                 ;KEY_COUNT=6
20      SNZ     KEY_PORT.6          ;KEY_COUNT=7,ROW 1 PRESSED?
21      JMP     END_KEY             ;YES.
22      INC     KEY                 ;NO, INCREASE KEY CODE
23      SNZ     KEY_PORT.5          ;ROW 2 PRESSED?
24      JMP     END_KEY             ;YES.
25      INC     KEY                 ;NO, INCREASE KEY CODE
26      SNZ     KEY_PORT.4          ;ROW 3 PRESSED?
27      JMP     END_KEY             ;YES.
```

28	INC	KEY	;NO, INCREASE KEY CODE
29	SNZ	KEY_PORT.3	;ROW 4 PRESSED?
30	JMP	END_KEY	;YES.
31	INC	KEY	;NO, INCREASE KEY CODE
32	SNZ	KEY_PORT.2	;ROW 5 PRESSED?
33	JMP	END_KEY	;YES.
34	INC	KEY	;NO, INCREASE KEY CODE
35	SNZ	KEY_PORT.1	;ROW 6 PRESSED?
36	JMP	END_KEY	;YES.
37	INC	KEY	;NO, INCREASE KEY CODE
38	SNZ	KEY_PORT.0	;ROW 7 PRESSED?
39	JMP	END_KEY	;YES.
40	INC	KEY	;NO, INCREASE KEY CODE
41	RR	COLUMN	;SCAN CODE FOR NEXT COLUMN
42	SDZ	KEY_COUNT	;HAVE ALL COULMN BEEN CHECKED?
43	JMP	SCAN_KEY	;NO, NEXT COLUMN
44	END_KEY:		
45	RET		
46	HALF_READ_KEY	ENDP	

HALF_READ_KEY 子程序说明

2~3　设置 COLUMN=01111111b；如之前的说明，半矩阵式键盘须依序扫描 PA7~PA1，因此以 COLUMN 作为扫描与 Port 定义数据寄存区。

4　　设置 KEY 值为 0。当执行完此子程序时，KEY 寄存器的值即为按键值；若按键都没有被按下，则 KEY=28。

5~6　设置 KEY_COUNT=7，作为扫描 7 行的控制寄存器。

8~10　因为扫描的行(即送出 0 的端口)必须定义为输出模式，而其余引脚必须为输入模式以便判断是否有按键被按下。因此这 3 行指令的目的是先依据 COLUMN 寄存器的值将 I/O 端口做适当的定义，然后再将扫描码送出。

11~40　依序检查各列是否有按键被按下，若无则将 KEY 寄存器值加 1，并继续检查下一列；若按键被按下则返回主程序(JMP END_KEY)。由于硬件结构的关系，每当送出新的扫描码时，并不见得要检查所有的输入引脚。例如当扫描第 1 行时(PA7=0)，必须循序检查 PA6~PA0；而扫描第 2 行时(PA6=0)，只须循序检查 PA5~PA0；依此类推。因此在程序第 11~19 行是以 KEY_COUNT 寄存器计算式的跳转(Computational Jump)，以避开不需检查的输入引脚。又因为 KEY_COUNT 寄存器的变化范围是 1~7，所以在第 13 行加入 NOP 指令作为补偿。

44　将扫描码右移一位，准备扫描下一行。

42~45　判断 KEY_COUNT 减 1 是否为 0，若成立表示 7 行均已完成扫描，返回主程序(RET)；否则继续检查下一行(JMP SCAN_KEY)。

HT46xx 单片机原理与实践

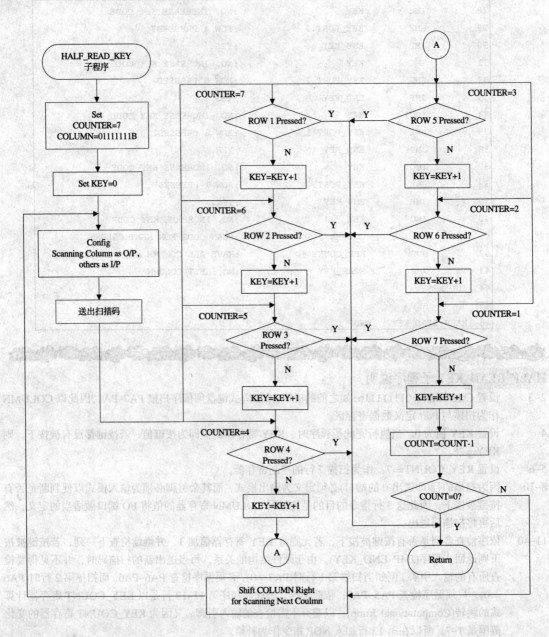

图5-12-4 流程图

5-12-4 流程图及程序

1. 流程图(见图 5-12-5)

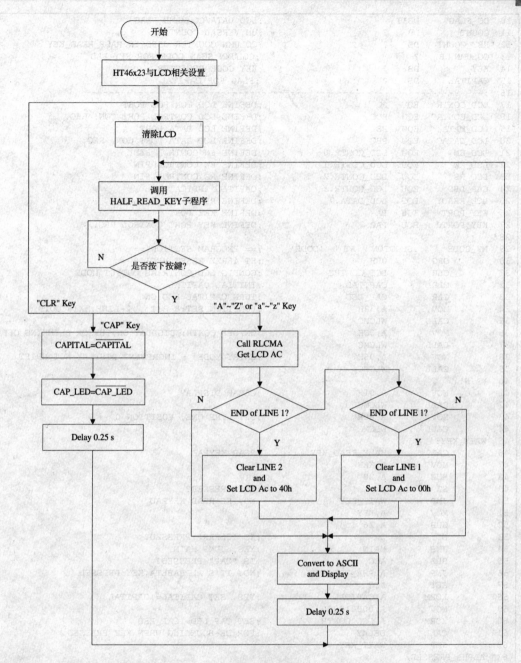

图5-12-5 流程图

2. 程序 5-12 半矩阵式键盘与 LCD 控制程序

```
1   ;PROGRAM : 5-12.ASM  (5-12.PRJ)                           BY STEVEN
2   ;FUNCTION: 16X2 LCD vs.HALF-MATRIX KEYPAD DEMO PROGRAM    2002.DEC.07.
3   #INCLUDE    HT46R23.INC
4           .CHIP HT46R23
5   ;--------------------------------------------------------------------
6   MY_DATA     .SECTION    'DATA'          ;== DATA SECTION ==
7   DEL1        DB      ?                   ;DELAY LOOP COUNT 1
8   DEL2        DB      ?                   ;DELAY LOOP COUNT 2
9   DEL3        DB      ?                   ;DELAY LOOP COUNT 3
```

277

```
10  DC_FLAG      DBIT                         ;LCD DATA/COMMAND FLAG
11  COUNT        DB      ?                    ;UNIVERSAL COUNTER
12  KEY_COUNT    DB      ?                    ;COLUMN COUNTER USED IN HALF_READ_KEY
13  COLUMN DB    ?                            ;COLUMN SCAN CODE FOR KEY PAD
14  KEY          DB      ?                    ;KEY CODE REGISTER
15  CAPITAL      DB      ?                    ;FLAG FOR CAPITAL
16  ;----------------------------------------------------------------
17  LCD_CONTR    EQU     PC                   ;DEFINE LCD CONTROL PORT
18  LCD_CONTRC   EQU     PCC                  ;DEFINE LCD CONTROL PORT CON. REG.
19  LCD_DATA     EQU     PB                   ;DEFINE LCD DATA PORT
20  LCD_DATAC    EQU     PBC                  ;DEFINE LCD DATA PORT CON. REG.
21  LCD_EN       EQU     LCD_CONTR.0          ;DEFINE EN CONTROL PIN
22  LCD_RW       EQU     LCD_CONTR.1          ;DEFINE RW CONTROL PIN
23  LCD_RS       EQU     LCD_CONTR.2          ;DEFINE RS CONTROL PIN
24  CAP_LED      EQU     LCD_CONTR.3          ;CAPITAL INDICATOR
25  LCD_READY    EQU     LCD_DATA.7           ;DEFINE READY BIT OF LCD
26  KEY_PORT     EQU     PA                   ;DEFINE KEY PORT
27  KEY_PORTC    EQU     PAC                  ;DEFINE KEY PORT CONTROL REG.
28
29  MY_CODE      .SECTION  AT 0 'CODE'        ;== PROGRAM SECTION ==
30       ORG         00H                      ;HT-46RXX RESET VECTOR
31       CLR         LCD_CONTRC               ;CONFIG LCD_CONTR AS OUTPUT MODE
32       CLR         CAPITAL                  ;INITIAL CAPITAL
33       CLR         CAP_LED                  ;TURN CAPITAL LED ON
34       MOV         A,38H                    ;FUNCTION SET:8-BIT  ,2-LINE,5X10 DOTS
35       CALL        WLCMC
36       MOV         A,0CH                    ;ON/OFF CONTR: DISPLAY ON,CURSOR BLINKING OFF
37       CALL        WLCMC
38       MOV         A,06H                    ;ENTRY MODE : INCREMENT,DISPLAY NOT SHIFT
39       CALL        WLCMC
40  MAIN:
41       MOV         A,01H                    ;CLEAR DISPLAY
42       CALL        WLCMC
43       MOV         A,80H                    ;SET LINE ONE, POSITION 0
44       CALL        WLCMC
45  WAIT_KEY:
46       CALL        HALF_READ_KEY            ;READ KEYPAD
47       MOV         A,KEY
48       SUB         A,28
49       SZ          Z                        ;KEY PRESSED?
50       JMP         WAIT_KEY                 ;NO, RE-READ KEYPAD
51       MOV         A,KEY
52       SUB         A,26
53       SZ          Z                        ;IS "CLEAR" PRESSED?
54       JMP         MAIN                     ;YES, JUMP MAIN
55       SDZ         ACC                      ;IS "CAP" PRESSED?
56       JMP         ALPHA_PRESSED            ;NO. IT'S ALPHABETA KEY PRESSED
57       SET         ACC
58       XORM        A,CAPITAL                ;YES, SET CAPITAL=!CAPITAL
59       MOV         A,00001000B
60       XORM        A,LCD_CONTR              ;SET CAP_LED=!CAP_LED
61       CALL        DELAY                    ;FOR DE-BOUNCING WHEN KEY PRESSED
62       JMP         WAIT_KEY
63  ALPHA_PRESSED:
64       CALL RLCMA                           ;READ LCM AC
65       SUB         A,16
66       SNZ         Z                        ;REACH LINE 1 ENDDING?
67       JMP         CHECK_LINE2              ;NO.
68       MOV         A,0C0H                   ;YES, SET LCD LINE 2, POSITION 0
69       CALL        WLCMC
70       MOV         A,16
71       MOV         DEL1,A
72       MOV         A,20H                    ;CLEAR LINE 2
73       CALL        WLCMD
74       SDZ         DEL1
75       JMP         $-2
76       MOV         A,0C0H                   ;SET LCD LINE 2, POSITION 0
```

```
77          CALL        WLCMC
78          JMP         DISPLAY
79  CHECK_LINE2:
80          SUB         A,40H
81          SNZ         Z                       ;REACH LINE 2 ENDDING?
82          JMP         DISPLAY                 ;NO.
83          MOV         A,80H                   ;YES, SET LCD LINE 1, POSITION 0
84          CALL        WLCMC
85          MOV         A,16
86          MOV         DEL1,A
87          MOV         A,20H                   ;CLEAR LINE 1
88          CALL        WLCMD
89          SDZ         DEL1
90          JMP         $-2
91          MOV         A,080H                  ;SET LCD LINE 1, POSITION 0
92          CALL        WLCMC
93  DISPLAY:
94          MOV         A,KEY                   ;GET KEY CODE
95          SZ          CAPITAL                 ;IS CAPITAL?
96          ADD         A,20H                   ;OFFSET OF ASCII "a" AND "A"
97          ADD         A,41H                   ;ASCII OF "A"
98          CALL        WLCMD                   ;DISPLAY KEY IN DATA
99          CALL        DELAY                   ;FOR DE-BOUNCING WHEN KEY PRESSED
100         JMP         WAIT_KEY                ;READ NEXT KEY
101 ;**********************************************************************
102 ;SCAN HALF MATRIX ON KEY PORT AND RETURN THE CODE IN KEY REGISTER
103 ; IF NO KEY BEEN PRESSED, KEY=28.
104 ;**********************************************************************
105 HALF_READ_KEY   PROC
106         MOV         A,01111111B
107         MOV         COLUMN,A                ;SET COLUMN FOR SCANNING
108         CLR         KEY                     ;INITIAL KEY REGISTER
109         MOV         A,7
110         MOV         KEY_COUNT,A             ;SET COUNTER=7 FOR 7 COLUMNS
111 SCAN_KEY:
112         MOV         A,COLUMN                ;GET COLUMN FOR SCANNING
113         MOV         KEY_PORTC,A             ;CONFIG KEY_PORT
114         MOV         KEY_PORT,A              ;SCANNING KEY PAD
115         MOV         A,KEY_COUNT             ;GET COLUMN COUNTER
116         ADDM        A,PCL                   ;COMPUTATIONAL JUMP
117         NOP                                 ;OFFSET FOR KEY_COUNT=0
118         JMP         $+24                    ;KEY_COUNT=1
119         JMP         $+20                    ;KEY_COUNT=2
120         JMP         $+16                    ;KEY_COUNT=3
121         JMP         $+12                    ;KEY_COUNT=4
122         JMP         $+8                     ;KEY_COUNT=5
123         JMP         $+4                     ;KEY_COUNT=6
124         SNZ         KEY_PORT.6              ;KEY_COUNT=7,ROW 1 PRESSED?
125         JMP         END_KEY                 ;YES.
126         INC         KEY                     ;NO, INCREASE KEY CODE
127         SNZ         KEY_PORT.5              ;ROW 2 PRESSED?
128         JMP         END_KEY                 ;YES.
129         INC         KEY                     ;NO, INCREASE KEY CODE
130         SNZ         KEY_PORT.4              ;ROW 3 PRESSED?
131         JMP         END_KEY                 ;YES.
132         INC         KEY                     ;NO, INCREASE KEY CODE
133         SNZ         KEY_PORT.3              ;ROW 4 PRESSED?
134         JMP         END_KEY                 ;YES.
135         INC         KEY                     ;NO, INCREASE KEY CODE
136         SNZ         KEY_PORT.2              ;ROW 5 PRESSED?
137         JMP         END_KEY                 ;YES.
138         INC         KEY                     ;NO, INCREASE KEY CODE
139         SNZ         KEY_PORT.1              ;ROW 6 PRESSED?
140         JMP         END_KEY                 ;YES.
141         INC         KEY                     ;NO, INCREASE KEY CODE
142         SNZ         KEY_PORT.0              ;ROW 7 PRESSED?
143         JMP         END_KEY                 ;YES.
```

```
144         INC     KEY                     ;NO, INCREASE KEY CODE
145         RR      COLUMN                  ;SCAN CODE FOR NEXT COLUMN
146         SDZ     KEY_COUNT               ;HAVE ALL COULMN BEEN CHECKED?
147         JMP     SCAN_KEY                ;NO, NEXT COLUMN
148 END_KEY:
149         RET
150 HALF_READ_KEY   ENDP
151 ;************************************************************************
152 ;           LCD DATA/COMMAND WRITE PROCEDURE
153 ;************************************************************************
154 WLCMD   PROC
155         SET     DC_FLAG                 ;SET DC_FLAG=1 FOR DATA WRITE
156         JMP     WLCM
157 WLCMC:
158         CLR     DC_FLAG                 ;SET DC_FLAG=0 FOR COMMAND WRITE
159 WLCM:
160         SET     LCD_DATAC               ;CONFIG LCD_DATA AS INPUT MODE
161         CLR     LCD_EN                  ;CLEAR ALL LCD CONTROL SIGNAL
162         CLR     LCD_RW
163         CLR     LCD_RS
164         SET     LCD_RW                  ;SET RW SIGNAL (READ)
165         NOP                             ;FOR TAS
166         SET     LCD_EN                  ;SET EN HIGH
167         NOP                             ;FOR TDDR
168 WF:
169         SZ      LCD_READY               ;IS LCD BUSY?
170         JMP     WF                      ;YES, JUMP TO WAIT
171         CLR     LCD_DATAC               ;NO, CONFIG LCD_DATA AS OUTPUT MODE
172         MOV     LCD_DATA,A              ;LATCH DATA/COMMAND ON PB(LCD DATA BUS)
173         CLR     LCD_EN                  ;CLEAR ALL LCD CONTROL SIGNAL
174         CLR     LCD_RW
175         CLR     LCD_RS
176         SZ      DC_FLAG                 ;IS COMMAND WRITE?
177         SET     LCD_RS                  ;NO, SET RS HIGH
178         SET     LCD_EN                  ;SET EN HIGH
179         NOP
180         CLR     LCD_EN                  ;SET EN LOW
181         RET
182 WLCMD   ENDP
183 ;************************************************************************
184 ;           LCD DATA/ADDRESS READ PROCEDURE
185 ;************************************************************************
186 RLCMD   PROC
187         SET     DC_FLAG                 ;SET DC_FLAG=1 FOR DATA READ
188         JMP     RLCM
189 RLCMA:
190         CLR     DC_FLAG                 ;SET DC_FLAG=0 FOR ADDRESS READ
191 RLCM:
192         SET     LCD_DATAC               ;CONFIG LCD_DATA AS INPUT MODE
193         CLR     LCD_EN                  ;CLEAR ALL LCD CONTROL SIGNAL
194         CLR     LCD_RW
195         CLR     LCD_RS
196         SET     LCD_RW                  ;SET RW=1
197         NOP                             ;FOR TAS
198         SET     LCD_EN                  ;SET EN=1
199         NOP                             ;FOR TDDR
200 RWF:
201         SZ      LCD_READY               ;IS LCD BUSY?
202         JMP     RWF                     ;YES,WAIT
203         MOV     A,LCD_DATA
204         SNZ     DC_FLAG                 ;ADDRESS READ?
205         RET                             ;YES.
206         CLR     LCD_EN                  ;CLEAR ALL LCD CONTROL SIGNAL
207         CLR     LCD_RW
208         CLR     LCD_RS
209         SET     LCD_RW                  ;SET LCD_RW
210         SET     LCD_RS                  ;SET RS=1 (DATA READ)
```

第 5 章　进阶实验篇

```
211             NOP
212             SET     LCD_EN              ;SET EN
213             NOP
214             MOV     A,LCD_DATA          ;READ LCD DATA
215             CLR     LCD_EN              ;CLEAR EN
216             RET
217 RLCMD   ENDP
218 ;***********************************************************************
219 ;                       Delay about DEL1*10ms
220 ;***********************************************************************
221 DELAY   PROC
222             MOV     A,25
223             MOV     DEL1,A              ;SET DEL1 COUNTER
224 DEL_1:  MOV     A,30
225             MOV     DEL2,A              ;SET DEL2 COUNTER
226 DEL_2:  MOV     A,110
227             MOV     DEL3,A              ;SET DEL3 COUNTER
228 DEL_3:  SDZ     DEL3                ;DEL3 DOWN COUNT
229             JMP     DEL_3
230             SDZ     DEL2                ;DEL2 DOWN COUNT
231             JMP     DEL_2
232             SDZ     DEL1                ;DEL1 DOWN COUNT
233             JMP     DEL_1
234             RET
235 DELAY   ENDP
236             END
```

程序说明

7~15 　依序定义变量地址。

17~20 　定义 LCD_DATA Port 与 LCD_CONTR Port 分别为 PB 与 PC。

21~25 　定义 LCD 的控制信号引脚，以及表示大小写的 LED 控制引脚。

26~27 　定义 KEY_PORT 为 PA。

30 　声明存储器地址由 000h 开始(HT46xx Reset Vector)。

31 　将 LCD_CONTR Port(PA)定义为输出模式。

32~33 　设置 CAPITAL＝0(Default 为大写字形)，并点亮代表大写字形的 LED。

34~35 　将 LCD 设置为双行显示(N = 1)、使用 8 位(DB7～DB0)控制模式(DL = 1)、5×7 点矩阵字形(F = 0)。

36~37 　将 LCD 设置为显示所有数据(D = 1)、显示光标(C = 1)、光标所在位置的字不会闪烁(B = 0)。

38~39 　将 LCD 的地址标志位(AC)设置为递加(I/D = 1)、显示器画面不因读/写数据而移动(S = 0)。

41~44 　将 LCD 整个显示器清空，并设置 DD RAM 地址为第 1 行第 0 个位置。

46~50 　调用 HALF_READ_KEY 子程序，并判定是否有按键被按下，若无则继续检查(JMP SCAN_KEY)。

51~54 　判断是不是按下 CLEAR 键，若是，则清除 LCD 并重新扫描按键(JMP MAIN)。

55~60 　判断是不是按下 CAP 键，若是，则将 CAPITAL 寄存器与 CAP_LED 反向，完成大/小写字形的切换。

61~62 　调用 DELAY 子程序，延时 0.25 s 后重新扫描按键(JMP WAIT_KEY)；延时时间的计算参考 4-1 节。

64~92 　此段程序主要是判断 LCD 显示器是否已经显示到第 1 行或第 2 行的最后一个位置。首先调用 RLCMA 子程序读取 LCD 目前的 AC 值，65~78 行是检查是否到达第 1 行的最后一个位置(本实验所采用的是 20×2 的 LCD)，若是则将第 2 行先予以清除，并设置 LCD DD RAM 的地址为第 2 行第 0 个位置，因此输入的字符会改由 LCD 的第 2 行开始显示。79~92 行则是检查是否到达第 2 行的最后一个位置(40h+16)，若是则将第 1 行先予以清除，并设置 LCD DD RAM 地址为第 1 行第 0 个位置，因此输入的字符会改由 LCD 的第 1 行开始显示。

94~98 　将按键值转换为 ASCII 码并调用 WLCMD 子程序予以显示；期间会以 CAPITAL 寄存器的值判断目前为大写模式(CAPITAL＝00h)或小写模式(CAPITAL＝FFh)，以便转换为大写或小写的 ASCII 码(注：大小写英文字形的 ASCII 码相差 20h)。

99~100　调用 DELAY 子程序，延时 0.25 s 后重新扫描按键（JMP WAIT_KEY）；延时时间的计算参考 4-1 节。
105~150　调用 HALF_READ_KEY 子程序，请参考 5-12-3 小节的说明。
154~182　WLCMD 与 WLCMC 子程序，请参考 5-5 节的说明。不过请读者留意此处的子程序与 5-5 节有一点不同，主要是因为 LCD_CONTR Port 不只用来控制 LCD，同时 LCD_CONTR.3 也用来控制代表大/小写模式的 LCD，为了不影响 LCD 的正常显示，将原来的"CLR CLD_CONTR"改以"CLR LCD_EN"、"CLR LCD_RW"以及"CLR LCD_RS"3 行指令加以取代。
186~217　RLCMD 与 RLCMD 子程序，请参考 5-5 节的时序图与说明。
221~235　DELAY 子程序，至于延时时间的计算请参考 4-1 节。

由于只有28个按键，为了能够区分出大小写字母，所以将按键S27定义为大/小写设置键（其功能就如PC键盘上的"Caps Lock"按键），并搭配CAPITAL寄存器的控制，决定S0~S25等26个按键究竟是大写还是小写字符。如此还剩下一个按键S26，就索性将它定义为LCD的屏幕清除功能按钮。

半矩阵式键盘与4×4矩阵式键盘相较其来，同样是运用一个I/O端口，但前者足足多出了12个按键数，程序代码虽然多出了几行，但并不是太复杂，读者历经了20余个实验单元的磨练，应该可以洞悉其工作原理才是。

5-12-5　动动脑＋动动手

　📖　请将 CLR 按键（S26）即"LCD 屏幕清除功能"更改为"Num Lock"功能。也就是说通过"Num Lock"键的控制，可以选择英文字/数字模式，在数字模式下除了可以显示 0~9 之外，并且还能显示 !、@、$ 等特殊符号。

5-13　HT46xx I²C Mater–Slave传输实验

5-13-1　目　的

本实验利用HT46x23的I²C串行传输接口，完成4×4键盘按键值的显示与传输动作。由Master端按下的键值，除了将显示在Master端LCD的第1行之外，并通过I²C串行传输接口传输到Slave端，同时显示在Slave端LCD的第2行反之不明确，可以删除。同时当显示超过20个字符时，自动清除该行并继续显示。

5-13-2　学习重点

通过本实验，读者对于HT46xx 的I²C串行传输接口的控制应透彻了解，同时对两个设备之间通过I²C串行接口传输数据的方式应该更加明了。

5-13-3　电路图

如图5-13-1所示为HT46x23 I2C传输Master控端制电路；图5-13-2为Slave端控制电路。

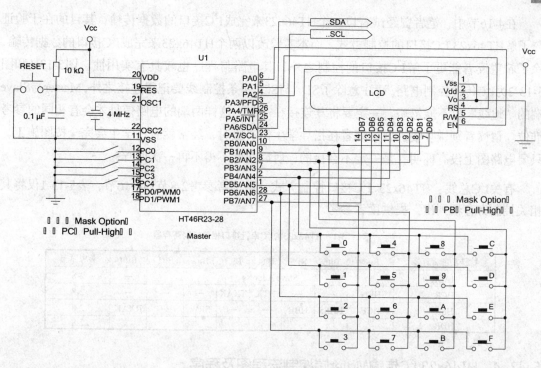

图5-13-1　HT46x23 I2C传输Mater端控制电路

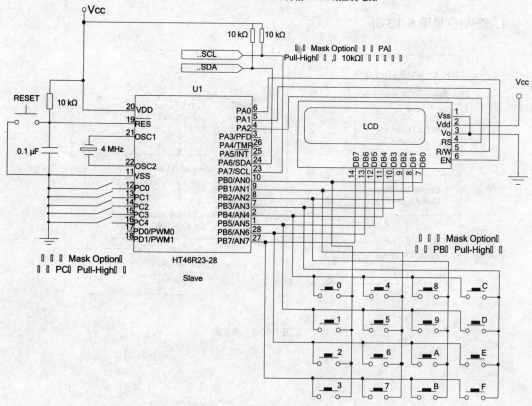

图5-13-2　HT46x23 I^{2}C传输Slave端控制电路

在4-16节中,笔者曾经试着以一个HT46x23来完成I²C接口的数据传输,其目的在于验证与了解HT46x23 I²C接口的控制方式。而本实验改以两个HT46x23来完成I²C接口的数据传输,除了希望读者能更了解I²C接口的控制之外,其电路与程序也较具有实用性。图5-13-1和图5-13-2为I²C传输控制电路,请注意除了SDA与SCL两条控制线要记得连接之外,Master与Slave端的"地线"也不能忽略,一定要将其连接在一起,这样两端的电路信号才会有相同的参考准位。请读者别嫌啰唆,因为笔者在指导学生做实验时,经常就有些学生会完全按图施工,只要电路图上没有标明,是一定不会接的,所以在此不得不再一次叮咛!

有关I²C总线、HT46x23 I²C接口控制方式,请读者参考2-8节与4-16节,表5-13-1仅将其相关控制寄存器列出,提供读者参考。

表5-13-1 HT46x23的I²C串行接口相关控制寄存器

特殊功能寄存器	Bit7	Bit6	Bit5	Bit4	Bit3	Bit2	Bit1	Bit0	参考章节
HADR	Slave Address							—	2-8
HCR	HEN	—	—	HTX	TXAK	—		—	2-8
HSR	HCF	HASS	HBB	—	—	SRW	—	RXAK	2-8
HDR									2-8

5-13-4 HT46x23 I²C传输Mater端控制流程图及程序

1. 流程图(见图 5-13-3)

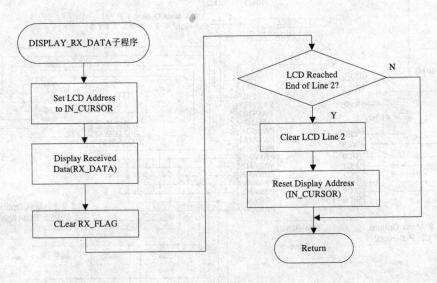

图5-13-3 流程图

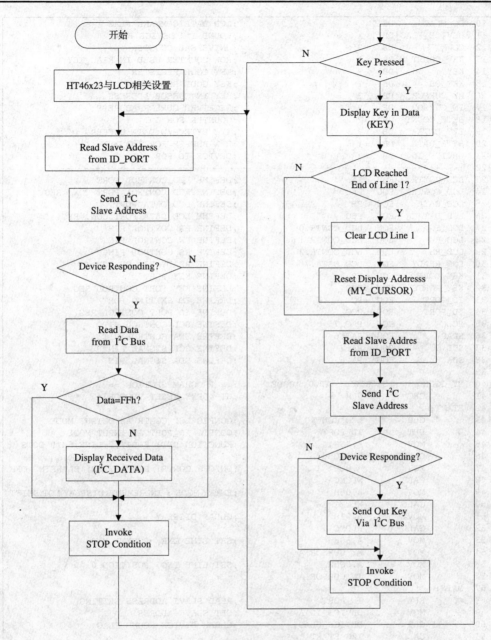

图5-13-3 流程图（续）

2. 程序 5-13-1 M HT46x23 I²C 传输实验 Mater 端控制程序

```
1    ;PROGRAM : 5-13M.ASM (5-13M.PRJ)                          BY STEVEN
2    ;FUNCTION: I2C SLAVE MODE DEMO PROGRAM (MASTER)           2002.DEC.07.
3    #INCLUDE    HT46R23.INC
4                .CHIP HT46R23
5    ;--------------------------------------------------------------------
6    MY_DATA     .SECTION    'DATA'          ;== DATA SECTION ==
7    DEL1        DB    ?                     ;DELAY LOOP COUNT 1
8    DEL2        DB    ?                     ;DELAY LOOP COUNT 2
9    DEL3        DB    ?                     ;DELAY LOOP COUNT 3
```

```
10      DC_FLAG      DBIT                       ;LCD DATA/COMMAND FLAG
11      DEVICE_FLAG  DBIT                       ;FOUND I²C DEVICE FLAG
12      COUNT        DB    ?                    ;UNIVERSAL COUNTER
13      KEY_COUNT    DB    ?                    ;ROW COUNTER USED IN READ_KEY
14      KEY          DB    ?                    ;KEY CODE REGISTER
15      KEY_PS       DB    ?                    ;KEY CODE REGISTER
16      MY_CURSOR    DB    ?                    ;KEY IN CURSOR ADDRESS
17      IN_CURSOR    DB    ?                    ;RECEIVED CURSOR ADDRESS
18      BYTE_COUNT   DB    ?                    ;COUNTER FOR I²C
19      I2C_DATA     DB    ?                    ;I2C BYTE R/W DATA BUFFER
20      BYTE_DATA    DB    ?                    ;R/W BUFFER FOR I²C DATA
21      DEVICE_ID    DB    ?                    ;DEVICE ID FOR I²C
22      ;----------------------------------------------------------------
23      LCD_CONTR    EQU   PA                   ;DEFINE LCD CONTROL PORT
24      LCD_CONTRC   EQU   PAC                  ;DEFINE LCD CONTROL PORT CON. REG.
25      LCD_DATA     EQU   PB                   ;DEFINE LCD DATA PORT
26      LCD_DATAC    EQU   PBC                  ;DEFINE LCD DATA PORT CON. REG.
27      LCD_EN       EQU   LCD_CONTR.0          ;DEFINE EN CONTROL PIN
28      LCD_RW       EQU   LCD_CONTR.1          ;DEFINE RW CONTROL PIN
29      LCD_RS       EQU   LCD_CONTR.2          ;DEFINE RS CONTROL PIN
30      LCD_READY    EQU   LCD_DATA.7           ;DEFINE READY BIT OF LCD
31      KEY_PORT     EQU   PB                   ;DEFINE KEY PORT
32      KEY_PORTC    EQU   PBC                  ;DEFINE KEY PORT CONTROL REG.
33      ID_PORT      EQU   PC                   ;DEFINE ID SETTING PORT
34      ID_PORTC     EQU   PCC                  ;DEFINE ID PORT CONTROL REG.
35      SCLC         EQU   PAC.7                ;DEFINE SCL CONTROL BIT
36      SDAC         EQU   PAC.6                ;DEFINE SDA CONTROL BIT
37      SCL          EQU   PA.7                 ;DEFINE SCL SIGNAL PIN
38      SDA          EQU   PA.6                 ;DEFINE SDA SIGNAL PIN
39
40      MY_CODE      .SECTION  AT 0 'CODE'      ;== PROGRAM SECTION ==
41              ORG         00H                 ;HT-46RXX RESET VECTOR
42      START:
43              CLR         LCD_CONTRC          ;CONFIG LCD_CONTR AS OUTPUT MODE
44              SET         ID_PORTC            ;CONFIG ID_PORT AS INPUT MODE
45              MOV         A,38H               ;FUNCTION SET: 8-BIT,2-LINE,5X10 DOTS
46              CALL        WLCMC
47              MOV         A,0CH               ;ON/OFF CONTR: DISPLAY ON, BLINKING OFF
48              CALL        WLCMC
49              MOV         A,06H               ;ENTRY MODE : INCREMENT,DISPLAY NOT SHIFT
50              CALL        WLCMC
51              MOV         A,01H               ;CLEAR DISPLAY
52              CALL        WLCMC
53              MOV         A,80H               ;SET LINE ONE, POSITION 0
54              MOV         MY_CURSOR,A
55              MOV         A,0C0H              ;SET LINE TWO, POSITION 0
56              MOV         IN_CURSOR,A
57      MAIN:
58              RLA         ID_PORT             ;READ SLAVE ADDRESS SETTING
59              MOV         DEVICE_ID,A         ;SET SLAVE ADDRES
60              CLR         DEVICE_FLAG         ;CLEAR DIVICE FOUND FLAG
61              CALL        I2C_START
62              CALL        I2C_DEV_SEL_R
63              SNZ         DEVICE_FLAG         ;DEVICE RESPOUNDING ?
64              JMP         NO_DEVICE           ;NO.
65              CALL        READ_BYTE           ;YES, READ DATA
66              SET         ACC                 ;SET ACC=0FFH
67              XOR         A,I2C_DATA
68              SZ          ACC
69              CALL        DISPLAY_RX_DATA     ;DEVICE GOT NEW DATA?
                                                ;YES
70      NO_DEVICE:
71              CALL        I2C_STOP
72              CALL        READ_KEY            ;READ KEYPAD
73              MOV         A,16
74              XOR         A,KEY
75              SZ          Z                   ;KEY PRESSED ?
76              JMP         MAIN                ;NO, RE-READ KEYPAD
```

```
77          MOV     A,KEY
78          MOV     KEY_PS,A            ;RESERVE KEY CODE
79  WAIT:
80          CALL    READ_KEY            ;CHECK KEYPAD RELEASED
81          MOV     A,16
82          XOR     A,KEY
83          SNZ     Z                   ;IS KEY RELEASED?
84          JMP     WAIT                ;NO,WAIT KEY RELEASE
85          CALL    DELAY               ;FOR DE-BOUNCING WHEN KEY RELEASE
86          MOV     A,MY_CURSOR         ;SET CURSOR ADDRESS
87          CALL    WLCMC
88          MOV     A,KEY_PS
89          SUB     A,10                ;CONVERT TO ASCII CODE
90          SZ      C
91          ADD     A,40H-30H-9
92          ADD     A,30H+10
93          MOV     KEY_PS,A
94          CALL    WLCMD               ;DISPLAY KEY IN DATA
95          INC     MY_CURSOR
96          MOV     A,MY_CURSOR
97          XOR     A,80H+20+1          ;20X2 LCD
98          SNZ     Z                   ;REACH LAST POSITION OF LINE 1?
99          JMP     TX_DATA             ;NO
100         MOV     A,80H                ;RESET LCD ADDRESS
101         CALL    WLCMC
102         MOV     A,20
103         MOV     DEL1,A
104         MOV     A,20H               ;CLEAR LINE 1
105         CALL    WLCMD
106         SDZ     DEL1
107         JMP     $-2
108         MOV     A,80H
109         MOV     MY_CURSOR,A         ;RESET LCD ADDRESS
110 TX_DATA:
111         RLA     ID_PORT             ;READ SLAVE ADDRESS SETTING
112         MOV     DEVICE_ID,A         ;SET SLAVE ADDRES
113         CLR     DEVICE_FLAG         ;CLEAR DIVICE FOUND FLAG
114         CALL    I2C_START
115         CALL    I2C_DEV_SEL_W
116         MOV     A,KEY_PS
117         SZ      DEVICE_FLAG         ;DEVICE RESPOUNDING ?
118         CALL    WRITE_BYTE          ;YES, WRITE DATA
119         CALL    I2C_STOP
120         JMP     MAIN                ;READ NEXT KEY
121 ;**********************************************************************
122 ;           DISPLAY RECEIVED DATA PROCEDURE
123 ;**********************************************************************
124 DISPLAY_RX_DATA     PROC
125         MOV     A,IN_CURSOR         ;SET CURSOR ADDRESS
126         CALL    WLCMC
127         MOV     A,I2C_DATA          ;COPY TO RX_DATA BUFFER
128         CALL    WLCMD               ;DISPLAY KEY IN DATA
129         INC     IN_CURSOR
130         MOV     A,IN_CURSOR
131         XOR     A,0C0H+20+1         ;20X2 LCD
132         SNZ     Z                   ;REACH LAST POSITION OF LINE 2
133         RET                         ;NO
134         MOV     A,0C0H              ;YES, RESET LCD ADDRESS
135         CALL    WLCMC
136         MOV     A,20
137         MOV     DEL1,A
138         MOV     A,20H               ;CLEAR LINE 2
139         CALL    WLCMD
140         SDZ     DEL1
141         JMP     $-2
142         MOV     A,0C0H
143         MOV     IN_CURSOR,A         ;RESET LCD ADDRESS
```

```
144         RET
145 DISPLAY_RX_DATA  ENDP
146 ;*************************************************************************
147 ;               GENERATE I2C START CONDITION
148 ;*************************************************************************
149 I2C_START   PROC
150         CLR     SDAC                ;CONFIG SDA AS OUTPUT MODE
151         CLR     SCL                 ;SET SCL=0
152         CLR     SDA                 ;SET SDA=0
153         CALL    DELAY_10            ;DELAY
154         SET     SDA                 ;SET SDA=1
155         SET     SCL                 ;SET SCL=1
156         CALL    DELAY_10            ;DELAY
157         CLR     SDA                 ;SET SDA=0
158         CALL    DELAY_10            ;DELAY
159         CLR     SCL                 ;SET SCL=0
160         CALL    DELAY_10            ;DELAY
161         RET
162 I2C_START   ENDP
163 ;*************************************************************************
164 ;          SEND I2C DEVICE SELECT CODE (R/W) TO DEVICE
165 ;*************************************************************************
166 I2C_DEV_SEL_W   PROC                ;WRITE MODE
167         MOV     A,DEVICE_ID         ;LOAD A WITH DEVICE SELECT CODE
168         CLR     ACC.0               ;SET SRW T0 0 FOR WRITING TO I$^2$C
169         JMP     DEV_SEL
170 I2C_DEV_SEL_R:                      ;READ MODE
171         MOV     A,DEVICE_ID         ;LOAD A WITH DEVICE SELECT CODE
172         SET     ACC.0               ;SET SRW T0 1 FOR READING FROM I$^2$C
173 DEV_SEL:
174         CALL    WRITE_BYTE          ;SEND OUT 1 BYTE DATA IN Acc
175         CALL    CHECK_ACK           ;WAIT FOR DEVICE ACK SIGNAL
176         RET
177 I2C_DEV_SEL_W   ENDP
178 ;*************************************************************************
179 ;               GENERATE I2C STOP CONDITION
180 ;*************************************************************************
181 I2C_STOP    PROC
182         CLR     SDAC                ;CONFIG SDA AS OUTPUT MODE
183         CLR     SCL                 ;SET SCL=0
184         CLR     SDA                 ;SET SDA=0
185         CALL    DELAY_10            ;DELAY
186         CLR     SDA                 ;SET SDA=0
187         CALL    DELAY_10            ;DELAY
188         SET     SCL                 ;SET SCL=1
189         CALL    DELAY_10            ;DELAY
190         SET     SDA                 ;SET SDA=1
191         RET
192 I2C_STOP    ENDP
193 ;*************************************************************************
194 ;          SERIAL OUT DATA IN Acc VIA SDA & SCL
195 ;*************************************************************************
196 WRITE_BYTE  PROC
197         CLR     SDAC                ;CONFIG SDA AS OUTPUT MODE
198         CLR     SCL                 ;SET SCL=0
199         CLR     SDA                 ;SET SDA=0
200         MOV     I2C_DATA,A          ;RESERVED DATA IN TX BUFFER
201         MOV     A,8                 ;SET 8 BIT COUNTER
202         MOV     BYTE_COUNT,A
203 WNB_0:
204         SZ      I2C_DATA.7          ;IS MSB = 0?
205         JMP     WRITE_1             ;NO, JUMP TO WRITE_1
206 WRITE_0:
207         CLR     SDA                 ;SET SDA=0
208         JMP     WNB_1               ;JUMP TO WNB_1
209 WRITE_1:
210         SET     SDA                 ;SET SDA=1
```

```
211 WNB_1:
212         CALL    DELAY_10            ;DELAY
213         SET     SCL                 ;SET SCL=1
214         CALL    DELAY_10            ;DELAY
215         CLR     SCL                 ;SET SCL=0
216         CALL    DELAY_10            ;DELAY
217         RL      I2C_DATA            ;SHIFT TX BUFFER
218         SDZ     BYTE_COUNT          ;BYTE_COUNT-1 = 0?
219         JMP     WNB_0               ;NO, WRITE NEXT BIT
220         RET                         ;YES, RETURN.
221 WRITE_BYTE  ENDP
222 ;************************************************************************
223 ;           SERIAL IN DATA TO Acc VIA SDA & SCL
224 ;************************************************************************
225 READ_BYTE   PROC
226         SET     SDAC                ;CONFIG SDA AS INTPUT MODE
227         MOV     A,8                 ;SET 8 BIT COUNTER
228         MOV     BYTE_COUNT,A
229 RNB_0:
230         SET     SCL                 ;SET SCL=0
231         CALL    DELAY_10            ;DELAY
232         RLC     I2C_DATA            ;SHIFT RX BUFFER
233         SZ      SDA                 ;SDA = 0?
234         JMP     READ_1              ;NO, JUMP TO READ_1
235 READ_0: CLR     I2C_DATA.0          ;YES, SET LSB=0
236         JMP     RNB_1               ;JUMP TO RNB_1
237 READ_1:
238         SET     I2C_DATA.0          ;SET LSB=1
239 RNB_1:
240         CALL    DELAY_10            ;DELAY
241         CLR     SCL                 ;SET SCL=0
242         CALL    DELAY_10            ;DELAY
243         SDZ     BYTE_COUNT          ;BYTE_COUNT-1 = 0?
244         JMP     RNB_0               ;NO, READ NEXT BIT
245         MOV     A,I2C_DATA          ;RELOAD RX DATA TO Acc
246         CLR     SDAC                ;YES, SEND NO_ACK_SIGNAL
247         CLR     SDA
248         SET     SCL
249         CALL    DELAY_10
250         SET     SDA
251         CLR     SCL
252         RET                         ;YES, RETURN
253 READ_BYTE   ENDP
254 ;************************************************************************
255 ;           WAIT FOR ACK SIGNAL FROM I2C DEVICE
256 ;************************************************************************
257 CHECK_ACK   PROC
258         SET     SDAC                ;CONFIG SDA AS INPUT MODE
259         CALL    DELAY_10            ;DELAY
260         SET     SCL                 ;SET SCL=1
261         CALL    DELAY_10            ;DELAY
262         CLR     DEL1
263         SET     DEVICE_FLAG         ;ASSUME DEVICE FOUND
264 WAIT_ACK:
265         SNZ     SDA                 ;ACK SINGLE SET ?
266         JMP     GET_ACK             ;YES.
267         SDZ     DEL1                ;NO. OUT OF WAIT ACK TIME
268         JMP     WAIT_ACK            ;NO. WAIT AGAIN
269         CLR     DEVICE_FLAG         ;NO DEVICE FOUND!
270 GET_ACK:
271         CALL    DELAY_10            ;DELAY
272         CLR     SCL                 ;SET SCL=0
273         CALL    DELAY_10            ;DELAY
274         RET
275 CHECK_ACK   ENDP
276 ;************************************************************************
277 ;           DELAY 10us FOR TIMING CONSIDERATION
```

```
278 ;********************************************************************
279 DELAY_10:
280         TABRDL      DEL1                ;NULL READ
281         TABRDL      DEL1                ;DELAY 18 INS. CYCLES
282         TABRDL      DEL1
283         TABRDL      DEL1
284         TABRDL      DEL1
285         TABRDL      DEL1
286         TABRDL      DEL1
287         TABRDL      DEL1
288         RET
289 ;********************************************************************
290 ;           LCD DATA/COMMAND WRITE PROCEDURE
291 ;********************************************************************
292 WLCMD   PROC
293         SET         DC_FLAG             ;SET DC_FLAG=1 FOR DATA WRITE
294         JMP         WLCM
295 WLCMC:
296         CLR         DC_FLAG             ;SET DC_FLAG=0 FOR COMMAND WRITE
297 WLCM:
298         SET         LCD_DATAC           ;CONFIG LCD_DATA AS INPUT MODE
299         CLR         LCD_EN              ;CLEAR ALL LCD CONTROL SIGNAL
300         CLR         LCD_RW
301         CLR         LCD_RS
302         SET         LCD_RW              ;SET RW SIGNAL (READ)
303         NOP                             ;FOR TAS
304         SET         LCD_EN              ;SET EN HIGH
305         NOP                             ;FOR TDDR
306 WF:
307         SZ          LCD_READY           ;IS LCD BUSY?
308         JMP         WF                  ;YES, JUMP TO WAIT
309         CLR         LCD_DATAC           ;NO, CONFIG LCD_DATA AS OUTPUT MODE
310         MOV         LCD_DATA,A          ;LATCH DATA/COMMAND ON PB(LCD DATA BUS)
311         CLR         LCD_EN
312         CLR         LCD_RW
313         CLR         LCD_RS
314         SZ          DC_FLAG             ;IS COMMAND WRITE?
315         SET         LCD_RS              ;NO, SET RS HIGH
316         SET         LCD_EN              ;SET EN HIGH
317         NOP
318         CLR         LCD_EN              ;SET EN LOW
319         RET
320 WLCMD   ENDP
321 ;********************************************************************
322 ;SCAN 4x4 MATRIX ON KEY PORT AND RETURN THE CODE IN KEY REGISTER
323 ; IF NO KEY BEEN PRESSED, KEY=16.
324 ;********************************************************************
325 READ_KEY    PROC
326         MOV         A,11110000B
327         MOV         KEY_PORTC,A         ;CONFIG KEY_PORT
328         SET         KEY_PORT            ;INITIAL KEY PORT
329         CLR         KEY                 ;INITIAL KEY REGISTER
330         MOV         A,04
331         MOV         KEY_COUNT,A         ;SET ROW COUNTER
332         CLR         C                   ;CLEAR CARRY FLAG
333 SCAN_KEY:
334         RLC         KEY_PORT            ;ROTATE SCANNING BIT
335         SET         C                   ;MAKE SURE C=1
336         SNZ         KEY_PORT.4          ;COLUMN 0 PRESSED?
337         JMP         END_KEY             ;YES.
338         INC         KEY                 ;NO, INCREASE KEY CODE.
339         SNZ         KEY_PORT.5          ;COLUMN 1 PRESSED?
340         JMP         END_KEY             ;YES.
341         INC         KEY                 ;NO, INCREASE KEY CODE.
342         SNZ         KEY_PORT.6          ;COLUMN 2 PRESSED?
343         JMP         END_KEY             ;YES.
344         INC         KEY                 ;NO, INCREASE KEY CODE.
```

```
345             SNZ       KEY_PORT.7            ;COLUMN 3 PRESSED?
346             JMP       END_KEY               ;YES.
347             INC       KEY                   ;NO, INCREASE KEY CODE.
348             SDZ       KEY_COUNT             ;HAVE ALL ROWs BEEN CHECKED?
349             JMP       SCAN_KEY              ;NO, NEXT ROW.
350   END_KEY:
351             RET
352   READ_KEY  ENDP
353   ;********************************************************************
354   ;                   Delay about DEL1*10ms
355   ;********************************************************************
356   DELAY     PROC
357             MOV       A,6
358             MOV       DEL1,A                ;SET DEL1 COUNTER
359   DEL_1:    MOV       A,30
360             MOV       DEL2,A                ;SET DEL2 COUNTER
361   DEL_2:    MOV       A,110
362             MOV       DEL3,A                ;SET DEL3 COUNTER
363   DEL_3:    SDZ       DEL3                  ;DEL3 DOWN COUNT
364             JMP       DEL_3
365             SDZ       DEL2                  ;DEL2 DOWN COUNT
366             JMP       DEL_2
367             SDZ       DEL1                  ;DEL1 DOWN COUNT
368             JMP       DEL_1
369             RET
370   DELAY     ENDP
371             END
```

MASTER 端控制程序说明

7~21   依序定义变量地址。

23~24   定义 LCD_CONTR 为 PA。

25~26   定义 LCD_DATA 为 PB。

27~30   定义 LCD 的控制信号引脚。

31~32   定义 KEY_PORT 为 PB。

33~34   定义 ID_PORT 为 PC，因此所送出的 Slave 地址范围为 00h~1Fh。

35~38   定义 SCL 为 PA7、SDA 为 PA6，作为控制 I^{2}C 串行传输的引脚。

41   声明存储器地址由 000h 开始(HT46x23 Reset Vector)。

43~44   定义 LCD_CONTR 为输出模式、ID_PORT 为输入模式。

45~46   将 LCD 设置为双行显示(N = 1)、使用 8 位(DB7 ~ DB0)控制模式(DL = 1)、5×7 点矩阵字形(F = 0)。

47~48   将 LCD 设置为显示所有数据(D = 1)、显示光标(C = 1)、光标所在位置的字不会闪烁(B = 0)。

49~50   将 LCD 的地址标志位(AC)设置为递加(I/D = 1)、显示器画面不因读/写数据而移动(S = 0)。

51~52   将 LCD 整个显示器清空，并设置 DD RAM 地址为第 1 行第 0 个位置。

53~54   设置 MY_CURSOR=80h，此寄存器是记录 Master 端按键值在 LCD 上的显示地址，以作为是否到达第 1 行最后一个位置的判断。

55~56   设置 IN_CURSOR=C0h，此寄存器是记录由 Slave 端送过来的按键值在 LCD 上的显示地址，以作为是否到达第 2 行最后一个位置的判断。

58~59   由 ID_PORT 读回指拨开关设置数值，并存放在 DEVICE_ID 寄存器，此即代表 I^{2}C Bus Master 所要存取的 Slave 地址。

60   清除 DEVICE_FLAG，程序中将以此位判断是否有与 DEVICE_ID 相符的设备回应(1 表示与 DEVICE_ID 相符)，因此先将其清除为 0。

61~62   送出 I^{2}C Start 条件与 Slave 地址(DEVICE_ID)。注意，此时 R/W 位＝1，表示 Master 要由 I^{2}C Device 读取数据，所以 Slave 端的 I^{2}C 接口必须准备送出数据(这里是指 Device ID 与 Slave 地址相符的设备)。

行号	说明
63~69	判断是否有 DEVICE_ID 相符的设备回应；若有则调用 READ_BYTE 子程序，通过 SDA 与 SCL 的控制从 I²C 接口读回 8 位的数据。接着判断读回的数据是否为 FFh，如果是的话则表示 Slave 端没有按键按下；反之，则表示有按键按下，所以必须调用 DISPLAY_RX_DATA 子程序显示接收到的数据。
71	调用 I2C_STOP 子程序产生 STOP 条件，结束此次 I²C 接口的数据传输。
71~76	调用 READ_KEY 子程序扫描键盘，完成后判断 KEY 寄存器内容是否为 16，若成立表示键盘无按键输入，则重新读取 ID_PORT，通过 I²C 接口读取 Slave 端的数据并再次扫描键盘(JMP MAIN)。反之，则代表键盘已被按下，准备开始将按键值显示并通过 I²C 接口发送出去的过程。
77~78	将按键值保留在 KEY_PS 寄存器。
80~85	因为 LCD 的数据端口与键盘是共用一个 I/O 端口，所以必须等按键放开之后才能送出 LCD 的显示数据，以避免造成数据被干扰的现象。此段程序的功能即是确认按键已经完全放开，以便继续往下执行显示的动作。
86~87	依据 MY_CURSOR 寄存器的值设置 LCD 显示地址。
88~94	将按键值转换为 ASCII Code 之后调用 WLCMD 子程序予以显示。
95~109	将负责记录 Master 端按键值在 LCD 上显示地址的寄存器(MY_CURSOR)加 1，判断是否到达第 1 行最后一个位置。若是则清除第 1 行，并重新设置 MY_CURSOR=80h。(注：本实验使用 20×2 LCD。)
111~112	由 ID_PORT 读回指拨开关设置数值，并存放在 DEVICE_ID 寄存器，此即代表 I²C Bus Master 所要执行写入动作的 Slave 地址。
113	清除 DEVICE_FLAG，程序中将以此位判断是否有与 DEVICE_ID 相符的设备回应(1 表示与 DEVICE_ID 相符)，因此先将其清除为 0。
114~115	送出 I²C Start 条件与 Slave 地址(DEVICE_ID)。注意，此时 R/W 位=0，表示 Master 要送数据给 I²C 设备，所以 Slave 端的 I²C 接口必须准备接收数据。
116	将按键值放在 Acc 寄存器，准备由 WRITE_BYTE 子程序发送。
117~119	当调用 I2C_DEV_SEL_W 子程序时，该程序会检查是否有 I²C 设备对所送出的 DEVICE_ID 产生回应(ACK=0)，若无设备回应则传回 DEVICE_FLAG=0；若有则传回 DEVICE_FLAG=1。此段程序是依 DEVICE_FLAG 为 0 或 1 分别完成以下动作： DEVICE_FLAG=1：表示有设备回应，调用 WRITE_BYTE 子程序，通过 SDA 与 SCL 的控制将按键值送到 I²C 接口，然后调用 I2C_STOP 子程序产生 STOP 条件，结束此次 I²C 接口的数据传输。 DEVICE_FLAG=0：表示没有设备回应，直接调用 I2C_STOP 子程序产生 STOP 条件，结束此次 I²C 接口传输。
120	重新执行程序(JMP MAIN)。
124~145	DISPLAY_RX_DATA 子程序，负责将接收到的数据显示在 LCD 上。首先会依据 IN_CURSOR 寄存器的值设置显示地址，其次将按键值转换为 ASCII Code 后调用 WLCMD 子程序予以显示。最后再将负责记录 Slave 端按键值在 LCD 上显示地址的寄存器(IN_CURSOR)加 1，判断是否到达第 2 行最后一个位置。若是则清除第 2 行，并重新设置 IN_CURSOR=C0h。
149~162	I2C_START 子程序，产生 START 条件。
166~177	I2C_DEV_SEL_W 与 I2C_DEV_SEL_R 子程序，将设备选择码(DEVICE_ID)输出到 I²C Bus。I2C_DEV_SEL_W 子程序会设置 R/W 位=0，表示是对 I²C 设备进行写入动作；I2C_DEV_SEL_R 子程序会设置 R/W 位=1，表示要对 I²C 设备进行读取动作。程序中会调用 WRITE_BYTE 子程序，将设备选择码以串行方式一一送出。此外，调用 CHECK_ACK 子程序的目的，是看看是否有 I²C 设备对此 DEVICE_ID 产生回应。
181~192	I2C_STOP 子程序，产生 STOP 条件。
196~221	WRITE_BYTE 子程序，将 Acc 寄存器的数据由 SDA 引脚循序移出(配合 SCL 信号)。

225~253 READ_BYTE 子程序，将 SDA 引脚上的信号循序移入(配合 SCL 信号)I2C_DATA 寄存器。由于本实验是一次只读取一个字节，因此第 244~249 行是设置 ACK=1(亦即送出 NO_ACK 信号，所以 HT46x23 的 RXAK 会为 0)。

257~275 CHECK_ACK 子程序，等待 I^{2}C 设备回应 ACK 信号，261~268 行是将检查的次数设置为 256 次。如果经过 256 次的检查 ACK 信号仍未送出，就代表 I^{2}C Bus 上没有与 DEVICE_ID 相符的设备，此时设置 DEVICE_FLAG=0。

279~288 DELAY_10 子程序，延时 20 个指令周期的时间，以确保 I^{2}C 接口来得及反应。

292~320 WLCMD 与 WLCMC 子程序，请参考 5-5 节的说明。不过请读者留意此处的子程序与 5-5 节有一点不同，主要是因为 LCD_CONTR 端口不只用来控制 LCD，同时 LCD_CONTR.3 也用来控制 I^{2}C Bus 的 SDA、SCL 信号，为了不影响 I^{2}C Bus 的正常动作，将原来的 "CLR CLD_CONTR" 改以 "CLR LCD_EN"、"CLR LCD_RW" 以及 "CLR LCD_RS" 3 行指令加以取代。

325~352 READ_KEY 子程序，请参考 4-7 节中的说明。

356~370 DELAY 子程序，延时时间的计算请参考 4-1 节。

5-13-5　HT46x23 I^{2}C传输Slave端控制流程图及程序

1. 流程图(见图 5-13-4)

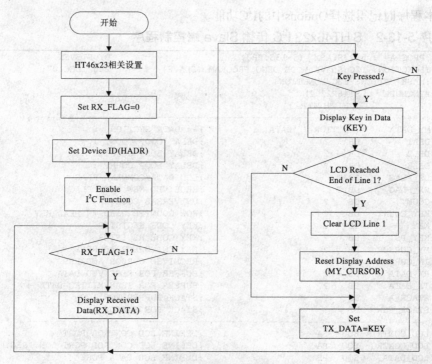

图5-13-4　流程图

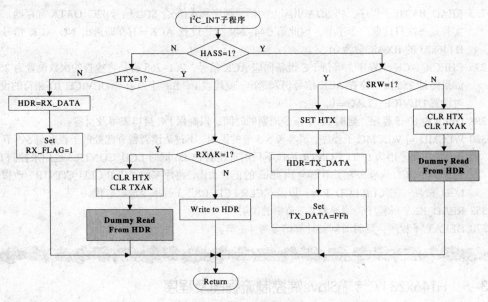

图5-13-4 流程图（续）

执行本程序时记得选择Options中的I²C功能。

2. 程序 5-13-2　S HT46x23 I²C 传输 Slave 端控制程序

```
1    ;PROGRAM : 5-13S.ASM (5-13S.PRJ)                   BY STEVEN
2    ;FUNCTION: I2C SLAVE MODE DEMO PROGRAM (SLAVE)     2002.DEC.07.
3    ;MASK OPTION: I2C FUNCTION
4    #INCLUDE    HT46R23.INC
5                .CHIP HT46R23
6    ;----------------------------------------------------------------
7    MY_DATA     .SECTION    'DATA'      ;== DATA SECTION ==
8    DEL1        DB      ?               ;DELAY LOOP COUNT 1
9    DEL2        DB      ?               ;DELAY LOOP COUNT 2
10   DEL3        DB      ?               ;DELAY LOOP COUNT 3
11   DC_FLAG     DBIT                    ;LCD DATA/COMMAND FLAG
12   RX_FLAG     DBIT                    ;RECEIVED NEW DATA FLAG
13   COUNT       DB      ?               ;UNIVERSAL COUNTER
14   KEY_COUNT   DB      ?               ;ROW COUNTER USED IN READ_KEY
15   KEY         DB      ?               ;KEY CODE REGISTER
16   KEY_PS      DB      ?               ;KEY CODE REGISTER
17   MY_CURSOR   DB      ?               ;KEY IN CURSOR ADDRESS
18   IN_CURSOR   DB      ?               ;RECEIVED CURSOR ADDRESS
19   RX_DATA     DB      ?               ;BUFFER FOR RECEIVED DATA
20   TX_DATA     DB      ?               ;BUFFER FOR TRANSMITTED DATA
21   STACK_A     DB      ?               ;STACK BUFFER FOR ACC
22   STACK_PSW   DB      ?               ;STACK FOR STATUS
23   ;----------------------------------------------------------------
24   LCD_CONTR   EQU     PA              ;DEFINE LCD CONTROL PORT
25   LCD_CONTRC  EQU     PAC             ;DEFINE LCD CONTROL PORT CON. REG.
26   LCD_DATA    EQU     PB              ;DEFINE LCD DATA PORT
27   LCD_DATAC   EQU     PBC             ;DEFINE LCD DATA PORT CON. REG.
28   LCD_EN      EQU     LCD_CONTR.0     ;DEFINE EN CONTROL PIN
29   LCD_RW      EQU     LCD_CONTR.1     ;DEFINE RW CONTROL PIN
30   LCD_RS      EQU     LCD_CONTR.2     ;DEFINE RS CONTROL PIN
31   LCD_READY   EQU     LCD_DATA.7      ;DEFINE READY BIT OF LCD
32   KEY_PORT    EQU     PB              ;DEFINE KEY PORT
33   KEY_PORTC   EQU     PBC             ;DEFINE KEY PORT CONTROL REG.
34   ID_PORT     EQU     PC              ;DEFINE ID SETTING PORT
35   ID_PORTC    EQU     PCC             ;DEFINE ID PORT CONTROL REG.
36   HCF         EQU     HSR.7           ;DEFINE HCF
```

```
37   HAAS         EQU    HSR.6              ;DEFINE HAAS
38   SRW          EQU    HSR.2              ;DEFINE SRW
39   RXAK         EQU    HSR.0              ;DEFINE RXAK
40   HEN          EQU    HCR.7              ;DEFINE HEN
41   HTX          EQU    HCR.4              ;DEFINE HTX
42   TXAK         EQU    HCR.3              ;DEFINE TXAK
43
44   MY_CODE      .SECTION   AT 0 'CODE'    ;== PROGRAM SECTION ==
45        ORG            00H                ;HT-46RXX RESET VECTOR
46        JMP            START
47        ORG            10H                ;HT-46XX H-BUS INTERRUPT VECTOR
48        JMP            I2C_INT
49   START:
50        CLR            LCD_CONTRC         ;CONFIG LCD_CONTR AS OUTPUT MODE
51        SET            ID_PORTC           ;CONFIG ID_PORT AS INPUT MODE
52        MOV            A,38H              ;FUNCTION SET: 8-BIT,2-LINE,5X10 DOTS
53        CALL           WLCMC
54        MOV            A,0CH              ;ON/OFF CONTR: DISPLAY ON, BLINKING OFF
55        CALL           WLCMC
56        MOV            A,06H              ;ENTRY MODE : INCREMENT,DISPLAY NOT SHIFT
57        CALL           WLCMC
58        MOV            A,01H              ;CLEAR DISPLAY
59        CALL           WLCMC
60        MOV            A,80H              ;SET LINE ONE, POSITION 0
61        MOV            MY_CURSOR,A
62        MOV            A,0C0H             ;SET LINE TWO, POSITION 0
63        MOV            IN_CURSOR,A
64        SET            EMI                ;ENABLE GLOBAL INTERRUPT
65        SET            EHI                ;ENABLE I2C INTERRUPT
66        RLA            ID_PORT            ;READ SLAVE ADDRESS SETTING
67        MOV            HADR,A             ;SET SLAVE ADDRES
68        MOV            A,10000000B
69        MOV            HCR,A              ;ENABLE I2C FUNCTION
70        CLR            RX_FLAG            ;CLEAR RECEIVED DATA FLAG
71   MAIN:
72        SZ             RX_FLAG
73        CALL           DISPLAY_RX_DATA
74        CALL           READ_KEY           ;READ KEYPAD
75        MOV            A,16
76        XOR            A,KEY
77        SZ             Z                  ;KEY PRESSED ?
78        JMP            MAIN               ;NO, RE-READ KEYPAD
79        MOV            A,KEY
80        MOV            KEY_PS,A           ;RESERVE KEY CODE
81   WAIT:
82        CALL           READ_KEY           ;CHECK KEYPAD RELEASED
83        MOV            A,16
84        XOR            A,KEY
85        SNZ            Z                  ;IS KEY RELEASED?
86        JMP            WAIT               ;NO,WAIT KEY RELEASE
87        CALL           DELAY              ;FOR DE-BOUNCING WHEN KEY RELEASE
88        MOV            A,MY_CURSOR        ;SET CURSOR ADDRESS
89        CALL           WLCMC
90        MOV            A,KEY_PS
91        SUB            A,10               ;CONVERT TO ASCII CODE
92        SZ             C
93        ADD            A,40H-30H-9
94        ADD            A,30H+10
95        MOV            TX_DATA,A
96        CALL           WLCMD              ;DISPLAY KEY IN DATA
97        INC            MY_CURSOR
98        MOV            A,MY_CURSOR
99        XOR            A,80H+20+1         ;20X2 LCD
100       SNZ            Z                  ;REACH LAST POSITION OF LINE 1?
101       JMP            MAIN               ;NO
102       MOV            A,80H              ;RESET LCD ADDRESS
```

```
103             CALL        WLCMC
104             MOV         A,20
105             MOV         DEL1,A
106             MOV         A,20H               ;CLEAR LINE 1
107             CALL        WLCMD
108             SDZ         DEL1
109             JMP         $-2
110             MOV         A,80H
111             MOV         MY_CURSOR,A         ;RESET LCD ADDRESS
112             JMP         MAIN                ;READ NEXT KEY
113 ;************************************************************************
114 ;           DISPLAY RECEIVED DATA PROCEDURE
115 ;************************************************************************
116 DISPLAY_RX_DATA  PROC
117             MOV         A,IN_CURSOR         ;SET CURSOR ADDRESS
118             CALL        WLCMC
119             MOV         A,RX_DATA           ;COPY TO RX_DATA BUFFER
120             CALL        WLCMD               ;DISPLAY KEY IN DATA
121             CLR         RX_FLAG
122             INC         IN_CURSOR
123             MOV         A,IN_CURSOR
124             XOR         A,0C0H+20+1         ;20x2 LCD
125             SNZ         Z                   ;REACH LAST POSITION OF LINE 2
126             RET                             ;NO
127             MOV         A,0C0H              ;YES, RESET LCD ADDRESS
128             CALL        WLCMC
129             MOV         A,20
130             MOV         DEL1,A
131             MOV         A,20H               ;CLEAR LINE 2
132             CALL        WLCMD
133             SDZ         DEL1
134             JMP         $-2
135             MOV         A,0C0H
136             MOV         IN_CURSOR,A         ;RESET LCD ADDRESS
137             RET
138 DISPLAY_RX_DATA  ENDP
139 ;************************************************************************
140 ;           I2C INTERRUPT SERVICE ROUTINE
141 ;************************************************************************
142 I2C_INT     PROC
143             MOV         STACK_A,A           ;PUSH ACC
144             MOV         A,STATUS
145             MOV         STACK_PSW,A         ;PUSH STATUS
146             SNZ         HAAS                ;HAAS = 1?
147             JMP         DATA_RT             ;NO, JUMP DATA RECEIVED/TRANSMITTED
148             SNZ         SRW                 ;ADRESS MATCH! TX MODE?
149             JMP         RX_MODE             ;NO, JUMP TO RX MODE
150 TX_MODE:
151             SET         HTX                 ;SET TX MODE
152             MOV         A,TX_DATA
153             MOV         HDR,A               ;WRITE DATA TO HDR
154             SET         TX_DATA
155             JMP         END_I2C_INT
156 RX_MODE:
157             CLR         HTX                 ;SET RX MODE
158             CLR         TXAK                ;ACKNOWLEDGE
159             MOV         A,HDR               ;DUMMY READ FROM HDR
160             JMP         END_I2C_INT
161 DATA_RT:
162             SNZ         HTX                 ;TX MODE?
163             JMP         DATA_RX             ;NO, JUMP TO DATA RX MODE
164 DATA_TX:
165             SZ          RXAK                ;RXAK = 0? CONTINUE DATA READ?
166             JMP         WITH_ACK            ;YES!
167 NO_ACK:
168             CLR         HTX
169             CLR         TXAK
```

```
170         MOV       A,HDR                   ;DUMMY READ FORM HDR
171         JMP       END_I2C_INT
172 WITH_ACK:
173         MOV       A,TX_DATA
174         MOV       HDR,A                   ;WRTIE TO HDR
175         SET       TX_DATA
176         JMP       END_I2C_INT
177 DATA_RX:
178         MOV       A,HDR                   ;SET CURSOR ADDRESS
179         MOV       RX_DATA,A               ;COPY TO RX_DATA BUFFER
180         SET       RX_FLAG
181 END_I2C_INT:
182         MOV       A,STACK_PSW
183         MOV       STATUS,A                ;POP PSW
184         MOV       A,STACK_A               ;POP Acc
185         RETI
186 I2C_INT    ENDP
187 ;************************************************************************
188 ;         LCD DATA/COMMAND WRITE PROCEDURE
189 ;************************************************************************
190 WLCMD   PROC
191         SET       DC_FLAG                 ;SET DC_FLAG=1 FOR DATA WRITE
192         JMP       WLCM
193 WLCMC:
194         CLR       DC_FLAG                 ;SET DC_FLAG=0 FOR COMMAND WRITE
195 WLCM:
196         SET       LCD_DATAC               ;CONFIG LCD_DATA AS INPUT MODE
197         CLR       LCD_EN                  ;CLEAR ALL LCD CONTROL SIGNAL
198         CLR       LCD_RW
199         CLR       LCD_RS
200         SET       LCD_RW                  ;SET RW SIGNAL (READ)
201         NOP                               ;FOR TAS
202         SET       LCD_EN                  ;SET EN HIGH
203         NOP                               ;FOR TDDR
204 WF:
205         SZ        LCD_READY               ;IS LCD BUSY?
206         JMP       WF                      ;YES, JUMP TO WAIT
207         CLR       LCD_DATAC               ;NO, CONFIG LCD_DATA AS OUTPUT MODE
208         MOV       LCD_DATA,A              ;LATCH DATA/COMMAND ON PB(LCD DATA BUS)
209         CLR       LCD_EN                  ;CLEAR ALL LCD CONTROL SIGNAL
210         CLR       LCD_RW
211         CLR       LCD_RS
212         SZ        DC_FLAG                 ;IS COMMAND WRITE?
213         SET       LCD_RS                  ;NO, SET RS HIGH
214         SET       LCD_EN                  ;SET EN HIGH
215         NOP
216         CLR       LCD_EN                  ;SET EN LOW
217         RET
218 WLCMD   ENDP
219 ;************************************************************************
220 ;   SCAN 4x4 MATRIX ON KEY PORT AND RETURN THE CODE IN KEY REGISTER
221 ; IF NO KEY BEEN PRESSED, KEY=16.
222 ;************************************************************************
223 READ_KEY    PROC
224         MOV       A,11110000B
225         MOV       KEY_PORTC,A             ;CONFIG KEY_PORT
226         SET       KEY_PORT                ;INITIAL KEY PORT
227         CLR       KEY                     ;INITIAL KEY REGISTER
228         MOV       A,04
229         MOV       KEY_COUNT,A             ;SET ROW COUNTER
230         CLR       C                       ;CLEAR CARRY FLAG
231 SCAN_KEY:
232         RLC       KEY_PORT                ;ROTATE SCANNING BIT
233         SET       C                       ;MAKE SURE C=1
234         SNZ       KEY_PORT.4              ;COLUMN 0 PRESSED?
235         JMP       END_KEY                 ;YES.
236         INC       KEY                     ;NO, INCREASE KEY CODE.
```

```
237         SNZ     KEY_PORT.5          ;COLUMN 1 PRESSED?
238         JMP     END_KEY             ;YES.
239         INC     KEY                 ;NO, INCREASE KEY CODE.
240         SNZ     KEY_PORT.6          ;COLUMN 2 PRESSED?
241         JMP     END_KEY             ;YES.
242         INC     KEY                 ;NO, INCREASE KEY CODE.
243         SNZ     KEY_PORT.7          ;COLUMN 3 PRESSED?
244         JMP     END_KEY             ;YES.
245         INC     KEY                 ;NO, INCREASE KEY CODE.
246         SDZ     KEY_COUNT           ;HAVE ALL ROWs BEEN CHECKED?
247         JMP     SCAN_KEY            ;NO, NEXT ROW.
248 END_KEY:
249         RET
250 READ_KEY    ENDP
251 ;*****************************************************************
252 ;               Delay about DEL1*10ms
253 ;*****************************************************************
254 DELAY   PROC
255         MOV     A,6
256         MOV     DEL1,A              ;SET DEL1 COUNTER
257 DEL_1:
258         MOV     A,30
259         MOV     DEL2,A              ;SET DEL2 COUNTER
260 DEL_2:
261         MOV     A,110
262         MOV     DEL3,A              ;SET DEL3 COUNTER
263 DEL_3:
264         SDZ     DEL3                ;DEL3 DOWN COUNT
265         JMP     DEL_3
266         SDZ     DEL2                ;DEL2 DOWN COUNT
267         JMP     DEL_2
268         SDZ     DEL1                ;DEL1 DOWN COUNT
269         JMP     DEL_1
270         RET
271 DELAY   ENDP
272         END
```

SLAVE 端控制程序说明

8~22　　依序定义变量地址。

24~25　　定义 LCD_CONTR 为 PA。

26~27　　定义 LCD_DATA 为 PB。

28~31　　定义 LCD 的控制信号引脚。

32~33　　定义 KEY_PORT 为 PB。

34~35　　定义 ID_PORT 为 PC，因此可以选择的 Device ID 范围为 00h~1Fh。

36~42　　依序定义 HT46x23 中与 I^2C 串行传输接口的相关控制位的名称，以增加程序的可读性与方便性。

45　　　声明存储器地址由 000h 开始(HT46x23 Reset Vector)。

47　　　声明存储器地址由 010h 开始(HT46x23 I^2C Interrupt Vector)。

50~51　　定义 LCD_CONTR 为输出模式、ID_PORT 为输入模式。

52~53　　将 LCD 设置为双行显示(N = 1)、使用 8 位(DB7 ~ DB0)控制模式(DL = 1)、5×7 点矩阵字形(F = 0)。

54~55　　将 LCD 设置为显示所有数据(D = 1)、显示光标(C = 1)、光标所在位置的字不会闪烁(B = 0)。

56~57　　将 LCD 的地址标志位(AC)设置为递增(I/D = 1)、显示器画面不因读/写数据而移动(S = 0)。

58~59　　将 LCD 整个显示器清空，并设置 DD RAM 地址为第 1 行第 0 个位置。

60~61　　设置 MY_CURSOR＝80h，此寄存器是记录 Master 端按键值在 LCD 上的显示地址，以作为是否到达第 1 行最后一个位置的判断。

62~63　　设置 IN_CURSOR＝C0h，此寄存器是记录由 Slave 端送过来的按键值在 LCD 上的显示地址，以作为是否到达第 2 行最后一个位置的判断。

行号	说明
64~65	中断总开关(EMI)与 I²C Interrupt(EHI)有效。
66~67	由 ID_PORT 读回指拨开关设置数值，并存放在 HADR 寄存器，此即 Slave Mode I²C 的 Device ID。自此之后 HADR 便不再更改，除非程序重新执行。
68~69	开启 I²C 串行接口的传输功能。
70	清除 RX_FLAG，程序中将以此位作为是否由 I²C 串行接口接收到 Master 端数据的判断(1 表示有接收到数据)，因此先将其清除为 0。
72~73	判断 I²C 串行传输接口是否接收到数据，若有(RX_FLAG=1)则进行数据显示的动作(CALL DISPLAY_RX_DATA)；若无(RX_FLAG=0)就继续扫描 4×4 键盘。
74~78	调用 READ_KEY 子程序扫描键盘，完成后判断 KEY 寄存器内容是否为 16，若成立表示键盘无按键输入，重新扫描键盘(JMP MAIN)。反之，则代表按键已被按下，准备开始将按键值显示并通过 I²C 接口发送出去。
79~80	将按键值保留在 KEY_PS 寄存器。
82~87	因为 LCD 的数据端口与键盘共用一个 I/O 端口，所以必须等按键放开之后才能送出 LCD 的显示数据，不会造成数据被干扰的现象。此段程序的功能即是确认按键已经完全放开，以便继续往下执行显示的动作。
88~89	依据 MY_CURSOR 寄存器的值设置显示地址。
90~96	将按键值转换为 ASCII Code 之后调用 WLCMD 子程序予以显示，同时并将该值存放在 TX_DATA 寄存器，以便 Bus Master 读取。
97~111	将负责记录 Master 端按键值在 LCD 上显示地址的寄存器(MY_CURSOR)加 1，判断是否到达第 1 行最后一个位置。若是则清除第 1 行，并重新设置 MY_CURSOR=80h。(注：本实验使用 20×2 LCD。)
112	重新执行程序(JMP MAIN)。
116~138	DISPLAY_RX_DATA 子程序，负责将接收到的数据显示在 LCD 上。首先会依据 IN_CURSOR 寄存器的值设置显示地址，其次将按键值转换为 ASCII Code 后调用 WLCMD 子程序予以显示。最后再将负责记录 Slave 端按键值在 LCD 上显示地址的寄存器(IN_CURSOR)加 1，判断是否到达第 2 行最后一个位置。若是则清除第 2 行，并重新设置 IN_CURSOR=C0h。
142~186	I²C_INT 中断服务子程序，此段程序主要是根据原厂数据手册中的"I²C Bus 中断服务子程序"的流程所编写，请读者参考图 5-13-5 所示流程图的说明。

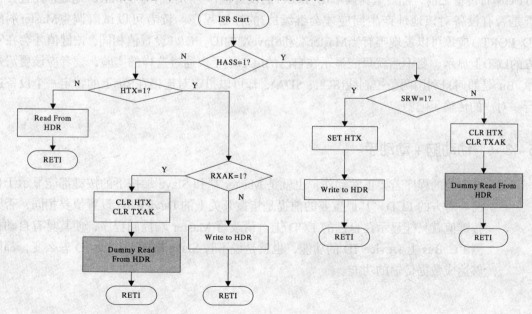

图5-13-5　I²C流程图

146~147 依据 HAAS 位分析是与 Device ID 相符所产生的中断(HASS=1)，或是 8 位数据发送/接收完毕所产生的中断(HASS=0)，若是则 JMP DATA_RT。

148~160 与 Device ID 相符所产生的中断：再依据 SRW 位分析为 Receive Mode(SRW=0)或 Transmit Mode(SRW=1)。若为 Transmit Mode，就将 HTX 设置为 1，并将要通过 I^2C 接口发送的数据 (TX_DATA)写至 HDR 寄存器。如果是 **Receive** Mode，就将 HTX 与 TXAK 位设置为 0，并依数据手册中的要求，对 HDR 寄存器执行一次 "Dummy Read"(MOV A,HDR)。

161~185 8 位数据发送/接收完毕所产生的中断：再依据 HTX 位分析为 Receive Mode(HTX=0)或 Transmit Mode(HTX=1)。

如果是 Transmit Mode，表示 8 位数据已经传输完毕，再根据 RXAK 位分析 Master 是否送出 ACK 信号，以决定是否要继续送出数据。如果 RXAK=1，表示 Master 端要继续读数据，所以再次将 TX_DATA 写至 HDR 寄存器；如果 RXAK=0，则清除 HTX 与 TXAK 位，并依数据手册中的要求，对 HDR 寄存器执行一次"Dummy Read"(MOV A,HDR)。本实验是一次只读取一个字节，因此 RXAK =0。

如果是 Receive Mode，表示 8 位数据已经接收完毕， 接收的数据复制到 RX_DATA 寄存器，并设置 RX_FLAG=1。

190~218 WLCMD 与 WLCMC 子程序，请参考 5-5 节的说明。不过请读者留意此处的子程序与 5-5 节有一点不同，主要是因为 LCD_CONTR 端口不只用来控制 LCD，同时 LCD_CONTR.3 也用来控制 I^2C Bus 的 SDA、SCL 信号，为了不影响 I^2C Bus 的正常动作，将原来的 "CLR CLD_CONTR" 改以 "CLR LCD_EN"、"CLR LCD_RW" 以及 "CLR LCD_RS" 3 行指令加以取代。

223~250 READ_KEY 子程序，请参考 4-7 节中的说明。

254~268 DELAY 子程序，延时时间的计算请参考 4-1 节。

由于 Slave 端在程序一开始执行时，就先读取 ID_PORT 的设置值作为本身的 Device ID，而且此后再也不会去更改 HADR，所以其 Device ID 是固定不变的。然而 Master 端在每次要进行 I^2C 串行传输之前，都会重新读取 ID_PORT 的设置值作为所要存取的 Slave 地址，并且会检查是否有设备对该地址产生回应才会继续 I^2C 的控制程序。读者可以试着调整 Master 端的 ID_PORT，应该可以发现唯有当 Master 端和 Slave 端的 ID_PORT 设置值相同，按键值才会在对方的 LCD 上显示，要不然就只能显示在自己的 LCD 上了。通过这样的实验，读者应该更清楚 I^2C Bus 是如何只利用两条控制线(SCL、SDA)，却可以指定对连接在 Bus 上的特定一个设备进行一对一的传输了吧？

5-13-6 动动脑+动动手

本实验的程序若要正常动作，也就是 Master 端和 Slave 端按下的按键都能显示于自己与对方的 LCD，其最重要的前提是指拨开关上的 Device ID 设置值要相同，否则按键值就只能显示在自己的 LCD 上。请改写 Master 端控制程序，使其具有自动检测 I^2C Bus 上 Device ID 的功能，也就是说不管 Slave 端的 Device ID 怎么变，都能够达成数据传输的功能。

第 6 章

实践应用篇

历经基础实验篇与进阶实验篇的磨练，读者对于 HT46xx 的硬件及程序编写技巧应该都已具备相当的能力。本章将继续介绍几个实用、有趣的专题制作，相信必定能够让读者对于单片机的应用能有更深一层的体会。不过本章的许多概念与子程序，都是延续前几个章节的学习成果，程序的编写技巧绝非一蹴而就，所以希望读者务必将之前的实验内容融会贯通，这样才能增加自己的学习成果。本章的实验内容包括：

- 6-1 专题一：数字温度计
- 6-2 专题二：密码锁
- 6-3 专题三：具记忆功能的密码锁(I^2C E^2PROM)
- 6-4 专题四：24小时时钟
- 6-5 专题五：猜数字游戏机
- 6-6 专题六：逻辑测试笔
- 6-7 专题七：频率计数器(Counter)的制作
- 6-8 专题八：简易信号产生器的制作
- 6-9 专题九：复频信号(DTMF)产生器的制作
- 6-10 专题十：简易低频电压—频率转换器(VCO)的制作
- 6-11 专题十一：简易声音调变器的制作
- 6-12 专题十二：RS-232串行传输

6-1 专题一：数字温度计

6-1-1 专题功能概述

以HT46x23配合温度传感器AD590，设计一个可以测量0℃~150℃的温度计，并将温度的变化以十进制显示在LCD上。

6-1-2 学习重点

通过本实验，读者应了解温度传感器AD590的工作原理，并且能充分运用HT46xx的ADC接口转换特性。有关HT46x23的模/数转换接口，请读者参阅2-9节与4-11节中的范例与说明。另外，在程序中需使用16位乘8位的乘法与16位的加、减法，这都是首次接触而且经常会使用到的编程技巧，读者应该了解其原理。

6-1-3 电路图（见图6-1-1）

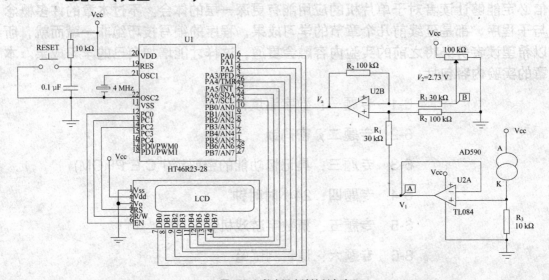

图6-1-1 数字温度计控制电路

AD590是一个PTAT(Proportional To Absolute Temperature)的电流调节器(Current Regulator)；也就是说它的输出电流正比于环境温度。AD590的电流温度系数为1μA/K；也就是说温度每上升1 K，电流就上升1μA。在绝对0 K时(开式温度)，其电流为0μA。AD590的主要特性如下：

线性的电流输出：1μA/K。

反应的温度范围：-55~150 ℃。

工作电压：+4~30 V。

高输出阻抗：>10 MΩ。

注意：开式(Kelvin)温度与惯用的华氏、摄氏温度间的关系为：

$$℃ = \frac{5}{9}(℉-32), \quad K = ℃+273.15; \quad ℉ = \frac{9}{5}(℃+32), \quad K = ℉+459.7$$

图6-1-2是AD590的工作电压、输出电流与温度之间的关系，读者可以发现其线性度相当不错，只要供给4 V以上的电压，它的输出电流几乎是只受温度的影响。AD590的完整数据已收录于本书所附的光盘中，有兴趣的读者可以了解其更多的特性。

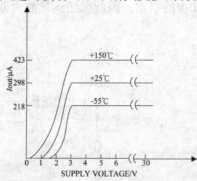

图6-1-2　AD590工作电压、输出电流与温度的关系

从以上关于AD590的特性介绍，以及图6-1-1的电路设计，其中R_3电阻的主要功能是将AD590的温度—电流(1 μA/K)的变化转换为温度—电压的变化(10 kΩ×1 μA/K = 10 mV/K)，以便HT46xx的ADC接口加以转换为数字输出并加以处理。U2A(TL084)是接成电压随耦器的形式，主要作为阻抗隔离之用，以避免HT46xx的ADC接口输入阻抗与R_3产生并联效应，影响温度与电压的变化关系(10 mV/K)。U2B的作用有二，其一是放大电压，以便提高ADC的分辨率；其二是作为最低温度调整或校正之用。

首先先分析HT46xx的ADC接口的输入电压(V_a)的动态变化范围(Dynamic Range)，由于要设计的是0~150 ℃的温度计，因此V_1电压的变化范围为：

$$(150℃-0℃)×10 \text{ mV/K} = 150 \text{ K}×10 \text{ mV/K} = 1.5 \text{ V}$$

而在0 ℃时，R_3上会有一个固定的压降(Offset Voltage)为：

$$(0+273.15)\text{K}×10 \text{ mV/K} = 2.73 \text{ V}$$

由此可以推断在0~150 ℃的温度变化范围时，HT46xx的ADC接口的输入电压变动大小为2.73~(2.73+1.5)V，如图6-1-3所示。

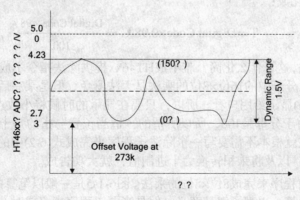

图6-1-3　温度—电压变化示意图

依照HT46x23数据手册有关A/D的叙述，其V_{AD}(A/D Input Voltage)的输入电压范围约为 0~V_{DD}之间，其间输入电压与转换后的数字数值呈线性关系。以本例HT46x23工作在$V_{DD}=5$ V的情况下，如果直接将V_1输入到模拟输入端(AN0)，势必造成有些数字输出位永远不会变化，这是因为电压的变动范围太小所造成的。因此必须将V_1的电压放大，最好使其动态范围是由0(0℃)变化到V_{DD}(150℃)之间，这样才能将HT46x23单片机的分辨率发挥到极致，但同时又得考虑在0℃时V_1的电压并非0 V的事实，因此采用运算放大器中差动放大特性来加以克服，如图6-1-4所示。

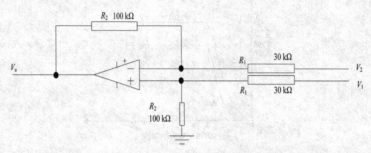

图6-1-4 运算放大器电路

图6-1-4中输入与输出的关系为：$V_a=(V_1-V_2)R_2/R_1$，若将V_2设置为2.73 V(0 ℃时的V_1电压)，再让$R_2/R_1=3.333$，就可以使得当温度由0 ℃变化到150 ℃时，V_a的电压恰巧约从0 V变化到5 V，让整个ADC的分辨率都能发挥。另外，在HT46x23数据手册中还提到ADC的分辨率为10位，但准确度只有9位，因此在程序中将只提取最高的9位加以处理，即ADRH(D9~D2)与ADRL(D1)，所以此时的分辨率为$5\text{ V}/2^9=9.765\ 625\text{ mV}\approx 9.8\text{ mV}$。这表示当输入电压($V_a$)变化超过9.8 mV时，ADC数字输出的最低位(LSB-D1)才会有所变化，所以一般又将其称之为LSB电压。在此设置之下，ADC数字输出的分辨率(或者说是LSB电压)为9.8 mV，而温度与输出数字数码(Digital Code)的关系为：

$$T=(\text{Digital Code}\times 9.8\text{ mV})/10\text{ mV}/℃\approx \text{Digital Code}\times 0.98℃ \tag{6-1}$$

硬件的分析到此为止，如何依据上式把Digital Code转换为实际的温度呢？

式(6-1)中有乘法又有小数，HT46xx又没有乘法指令、汇编语言又要如何处理小数呢？

式(6-1)可以再简化成：

$$T=\text{Digital Code}\times 0.98℃=\frac{\text{Digital Code}\times 98}{100} \tag{6-2}$$

小数变成了除法,这算什么化简？难不成HT46xx提供除法指令不成？答案是：当然没有。但仔细观察式(6-2)，分子中的16位×8位的乘法是无法避免，但是乘完后的结果终究要转成十进制(或说是ASCII码)显示给用户看。所以，只要在显示的时候将小数点插在"适当的位置"上不就可以了吗？由于分母是100，所以所谓的"适当位置"指的是十进制数值的倒数第二个位数，通过此方式就根本不需要写一个除法子程序来完成式(6-2)中的除法了。现在只要解决16位与8位的乘法，以及将乘积转换为十进制值，就大功告成。

首先分析如何由程序来完成8位×8位的乘法。图6-1-5是一般以笔算的方式完成8位乘法的过程，相信读者都会算，也都了解其意义。但是当要以程序来完成时，就有点裹足不前、不知如何下手才好。我们尝试以写程序的逻辑观点来解释此运算过程，希望能对读者在程序的了解上有所帮助。先看乘数(Multiplier)的B_0(Bit0)，因为其为1，所以必须将被乘数(Multiplicand)累加至乘积(Product)项，然后再将被乘数左移一位(相当于乘2)，为下一位的运算预先做好准备。其次再看乘数的B_1，因为其为1，所以必须将移位后的被乘数累加至乘积项，然后再次将被乘数左移一位，为B_2的运算做好准备。乘数的$B_2=0$，所以不须将被乘数累加至乘积项，但仍须将被乘数左移一位，继续下一位的运算，其余以此类推。

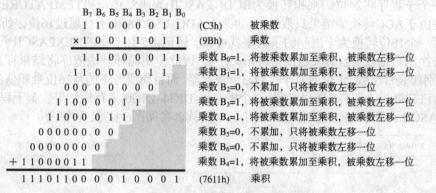

图6-1-5 8位×8-位的乘法运算范例

其实说穿了，就程序而言，乘法子程序不外乎就是一连串的相加与移位的动作而已。只要能熟练运用HT46xx的ADD、RRC、RLC与位判断指令，应该就不是难事了。请读者参考以下的流程图(见图6-1-6)及程序：

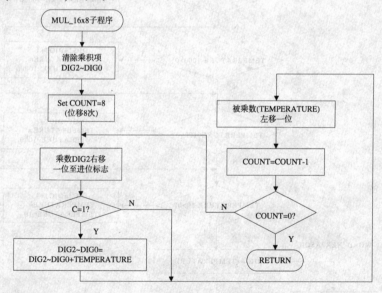

图6-1-6 流程图

只要先将ADC的数字输出码先读到TEMPERATURE寄存器(由于转换结果为9位，因此分别存放在TEMPERATURE_H与TEMPERATURE_L两个寄存器)，再调用此程序即可在DIG2~DIG0(24位)寄存器得到乘积值(DIGITAL CODE×98；请参考式(6-2))。 读者只要将第9、10两行稍加修改，就可以将此程序变成通用的16位×8位子程序(被乘数不再固定为常量即ADC_LSB)，以后凡是需要用到乘法运算时，都可以拿来运用。程序中有一点必须提醒读者，当被乘数(TEMPERATURE_H与TEMPERATURE_L)左移之后会变成大于16位的数值，在此是以DIG4与其配合使用，所以在第24~26行将被乘数左移时，是3个寄存器一起移动。也由于

305

移位后的被乘数大于8位(DIG4、TEMPERATURE_H与TEMPERATURE_L)，所以当累加乘积项(DIG2~DIG0)时，必须使用24位的加法，也就是必须先将低8位相加之后，再执行高8位的加法运算，而在执行高8位运算时必须将低8位相加的进位一起考虑，这就是在第17~22行程序所负责的任务。

解决了乘法子程序，最后就只剩下将乘积转换为十进制的工作。但是乘积为24位，是不是要写一个子程序将24位的乘积转换为BCD(或ASCII)码吗？还记得TEMPATURE虽然是16位，但是由于ADC转换的结果只取9位，再乘上ADC_LSB(98)也不会超过16位，所以实际上只需要写一个16位转换为十进制的子程序就可以。还记得5-10节中的HEX2ASCII子程序吗？因为要转换的数据仅8位(最大数值为FFh＝255)，因此使用循环减法依序将结果存到DIG2～DIG0(分别代表百位数、十位数及个位数)3个寄存器。而目前所考虑的是16位乘积值的转换(最大数值为FFFFh＝65 535)，所以必须以5个寄存器(DIG4~DIG0)来存放结果，至于程序的写法与HEX2ASCII子程序的原理及作法大同小异，请读者参阅图6-1-7及程序。

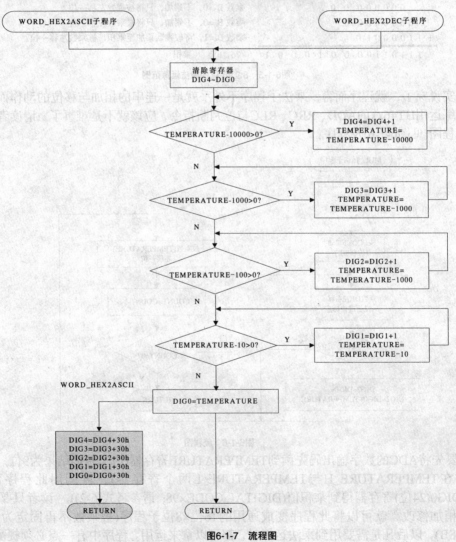

图6-1-7　流程图

```
1    ;****************************************************************
2    ;            CONVERT 16-BIT HEX TO 5-DIGIT DECIMAL
3    ;      I/P: TEMPERATURE_H & TEMPERATURE_L   O/P: DIG4 ~ DIG0
4    ;****************************************************************
5    WORD_HEX2ASCII PROC
6         CLR       DIG4                   ;INITIAL RESULT BUFFER
7         CLR       DIG3
8         CLR       DIG2
9         CLR       DIG1
10        CLR       DIG0
11   DIG_4:
12        MOV       A,TEMPERATURE_L        ;TEMPERATURE-10000
13        SUB       A,LOW 10000
14        MOV       TEMPERATURE_L,A
15        MOV       A,(NOT (HIGH 10000))+1
16        SNZ       C
17        DEC       Acc
18        ADDM      A,TEMPERATURE_H
19        SNZ       C                      ;TEMPERATURE-10000 > 0?
20        JMP       DIG_4_1                ;NO.
21        INC       DIG4                   ;YES, DIG4=DIG4+1
22        JMP       DIG_4                  ;DO AGAIN
23   DIG_4_1:
24        MOV       A,LOW 10000            ;RESTORE TEMPERATURE
25        ADDM      A,TEMPERATURE_L
26        MOV       A,HIGH 10000
27        ADCM      A,TEMPERATURE_H
28   DIG_3:
29        MOV       A,TEMPERATURE_L        ;TEMPERATURE-1000
30        SUB       A,LOW 1000
31        MOV       TEMPERATURE_L,A
32        MOV       A,(NOT (HIGH 1000))+1
33        SNZ       C
34        DEC       Acc
35        ADDM      A,TEMPERATURE_H
36        SNZ       C                      ;TEMPERATURE-1000 > 0?
37        JMP       DIG_3_1                ;NO.
38        INC       DIG3                   ;YES,DIG3=DIG3+1
39        JMP       DIG_3                  ;DO AGAIN
40   DIG_3_1:
41        MOV       A,LOW 1000             ;RESTORE TEMPERATURE
42        ADDM      A,TEMPERATURE_L
43        MOV       A,HIGH 1000
44        ADCM      A,TEMPERATURE_H
45   DIG_2:
46        MOV       A,TEMPERATURE_L        ;TEMPERATURE-100
47        SUB       A,100
48        MOV       TEMPERATURE_L,A
49        MOV       A,TEMPERATURE_H
50        SNZ       C
51        SUB       A,1
52        MOV       TEMPERATURE_H,A
53        SNZ       C                      ;TEMPERATURE-100 > 0?
54        JMP       DIG_2_1                ;NO.
55        INC       DIG2                   ;YES, DIG2=DIG2+1
56        JMP       DIG_2                  ;DO AGAIN
57   DIG_2_1:
58        MOV       A,100                  ;RESTORE TEMPERATURE
59        ADDM      A,TEMPERATURE_L
60        CLR       Acc
61        ADCM      A,TEMPERATURE_H
62   DIG_1:
63        MOV       A,TEMPERATURE_L        ;TEMPERATURE-10
64   DIG_1_1:
65        SUB       A,10
66        SNZ       C                      ;TEMPERATURE-10 > 0?
67        JMP       DIG_0                  ;NO.
```

```
68        INC     DIG1            ;YES, DIG1=DIG1+1
69        JMP     DIG_1_1         ;DO AGAIN
70  DIG_0:
71        ADD     A,10            ;RESTORE TEMPERATURE
72        MOV     DIG0,A
73        MOV     A,30H           ;BEFORE THIS LINE HEX TO UN-PACK BCD
74        ADDM    A,DIG4          ;CONVERT DIG4 TO ASCII
75        ADDM    A,DIG3          ;CONVERT DIG3 TO ASCII
76        ADDM    A,DIG2          ;CONVERT DIG2 TO ASCII
77        ADDM    A,DIG1          ;CONVERT DIG1 TO ASCII
78        ADDM    A,DIG0          ;CONVERT DIG0 TO ASCII
79        RET
80  WORD_HEX2ASCII ENDP
```

注：代表乘积项的变量为16位，在程序中是以TEMPERATURE_H与TEMPERATURE_L两个寄存器来存放，但为了方便解释，在以下的说明中就以TEMPERATURE来代表这两个寄存器。

如果读者细心浏览此流程图及程序，应该发现写法其实很单纯。首先把TEMPERATURE减10 000，若结果大于或等于0，就将代表万位数的DIG4加1，再继续减。当发现减10 000后的结果小于0，就先将TEMPERATURE还原(加回10 000)之后，再以相同步骤处理千位数、百位数、十位数与个位数。相减的过程中需用到16位的减法，这与16位加法运算的观念差不多，所以不再赘述。读者要特别留意SUB指令对进位标志位的影响规则是："有借位时C=0，没有借位时C=1"。程序中使用了两个编译器提供的运算LOW与HIGH，因为10 000与1 000都是大于8位的常量，所以必须分为两个字节来处理，一般人会在写程序前先将其转换好(如10000=2710h)。虽然现在计算机按一按很方便，但是既然编译器提供了相关指令，何不让它发挥功效呢？HIGH、LOW虚指令，不用再多做说明了吧？其分别代表取得操作数的低字节与高字节。完成此转换程序之后，别忘了输出结果时必须在百位数与十位数插入一个小数点。

6-1-4 流程图及程序

1. 流程图(见图6-1-8)

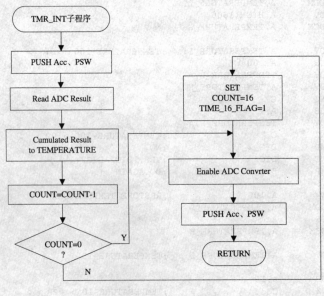

图6-1-8 流程图

第6章 实践应用篇

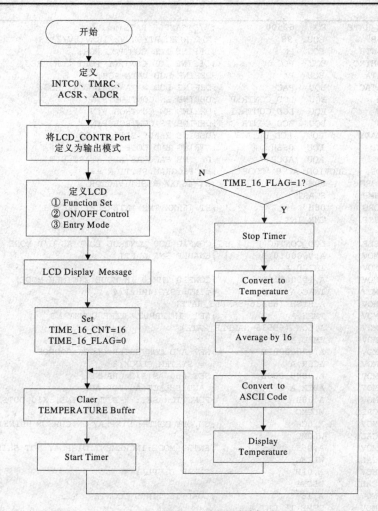

图6-1-8 流程图（续）

2. 程序 6-1 数字温度计

```
1    ;PROGRAM : 6-1.ASM  (6-1.PRJ)                              BY STEVEN
2    ;FUNCTION: DIGITAL TEMPERATURER (AD590) DEMO PROGRAM     2002.DEC.07.
3    ;  NOTE  : 10-BIT ADC IS TAKE THE 9 BITS ACCURACY
4    #INCLUDE  HT46R23.INC
5              .CHIP    HT46R23
6    ;----------------------------------------------------------------
7    MY_DATA       .SECTION  'DATA'        ;== DATA SECTION ==
8    DC_FLAG          DBIT                 ;LCD DATA/COMMAND FLAG
9    TIME_16_FLAG     DBIT                 ;16 TIMES CUMULATED FLAG
10   TIME_16_CNT      DB    ?              ;16 TIMES CUMULATED COUNTER
11   DIG0             DB    ?              ;DECIMAL TEMPERATURE DIGIT 0
12   DIG1             DB    ?              ;DECIMAL TEMPERATURE DIGIT 1
13   DIG2             DB    ?              ;DECIMAL TEMPERATURE DIGIT 2
14   DIG3             DB    ?              ;DECIMAL TEMPERATURE DIGIT 3
15   DIG4             DB    ?              ;DECIMAL TEMPERATURE DIGIT 4
16   TEMP_L           DB    ?              ;TEMPORY BUFFER
17   TEMP_H           DB    ?              ;TEMPORY BUFFER
18   TEMPERATURE_L    DB    ?              ;TEMPERATURE BUFFER LO-BYTE
19   TEMPERATURE_H    DB    ?              ;TEMPERATURE BUFFER HI-BYPE
20   COUNT            DB    ?              ;DISPLAY COUNT
21   STACK_A          DB    ?              ;STACK BUFFER FOR ACC
22   STACK_PSW        DB    ?              ;STACK BUFFER FOR STATUS
23   ;----------------------------------------------------------------
```

```
24   SAMPLE_RATE    EQU    62500              ;A/D SAMPLE RATE(μs)
25   ADC_LSB        EQU    98                 ;A/D RESOLUTION,98/10mV(5V/2**9)
26   LCD_CONTR      EQU    PC                 ;DEFINE LCD CONTROL PORT
27   LCD_CONTRC     EQU    PCC                ;DEFINE LCD CONTROL PORT CON. REG.
28   LCD_DATA       EQU    PA                 ;DEFINE LCD DATA PORT
29   LCD_DATAC      EQU    PAC                ;DEFINE LCD DATA PORT CON. REG.
30   LCD_EN         EQU    LCD_CONTR.0        ;DEFINE EN CONTROL PIN
31   LCD_RW         EQU    LCD_CONTR.1        ;DEFINE RW CONTROL PIN
32   LCD_RS         EQU    LCD_CONTR.2        ;DEFINE RS CONTROL PIN
33   LCD_READY      EQU    LCD_DATA.7         ;DEFINE READY BIT OF LCD
34   STR_END        EQU    0FEH               ;STRING END CODE
35   SADC           EQU    ADCR.7             ;DEFINE SADC AS ADC START BIT
36   MY_CODE   .SECTION AT 0 'CODE'           ;== PROGRAM SECTION ==
37        ORG       00H                       ;HT-46RXX RESET VECTOR
38        JMP       START
39        ORG       08H                       ;HT-46RXX TMR INTERRUPT VECTOR
40        JMP       TMR_INT
41   START:
42        CLR       LCD_CONTRC                ;CONFIG LCD CONTROL PORT AS O/P MODE
43        MOV       A,00000101B               ;ENABLE EMI & ETI
44        MOV       INTC0,A
45        MOV       A,10000010B               ;CONFIG TMR 0 IN MODE 2(TIMER MODE)
46        MOV       TMRC,A                    ;fINT=fSYS(4MHz)/4
47        MOV       A,HIGH (65536-SAMPLE_RATE)
48        MOV       TMRH,A                    ;SET INTERRUTP TIME CONSTANT
49        MOV       A,LOW (65536-SAMPLE_RATE)
50        MOV       TMRL,A
51        MOV       A,00000001B               ;SET A/D CONVERTER CLOCK SOURCE
52        MOV       ACSR,A                    ; = SYSTEM CLOCK/8
53        MOV       A,00001000B               ;SET PB0 AS A/D CHANNEL
54        MOV       ADCR,A                    ; AND SELECT A0 FOR ADC
55        MOV       A,38H                     ;FUNCTION SET: 8-BIT,2-LINE,5X10 DOTS
56        CALL      WLCMC
57        MOV       A,0CH                     ;ON/OFF CONTR: DISPLAY ON,CURSOR BLINKING OFF
58        CALL      WLCMC
59        MOV       A,06H                     ;ENTRY MODE:INCREMENT,DISPLAY NOT SHIFT
60        CALL      WLCMC
61        MOV       A,01H                     ;CLEAR DISPLAY
62        CALL      WLCMC
63        MOV       A,16                      ;LOAD 16 TIMES COUNTE
64        MOV       TIME_16_CNT,A
65        SET       SADC                      ;RESET ADC
66        CLR       SADC                      ;START ADC CONVERSION
67        CLR       TIME_16_FLAG              ;SET 16_TIME_FLAG
68        MOV       A,80H                     ;SET LCD LINE 1, POSITION 0
69        CALL      WLCMC
70        MOV       A,OFFSET STR1
71        CALL      PRINT
72   RE_START:
73        CLR       TEMPERATURE_L             ;CLEAR TEMPERATURE BUFFER
74        CLR       TEMPERATURE_H
75        CLR       TIME_16_FLAG              ;RESET 16_TIME_FLAG
76        SET       TON                       ;START TIMER
77   WAIT_16_TIME:
78        SNZ       TIME_16_FLAG              ;IS 16 TIMES CUMULATED?
79        JMP       WAIT_16_TIME              ;NO, WAIT
80        CLR       TON                       ;STOP CONVERSION
81        CALL      MUL_16X8                  ;CONVERT DIGITAL CODE TO TEMPERATURE
82        MOV       A,4                       ;CALCULATE AVARAGE TEMPERATURE DURING 1 SEC
83   AVERAGE:
84        CLR       C                         ;DIVIDED BY 16 IS EQUVALENT SHIFT RIGHT 4 TIMES
85        RRC       DIG2
86        RRC       DIG1
87        RRC       DIG0
88        SDZ       ACC                       ;IS SHIFT 4 TIMES ?
89        JMP       AVERAGE                   ;NO, NEXT SHIFTING
90        MOV       A,DIG0                    ;COPY RESULT TO TEMPURATURE BUFFER
```

```
 91         MOV      TEMPERATURE_L,A
 92         MOV      A,DIG1
 93         MOV      TEMPERATURE_H,A
 94         CALL     WORD_HEX2ASCII       ;CONVERT HEXDECIMAL TO ASCII CODE
 95         MOV      A,0C4H               ;SET LCD LINE 2, POSITION 4
 96         CALL     WLCMC
 97         CALL     DISPLAY_DIGIT        ;DISPLAY TEMPERATURE
 98         JMP      RE_START
 99 ;****************************************************************
100 ;              TIMER INTERRUPT SERVICE ROUTINE
101 ;****************************************************************
102 TMR_INT    PROC
103         MOV      STACK_A,A            ;PUSH A
104         MOV      A,STATUS
105         MOV      STACK_PSW,A          ;PUSH STATUS
106         CLR      TEMP_H               ;CLEAR TEMP_H
107         MOV      A,ADRH
108         MOV      TEMP_L,A             ;TEMP_L=ADC HI-BYTE
109         RLC      ADRL                 ;ROTATE LEFT TO GET BIT 1
110         RLCA     TEMP_L               ;SAVE DATA IN Acc
111         RLC      TEMP_H
112         ADDM     A,TEMPERATURE_L      ;CUMULATED THE TEMPEATURE
113         MOV      A,TEMP_H
114         ADCM     A,TEMPERATURE_H
115         SDZ      TIME_16_CNT          ;CUMULATED TEMPERATURE 16 TIMES?
116         JMP      TMR_INT_END          ;NO, RETURN
117         SET      TIME_16_FLAG         ;SET 16_TIME_FLAG
118         MOV      A,16                 ;RELOAD 16 TIMES COUNTER
119         MOV      TIME_16_CNT,A
120 TMR_INT_END:
121         SET      SADC                 ;RESET ADC
122         CLR      SADC                 ;START ADC CONVERSION
123         MOV      A,STACK_PSW
124         MOV      STATUS,A             ;POP STATUS
125         MOV      A,STACK_A            ;POP A
126         RETI
127 TMR_INT    ENDP
128 ;****************************************************************
129 ;              UNSIGN 16-BIT x 8-BIT SUBROUTINE
130 ;****************************************************************
131 MUL_16X8   PROC
132         CLR      DIG0                 ;CLEAR RESULT BUFFER LO
133         CLR      DIG1                 ;CLEAR RESULT BUFFER MID
134         CLR      DIG2                 ;CLEAR RESULT BUFFER HI
135         CLR      DIG4                 ;CLEAR TEMP BUFFER
136         MOV      A,ADC_LSB
137         MOV      DIG3,A               ;LOAD MULTIPLIER WITH ADC LSB VOLTAGE
138         MOV      A,8                  ;SET COUNTER
139         MOV      COUNT,A              ;SET DIG3 AS COUNTER
140 LOOP:
141         RRC      DIG3                 ;CHECK THE LSB OF MULITPILIER
142         SNZ      C                    ;SKIP IF BIT=1
143         JMP      C_EQU_0              ;JUMP IF BIT=0
144         MOV      A,TEMPERATURE_L      ;ADDITION
145         ADDM     A,DIG0               ;ADD TO RESULT BUFFER LO-BYTE
146         MOV      A,TEMPERATURE_H      ;ADDITION
147         ADCM     A,DIG1               ;ADD TO RESULT BUFFER MID-BYTE
148         MOV      A,DIG4
149         ADCM     A,DIG2               ;ADD TO RESULT BUFFER MID-BYTE
150 C_EQU_0:
151         RLC      TEMPERATURE_L        ;MULTIPLICANT SHIFT LEFT 1 BIT
152         RLC      TEMPERATURE_H
153         RLC      DIG4
154         SDZ      COUNT                ;SKIP IF COUNTER = 0
155         JMP      LOOP
156         RET
157 MUL_16X8   ENDP
```

```
158 ;****************************************************************
159 ;                    DISPLAY_DIGIT SUBROUTINE
160 ;****************************************************************
161 DISPLAY_DIGIT:
162         MOV         A,DIG4              ;DISPLAY DIGIT 4
163         CALL        WLCMD
164         MOV         A,DIG3              ;DISPLAY DIGIT 3
165         CALL        WLCMD
166         MOV         A,DIG2              ;DISPLAY DIGIT 2
167         CALL        WLCMD
168         MOV         A,2EH               ;'.'
169         CALL        WLCMD
170         MOV         A,DIG1              ;DISPLAY DIGIT 1
171         CALL        WLCMD
172         MOV         A,DIG0              ;DISPLAY DIGIT 0
173         CALL        WLCMD
174         MOV         A,0DFH              ;'°'
175         CALL        WLCMD
176         MOV         A,43H               ;'C'
177         CALL        WLCMD
178         RET
179 ;****************************************************************
180 ;           CONVERT 16-BIT HEX TO 5-DIGIT DECIMAL
181 ;   I/P: TEMPERATURE_H & TEMPERATURE_L   O/P: DIG4 ~ DIG0
182 ;****************************************************************
183 WORD_HEX2ASCII PROC
184         CLR         DIG4                ;INITIAL RESULT BUFFER
185         CLR         DIG3
186         CLR         DIG2
187         CLR         DIG1
188         CLR         DIG0
189 DIG_4:
190         MOV         A,TEMPERATURE_L     ;TEMPERATURE-10000
191         SUB         A,LOW 1000
192         MOV         TEMPERATURE_L,A
193         MOV         A,(NOT (HIGH 10000))+1
194         SNZ         C
195         DEC         ACC
196         ADDM        A,TEMPERATURE_H
197         SNZ         C                   ;TEMPERATURE-10000 > 0?
198         JMP         DIG_4_1             ;NO.
199         INC         DIG4                ;YES, DIG4=DIG4+1
200         JMP         DIG_4               ;DO AGAIN
201 DIG_4_1:
202         MOV         A,LOW 10000         ;RESTORE TEMPERATURE
203         ADDM        A,TEMPERATURE_
204         MOV         A,HIGH 10000
205         ADCM        A,TEMPERATURE_H
206 DIG_3:
207         MOV         A,TEMPERATURE_L     ;TEMPERATURE-1000
208         SUB         A,LOW 1000
209         MOV         TEMPERATURE_L,A
210         MOV         A,(NOT (HIGH 1000))+1
211         SNZ         C
212         DEC         ACC
213         ADDM        A,TEMPERATURE_H
214         SNZ         C                   ;TEMPERATURE-1000 > 0?
215         JMP         DIG_3_1             ;NO.
216         INC         DIG3                ;YES,DIG3=DIG3+1
217         JMP         DIG_3               ;DO AGAIN
218 DIG_3_1:
219         MOV         A,LOW 1000          ;RESTORE TEMPERATURE
220         ADDM        A,TEMPERATURE_L
221         MOV         A,HIGH 1000
222         ADCM        A,TEMPERATURE_H
223 DIG_2:
224         MOV         A,TEMPERATURE_L     ;TEMPERATURE-100
```

```
225         SUB     A,100
226         MOV     TEMPERATURE_L,A
227         MOV     A,TEMPERATURE_H
228         SNZ     C
229         SUB     A,1
230         MOV     TEMPERATURE_H,A
231         SNZ     C                     ;TEMPERATURE-100 > 0?
232         JMP     DIG_2_1               ;NO.
233         INC     DIG2                  ;YES, DIG2=DIG2+1
234         JMP     DIG_2                 ;DO AGAIN
235 DIG_2_1:
236         MOV     A,100                 ;RESTORE TEMPERATURE
237         ADDM    A,TEMPERATURE_L
238         CLR     ACC
239         ADCM    A,TEMPERATURE_H
240 DIG_1:
241         MOV     A,TEMPERATURE_L       ;TEMPERATURE-10
242 DIG_1_1:
243         SUB     A,10
244         SNZ     C                     ;TEMPERATURE-10 > 0?
245         JMP     DIG_0                 ;NO.
246         INC     DIG1                  ;YES, DIG1=DIG1+1
247         JMP     DIG_1_1               ;DO AGAIN
248 DIG_0:
249         ADD     A,10                  ;RESTORE TEMPERATURE
250         MOV     DIG0,A
251         MOV     A,30H                 ;BEFORE THIS LINE HEX TO UN-PACK BCD
252         ADDM    A,DIG4                ;CONVERT DIG4 TO ASCII
253         ADDM    A,DIG3                ;CONVERT DIG3 TO ASCII
254         ADDM    A,DIG2                ;CONVERT DIG2 TO ASCII
255         ADDM    A,DIG1                ;CONVERT DIG1 TO ASCII
256         ADDM    A,DIG0                ;CONVERT DIG0 TO ASCII
257         RET
258 WORD_HEX2ASCII ENDP
259 ;*************************************************************
260 ;       LCD PRINT PROCEDURE (START ADDRS=Acc)
261 ;*************************************************************
262 PRINT PROC
263         MOV     TBLP,A                ;LOAD TABLE POINTER
264 PRINT_1:
265         TABRDL  ACC                   ;READ STRING
266         XOR     A,STR_END
267         SZ      Z                     ;END OF STRING CHARACTER?
268         JMP     END_PRINT             ;YES, STOP PRINT
269         TABRDL  ACC                   ;NO, RE-READ
270         CALL    WLCMD                 ;SEND TO LCD
271         INC     TBLP                  ;INCREASE DATA POINTER
272         JMP     PRINT_1               ;NEXT CHARACTER
273 END_PRINT:
274         RET                           ;YES, RETURN
275 PRINT ENDP
276 ;*************************************************************
277 ;       LCD DATA/COMMAND WRITE PROCEDURE
278 ;*************************************************************
279 WLCMD PROC
280         SET     DC_FLAG               ;SET DC_FLAG=1 FOR DATA WRITE
281         JMP     WLCM
282 WLCMC:
283         CLR     DC_FLAG               ;SET DC_FLAG=0 FOR COMMAND WRITE
284 WLCM:
285         SET     LCD_DATAC             ;CONFIG LCD_DATA AS INPUT MODE
286         CLR     LCD_CONTR             ;CLEAR ALL LCD CONTROL SIGNAL
287         SET     LCD_RW                ;SET RW SIGNAL (READ)
288         NOP                           ;FOR TAS
289         SET     LCD_EN                ;SET EN HIGH
290         NOP                           ;FOR TDDR
291 WF:
292         SZ      LCD_READY             ;IS LCD BUSY?
```

```
293         JMP     WF                  ;YES, JUMP TO WAIT
294         CLR     LCD_DATAC           ;NO, CONFIG LCD_DATA AS OUTPUT MODE
295         MOV     LCD_DATA,A              ;LATCH DATA/COMMAND ON PB(LCD DATA BUS)
296         CLR     LCD_CONTR           ;CLEAR ALL LCD CONTROL SIGNAL
297         SZ      DC_FLAG             ;IS COMMAND WRITE?
298         SET     LCD_RS              ;NO, SET RS HIGH
299         SET     LCD_EN              ;SET EN HIGH
300         NOP
301         CLR     LCD_EN              ;SET EN LOW
302         RET
303  WLCMD  ENDP
304         ORG     LASTPAGE
305  STR1:  DC      'TEMPERATURE IS :',STR_END
306         END
```

程序说明

8~22   依序定义变量地址。

24   定义采样率为 $62\,500\,\mu s$。

25   定义 ADC 的分辨率为 $9.8\,mV(5\,V/2^9)$。

26~29 定义 LCD_DATA Port 与 LCD_CONTR Port 分别为 PA 与 PC。

30~33 定义 LCD 的控制信号引脚。

34   定义字符串结束码 STR_END=FEh。

38   声明存储器地址由 000h 开始(HT46xx Reset Vector)。

40   声明存储器地址由 008h 开始(HT46xx Timer/Event Counter Interrupt Vector)。

43   将 LCD_CONTR Port(PA)定义为输出模式。

44~45 EMI(Global Interrupt)与 ETI(Timer/Event Counter Interrupt)有效。

46~47 定义 Timer/Event Counter 的工作模式：Timer Mode、计数比例 1:4($f_{INT}=f_{SYS}/4$)。

48~51 将取样率载入 TMRH 与 TMRL 寄存器。

52~53 定义 ACSR 控制寄存器：ADC Clock=$f_{SYS}/8$。

54~55 定义 ADCR 控制寄存器：定义 PB0=A0，选择 A0 做 ADC 转换的信号来源。

56~57 将 LCD 设置为双行显示(N = 1)、使用 8 位(DB7 ~ DB0)控制模式(DL = 1)、5×7 点矩阵字形(F = 0)。

58~59 将 LCD 设置为显示所有数据(D = 1)、显示光标(C = 1)、光标所在位置的字不会闪烁(B = 0)。

60~61 将 LCD 的地址标志位(AC)设置为递加(I/D = 1)、显示器画面不因读/写数据而移动(S = 0)。

62~63 将 LCD 整个显示器清空。

64~65 设置 TIME_16_CNT=16。为求温度测量的稳定性及准确性，程序中是将连续测量 16 次的结果平均之后再予以显示；TIME_16_CNT 即记录是否已完成 16 次测量的计数器。

66~67 启动 ADC 开始转换。

68   清除代表已经完成 16 次测量的标志位 TIME_16_FLAG。

69~72 调用 PRINT 子程序，在 LCD 上显示 "TEMPERATURE IS:" 字符串。

74~77 清除相关寄存器，启动 Timer/Event counter 开始计数。

79~80 检查 TIME_16_FLAG 是否为 1，若不是，代表还未完成 16 次测量，继续等待。

81   Timer/Event counter 停止计数。

82   调用 MUL_16×8 子程序，将数字码转成温度值。

83~90 将温度值除以 16，求取平均值。因为右移一位等于除二的功能，因此将温度值连续右移 4 次，达到除以 16 的目的。

91~95　调用 WORD_HEX2ASCII 子程序将数值转换为 ASCII 码。

96~98　设置 DD RAM 地址，并调用 DISPLAY_DIGIT 子程序将数据显示在 LCD 上。

103~128　Timer/Event counter 中断服务子程序。第 107~115 行程序是取出 9 位(D9~D1)的转换结果，并累加于 TEMPERATURE_H(High Byte)与 TEMPERATURE_L(Low Byte)寄存器。第 116~117 行程序是判断 TIME_16_CNT-1 是否j为 0，若是代表已经测量了 16 次，接着再设置 TIME_16_FLAG=1、TIME_16_CNT=16；否则代表还未完成 16 次测量。第 122~123 行程序则是再次启动 ADC 转换。

132~158　MUL_16×8 子程序，请参考本实验的说明。

162~179　DISPLAY_DIGIT 子程序，记得在适当位置插入小数点，请参考本实验的说明。

184~259　WORD_HEX2ASCII 子程序，请参考本实验的说明。

263~276　PRINT 子程序，此子程序负责将定义好的字符串依序显示在 LCD 上。在调用此子程序之前，除了必须先设置好 LCD 的位置之外，还须先在 Acc 寄存器中指定字符串的起始地址(字符串必须存放在最末程序页 Last Page)，并请在字符串的最后一个字符插入 STR_END，代表字符串结束。

280~304　WLCMD 与 WLCMC 子程序，请参考 5-5 节的说明。

306　字符串定义区。

程序执行时，读者可以用烙铁在 AD590 的外壳上加热，验收本次小专题的成果。ADC0 的分辨率为 9.8 mV，而其输入的温度变化为 10 mV/℃，所以这个数字温度计的分辨率约在 0.9 ℃左右，至于准确度则取决于电路所使用的元件特性以及接线的阻抗。关于两级 OP-AMP 的调整及处理，虽然在实验一开始便提出了完整的理论分析以决定电路的各项输入参数，如 V_2 的输入电压、放大倍率等，但是读者仍须视实际的状况予以调整，以求获得更准确的温度测量结果。最好是在一个已知的标准温度环境中，来调校 V_2 的数值，方能获得较准确的测量结果。另外，在程序中利用 Timer/Event Counter 每隔 62.5 ms 采集一次温度转换的数值，连续采集 16 次(16×62.5 ms=1 s)之后再由主程序计算其平均值，其目的只是希望温度显示值不要跳动太快，而取其平均值将使显示的温度准确度更高。或许读者要问："求取平均值为何不需用到除法？"，这是因为在 Binary 系统中，右移一个位即是除二的特性，所以在程序中将累积 16 次的温度连续向右移位 4 次，就达到求取平均值的目的。

电路中的 TL084 是一般常用的运算放大器，内含 4 组 OP-AMP，但其截止与饱和的输出并无法达到 VSS 与 VDD。读者可以改用具有 "Rail to Rail" 特性的 IC(如盛群半导体公司的 HT82V732；详见随书光盘中的 Datasheet)，将测量的准确度再向上提升。

6-1-5　动动脑 + 动动手

- 如果将测量的温度范围限制在 0~100 ℃，请重新分析 V_2 的输入电压、U2B 放大倍率应调整为多少，才能充分发挥 HT46xx 的 ADC 分辨率特性。此时温度计的分辨率为多少？程序需要修改吗？(Hint：ADC_LSB)

- 本程序在显示温度时，会有 "前导零(Leading Zero)" 的显示，如 030 ℃、001 ℃，感觉上十分不自然，是不是可改成只显示 30 ℃、1 ℃呢？请修改 DISPLAY_DIGIT 子程序，改进此项缺点。

📖 如果将测量的温度范围限制在−10~100 ℃，请重新分析 V_2 的输入电压、U2B 放大倍率该调整为多少，才能充分发挥 HT46xx 的 ADC 分辨率特性。此时温度计的分辨率为多少？为便于显示，要如何修改程序才能分辨小于 0 ℃以下的温度呢？

📖 修改程序，使温度计显示的温度单位由摄氏(℃)改为华氏(℉)。

📖 如果测量的温度范围只为 0~50 ℃，可以考虑增大 R3 的阻值。请重新分析 V_2 的输入电压、U2B 放大倍率该调整为多少，才能充分发挥 HT46xx 的 ADC 分辨率特性。此时温度计的分辨率为多少？

📖 如果再配合上喇叭控制电路(见图 6-1-8)，可以很容易地将本专题程序改写成"温度警报系统"，请插入部分程序使温度超过 100 ℃时就发出警报声。

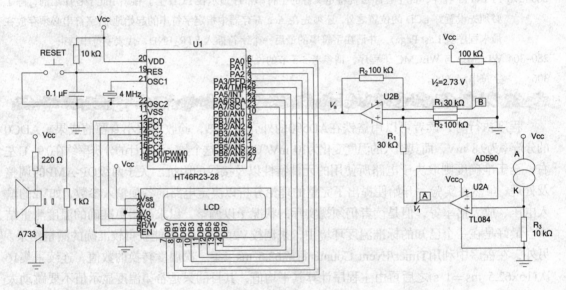

图6-1-8 温度警报系统

6-2 专题二：密码锁

6-2-1 专题功能概述

密码锁的功能和用途，想必不需多做介绍读者应该十分清楚。而本专题的特色是要求密码的长度为可变(4~10 Digits)，且当输入密码错误次数达4次以上时，系统将产生警告声。

6-2-2 学习重点

本专题的实现电路，可说是几个实验的综合，在硬件方面对读者而言应该是驾轻就熟，因此本专题的重点乃着重于程序技巧的磨练。

6-2-3 电路图（见图6-2-1）

第 6 章 实践应用篇

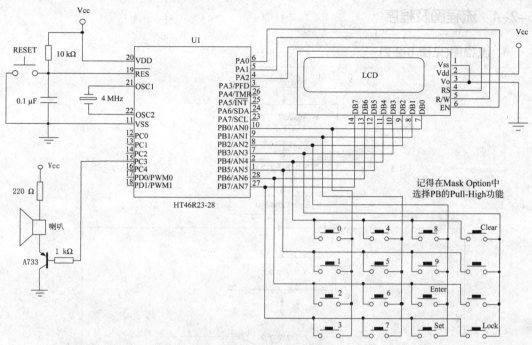

图6-2-1 密码锁控制电路

首先说明各按键所定义的功能：

0~9： 密码输入键，可输入4～9位数的密码。

Enter： 密码输入结束键，按下此键即决定了密码以及其长度。

Set： 更改密码键，按下此键表示要更改密码，但系统会要求用户先输入旧密码(Old Password)；输入正确后方能设置新密码(New Password)。(注：必须先进入开锁模式才能更改密码。)

Clear： 清除目前输入的密码，当用户发现按错输入的密码时，可按下此键清除，并重新输入密码。

Lock： 上锁键，当用户输入系统的密码正确时立即开锁(开锁模式)，此时用户可按下Lock键，将锁重新锁上。

密码锁的基本做法应该很容易，不外乎就是将用户输入的密码与系统密码加以比较，再判断是否相同而已。但是为了要让用户知道现在该输入什么，就必须要花费大半的功夫在人机接口的设计上，也就是说，该显示什么消息让用户一看就知道该按什么按键呢？这就是为什么程序会如此冗长的原因。

为方便读者阅读程序了解逻辑思考，故意将所有与LCD显示的相关部分以阴影方式表示，读者可以先不予以理会，而完全专注于程序的逻辑概念。配合流程图以及程序中的叙述，相信读者应该可以理解此程序的写法。

6-2-4 流程图及程序

1. 流程图(见图 6-2-2)

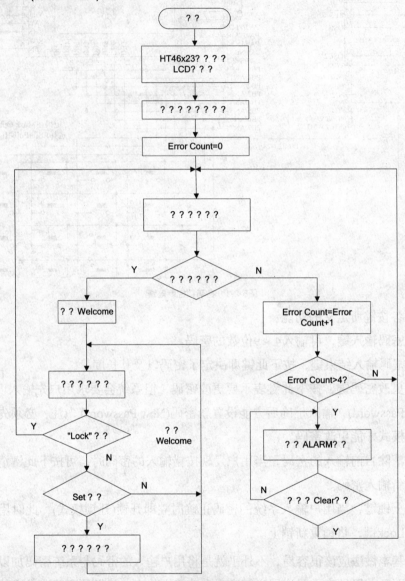

图6-2-2 流程图

2. 程序 6-2 密码锁

```
1    ;PROGRAM : 6-2.ASM  (6-2.PRJ)                              BY STEVEN
2    ;FUNCTION: DIGITAL CODE KEY DEMO PROGRAM                   2002.DEC.07.
3    #INCLUDE   HT46R23.INC
4              .CHIP    HT46R23
5    ;------------------------------------------------------------------
6    MY_DATA    .SECTION 'DATA'            ;== DATA SECTION ==
7    DEL1       DB    ?                    ;DELAY LOOP COUNT 1
8    DEL2       DB    ?                    ;DELAY LOOP COUNT 2
9    DEL3       DB    ?                    ;DELAY LOOP COUNT 3
10   DC_FLAG    DBIT                       ;LCD DATA/COMMAND FLAG
```

```
11      COUNT       DB      ?                       ;DISPLAY COUNT
12      PASS_WORD DB        10 DUP(?)               ;BUFFER FOR PASS WORD
13      KEY         DB      ?                       ;REGISTER FOR READ_KEY
14      KEY_PS      DB      ?                       ;BUFFER FOR READ_A_KEY
15      KEY_COUNT DB        ?                       ;ROW COUNTER USED IN READ_KEY
16      DIG_COUNT DB        ?                       ;COUNTER FOR INPUT DIGIT
17      PWD_COUNT DB        ?                       ;COUNTER FOR PASS WORD DIGIT
18      ERR_N       DB      ?                       ;ERROR RECORD COUNTER
19      STACK_A     DB      ?                       ;STACK BUFFER FOR ACC
20      STACK_PSW DB        ?                       ;STACK BUFFER FOR STATUS
21      KEY_BUFFERDB        10 DUP(?)               ;BUFFER FOR INPUT PASS WORD
22      ;----------------------------------------------------------    -----
23      LCD_CONTR EQU       PA                      ;DEFINE LCD CONTROL PORT
24      LCD_CONTRC  EQU     PAC                     ;DEFINE LCD CONTROL PORT CON. REG.
25      LCD_DATA EQU        PB                      ;DEFINE LCD DATA PORT
26      LCD_DATAC EQU       PBC                     ;DEFINE LCD DATA PORT CON. REG.
27      LCD_EN      EQU     LCD_CONTR.0             ;DEFINE EN CONTROL PIN
28      LCD_RW      EQU     LCD_CONTR.1             ;DEFINE RW CONTROL PIN
29      LCD_RS      EQU     LCD_CONTR.2             ;DEFINE RS CONTROL PIN
30      LCD_READY EQU       LCD_DATA.7              ;DEFINE READY BIT OF LCD
31      KEY_PORT    EQU     PB                      ;DEFINE KEY PORT
32      KEY_PORTC EQU       PBC                     ;DEFINE KEY PORT CONTROL REG.
33      END_CODE    EQU     0AH                     ;DEFINE PASS WORD END CODE
34      STR_END     EQU     0FEH                    ;STRING END CODE
35      SPK_PORT    EQU     PC                      ;BUZZER CONNECT PORT
36      SPK_PORTC EQU       PCC                     ;BUZZER CONNECT PORT CON. REG.
37      MY_CODE     .SECTION AT 0 'CODE'            ;== PROGRAM SECTION ==
38              ORG         00H                     ;HT-46RXX RESET VECTOR
39              JMP         START
40              ORG         08H                     ;HT-46RXX TMR INTERRUPT VECTOR
41      ;**********************************************************************
42      ;               TIMER INTERRUPT SERVICE ROUTINE
43      ;**********************************************************************
44      TMR_INT:
45              MOV         STACK_A,A               ;PUSH A
46              MOV         A,STATUS
47              MOV         STACK_PSW,A             ;PUSH STATUS
48              MOV         A,00001000B
49              XORM        A,SPK_PORT              ;COMPLEMENT
50              MOV         A,STACK_PSW
51              MOV         STATUS,A                ;POP STATUS
52              MOV         A,STACK_A               ;POP A
53              RETI
54      START:
55              CLR         LCD_CONTRC              ;CONFIG PA AS OUTPUT MODE
56              CLR         SPK_PORTC.3             ;CONFIG PC.0 AS OUTPUT MODE
57              MOV         A,00000101B
58              MOV         INTC0,A                 ;ENABLE EMI & ETI
59              MOV         A,38H                   ;FUNCTION SET: 8-BIT,2-LINE,5X10 DOTS
60              CALL        WLCMC
61              MOV         A,0FH                   ;ON/OFF CONTR: DISPLAY ON,CURSOR BLINKING ON
62              CALL        WLCMC
63              MOV         A,06H                   ;ENTRY MODE  : INCREMENT,DISPLAY NOT SHIFT
64              CALL        WLCMC
65              MOV         A,01H                   ;CLEAR DISPLAY
66              CALL        WLCMC
67              MOV         A,80H                   ;SET LCD LINE 1, POSITION 0
68              CALL        WLCMC
69              MOV         A,OFFSET PASS_WORD      ;INITIAL BUFFER ADDRESS
70              MOV         MP0,A
71              MOV         A,OFFSET DEFAULT_PWD    ;MOVE DEFAULT PASSWORD TO RAM BUFFER
72              MOV         TBLP,A                  ;INITIAL PASSWORD POINTER
73              CLR         PWD_COUNT               ;PWD_COUNT=0
74      READ_PWD:
75              TABRDL      IAR0                    ;READ DEFAULT PWD TO RAM BUFFER
76              MOV         A,END_CODE
77              XOR         A,IAR0
```

```
78        SZ      Z                       ;END_CODE READ?
79        JMP     AGAIN                   ;YES!
80        INC     MP0                     ;NO, INCREASE POINTER BY 1
81        INC     TBLP                    ;UPDATE TABLE POINTER
82        INC     PWD_COUNT               ;INCREASE PWD_COUNTER BY 1
83        JMP     READ_PWD                ;NEXT DEFAULT DIGIT
84 AGAIN:
85        CLR     ERR_N                   ;CLEAR ERROR COUNT
86 MAIN:
87        MOV     A,01                    ;CLEAR LCD
88        CALL    WLCMC                   ;PRINT "Password:(Err_N)" MESSAGE
89        MOV     A,OFFSET STR1           ;INITIAL STRING START ADDRS
90        CALL    PRINT
91        MOV     A,80H+13                ;SET LCD LINE 1 POSITION 13
92        CALL    WLCMC
93        MOV     A,ERR_N                 ;GET ERROR COUNTER
94        ADD     A,30H                   ;CONVERT TO ASCII
95        CALL    WLCMD                   ;DISPLAY ON LCD
96        CALL    BUFFER                  ;GET INPUT PASSWORD
97        CALL    CHECK                   ;CHECK PASSWORD FORMAT
98        XOR     A,00H
99        SZ      Z                       ;IS FORMAT CORRECT?
100       JMP     O_ERROR                 ;NO,FORMAT ERROR
101       MOV     A,OFFSET STR7           ;YES,PRINT "Pass"
102       CALL    PRINT
103       MOV     A,180                   ;DELAY 180ms
104       CALL    DELAY
105 OPEN:
106       MOV     A,01                    ;CLEAR LCD
107       CALL    WLCMC
108       MOV     A,OFFSET STR2           ;PRINT "Welcome (Open)"
109       CALL    PRINT
110       CALL    READ_KEY_PRESSED        ;WAIT KEY INPUT
111       MOV     A,KEY_PS
112       XOR     A,0BH
113       SZ      Z                       ;IS 'SET(B)' PRESSED
114       CALL    SET_PAS                 ;CALL IF "SET" KEY PRESSED
115       MOV     A,KEY_PS
116       XOR     A,0FH
117       SNZ     Z                       ;IS 'LOCK(F)' PRESSED
118       JMP     OPEN
119       JMP     AGAIN                   ;JUMP IF "LOCK" PRESSED
120 O_ERROR:
121       MOV     A,OFFSET STR3           ;PRINT "Error"
122       CALL    PRINT
123       MOV     A,100                   ;DELAY 100ms
124       CALL    DELAY
125       INC     ERR_N                   ;INCREASE ERROR COUNT BY 1
126       MOV     A,ERR_N
127       XOR     A,04
128       SNZ     Z                       ;IS ERROR NUMBER =4 ?
129       JMP     MAIN                    ;NO, JUMP MAIN
130       CALL    PHONE                   ;YES,CALL ALARM!!
131       JMP     AGAIN                   ;RE-START
132 ;****************************************************************
133 ;                  SET_PASS SUBROUTINE
134 ;****************************************************************
135 SET_PAS    PROC
136       MOV     A,01                    ;CLEAR LCD
137       CALL    WLCMC
138       MOV     A,OFFSET STR4           ;PRINT "O_Password:"
139       CALL    PRINT
140       MOV     A,0C0H                  ;SET LINE 2
141       CALL    WLCMC
142       CALL    BUFFER                  ;INPUT PASSWORD
143       CALL    CHECK                   ;CHECK PASSWORD FORMAT
144       XOR     A,00
```

```
145         SZ          Z                       ;IS FORMAT CORRECT?
146         JMP         S_ERROR                 ;NO,JUMP TO FORMAT ERROR
147         MOV         A,OFFSET STR7           ;YES, PRINT "Pass"
148         CALL        PRINT
149         MOV         A,100                   ;DELAY 100ms
150         CALL        DELAY
151         CALL        KEY_PAS                 ;MOVE NEW PASSWORD TO BUFFER
152         XOR         A,01
153         SNZ         Z                       ;IS FORMAT CORRECT?
154         JMP         S_ERROR                 ;NO, JUMP TO FORMAT ERROR
155         MOV         A,4FH                   ;YES, PRINT "OK"
156         CALL        WLCMD
157         MOV         A,4BH
158         CALL        WLCMD
159         MOV         A,200                   ;DELAY 200ms
160         CALL        DELAY
161         CALL        CHA_PAS                 ;MOVE NEW PWD TO PAS_BUF
162         RET
163 SET_PAS ENDP
164 ;*************************************************************************
165 ;                 S_ERROR SUBROUTINE
166 ;*************************************************************************
167 S-ERROR     PROC
168         MOV         A,OFFSET STR3           ;PRINT "Error"
169         CALL        PRINT
170         MOV         A,150                   ;DELAY 150ms
171         CALL        DELA
172         RET
173 S-ERROR ENDP
174 ;*************************************************************************
175 ;                   KEY_PASS SUBROUTINE
176 ;*************************************************************************
177 KEY_PAS     PROC
178         MOV         A,01                    ;CLEAR LCD
179         CALL        WLCMC
180         MOV         A,OFFSET STR5           ;PRINT "New Password:"
181         CALL        PRINT
182         MOV         A,0C0H                  ;SET LCD LINE 2
183         CALL        WLCMC
184         CALL        BUFFER                  ;GET PASSWORD
185         MOV         A,DIG_COUNT
186         SUB         A,10
187         SZ          C                       ;A <=10?
188         RET         A,0                     ;PASSWROD FORMAT ERROR
189         MOV         A,DIG_COUNT
190         SUB         A,4
191         SNZ         C                       ;A <=4?
192         RET         A,0                     ;PASSWROD FORMAT ERROR
193         RET         A,1                     ;PASSWROD FORMAT CORRECT
194 KEY_PAS ENDP
195 ;*************************************************************************
196 ;                 CHANGE PASSWORD SUBROUTINE
197 ;*************************************************************************
198 CHA_PAS     PROC
199         MOV         A,OFFSET KEY_BUFFER     ;SET DATA POINTER 0
200         MOV         MP0,A
201         MOV         A,OFFSET PASS_WORD      ;SET DATA POINTER 1
202         MOV         MP1,A
203         MOV         A,DIG_COUNT
204         MOV         PWD_COUNT,A             ;UPDATE PASSWORD DIGIT COUNT
205 CHA_PAS_1:
206         MOV         A,IAR0                  ;GET NEW PASSWORD DIGIT
207         MOV         IAR1,A                  ;UPDATE PASSWORD BUFFER
208         INC         MP0                     ;INCREASE DATA POINTER 0
209         INC         MP1                     ;INCREASE DATA POINTER 1
210         SDZ         DIG_COUNT               ;DIG_COUNT-1 = 0?
211         JMP         CHA_PAS_1               ;NO, NEXT DIGIT
```

```
212        RET                                 ;YES.
213 CHA_PAS  ENDP
214 ;************************************************************************
215 ;            PASSWORD CHECK SUBROUTINE
216 ;************************************************************************
217 CHECK PROC
218        MOV      A,PWD_COUNT              ;CHECK THE DIGIT NUMER
219        XOR      A,DIG_COUNT
220        SNZ      Z                        ;IS DIGIT NUMBER CORRRECT?
221        JMP      CHE_ERR                  ;NO, JUMP TO CHE_ERROR
222        MOV      A,OFFSET KEY_BUFFER      ;CHECK PASSWORD WITH KEY BUFFER
223        MOV      MP0,A
224        MOV      A,OFFSET PASS_WORD
225        MOV      MP1,A
226 CHE_PAS:
227        MOV      A,IAR0
228        XOR      A,IAR1
229        SNZ      Z                        ;IS THE DIGIT THE SAME?
230        JMP      CHE_ERR                  ;NO, JUMP TO CHE_ERROR
231        INC      MP0                      ;INCREASE DATA POINTER 0
232        INC      MP1                      ;INCREASE DATA POINTER 1
233        SDZ      DIG_COUNT                ;IS DIG_COUNT-1 = 0?
234        JMP      CHE_PAS                  ;NO, NEXT DIGIT
235 CHE_OK:
236        RET      A,1                      ;YES, RETURN WITH OK
237 CHE_ERR:
238        RET      A,0                      ;RETURN WITH ERROR
239 CHECK ENDP
240 ;************************************************************************
241 ;            GET PASSWORD TO BUFFER SUBROUTINE
242 ;************************************************************************
243 BUFFER    PROC
244        MOV      A,0C0H                   ;SET LCD LINE 2
245        CALL     WLCMC
246        MOV      A,16
247        MOV      DIG_COUNT,A
248        MOV      A,20H                    ;CLEAR LINE 2
249 LOOP:
250        CALL     WLCMD                    ;CLEAR LCD LINE 2
251        SDZ      DIG_COUNT
252        JMP      LOOP
253        MOV      A,0C0H                   ;SET LCD LINE 2, POSITION 0
254        CALL     WLCMC
255        MOV      A,OFFSET KEY_BUFFER      ;SET INPUT BUFFER
256        MOV      MP0,A
257        CLR      DIG_COUNT                ;SET GIGIT COUNT TO 0
258 SAV_BUF:
259        CALL     READ_KEY_PRESSED         ;READ UNTIL KEY PRESSED
260        MOV      A,KEY_PS
261        XOR      A,0CH
262        SZ       Z                        ;IS "CLEAR(C)" PRESSED
263        JMP      BUFFER                   ;YES, RESTART
264        MOV      A,0AH
265        XOR      A,KEY_PS
266        SZ       Z                        ;IS "ENTER(A)" KEY PRESSED?
267        JMP      END_BUF                  ;YES, RETUTN
268        MOV      A,KEY_PS
269        SUB      A,10                     ;A-10
270        SZ       C                        ;IS KEY CODE < 10?
271        JMP      SAV_BUF                  ;NO, SKIP UNDEFINED KEY
272        MOV      A,2AH
273        CALL     WLCMD                    ;PRINT "*" ON LCD
274        MOV      A,KEY_PS
275        MOV      R0,A                     ;SAVE PRESSED KEY CODE IN BUFFER
276        INC      MP0                      ;INCREASE DATA POINTER
277        INC      DIG_COUNT                ;INCREASE DIGIT COUNT
278        JMP      SAV_BUF                  ;NEXT KEY
```

```
279 END_BUF:
280         RET
281 BUFFER      ENDP
282 ;***********************************************************************
283 ;     READ KEYPAD AND WAIT UNTIL KEY IS PRESSED THEN RELEASED
284 ;***********************************************************************
285 READ_KEY_PRESSED    PROC
286         CALL    READ_KEY                ;SCAN KEYPAD
287         MOV     A,16
288         XOR     A,KEY
289         SZ      Z                       ;IS KEY BEEN PRESSED?
290         JMP     READ_KEY_PRESSED        ;NO, RE-SCAN KEYPAD
291         MOV     A,KEY
292         MOV     KEY_PS,A                ;RESERVED KEY CODE
293 WAIT_KEY_RELEASED:
294         CALL    READ_KEY                ;SCAN KEYPAD
295         MOV     A,16
296         XOR     A,KEY
297         SZ      ACC
298         JMP     WAIT_KEY_RELEASED       ;IS KEY BEEN RELEASED?
                                            ;NO, RE-SCAN KEYPAD
299         MOV     A,4
300         CALL    DELAY                   ;DELAY 40ms
301         RET
302 READ_KEY_PRESSED    ENDP
303 ;***********************************************************************
304 ;                    ALARM SUBROUTINE
305 ;***********************************************************************
306 PHONE PROC
307         MOV     A,10000000B             ;TIMER MODE, Fint=Fsys, TON=0
308         MOV     TMRC,A
309         MOV     A,0F4H
310         MOV     DIG_COUNT,A             ;TIMER CONSTANT
311         MOV     A,0A0H
312         MOV     PWD_COUNT,A             ;TIMER CONSTAND
313 PHONE_1:
314         MOV     A,01                    ;CLEAR LCD
315         CALL    WLCMC
316         MOV     A,OFFSET STR6           ;PRINT "Call 110......"
317         CALL    PRINT
318         MOV     A,8
319         MOV     COUNT,A
320 QWE:
321         MOV     A,2EH                   ;PRINT "."
322         CALL    WLCMD
323         MOV     A,DIG_COUNT
324         MOV     TMRH,A                  ;INITIAL TMRH
325         MOV     A,PWD_COUNT
326         MOV     TMRL,A                  ;INITIAL TMRH
327         SET     TON                     ;START TIMER
328         MOV     A,20                    ;DELAY 200ms
329         CALL    DELAY
330         CLR     TON                     ;STOP TIMER
331         MOV     A,07H
332         XORM    A,DIG_COUNT             ;CHANGE TIME CONSTANT
333         CALL    READ_KEY
334         MOV     A,0CH                   ;WAIT "C" CODE
335         XOR     A,KEY
336         SZ      Z                       ;IS "C" PRESSED?
337         JMP     STOP                    ;YES, JUMP TO STOP
338         SDZ     COUNT                   ;COUNT-1 = 0?
339         JMP     QWE                     ;NO!!
340         JMP     PHONE_1                 ;YES
341 STOP:
342         CALL    READ_KEY_PRESSED        ;WAIT KEY RELEASE
343         RET
344 PHONE ENDP
345 ;***********************************************************************
```

```
346 ;          LCD PRINT PROCEDURE (START ADDRS=TBLP)
347 ;**********************************************************************
348 PRINT PROC
349       MOV       TBLP,A                  ;LOAD TABLE POINTER
350 PRINT_1:
351       TABRDL    ACC                     ;READ STRING
352       XOR       A,STR_END
353       SZ        Z                       ;END OF STRING CHARACTER?
354       JMP       END_PRINT               ;YES, STOP PRINT
355       TABRDL    ACC                     ;NO, RE-READ
356       CALL      WLCMD                   ;SEND TO LCD
357       INC       TBLP                    ;INCREASE DATA POINTER
358       JMP       PRINT_1                 ;NEXT CHARACTER
359 END_PRINT:
360       RET                               ;YES, RETURN
361 PRINT ENDP
362 ;**********************************************************************
363 ;          LCD DATA/COMMAND WRITE PROCEDURE
364 ;**********************************************************************
365 WLCMD PROC
366       SET       DC_FLAG                 ;SET DC_FLAG=1 FOR DATA WRITE
367       JMP       WLCM
368 WLCMC:
369       CLR       DC_FLAG                 ;SET DC_FLAG=0 FOR COMMAND WRITE
370 WLCM:
371       SET       LCD_DATAC               ;CONFIG LCD_DATA AS INPUT MODE
372       CLR       LCD_CONTR               ;CLEAR ALL LCD CONTROL SIGNAL
373       SET       LCD_RW                  ;SET RW SIGNAL (READ)
374       NOP                               ;FOR TAS
375       SET       LCD_EN                  ;SET EN HIGH
376       NOP                               ;FOR TDDR
377 WF:
378       SZ        LCD_READY               ;IS LCD BUSY?
379       JMP       WF                      ;YES, JUMP TO WAIT
380       CLR       LCD_DATAC               ;NO, CONFIG LCD_DATA AS OUTPUT MODE
381       MOV       LCD_DATA,A              ;LATCH DATA/COMMAND ON PB(LCD DATA BUS)
382       CLR       LCD_CONTR               ;CLEAR ALL LCD CONTROL SIGNAL
383       SZ        DC_FLAG                 ;IS COMMAND WRITE?
384       SET       LCD_RS                  ;NO, SET RS HIGH
385       SET       LCD_EN                  ;SET EN HIGH
386       NOP
387       CLR       LCD_EN                  ;SET EN LOW
388       RET
389 WLCMD ENDP
390 ;**********************************************************************
391 ; SCAN 4x4 MATRIX ON KEY PORT AND RETURN THE CODE IN KEY REGISTER
392 ; IF NO KEY BEEN PRESSED, KEY=16.
393 ;**********************************************************************
394 READ_KEY  PROC
395       MOV       A,11110000B
396       MOV       KEY_PORTC,A             ;CONFIG KEY_PORT
397       SET       KEY_PORT                ;INITIAL KEY PORT
398       CLR       KEY                     ;INITIAL KEY REGISTER
399       MOV       A,04
400       MOV       KEY_COUNT,A             ;SET ROW COUNTER
401       CLR       C                       ;CLEAR CARRY FLAG
402 SCAN_KEY:
403       RLC       KEY_PORT                ;ROTATE SCANNING Bit
404       SET       C                       ;MAKE SURE C=1
405       SNZ       KEY_PORT.4              ;COLUMN 0 PRESSED?
406       JMP       END_KEY                 ;YES.
407       INC       KEY                     ;NO, INCREASE KEY CODE.
408       SNZ       KEY_PORT.5              ;COLUMN 1 PRESSED?
409       JMP       END_KEY                 ;YES.
410       INC       KEY                     ;NO, INCREASE KEY CODE.
411       SNZ       KEY_PORT.6              ;COLUMN 2 PRESSED?
412       JMP       END_KEY                 ;YES.
```

```
413         INC     KEY                     ;NO, INCREASE KEY CODE.
414         SNZ     KEY_PORT.7              ;COLUMN 3 PRESSED?
415         JMP     END_KEY                 ;YES.
416         INC     KEY                     ;NO, INCREASE KEY CODE.
417         SDZ     KEY_COUNT               ;HAVE ALL ROWs BEEN CHECKED?
418         JMP     SCAN_KEY                ;NO, NEXT ROW.
419 END_KEY:
420         RET
421 READ_KEY  ENDP
422 ;***********************************************************************
423 ;                    Delay about Acc*10ms
424 ;***********************************************************************
425 DELAY    PROC
426         MOV     DEL1,A                  ;SET DEL1 COUNTER
427 DEL_1:
428         MOV     A,30
429         MOV     DEL2,A                  ;SET DEL2 COUNTER
430 DEL_2:
431         MOV     A,110
432         MOV     DEL3,A                  ;SET DEL3 COUNTER
433 DEL_3:
434         SDZ     DEL3                    ;DEL3 DOWN COUNT
435         JMP     DEL_3
436         SDZ     DEL2                    ;DEL2 DOWN COUNT
437         JMP     DEL_2
438         SDZ     DEL1                    ;DEL1 DOWN COUNT
439         JMP     DEL_1
440         RET
441 DELAY ENDP
442         ORG     LASTPAGE
443 DEFAULT_PWD:                             ;DEFAULT SYSTEM PASSWORD
444         DC      0,1,2,3                 ;0~9 AND MAXIMUM 10 DIGITS
445         DC      END_CODE                ;NOTE: MUST END WITH END_CODE
446 STR1: DC   'Password(ERR_ ):',STR_END   ;STRING DEFINE 1
447 STR2: DC   'Welcome (Open)',STR_END     ;STRING DEFINE 2
448 STR3: DC   ' Err!!',STR_END             ;STRING DEFINE 3
449 STR4: DC   'Old Password:',STR_END      ;STRING DEFINE 4
450 STR5: DC   'New Password:',STR_END      ;STRING DEFINE 5
451 STR6: DC   'Call 110',STR_END           ;STRING DEFINE 6
452 STR7: DC   ' pass !!',STR_END           ;STRING DEFINE 7
453 END
```

程序说明

- 7~21　依序定义变量地址。
- 23~26　定义 LCD_DATA Port 与 LCD_CONTR Port 分别为 PB 与 PA。
- 27~30　定义 LCD 的控制信号引脚。
- 31~32　定义 KEY_PORT 为 PB。
- 33~34　定义常量 END_CODE 与 STR_END 分别为 0Ah、FEh；其中 END_CODE 作为系统自设密码结束之用，STR_END 则代表字符串的结束字符。
- 35~36　定义 SPK_PORT 为 PC。
- 39　声明存储器地址由 000h 开始。
- 41　声明存储器地址由 008h 开始。
- 45~54　TMR_INT 子程序，此段子程序主要是将 SPK_PORT.3 反向一次。结合 Timer/Event Counter 的计数溢位时间，可产生不同频率的方波输出。由于 SPK_PORT.3 用来触发喇叭，因此可以造成不同的输出声音。
- 56~57　定义 LCD_CONTR Port 与 SPK_PORT 为输出模式。
- 58~59　Global Interrupt 与 Timer/Event Counter Interrupt 有效。
- 60~61　将 LCD 设置为双行显示(N = 1)、使用 8 位(DB7 ~ DB0)控制模式(DL = 1)、5×7 点矩阵字形(F = 0)。

行号	说明
62~63	将 LCD 设置为显示所有数据(D = 1)、显示光标(C = 1)、光标所在位置的字会闪烁(B = 1)。
64~65	将 LCD 的地址标志位(AC)设置为递加(I/D = 1)、显示器画面不因读/写数据而移动(S = 0)。
66~67	将 LCD 整个显示器清空。
68~69	将光标移到第 1 行的第 1 个位置。
70~84	运用间接寻址法,将系统原始密码(Default Password)传送至系统密码区(由地址 PASS_WORD 开始)。
86	设置 ERR=0,ERR 是记录密码输入错误次数的寄存器。
88~96	显示 Password:(Err_n)字符串(n 代表输入错误次数)。
97	调用 BUFFER 子程序,读取用户输入的密码并将其显示在 LCD 第 2 行。
98	调用 CHECK 子程序,检查用户输入的密码是否正确。若正确则返回 Acc=1,否则 Acc=0。
99~101	密码若正确则继续执行,否则跳至 O_ERROR。
102~105	密码正确,显示 Pass 字符串并调用 DELAY 子程序延时 1.8 s。
107~110	清除 LCD 后,显示 Welcome(Open)字符串,Open 表示锁已打开(此即所谓开门模式)。
111~120	调用 READ_KEY_PRESSED 子程序,读取用户输入的按键。若为 Set 键,则进入重新设置密码的模式(SET_PAS 子程序);若为 Lock 键,则跳出开门模式。
121~125	密码不正确,显示 Error 字符串。
126~132	错误次数累加一次(ERR=ERR+1)。检查错误次数是否大于 4 次,若是则调用 PHONE 子程序,启动警报系统;否则重新要求输入密码。
136~164	SET_PASS 子程序,此程序会先调用 BUFFER 子程序,要求用户输入旧密码(Old_Password)。若密码正确(调用 CHECK 子程序检查),会再要求用户输入新密码(New_Password,调用 KEY_PAS 子程序),并将新密码传送至系统密码区(由地址 PASS_WORD 开始)。
168~174	S_ERROR 子程序,显示 ERROR 字符串并调用 DELAY 子程序延时 1.5 s。
178~195	KEY_PAS 子程序,此程序会调用 BUFFER 子程序读回用户所输入的密码,然后检查其位数(Digit)是否介于 4~10 之间。若位数超过此范围,则返回 0,表示错误;否则返回 1。
199~214	CHA_PAS 子程序,运用间接寻址法将新输入的密码由 KEY_BUFFER 传送到系统密码区(由地址 PASS_WORD 开始)。
218~240	CHECK 子程序,运用间接寻址法检查用户输入的密码(KEY_BUFFER)与系统密码(PASS_WORD)是否相同。如果相同则返回 1;否则返回 0,表示密码错误。
244~282	BUFFER 子程序,负责读取用户输入的密码(由地址 KEY_BUFFER 开始存放),并返回密码的长度(DIG_COUNT)。若输入 Clear 键,则清除之前所有输入的按键值并将 DIG_COUNT 归零;若输入 Enter 键,表示密码输入结束。另外,若输入的按键值不是 0~9,将不予理会。
286~303	READ_KEY_PRESSED 子程序,此子程序会调用 READ_KEY 子程序读取按键值,但必须确定用户按下按键,且等到按键放开后才会跳离此子程序。此举的主要目的,是避免 LCD 受 4×4 Key Pad 的干扰(因为 KEY_PORT 与 LCD_DATA Port 是共用的)。
307~344	PHONE 子程序,显示 "Call 110 ……"字符串并发出警报声。待按下 Clear 键时,重新启动系统。PHONE 子程序会启动 Timer/Event Counter 开始计数,通过 TMRH、TMRL 计数值的改变结合 Timer/Event Counter 中断服务子程序产生音调的变化。
349~362	PRINT 子程序,此子程序负责将定义好的字符串依序显示在 LCD 上。在调用此子程序之前,除了必须先设置好 LCD 的位置之外,还须先在 Acc 寄存器中指定字符串的起始地址(字符串必须存放在最末程序页 Last Page),并请在字符串的最后一个字符放入 STR_END,代表字符串结束。
366~390	WLCMD 与 WLCMC 子程序,请参考 5-5 节中的说明。
395~422	READ_KEY 子程序,请参考 4-7 节中的说明。
426~439	DELAY 子程序,延迟时间的计算请参考实验 4-1 中的说明。。
442~443	系统原始密码存放区,请记得以 END_CODE 代表密码结束。
444~450	字符串定义区,请注意每一个字符串必须以 STR_END 作为结束。

程序一开始执行时，LCD会显现如下的画面(此称为关门模式)：

P	a	s	s	w	o	r	d	:	(	E	r	r	_	0	)

要求用户输入系统的缺省密码(Default Password；目前设为0123)，(Err_0)表示输入密码错误的次数，输入正确的密码之后，会显现如下的画面：

W	e	l	c	o	m	e		(	O	p	e	n	)		

此画面表示锁已打开(此即所谓开门模式)。此时用户可以按下Lock键，将锁再锁上；或者按下Set键，重新输入密码。显现的画面如下：

O	_	P	a	s	s	w	o	r	d	:					
*	*	*	*	*	P	a	s	s							

要求用户先输入旧的密码，待输入正确的密码后，再要求用户输入新的密码。显现的画面如下：

N	_	P	a	s	s	w	o	r	d	:					

若发现输入的密码错误或想修改密码，只要在按下Enter键之前按Clear键即可重新输入。本系统预定输入的密码长度在4～10位数间，若不在此范围内则视为密码"格式"错误。LCD上会显示以下的消息：

N	_	P	a	s	s	w	o	r	d	:					
*	*	*	*	E	r	r	o	r							

并回到上述的开门模式。

当在关门模式下要求输入密码时，若用户输入的密码内容错误次数达4次以上，系统会启动警报系统(Buzzer发出声音)，并显示以下的画面：

C	a	l	l		1	1	0	.	.	.	.	.	.	.	.

此时唯有按下Clear键使系统回到关门模式，重新要求输入密码；否则警报器将持续发声。虽然显示"Call 110"的消息在此只是个噱头，但在实践运用上却确定可行，读者可以配合电话控制专用IC加以实现。

6-2-5 动动脑 + 动动手

  📖 在 BUFFER 子程序中，并未将输入的密码长度予以限制，万一用户一直输入密码而不按下 Enter 键，可能会导致系统动作不正常。请在 BUFFER 子程序中加上限制，当用户输入的密码超过 20 个位数就跳出此子程序。

6-3 专题三：具记忆功能的密码锁(I²C E²PROM)

6-3-1 专题功能概述

前一个密码锁专题中每次关机之后，必须以预设密码进入系统。本专题利用HT46x23配合I²C E²PROM(24Cxx)以及4×4键盘、LCD等控制电路，制作一个断电之后仍可记忆密码的密码锁。

6-3-2 学习重点

通过本专题，读者应充分了解I²C接口的控制方式。

6-3-3 电路图（见图6-3-1）

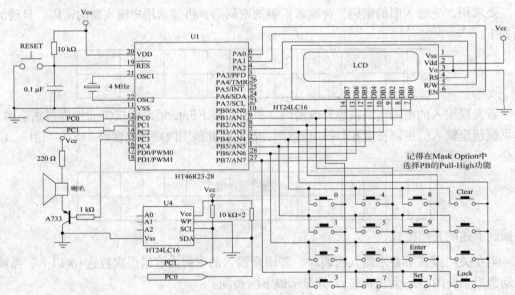

图6-3-1 具记忆功能的密码锁控制电路

I²C(Inter IC)又可称为I²C-Bus，是由飞利浦(Philips)公司研发出来的串行存取方式。目前市面上以串行存取电气擦除式可编程只读存储器(Serial Access E²PROM)，常见的控制接口有以下4种方式：I²C-Bus、XI²C- Bus、SPI- Bus、MicroWire- Bus，特将其特性简述如下：

➢ **I²C-Bus**：为飞利浦公司所设计的串行总线接口，当初是针对消费应用产品所设计的标准界面，只需两条控制线就可达成数据传输的目的。除了存储器设备之外，目前许多设备都配备了I²C-Bus 的传输接口。I²C-Bus 的寻址能力只有 16 Kbit，总线的传输速度最高只到 100 kHz。而 XI²C-Bus(Extended I²C-Bus)将寻址能力提高到了 4 Mbit，总线传输速度也上拉到 400 kHz，新的版本更已经上拉到 3.4 MHz。

- **SPI-Bus**：Serial Port Interface Bus，最早是由 Motorola 公司所提出的三线式(Data Out、Data In、Clock)串行总线传输接口，但后来是由 ST 以及其他公司将其集成到单片机内部，作为数据传输的独立接口单元，总线传输速度上限为 5 MHz。
- **MicroWire-Bus**：由 National Semiconductor 所开发的串行总线传输接口，如同 SPI-Bus 也是以三条控制线(Data OUT、Data In、Clock)完成数据传输的动作，其总线传输速度上限为 1 MHz。

这几种串行传输接口主要差异在于总线尺寸、总线通信协协议(Bus Protocol)、噪音抗扰性(Noise Immunity)以及存取时间。其中又以I^2C(Inter IC)的应用最为广泛，因此特别将I^2C E^2PROM运用在此次的专题电路中，也通过此机会让读者认识I^2C接口的控制技巧。串行数据存取比并行数据存取的传输数据速度慢，但是因为所使用的I/O引脚数极少，因此当传输速度要求不是太快而I/O资源又弥足珍贵时(如小型的单片机)，采用此类接口可说是最正确的选择。本专题就以具有I^2C传输接口的E^2PROM——HT24LCxx(请参考表6-3-1，盛群半导体公司I^2C E^2PROM产品型号)为例，说明I^2C接口的控制方式。

表6-3-1 盛群半导体公司I^2C E^2PROM产品型号

Part No.	Clock Rate /kHz	Write Speed @2.4V /ms	Operating Current @5V/mA	Standby Current @5V /μA	Capacity
HT24LC02	400	5	5	5	256×8
HT24LC04	400	5	5	5	512×8
HT24LC08	400	5	5	5	1024×8
HT24LC16	400	5	5	5	2048×8
HT24LC32	400	5	5	5	4096×8
HT24LC64	400	5	5	5	8192×8
HT24LC128	400	5	5	5	16384×8
HT24LC256	400	5	5	5	32768×8

图6-3-2为HT24LCxx的引脚及功能概述，完整的电气特性还是请读者参考光盘中的IC数据手册。由引脚的特性来看，所有的数据控制及读/写动作，都是通过SDA、SCL引脚来完成，因此有关学习I^2C接口的传输方式，其实等于是在研究这两个信号间的关系。不过首先要注意，I^2C接口SDA、SCL两个引脚必须接上拉电阻，如图6-3-3所示，其引脚功能如表6-3-2所列。

表6-3-2 HT24LCxx引脚功能列表

引脚名称	功能	概述
A0 A1 A2	Chip Enable	当增加存储器时的控制信号，也就是说利用此3支引脚的信号号码可以同时并联多个I^2C存储器。本专题使用 HT24LC16(2k×8)，已达I^2C之最大容许范围，因此无须使用这3支有效脚位
SDA	Serial Data	此引脚是用来传输连续的数据，但是要注意此引脚他是属于双向传输的方式。当在执行写入数据时，若接收完一笔资料时此引脚将会输出一个 Acknowledge 信号表示已接收到数据。当在执行读取数据时，若送出完一笔数据，要由接收数据端送出一个 Acknowledge 已接收到的信号
SCL	Serial Clock	串行数据传输时钟脉冲输入引脚。
WP	Write Protect	引脚若接为高准位，此时 IC 将禁止写入数据的动作，但是可以读取数据；若引脚为低准位时，则可以进行写入及读取数据的动作
V_{CC}	Supply Voltage	工作电压输入端，工作电压 2.2~5.5V
V_{SS}	Ground	接地

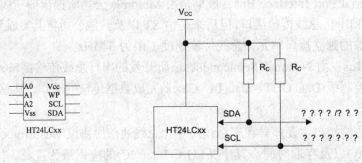

图6-3-2 HT24LCxx的引脚图　　图6-3-3 SDA、SCL两个引脚的上拉电阻

接下来就需要了解SDA、SCL的控制时序，I^2C Bus的任何读/写控制动作，都必须以"START Condition"与"STOP Condition"作为起始与结束，因为只有使用SCL、SDA两根信号线，所以就以其电位的高低状态来区分，请参考图6-3-4。

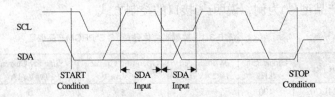

图6-3-4 START与STOP Condition

也就是说：当SCL=1时，如果SDA由1→0，则为START Condition，I^2C设备将准备进行数据的读/写动作；而当SCL=1时，如果SDA由0→1，则为STOP Condition，表示结束数据的读/写动作。因此在I^2C Bus的传输过程中，数据的改变一定要在SCL=0的状态时进行，否则很容易被I^2C设备误认为是START或STOP Condition。这样可能不正常地终止目前的传输，而导致存取的数据错误，读者不得不小心。在程序中分别以I2C_START与I2C_STOP子程序来产生这两种状态，特将其一并列出，请读者结合上述的时序图来分析，其中的DELAY_10子程序大约延时10个指令周期，主要是为了让此子程序可以适用于更高的工作频率(例如当HT46xx工作在8 MHz时)。

```
1   ;************************************************************
2   ;           GENERATE I2C START CONDITION
3   ;************************************************************
4   I2C_START  PROC
5              CLR     SCLC            ;CONFIG SLC AS OUTPUT MODE
6              CLR     SDAC            ;CONFIG SDA AS OUTPUT MODE
7              CLR     SCL             ;SET SCL=0
8              CLR     SDA             ;SET SDA=0
9              CALL    DELAY_10        ;DELAY
10             SET     SCL             ;SET SCL=1
11             SET     SDA             ;SET SDA=1
12             CALL    DELAY_10        ;DELAY
13             CLR     SDA             ;SET SDA=0
14             CALL    DELAY_10        ;DELAY
15             CLR     SCL             ;SET SCL=0
16             CALL    DELAY_10        ;DELAY
17             RET
18  I2C_START  ENDP
19  ;************************************************************
20  ;           GENERATE I2C STOP CONDITION
21  ;************************************************************
22  I2C_STOP   PROC
```

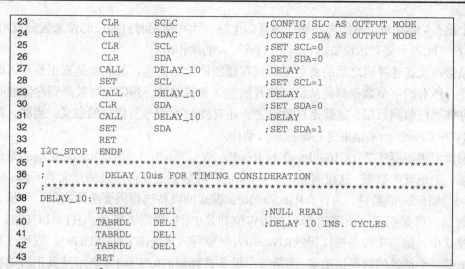

接下来就是如何让I²C设备分辨读或写的动作了，请参考图6-3-5的时序。

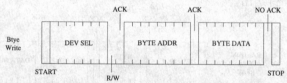

图6-3-5 HT24LCxxByte Write模式

图6-3-5是HT24LCxx Byte Write模式的示意图，在送出START状态之后，接着要送设备选择码(DEV SEL)，请参考表6-3-3。

表6-3-3 HT24LCxx设备选择码

Part No.	Device Type ID				Chip Enable			R/W
	b7	b6	b5	b4	b3	b2	b1	b0
HT24LC02	1	0	1	0	E2	E1	E0	R/W
HT24LC04	1	0	1	0	E2	E1	A8	R/W
HT24LC08	1	0	1	0	E2	A9	A8	R/W
HT24LC16	1	0	1	0	A10	A9	A8	R/W

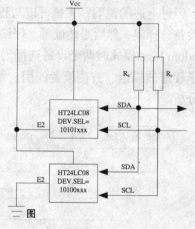

图6-3-6 Device Select Code图例

I2C Bus对不同的设备定义了不同的辨识码(Device Type ID)，目前大约有150几种不同的I2C设备，每一种设备都定义了不同的设备辨识码，而存储器设备的辨识码为1010。b3~b1为有效控制引脚，当送出的有效信号与引脚上所设置状态相同时，该设备才会动作。因为I2C Bus所能接受的最大存储容量为2 KB，而本书所使用的HT24LC16恰为2 KB的存储容量，因此A2、A1、A0硬件引脚实际上是没有用的，而此时的b3~b1就代表A10~A8的地址选择信号。如果在电路中使用两个HT24LC08(每个存储容量为1 KB)，此时就可以将其中一个A2的接至VCC，

331

另一个接至VSS即可，不需再外加任何译码电路，而在使用时只要以b3位来区分就可以了(见图6-3-6)。b0表示要对该设备进行读出(1)或写入的动作(0)。

紧接在设备选择码之后的就是要写到存储器的地址信息，其次就是真正要写入的数据。请注意，所有的字节数据都是从最高位开始发送和接收的，当CPU写数据到存储器时，必须确认存储器已收到数据，这就是在数据之外还有些ACK信号所代表的意义。当然，最后别忘记以STOP Condition来结束这一次的写入动作。

请读者再观察HT24LC16的Random Read模式(见图6-3-7)，所谓Random Read是指可以读取任意一个地址的数据，这跟刚刚Byte Write可以写数据到任何地址的意义差不多，但请比较图6-3-5与图6-3-7的差异，为什么Random Read模式的设备选择码要送两次呢？为什么明明要读出数据，可是第一个设备选择码的R/W位却是0呢？请读者留意，HT24LC16的 Random Read模式中，第一个设备选择码的R/W＝0表示要写入地址(BYTE ADDR)，以便让存储器知道该把哪一个地址的数据送出来，而第二个设备选择码的R/W＝1才表示是真正启动存储器将指定地址内的数据串行送出。

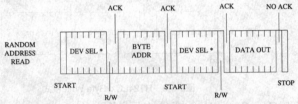

图6-3-7　HT24LC16 Random Read模式

经过上述的说明，大概可以写程序来控制M24C16。首先在START Condition之后必须送出设备选择码，然后是地址等。仔细观察图6-3-5与图6-3-7，前面两组送的消息都一样，而在Random Read模式时两个设备选择码也只有最后一个位(R/W)不相同而已。因此在此想把它写成子程序的模组，方便读者使用。请参考图6-3-8所示的流程图，并结合图6-3-5与图6-3-7进行分析。

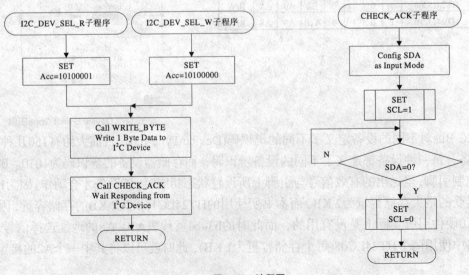

图6-3-8　流程图

当一笔数据由HT46x23写到HT24LC16之后,必须确定对方已经收到数据才能再继续做下一步的工作,因此当最后一个位写完之时,必须等待HT24LC16送出SDA=0的信号,即所谓的ACK(Acknowledge)回应信号(请参考CHECK_ACK子程序的流程图)。

不管是写入Device Select Code、设置地址,乃至于存入数据,都必须在SCL时钟脉冲的控制之下,将数据通过SDA信号线串行送出,因此就将其写成WRITE_BYTE子程序,以方便读者使用,请参考如图6-3-9所示的流程图。

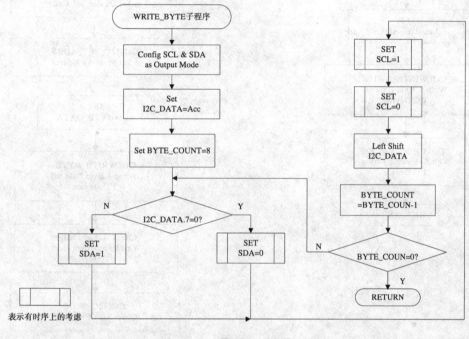

图6-3-9 流程图

在HT24LC16数据手册中,对于SCL与SDA变化之间有一定的时间需求(即时序关系),在程序中是以DELAY10子程序来完成,但是为了避免使流程图看起来过于繁杂,在图中并未特别表示出来,请读者留意!接下来就是送出地址,让HT24LC16知道该针对哪一个存储位置做动作,其实这与上述的流程大同小异。

最后就是数据读或写的动作,写数据的动作和写入地址是一样的,读者可以把I2C_SET_ADRS子程序中的地址(I2C_ADRS)改成数据即可。不过在写入数据之前,是不是要让HT24LC16知道该把数据放到哪一个位置呢?所以必须由HT46x23先把地址信息送给它,读者应该了解其流程图所代表的意义了吧。

HT24LC16提供4种不同的读取模式:随机读取(Random Read)、连续随机读取(Sequential Random Read)、目前地址读取(Current Address Read)与连续目前地址读取(Sequential Current Read)。而不论是哪一种读取模式,都必须在SCL时钟脉冲的控制之下,将数据通过SDA信号线串行读入,因此就将其写成READ_BYTE子程序,以方便读者使用,请参考图6-3-10所示的流程图。

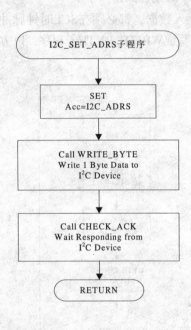 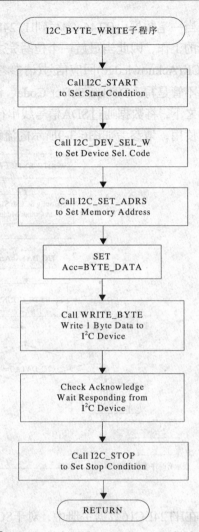

图6-3-10 流程图

其实数据也只是以串行方式,配合SCL信号一个位接一个位循序地读入DATA寄存器中。不过请读者注意,此时HT46x23为数据的接收者,所以在收完8个位的数据之后,必须在第9个位的时间由SDA送出1(NO ACK)或0(ACK)的信号,以通知HT24LC16是否需要继续送出下一个地址的数据。这就是上述的4种模式中,非连续读取与连续读取的最大差异。以I2C_RM_READ子程序为例,它是负责由指定的地址(I2C_ADRS)读回一个字节数据的子程序,其流程图如图6-3-11所示。

由于只读取一个字节,所以在读完8位的数据之后,由HT46x23在第9位送出ACK＝1,来通知HT24LC16不需继续再送出下一个地址的数据。上图中也一并把I2C_CT_READ(Current Address Read)子程序的流程图列出,其实这两个子程序的差异只在于I2C_CT_READ子程序是负责将目前地址计数器(在HT24LCxx内部)所指示的地址数据读出,所以不需要再重新设置地址。

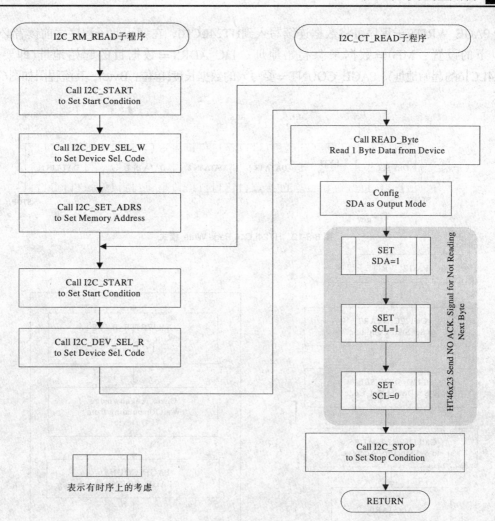

图6-3-11 流程图

通过上述的过程可发现，每次的读/写动作都要送出地址通知HT24LC16，不是既浪费时间又麻烦吗？其实在HT24LCxx内部有一个地址计数器(Internal Address Counter；AC)和LCD有些类似，每执行一次读/写的动作时，AC就会自动加1。所以如果是读/写连续的存储器地址，其实是可以不用每次都重新设置AC值的。这也是读者在HT24LC16数据手册中发现可以有好几种不同读/写模式的原因，请参考以下的说明。

如图6-3-12所示，所谓的Page Write就是在通知HT24LC16地址之后连续写N个字节的数据，而且只需在最后一笔数据写完之后，再以STOP Condition结束Page Write的动作即可。这样可以省去每次都送地址的时间，不过要注意的是N不得大于16。由于E^2PROM的写入速度较慢，所以在HT24LC16内部有一块存取速度较快的RAM Buffer，外界写入的数据其实都是先暂时存放在RAM Buffer内，待进入STOP Condition后，HT24LC16再把RAM Buffer内的数据复制到E^2PROM上。这样的作法是为了提升与外部传输的速度，但是因为内部的RAM Buffer只有16字节，所以外界一次最多只能连续写入16字节的数据就必须进入STOP Condition，否则将会造成数据错误。而E^2PROM的读取速度与RAM相当，所以在连续读取时就无此限制。

I2C_PAGE_WRITE子程序可将数据连续写入到HT24LC16，在调用此子程序之前读者必须先做以下的设置：MP0＝数据来源起始地址、I2C_ADRS＝数据目的起始地址(即要写入HT24LC16的起始地址)、PAGE_COUNT＝要写入的数据长度(单位：Byte)，其流程图如图6-3-13所示。

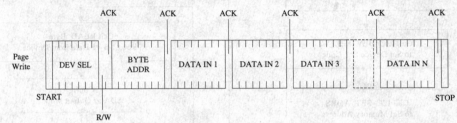

图 6-3-12　HT24LC16 Page Write 模式

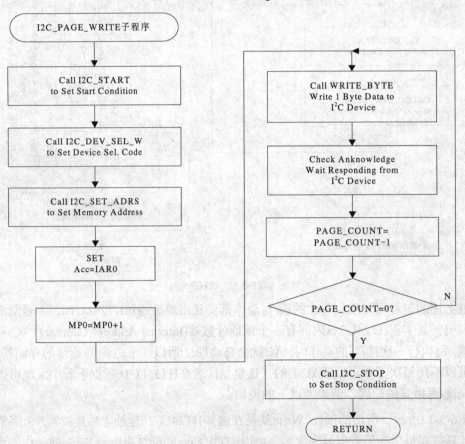

图6-3-13　流程图

Sequential Random Read就是在通知HT24LC16地址之后，连续读N个字节的数据，而且只需在最后一笔数据写完之后，再以STOP Condition结束Read的动作即可。Read时N并没有限制，不过要注意的是当读到最后一个地址时，AC会重新归0，所以又会从第0个地址开始读起。问题是HT24LC16怎么知道究竟何时停止呢？请注意图6-3-14中的ACK信号，因为是读取数据，所以收到数据之后HT46xx应该送出ACK信号通知HT24LC16，当HT46x23的ACK信号=0时，代表还要继续读；如果ACK信号=1，就表示不再读了。

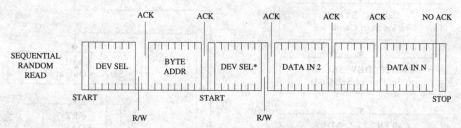

图6-3-14 HT24LC16 Sequential Random Read 模式

如图6-3-15所示，Current Address Read就是读取目前AC所指的地址内容，由于不需通知HT24LC16地址，因此在单位时间内数据的传送量较大，请参考前述有关I2C_CT_READ与I2C_SQ_READ子程序的流程图(见图6-3-11)。

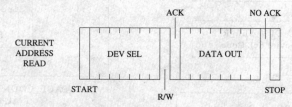

图6-3-15 HT24LC16 Current Address Read 模式

如图6-3-16所示，Sequential Current Read就是依目前AC所指的地址，连续读N个字节的数据，而且只需在最后一笔数据写完之后，再以STOP Condition结束Read的动作。与Sequential Random Read一样，结束读取是以HT46x23回应ACK信号=1，就表示不再读了。请参考I2C_SQ_CT_READ与I2C_SQ_RM_READ子程序的流程图(见图6-3-17)。

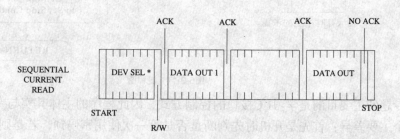

图6-3-16 HT24LC16 Sequential Current Read 模式

关于HT24LC16的说明到此告一段落，若有不清楚的地方还请读者参考光盘中的完整数据。比较重要的是信号间的"时序(Timing)"关系，在流程图中并没有详细地表达出来，读者在阅读程序时会发现每设置完一个信号时(SCL、SDA)，就会调用DELAY_10子程序延时一段时间，主要就是为了符合数据手册中的要求。

337

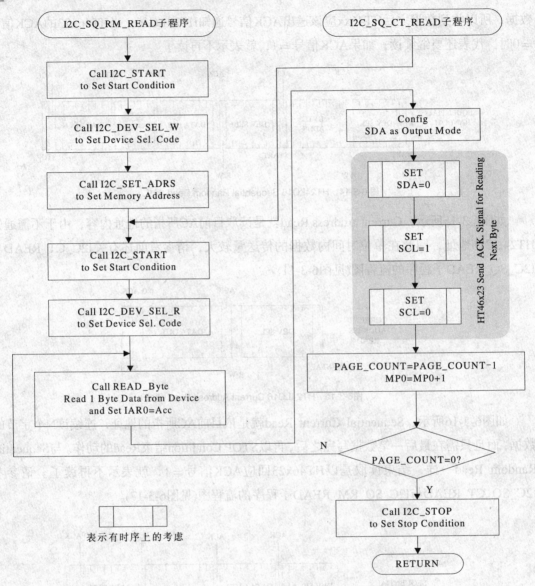

图6-3-17 流程图

本专题的主要目的是学习I²C接口的控制方式,因此程序的主体其实与专题二差异甚少,只有两个主要差异:首先是开机时先判断是否是第一次使用密码锁,若是则以系统预设值为密码(0123);否则就以HT24LC16中所存放的数据为密码(上一次用户所设置的密码)。第二个差异就是当用户重新设置完密码后,必须将新的密码存放在HT24LC16中,以便断电之后仍旧可以用用户重新设置的密码进入系统,而不需以系统预设密码进入,请参考以下的流程图(见图6-3-18)及程序。

6-3-4 流程图及程序

1. 流程图(见图 6-3-18)

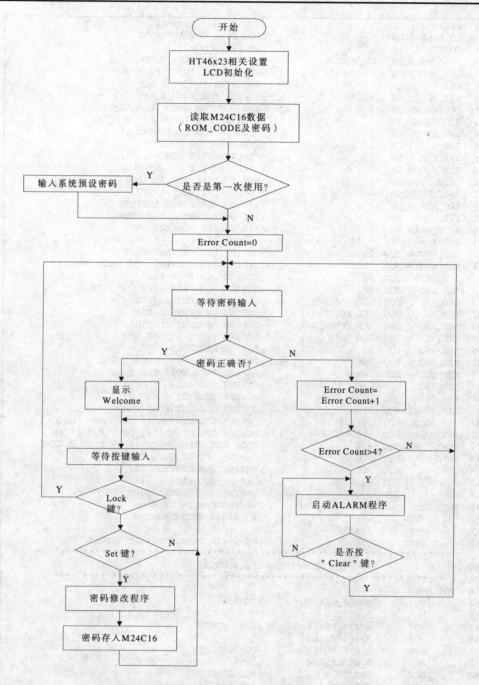

图6-3-18 流程图

2. 程序 6-3 具密码记忆功能密码锁控制程序

```
1   ;PROGRAM : 6-3.ASM (6-3.PRJ)                         BY STEVEN
2   ;FUNCTION: DIGITAL CODE KEY WITH EEPROM DEMO PROGRAM 2002.DEC.07.
3   #INCLUDE  HT46R23.INC
4             .CHIP     HT46R23
5   ;-----------------------------------------------------------------
6   MY_DATA   .SECTION  'DATA'        ;== DATA SECTION ==
7   DEL1      DB        ?             ;DELAY LOOP COUNT 1
8   DEL2      DB        ?             ;DELAY LOOP COUNT 2
```

```
9       DEL3        DB      ?                   ;DELAY LOOP COUNT 3
10      DC_FLAG     DBIT                        ;LCD DATA/COMMAND FLAG
11      COUNT       DB      ?                   ;DISPLAY COUNT
12      PASS_WORD   DB      10 DUP(?)           ;BUFFER FOR PASS WORD
13      KEY_BUFFER  DB      10 DUP(?)           ;BUFFER FOR INPUT PASS WORD
14      KEY         DB      ?                   ;REGISTER FOR READ_KEY
15      KEY_PS      DB      ?                   ;BUFFER FOR READ_A_KEY
16      KEY_COUNT   DB      ?                   ;ROW COUNTER USED IN READ_KEY
17      DIG_COUNT   DB      ?                   ;COUNTER FOR INPUT DIGIT
18      PWD_COUNT   DB      ?                   ;COUNTER FOR PASS WORD DIGIT
19      ERR_N       DB      ?                   ;ERROR RECORD COUNTER
20      STACK_A     DB      ?                   ;STACK BUFFER FOR ACC
21      STACK_PSW   DB      ?                   ;STACK BUFFER FOR STATUS
22      BYTE_COUNT  DB      ?                   ;BIT COUNTER FOR I2C R/W
23      PAGE_COUNT  DB      ?                   ;COUNT FOR I2C R/W BIT
24      I2C_ADRS    DB      ?                   ;I2C DEVICE START ADDRESS
25      I2C_DATA    DB      ?                   ;I2C BYTE R/W DATA BUFFER
26      BYTE_DATA   DB      ?                   ;R/W BUFFER FOR I2C DATA
27      ;-------------------------------------------------------------
28      LCD_CONTR   EQU     PA                  ;DEFINE LCD CONTROL PORT
29      LCD_CONTRC  EQU     PAC                 ;DEFINE LCD CONTROL PORT CON. REG.
30      LCD_DATA    EQU     PB                  ;DEFINE LCD DATA PORT
31      LCD_DATAC   EQU     PBC                 ;DEFINE LCD DATA PORT CON. REG.
32      LCD_EN      EQU     LCD_CONTR.0         ;DEFINE EN CONTROL PIN
33      LCD_RW      EQU     LCD_CONTR.1         ;DEFINE RW CONTROL PIN
34      LCD_RS      EQU     LCD_CONTR.2         ;DEFINE RS CONTROL PIN
35      LCD_READY   EQU     LCD_DATA.7          ;DEFINE READY BIT OF LCD
36      KEY_PORT    EQU     PB                  ;DEFINE KEY PORT
37      KEY_PORTC   EQU     PBC                 ;DEFINE KEY PORT CONTROL REG.
38      END_CODE    EQU     0AH                 ;DEFINE PASS WORD END CODE
39      STR_END     EQU     0FEH                ;STRING END CODE
40      SPK_PORT    EQU     PC                  ;BUZZER CONNECT PORT
41      SPK_PORTC   EQU     PCC                 ;BUZZER CONNECT PORT CONTR. REG.
42      DEVSEL_R    EQU     10100001B           ;DEVICE SELECT(READ)  CODE FOR 24CXX
43      DEVSEL_W    EQU     10100000B           ;DEVICE SELECT(WRITE) CODE FOR 24CXX
44      POWER_CODE  EQU     55H                 ;CODE FOR DESCRIMATE FIRST TIME USEING
45      SCLC        EQU     PCC.1               ;DEFINE SCL CONTROL BIT
46      SDAC        EQU     PCC.0               ;DEFINE SDA CONTROL BIT
47      SCL         EQU     PC.1                ;DEFINE SCL SIGNAL PIN
48      SDA         EQU     PC.0                ;DEFINE SDA SIGNAL PIN
49      MY_CODE     .SECTION AT 0 'CODE'        ;== PROGRAM SECTION ==
50              ORG         00H                 ;HT-46RXX RESET VECTOR
51              JMP         START
52              ORG         08H                 ;HT-46RXX TMR INTERRUPT VECTOR
53      ;***************************************************************
54      ;           TIMER INTERRUPT SERVICE ROUTINE
55      ;***************************************************************
56      TMR_INT :
57              MOV         STACK_A,A           ;PUSH A
58              MOV         A,STATUS
59              MOV         STACK_PSW,A         ;PUSH STATUS
60              MOV         A,00000100B
61              XORM        A,SPK_PORT          ;COMPLEMENT
62              MOV         A,STACK_PSW
63              MOV         STATUS,A            ;POP STATUS
64              MOV         A,STACK_A           ;POP A
65              RETI
66      START:
67              CLR         LCD_CONTRC          ;CONFIG LCD_CONTR AS OUTPUT MODE
68              CLR         SPK_PORTC.3         ;CONFIG SPK_PORT.3 AS OUTPUT MODE
69              MOV         A,00000101B         ;ENABLE EMI & ETI
70              MOV         INTC0,A
71              MOV         A,38H               ;FUNCTION SET: 8-BIT,2-LINE,5X10 DOTS
72              CALL        WLCMC
73              MOV         A,0FH               ;ON/OFF CONTR: DISPLAY ON,CURSOR BLINKING ON
74              CALL        WLCMC
75              MOV         A,06H               ;ENTRY MODE : INCREMENT,DISPLAY NOT SHIFT
```

```
76           CALL    WLCMC
77           MOV     A,01H              ;CLEAR DISPLAY
78           CALL    WLCMC
79           MOV     A,80H              ;SET LCD LINE 1, POSITION 0
80           CALL    WLCMC
81           CLR     I2C_ADRS           ;SET I2C START ADDRESS
82           CALL    I2C_RM_READ        ;READ POWED_CODE
83           MOV     A,POWER_CODE
84           XOR     A,BYTE_DATA        ;IS SYSTEM INITIAL CONDITION?
85           SZ      ACC                ;NO, READ PWD FROM I2C DEV.
86           JMP     SYSTEM_PWD         ;YES, READ DEFAULT PASSWORD
87           CALL    I2C_CT_READ        ;READ PWD NUMBER
88           MOV     A,BYTE_DATA
89           MOV     PWD_COUNT,A
90           MOV     A,OFFSET PASS_WORD
91           MOV     MP0,A              ;SET BUFFER ADDRESS
92           MOV     A,10
93           MOV     PAGE_COUNT,A       ;SET PAGE READ COUNT
94           CALL    I2C_SQ_CT_READ     ;READ PWD ON EEPROM
95           JMP     AGAIN
96  SYSTEM_PWD:
97           MOV     A,80H
98           CALL    WLCMC
99           MOV     A,OFFSET STR8      ;DISPLAY FIRST TIME MESSAGE
100          CALL    PRINT
101          MOV     A,0C0H
102          CALL    WLCMC
103          MOV     A,OFFSET STR9
104          CALL    PRINT
105          MOV     A,250              ;DELAY 2.5 SEC
106          MOV     DEL1,A
107          CALL    DELAY
108          MOV     A,OFFSET PASS_WORD ;INITIAL BUFFER ADDRESS
109          MOV     MP0,A              ;MOVE DEFAULT PASSWORD TO RAM BUFFER
110          MOV     A,OFFSET DEFAULT_PWD
111          MOV     TBLP,A             ;INITIAL PASSWORD POINTER
112          CLR     PWD_COUNT          ;PWD_COUNT=0
113 READ_PWD:
114          TABRDL  IAR0               ;READ DEFAULT PWD TO RAM BUFFER
115          MOV     A,END_CODE
116          XOR     A,IAR0
117          SZ      Z                  ;END_CODE READ?
118          JMP     AGAIN              ;YES!
119          INC     MP0                ;NO, INCREASE POINTER BY 1
120          INC     TBLP               ;UPDATE TABLE POINTER
121          INC     PWD_COUNT          ;INCREASE PWD_COUNTER BY 1
122          JMP     READ_PWD           ;NEXT DEFAULT DIGIT
123 AGAIN:
124          CLR     ERR_N              ;CLEAR ERROR COUNT
125 MAIN:
126          MOV     A,01               ;CLEAR LCD
127          CALL    WLCMC              ;PRINT "Password:(Err_N)" MESSAGE
128          MOV     A,OFFSET STR1
129          CALL    PRINT
130          MOV     A,80H+13           ;SET LCD LINE 1 POSITION 13
131          CALL    WLCMC
132          MOV     A,ERR_N            ;GET ERROR COUNTER
133          ADD     A,30H              ;CONVERT TO ASCII
134          CALL    WLCMD              ;DISPLAY ON LCD
135          CALL    BUFFER             ;GET INPUT PASSWORD
136          CALL    CHECK              ;CHECK PASSWORD FORMAT
137          XOR     A,00H
138          SZ      Z                  ;IS FORMAT CORRECT?
139          JMP     O_ERROR            ;NO,FORMAT ERROR
140          MOV     A,OFFSET STR7      ;YES,PRINT "Pass"
141          CALL    PRINT
142          MOV     A,180              ;DELAY 1.8S
```

```
143         CALL    DELAY
144 OPEN:
145         MOV     A,01                ;CLEAR LCD
146         CALL    WLCMC
147         MOV     A,OFFSET STR2       ;PRINT "Welcome (Open)"
148         CALL    PRINT
149         CALL    READ_KEY_PRESSED    ;WAIT KEY INPUT
150         MOV     A,KEY_PS
151         XOR     A,0BH
152         SZ      Z                   ;IS 'SET(B)' PRESSED
153         CALL    SET_PAS             ;CALL IF "SET" KEY PRESSED
154         MOV     A,KEY_PS
155         XOR     A,0FH
156         SNZ     Z                   ;IS 'LOCK(F)' PRESSED
157         JMP     OPEN
158         JMP     AGAIN               ;JUMP IF "LOCK" PRESSED
159 O_ERROR:
160         MOV     A,OFFSET STR3       ;PRINT "Error"
161         CALL    PRINT
162         MOV     A,100               ;DELAY 100mS
163         CALL    DELAY
164         INC     ERR_N               ;INCREASE ERROR COUNT BY 1
165         MOV     A,ERR_N
166         XOR     A,04
167         SNZ     Z                   ;IS ERROR NUMBER =4 ?
168         JMP     MAIN                ;NO, JUMP MAIN
169         CALL    PHONE               ;YES,CALL ALARM!!
170         JMP     AGAIN               ;RE-START
171 ;********************************************************************
172 ;           GENERATE I2C START CONDITION
173 ;********************************************************************
174 I2C_START PROC
175         CLR     SCLC                ;CONFIG SLC AS OUTPUT MODE
176         CLR     SDAC                ;CONFIG SDA AS OUTPUT MODE
177         CLR     SCL                 ;SET SCL=0
178         CLR     SDA                 ;SET SDA=0
179         CALL    DELAY_10            ;DELAY
180         SET     SCL                 ;SET SCL=1
181         SET     SDA                 ;SET SDA=1
182         CALL    DELAY_10            ;DELAY
183         CLR     SDA                 ;SET SDA=0
184         CALL    DELAY_10            ;DELAY
185         CLR     SCL                 ;SET SCL=0
186         CALL    DELAY_10            ;DELAY
187         RET
188 I2C_START ENDP
189 ;********************************************************************
190 ;       SEND I2C DEVICE SELECT CODE (R/W) TO DEVICE
191 ;********************************************************************
192 I2C_DEV_SEL_W  PROC                 ;WRITE MODE
193         MOV     A,DEVSEL_W          ;LOAD A WITH DEVICE SELECT CODE
194         JMP     DEV_SEL
195 I2C_DEV_SEL_R:                      ;READ MODE
196         MOV     A,DEVSEL_R          ;LOAD A WITH DEVICE SELECT CODE
197 DEV_SEL:
198         CALL    WRITE_BYTE          ;SEND OUT 1 BYTE DATA IN Acc
199         CALL    CHECK_ACK           ;WAIT FOR DEVICE ACK SIGNAL
200         RET
201 I2C_DEV_SEL_W  ENDP
202 ;********************************************************************
203 ;       SET STARTING(I2C_ADRS) ADDRESS OF I2C DEVICE
204 ;********************************************************************
205 I2C_SET_ADRS  PROC
206         MOV     A,I2C_ADRS
207         CALL    WRITE_BYTE          ;WRITE ADRESS TO I2C DEVICE
208         CALL    CHECK_ACK
209         RET
```

```
210     I2C_SET_ADRS    ENDP
211 ;**********************************************************************
212 ;                   I2C RANDOM/CURRENT ADDRESS READ
213 ;       READ 1 BYTE(BYTE_DATA) FROM I2C DEVICE ADRS SPECIFIED BY I2C_ADRS
214 ;**********************************************************************
215 I2C_RM_READ     PROC
216         CALL    I2C_START           ;START SIGNAL
217         CALL    I2C_DEV_SEL_W       ;SET DEVICE CODE AND WRITE MODE
218         CALL    I2C_SET_ADRS        ;I2C START ADDRESS SET IN I2C_ADRR
219 I2C_CT_READ:
220         CALL    I2C_START           ;START SIGNAL
221         CALL    I2C_DEV_SEL_R       ;SET DEVICE CODE AND READ MODE
222         CALL    READ_BYTE           ;READ 1 BYTE
223         MOV     BYTE_DATA,A         ;SAVE TO DATA BUFFER
224         CLR     SCLC                ;CONFIG SLC AS OUTPUT MODE
225         CLR     SDAC                ;CONFIG SDA AS OUTPUT MODE
226         SET     SDA                 ;SET SDA=1
227         CALL    DELAY_10            ;DELAY
228         SET     SCL                 ;SET SCL1, NO_ACK SIGNAL TO I2C DEVICE
229         CALL    DELAY_10            ;DELAY
230         CLR     SCL                 ;SET SCL=0
231         CALL    DELAY_10            ;DELAY
232         CALL    I2C_STOP            ;STOP SIGNAL
233         RET
234 I2C_RM_READ     ENDP
235 ;**********************************************************************
236 ;                   I2C SEQUENTIQL RANDOM/CURRENT READ
237 ;       READ CONSITIVE BYTES FROM I2C DEVICE TO BUFFER (DIRECTED BY MP0)
238 ;       NOTE: BYTE NUMBER IS DEFINED BY PAGE_COUNT (<=16)
239 ;**********************************************************************
240 I2C_SQ_RM_READ PROC
241         CALL    I2C_START           ;START SIGNAL
242         CALL    I2C_DEV_SEL_W       ;SET DEVICE CODE AND WRITE MODE
243         CALL    I2C_SET_ADRS        ;I2C START ADDRESS SET IN I2C_ADRS
244 I2C_SQ_CT_READ:
245         CALL    I2C_START           ;START SIGNAL
246         CALL    I2C_DEV_SEL_R       ;SET DEVICE CODE AND READ MODE
247 READ_NEXT_BYTE:
248         CALL    READ_BYTE           ;READ 1 BYTE
249         MOV     IAR0,A              ;SAVE TO DATA BUFFER
250         INC     MP0                 ;INCREASE BUFFER POINTER BY 1
251         CLR     SCLC                ;CONFIG SLC AS OUTPUT MODE
252         CLR     SDAC                ;CONFIG SDA AS OUTPUT MODE
253         CLR     SDA                 ;SET SDA=0
254         CALL    DELAY_10            ;DELAY
255         SET     SCL                 ;SET SCL=1, ACK SIGNAL TO I2C DEVICE
256         CALL    DELAY_10            ;DELAY
257         CLR     SCL                 ;SET SCL=0
258         CALL    DELAY_10            ;DELAY
259         SDZ     PAGE_COUNT          ;PAGE_COUNT-1 = 0?
260         JMP     READ_NEXT_BYTE      ;NO, READ NEXT BYTE
261         CALL    I2C_STOP            ;STOP SIGNAL
262         RET
263 I2C_SQ_RM_READ ENDP
264 ;**********************************************************************
265 ;       WRITE 1 BYTE(BYTE_DATA) TO I2C DEVICE ADRS SPECIFIED BY I2C_ADRS
266 ;**********************************************************************
267 I2C_BYTE_WRITE PROC
268         CALL    I2C_START           ;START SIGNAL
269         CALL    I2C_DEV_SEL_W       ;DEV. SELECT CODE WITH WRITE
270         CALL    I2C_SET_ADRS        ;SET DEVICE START ADDRESS
271         MOV     A,BYTE_DATA
272         CALL    WRITE_BYTE
273         CLR     SCLC                ;CONFIG SLC AS OUTPUT MODE
274         SET     SDAC                ;CONFIG SDA AS INPUT MODE
275         CALL    DELAY_10            ;DELAY
276         SET     SCL                 ;SCL=1
```

```
277           CALL    DELAY_10            ;DELAY
278   RP_WAIT:
279           SZ      SDA                 ;SDA = 0(DEV. ACK)?
280           JMP     RP_WAIT             ;NO, WAIT NO_ACK SIGNAL FROM I2C_DEVICE
281           CALL    DELAY_10            ;DELAY
282           CLR     SCL                 ;SCL=0
283           CALL    I2C_STOP            ;STOP CONDITION
284           MOV     A,1
285           MOV     DEL1,A
286           CALL    DELAY               ;DELAY 10MS FOR tW
287           RET
288   I2C_BYTE_WRITE ENDP
289   ;**********************************************************************
290   ;   WRITE CONSITIVE BYTES TO I2C DEVICE FORM BUFFER(DIRECTED BY MP0)
291   ;   NOTE: BYTE NUMBER IS DEFINED BY PAGE_COUNT (<=16)
292   ;**********************************************************************
293   I2C_PAGE_WRITE PROC
294           CALL    I2C_START           ;START SIGNAL
295           CALL    I2C_DEV_SEL_W       ;DEV. SELECT CODE WITH WRITE
296           CALL    I2C_SET_ADRS        ;SET DEVICE START ADDRESS
297   WRITE_NEXT_BYTE:
298           MOV     A,IAR0              ;READ BUFFER DATA
299           INC     MP0                 ;INCREASE DATA POINTER BY 1
300           CALL    WRITE_BYTE          ;SERIAL OUT 1 BYTE IN Acc
301           CLR     SCLC                ;CONFIG SLC AS OUTPUT MODE
302           SET     SDAC                ;CONFIG SDA AS INPUT MODE
303           CALL    DELAY_10            ;DELAY
304           SET     SCL                 ;SCL=1
305           CALL    DELAY_10            ;DELAY
306   WP_WAIT:
307           SZ      SDA                 ;SDA = 0(DEV. ACK)?
308           JMP     WP_WAIT             ;NO, WAIT NO_ACK SIGNAL FROM I2C_DEVICE
309           CALL    DELAY_10            ;DELAY
310           CLR     SCL                 ;SCL=0
311           CALL    DELAY_10            ;DELAY
312           SDZ     PAGE_COUNT          ;PAGE_COUNT-1 = 0?
313           JMP     WRITE_NEXT_BYTE     ;NO, WRITE NEXT BYTE
314           CALL    I2C_STOP            ;STOP CONDITION
315           RET
316   I2C_PAGE_WRITE ENDP
317   ;**********************************************************************
318   ;              GENERATE I2C STOP CONDITION
319   ;**********************************************************************
320   I2C_STOP   PROC
321           CLR     SCLC                ;CONFIG SLC AS OUTPUT MODE
322           CLR     SDAC                ;CONFIG SDA AS OUTPUT MODE
323           CLR     SCL                 ;SET SCL=0
324           CLR     SDA                 ;SET SDA=0
325.          CALL    DELAY_10            ;DELAY
326           SET     SCL                 ;SET SCL=1
327           CALL    DELAY_10            ;DELAY
328           CLR     SDA                 ;SET SDA=0
329           CALL    DELAY_10            ;DELAY
330           SET     SDA                 ;SET SDA=1
331           RET
332   I2C_STOP   ENDP
333   ;**********************************************************************
334   ;           SERIAL OUT DATA IN Acc VIA SDA & SCL
335   ;**********************************************************************
336   WRITE_BYTE     PROC
337           CLR     SCLC                ;CONFIG SLC AS OUTPUT MODE
338           CLR     SDAC                ;CONFIG SDA AS OUTPUT MODE
339           CLR     SCL                 ;SET SCL=0
340           CLR     SDA                 ;SET SDA=0
341           MOV     I2C_DATA,A          ;RESERVED DATA IN TX BUFFER
342           MOV     A,8                 ;SET 8 BIT COUNTER
343           MOV     BYTE_COUNT,A
```

```
344 WNB_0:
345       SZ        I2C_DATA.7        ;IS MSB = 0?
346       JMP       WRITE_1           ;NO, JUMP TO WRITE_1
347 WRITE_0:
348       CLR       SDA               ;SET SDA=0
349       JMP       WNB_1             ;JUMP TO WNB_1
350 WRITE_1:
351       SET       SDA               ;SET SDA=1
352 WNB_1:
353       CALL      DELAY_10          ;DELAY
354       SET       SCL               ;SET SCL=1
355       CALL      DELAY_10          ;DELAY
356       CLR       SCL               ;SET SCL=0
357       CALL      DELAY_10          ;DELAY
358       RL        I2C_DATA          ;SHIFT TX BUFFER
359       SDZ       BYTE_COUNT        ;BYTE_COUNT-1 = 0?
360       JMP       WNB_0             ;NO, WRITE NEXT BIT
361       RET                         ;YES, RETURN.
362 WRITE_BYTE    ENDP
363 ;************************************************************************
364 ;           SERIAL IN DATA TO Acc VIA SDA & SCL
365 ;************************************************************************
366 READ_BYTE PROC
367       CLR       SCLC              ;CONFIG SLC AS OUTPUT MODE
368       SET       SDAC              ;CONFIG SDA AS INTPUT MODE
369       MOV       A,8               ;SET 8 BIT COUNTER
370       MOV       BYTE_COUNT,A
371 RNB_0:
372       SET       SCL               ;SET SCL=0
373       CALL      DELAY_10          ;DELAY
374       RLC       I2C_DATA          ;SHIFT RX BUFFER
375       SZ        SDA               ;SDA = 0?
376       JMP       READ_1            ;NO, JUMP TO READ_1
377 READ_0: CLR     I2C_DATA.0        ;YES, SET LSB=0
378       JMP       RNB_1             ;JUMP TO RNB_1
379 READ_1:
380       SET       I2C_DATA.0        ;SET LSB=1
381 RNB_1:
382       CALL      DELAY_10          ;DELAY
383       CLR       SCL               ;SET SCL=0
384       CALL      DELAY_10          ;DELAY
385       SDZ       BYTE_COUNT        ;BYTE_COUNT-1 = 0?
386       JMP       RNB_0             ;NO, READ NEXT BIT
387       MOV       A,I2C_DATA        ;RELOAD RX DATA TO Acc
388       RET                         ;YES, RETURN
389 READ_BYTE ENDP
390 ;************************************************************************
391 ;           WAIT FOR ACK SIGNAL FROM I2C DEVICE
392 ;************************************************************************
393 CHECK_ACK PROC
394       SET       SDAC              ;CONFIG  SDA AS INPUT MODE
395       CALL      DELAY_10          ;DELAY
396       SET       SCL               ;SET SCL=1
397       CALL      DELAY_10          ;DELAY
398 WAIT_ACK:
399       SZ        SDA               ;ACK SINGLE SET ?
400       JMP       WAIT_ACK          ;NO, WAIT ACK
401       CALL      DELAY_10          ;YES, DELAY
402       CLR       SCL               ;SET SCL=0
403       CALL      DELAY_10          ;DELAY
404       RET
405 CHECK_ACK ENDP
406 ;************************************************************************
407 ;           DELAY 10us FOR TIMING CONSIDERATION
408 ;************************************************************************
409 DELAY_10:
410       TABRDL    DEL1              ;NULL READ
```

```
411         TABRDL      DEL1                ;DELAY 10 INS. CYCLES
412         TABRDL      DEL1
413         TABRDL      DEL1
414         RET
415 ;**********************************************************************
416 ;                SET_PASS SUBROUTINE
417 ;**********************************************************************
418 SET_PAS    PROC
419         MOV         A,01                ;CLEAR LCD
420         CALL        WLCMC
421         MOV         A,OFFSET STR4       ;PRINT "O_Password:"
422         CALL        PRINT
423         MOV         A,0C0H              ;SET LINE 2
424         CALL        WLCMC
425         CALL        BUFFER              ;INPUT PASSWORD
426         CALL        CHECK               ;CHECK PASSWORD FORMAT
427         XOR         A,00
428         SZ          Z                   ;IS FORMAT CORRECT?
429         JMP         S_ERROR             ;NO,JUMP TO FORMAT ERROR
430         MOV         A,OFFSET STR7       YES, PRINT "Pass"
431         CALL        PRINT
432         MOV         A,100               ;DELAY 100ms
433         CALL        DELAY
434         CALL        KEY_PAS             ;MOVE NEW PASSWORD TO BUFFER
435         XOR         A,01
436         SNZ         Z                   ;IS FORMAT CORRECT?
437         JMP         S_ERROR             ;NO, JUMP TO FORMAT ERROR
438         MOV         A,4FH               ;YES, PRINT "OK"
439         CALL        WLCMD
440         MOV         A,4BH
441         CALL        WLCMD
442         MOV         A,200               ;DELAY 200ms
443         CALL        DELAY
444         CALL        CHA_PAS             ;MOVE NEW PWD TO PAS_BUF
445         RET
446 SET_PAS    ENDP
447 ;**********************************************************************
448 ;              S_ERROR SUBROUTINE
449 ;**********************************************************************
450 S_ERROR    PROC
451         MOV         TBLP,A              ;INITIAL STRING START ADDRS
452         CALL        PRINT
453         MOV         A,150               ;DELAY 1.50s
454         CALL        DELAY
455         RET
456 S_ERROR    ENDP
457 ;**********************************************************************
458 ;                KEY_PASS SUBROUTINE
459 ;**********************************************************************
460 KEY_PAS    PROC
461         MOV         A,01                ;CLEAR LCD
462         CALL        WLCMC
463         MOV         A,OFFSET STR5       ;PRINT "New Password:"
464         CALL        PRINT
465         MOV         A,0C0H              ;SET LCD LINE 2
466         CALL        WLCMC
467         CALL        BUFFER              ;GET PASSWORD
468         MOV         A,DIG_COUNT
469         SUB         A,10
470         SZ          C                   ;A <=10?
471         RET         A,0                 ;PASSWROD FORMAT ERROR
472         MOV         A,DIG_COUNT
473         SUB         A,4
474         SNZ         C                   ;A <=4?
475         RET         A,0                 ;PASSWROD FORMAT ERROR
476         RET         A,1                 ;PASSWROD FORMAT CORRECT
477 KEY_PAS    ENDP
```

```
478 ;************************************************************************
479 ;                CHANGE PASSWORD SUBROUTINE
480 ;************************************************************************
481 CHA_PAS   PROC
482       MOV     A,OFFSET KEY_BUFFER;SET DATA POINTER 0
483       MOV     MP0,A
484       MOV     A,OFFSET PASS_WORD ;SET DATA POINTER 1
485       MOV     MP1,A
486       MOV     A,DIG_COUNT
487       MOV     PWD_COUNT,A         ;UPDATE PASSWORD DIGIT COUNT
488 CHA_PAS_1:
489       MOV     A,IAR0              ;GET NEW PASSWORD DIGIT
490       MOV     IAR1,A              ;UPDATE PASSWORD BUFFER
491       INC     MP0                 ;INCREASE DATA POINTER 0
492       INC     MP1                 ;INCREASE DATA POINTER 1
493       SDZ     DIG_COUNT           ;DIG_COUNT-1 = 0?
494       JMP     CHA_PAS_1           ;NO, NEXT DIGIT
495       CLR     I2C_ADRS            ;SET I2C_ADRS=0
496       MOV     A,POWER_CODE
497       MOV     BYTE_DATA,A
498       CALL    I2C_BYTE_WRITE      ;WRITE POWER CODE TO I2C DEV.
499       INC     I2C_ADRS            ;INCREASE I2C_ADRS BY 1
500       MOV     A,PWD_COUNT
501       MOV     BYTE_DATA,A
502       CALL    I2C_BYTE_WRITE      ;WRITE PWD NUMBER TO I2C DEV.
503       INC     I2C_ADRS            ;INCREASE I2C_ADRS BY 1
504       MOV     A,OFFSET PASS_WORD ;SET DATA POINTER 0
505       MOV     MP0,A
506       MOV     A,10
507       MOV     PAGE_COUNT,A        ;SET OAGE COUNT
508       CALL    I2C_PAGE_WRITE      ;SAVE PASSWORD TO I2C DEV.
509       RET                         ;YES.
510 CHA_PAS   ENDP
511 ;************************************************************************
512 ;                PASSWORD CHECK SUBROUTINE
513 ;************************************************************************
514 CHECK PROC
515       MOV     A,PWD_COUNT         ;CHECK THE DIGIT NUMER
516       XOR     A,DIG_COUNT
517       SNZ     Z                   ;IS DIGIT NUMBER CORRRECT?
518       JMP     CHE_ERR             ;NO, JUMP TO CHE_ERROR
519       MOV     A,OFFSET KEY_BUFFER;CHECK PASSWORD WITH KEY BUFFER
520       MOV     MP0,A
521       MOV     A,OFFSET PASS_WORD
522       MOV     MP1,A
523 CHE_PAS:
524       MOV     A,IAR0
525       XOR     A,IAR1
526       SNZ     Z                   ;IS THE DIGIT THE SAME?
527       JMP     CHE_ERR             ;NO, JUMP TO CHE_ERROR
528       INC     MP0                 ;INCREASE DATA POINTER 0
529       INC     MP1                 ;INCREASE DATA POINTER 1
530       SDZ     DIG_COUNT           ;IS DIG_COUNT-1 = 0?
531       JMP     CHE_PAS             ;NO, NEXT DIGIT
532 CHE_OK:
533       RET     A,1                 ;YES, RETURN WITH OK
534 CHE_ERR:
535       RET     A,0                 ;RETURN WITH ERROR
536 CHECK ENDP
537 ;************************************************************************
538 ;                GET PASSWORD TO BUFFER SUBROUTINE
539 ;************************************************************************
540 BUFFER    PROC
541       MOV     A,0C0H              ;SET LCD LINE 2
542       CALL    WLCMC
543       MOV     A,16
544       MOV     DIG_COUNT,A
```

```
545         MOV         A,20H               ;CLEAR LINE 2
546 LOOP:
547         CALL        WLCMD               ;CLEAR LCD LINE 2
548         SDZ         DIG_COUNT
549         JMP         LOOP
550         MOV         A,0C0H              ;SET LCD LINE 2, POSITION 0
551         CALL        WLCMC
552         MOV         A,OFFSET KEY_BUFFER ;SET INPUT BUFFER
553         MOV         MP0,A
554         CLR         DIG_COUNT           ;SET GIGIT COUNT TO 0
555 SAV_BUF:
556         CALL        READ_KEY_PRESSED    ;READ UNTIL KEY PRESSED
557         MOV         A,KEY_PS
558         XOR         A,0CH
559         SZ          Z                   ;IS "CLEAR(C)" PRESSED
560         JMP         BUFFER              ;YES, RESTART
561         MOV         A,0AH
562         XOR         A,KEY_PS
563         SZ          Z                   ;IS "ENTER(A)" KEY PRESSED?
564         JMP         END_BUF             ;YES,RETUTN
565         MOV         A,KEY_PS
566         SUB         A,10                ;A-10
567         SZ          C                   ;IS KEY CODE < 10?
568         JMP         SAV_BUF             ;NO, SKIP UNDEFINED KEY
569         MOV         A,2AH
570         CALL        WLCMD               ;PRINT "*" ON LCD
571         MOV         A,KEY_PS
572         MOV         R0,A                ;SAVE PRESSED KEY CODE IN BUFFER
573         INC         MP0                 ;INCREASE DATA POINTER
574         INC         DIG_COUNT           ;INCREASE DIGIT COUNT
575         JMP         SAV_BUF             ;NEXT KEY
576 END_BUF:
577         RET
578 BUFFER  ENDP
579 ;********************************************************************
580 ;    READ KEYPAD AND WAIT UNTIL KEY IS PRESSED THEN RELEASED
581 ;********************************************************************
582 READ_KEY_PRESSED    PROC
583         CALL        READ_KEY            ;SCAN KEYPAD
584         MOV         A,16
585         XOR         A,KEY
586         SZ          Z                   ;IS KEY BEEN PRESSED?
587         JMP         READ_KEY_PRESSED    ;NO, RE-SCAN KEYPAD
588         MOV         A,KEY
589         MOV         KEY_PS,A            ;RESERVED KEY CODE
590 WAIT_KEY_RELEASED:
591         CALL        READ_KEY            ;SCAN KEYPAD
592         MOV         A,16
593         XOR         A,KEY
594         SZ          ACC                 ;IS KEY BEEN RELEASED?
595         JMP         WAIT_KEY_RELEASED   ;NO, RE-SCAN KEYPAD
596         MOV         A,4
597         CALL        DELAY               ;DELAY 40ms
598         RET
599 READ_KEY_PRESSED    ENDP
600 ;********************************************************************
601 ;                   ALARM SUBROUTINE
602 ;********************************************************************
603 PHONE PROC
604         MOV         A,10000000B         ;TIMER MODE, Fint=Fsys, TON=0
605         MOV         TMRC,A
606         MOV         A,0F4H
607         MOV         DIG_COUNT,A         ;TIMER CONSTANT
608         MOV         A,0A0H
609         MOV         PWD_COUNT,A         ;TIMER CONSTAND
610 PHONE_1:
611         MOV         A,01                ;CLEAR LCD
```

```
612         CALL        WLCMC
613         MOV         A,OFFSET STR6       ;PRINT "Call 110..."
614         CALL        PRINT
615         MOV         A,8
616         MOV         COUNT,A
617 QWE:
618         MOV         A,2EH               ;PRINT "."
619         CALL        WLCMD
620         MOV         A,DIG_COUNT
621         MOV         TMRH,A              ;INITIAL TMRH
622         MOV         A,PWD_COUNT
623         MOV         TMRL,A              ;INITIAL TMRH
624         SET         TON                 ;START TIMER
625         MOV         A,20                ;DELAY 200mS
626         CALL        DELAY
627         CLR         TON                 ;STOP TIMER
628         MOV         A,07H
629         XORM        A,DIG_COUNT         ;CHANGE TIME CONSTANT
630         CALL        READ_KEY
631         MOV         A,0CH               ;WAIT "C" CODE
632         XOR         A,KEY
633         SZ          Z                   ;IS "C" PRESSED?
634         JMP         STOP                ;YES, JUMP TO STOP
635         SDZ         COUNT               ;COUNT-1 = 0?
636         JMP         QWE                 ;NO!!
637         JMP         PHONE_1             ;YES
638 STOP:
639         CALL        READ_KEY_PRESSED    ;WAIT KEY RELEASE
640         RET
641 PHONE ENDP
642 ;****************************************************************************
643 ;           LCD PRINT PROCEDURE (START ADDRS=TBLP)
644 ;****************************************************************************
645 PRINT PROC
646         MOV         TBLP,A              ;LOAD TABLE POINTER
647 PRINT_1:
648         TABRDL      ACC                 ;READ STRING
649         XOR         A,STR_END
650         SZ          Z                   ;END OF STRING CHARACTER?
651         JMP         END_PRINT           ;YES, STOP PRINT
652         TABRDL      ACC                 ;NO, RE-READ
653         CALL        WLCMD               ;SEND TO LCD
654         INC         TBLP                ;INCREASE DATA POINTER
655         JMP         PRINT_1             ;NEXT CHARACTER
656 END_PRINT:
657         RET                             ;YES, RETURN
658 PRINT ENDP
659 ;****************************************************************************
660 ;           LCD DATA/COMMAND WRITE PROCEDURE
661 ;****************************************************************************
662 WLCMD PROC
663         SET         DC_FLAG             ;SET DC_FLAG=1 FOR DATA WRITE
664         JMP         WLCM
665 WLCMC:
666         CLR         DC_FLAG             ;SET DC_FLAG=0 FOR COMMAND WRITE
667 WLCM:
668         SET         LCD_DATAC           ;CONFIG LCD_DATA AS INPUT MODE
669         CLR         LCD_CONTR           ;CLEAR ALL LCD CONTROL SIGNAL
670         SET         LCD_RW              ;SET RW SIGNAL (READ)
671         NOP                             ;FOR TAS
672         SET         LCD_EN              ;SET EN HIGH
673         NOP                             ;FOR TDDR
674 WF:
675         SZ          LCD_READY           ;IS LCD BUSY?
676         JMP         WF                  ;YES, JUMP TO WAIT
677         CLR         LCD_DATAC           ;NO, CONFIG LCD_DATA AS OUTPUT MODE
678         MOV         LCD_DATA,A          ;LATCH DATA/COMMAND ON PB(LCD DATA BUS)
```

```
679         CLR     LCD_CONTR          ;CLEAR ALL LCD CONTROL SIGNAL
680         SZ      DC_FLAG            ;IS COMMAND WRITE?
681         SET     LCD_RS             ;NO, SET RS HIGH
682         SET     LCD_EN             ;SET EN HIGH
683         NOP
684         CLR     LCD_EN             ;SET EN LOW
685         RET
686 WLCMD ENDP
687 ;**********************************************************************
688 ;    SCAN 4x4 MATRIX ON KEY PORT AND RETURN THE CODE IN KEY REGISTER
689 ;    IF NO KEY BEEN PRESSED, KEY=16.
690 ;**********************************************************************
691 READ_KEY  PROC
692         MOV     A,11110000B
693         MOV     KEY_PORTC,A        ;CONFIG KEY_PORT
694         SET     KEY_PORT           ;INITIAL KEY PORT
695         CLR     KEY                ;INITIAL KEY REGISTER
696         MOV     A,04
697         MOV     KEY_COUNT,A        ;SET ROW COUNTER
698         CLR     C                  ;CLEAR CARRY FLAG
699 SCAN_KEY:
700         RLC     KEY_PORT           ;ROTATE SCANNING BIT
701         SET     C                  ;MAKE SURE C=1
702         SNZ     KEY_PORT.4         ;COLUMN 0 PRESSED?
703         JMP     END_KEY            ;YES.
704         INC     KEY                ;NO, INCREASE KEY CODE.
705         SNZ     KEY_PORT.5         ;COLUMN 1 PRESSED?
706         JMP     END_KEY            ;YES.
707         INC     KEY                ;NO, INCREASE KEY CODE.
708         SNZ     KEY_PORT.6         ;COLUMN 2 PRESSED?
709         JMP     END_KEY            ;YES.
710         INC     KEY                ;NO, INCREASE KEY CODE.
711         SNZ     KEY_PORT.7         ;COLUMN 3 PRESSED?
712         JMP     END_KEY            ;YES.
713         INC     KEY                ;NO, INCREASE KEY CODE.
714         SDZ     KEY_COUNT          ;HAVE ALL ROWs BEEN CHECKED?
715         JMP     SCAN_KEY           ;NO, NEXT ROW.
716 END_KEY:
717         RET
718 READ_KEY  ENDP
719 ;**********************************************************************
720 ;                   Delay about Acc*10ms
721 ;**********************************************************************
722 DELAY   PROC
723         MOV     DEL1,A             ;SET DEL1 COUNTER
724 DEL_1:
725         MOV     A,30
726         MOV     DEL2,A             ;SET DEL2 COUNTER
727 DEL_2:
728         MOV     A,110
729         MOV     DEL3,A             ;SET DEL3 COUNTER
730 DEL_3:
731         SDZ     DEL3               ;DEL3 DOWN COUNT
732         JMP     DEL_3
733         SDZ     DEL2               ;DEL2 DOWN COUNT
734         JMP     DEL_2
735         SDZ     DEL1               ;DEL1 DOWN COUNT
736         JMP     DEL_1
737         RET
738 DELAY ENDP
739         ORG     LASTPAGE
740 DEFAULT_PWD:                       ;DEFAULT SYSTEM PASSWORD
741         DC      0,1,2,3            ;0~9 AND MAXIMUM 10 DIGITS
742         DC      END_CODE           ;NOTE: MUST END WITH END_CODE
743 STR1: DC       'Password(ERR_ ):',STR_END    ;STRING DEFINE 1
744 STR2: DC       'Welcome (Open)',STR_END      ;STRING DEFINE 2
745 STR3: DC       ' Err!!',STR_END              ;STRING DEFINE 3
746 STR4: DC       'Old Password:',STR_END       ;STRING DEFINE 4
747 STR5: DC       'New Password:',STR_END       ;STRING DEFINE 5
748 STR6: DC       'Call 110',STR_END            ;STRING DEFINE 6
```

```
749  STR7: DC        ' pass !!',STR_END           ;STRING DEFINE 7
750  STR8: DC        ' FIRST TIME UES ',STR_END   ;STRING DEFINE 8
751  STR9: DC        'PWD=DEFAULT CODE',STR_END   ;STRING DEFINE 8
752        END
```

程序说明

行	说明
7~26	依序定义变量地址。
28~31	定义 LCD_DATA Port 与 LCD_CONTR Port 分别为 PB 与 PA。
32~35	定义 LCD 的控制信号引脚。
36~37	定义 KEY_PORT 为 PB。
38~39	定义常量 END_CODE 与 STR_END 分别为 0Ah、FEh;其中 END_CODE 作为系统自设密码结束之用,STR_END 则代表字符串的结束字符。
40~41	定义 SPK_PORT 为 PC。
42~43	定义 HT24LC16 的读取(DEVSEL_W)与写入(DEVSEL_R)设备选择码,请参考表 6-3-3。
44	定义 POWER_CODE=55,POWER_CODE 的主要目的,是让 HT46x23 判断系统是否是第一次使用,若是则由程序存储器(ROM)读取系统密码;否则,就从 HT24LC16 中读取用户最后一次输入的自设密码。
45~48	定义 SCL 与 SDA 分别由 PC1、PC0 控制。
51	声明存储器地址由 000h 开始。
58	声明存储器地址由 008h 开始。
57~66	TMR_INT 子程序,此段子程序主要是将 SPK_PORT.3 反向一次。配合 Timer/Event Counter 的计数溢位时间,可产生不同频率的方波输出。由于 SPK_PORT.3 用来触发喇叭,因此可以造成不同的输出声音。
68~69	定义 LCD_CONTR Port 与 SPK_PORT 为输出模式。
70~71	Global Interrupt 与 Timer/Event Counter Interrupt 有效。
72~73	将 LCD 设置为双行显示(N = 1)、使用 8 位(DB7 ~ DB0)控制模式(DL = 1)、5×7 点矩阵字形(F = 0)。
74~75	将 LCD 设置为显示所有数据(D = 1)、显示光标(C = 1)、光标所在位置的字会闪烁(B = 1)。
76~77	将 LCD 的地址标志位(AC)设置为递加(I/D = 1)、显示器画面不因读/写数据而移动(S = 0)。
78~79	将 LCD 整个显示器清空。
80~81	将光标移到第一行的第一个位置。
82~83	调用 I2C_RM_READ 子程序,读回 HT24LC16 地址 0 内的数据。
84~96	判断是否是第一次使用,如果不是则调用 I2C_CT_READ 与 I2C_SQ_CT_READ 子程序,由 HT24LC16 读取密码长度及密码,然后跳到 AGAIN 执行;否则跳到 SYSTEM_PWD 读取程序存储器内的系统密码。
97~123	首先调用 PRINT 子程序,显示第一次使用的消息。然后运用间接寻址法,将系统原始密码(Default Password)传送到系统密码区(由地址 PASS_WORD 开始)。
125	设置 ERR=0,ERR 是记录密码输入错误次数的寄存器。
127~135	显示 Password:(Err_n)字符串(n 代表输入错误次数)。
136	调用 BUFFER 子程序,读取用户输入的密码并将其显示在 LCD 第 2 行。
137	调用 CHECK 子程序,检查用户输入的密码是否正确。若正确则返回 Acc=1,否则 Acc=0。
138~140	密码若正确则继续执行,否则跳到 O_ERROR。
141~144	密码正确,显示 Pass 字符串并调用 DELAY 子程序延时 1.8 s。
146~149	清除 LCD 后,显示 Welcome(Open)字符串,Open 表示锁已打开(此即所谓开门模式)。
150~159	调用 READ_KEY_PRESSED 子程序,读取用户输入的按键。若为 Set 键,则进入重新设置密码的模式(SET_PAS 子程序);若为 Lock 键,则跳出开门模式。

161~164　密码不正确，显示 Error 字符串。
165~171　错误次数累加一次(ERR＝ERR＋1)。检查错误次数是否大于 4 次，若是则调用 PHONE 子程序，启动警报系统；否则重新要求输入密码。
175~189　I2C_START 子程序，产生 START Condition。
193~202　I2C_DEV_SEL_W 与 I2C_DEV_SEL_R 子程序，将设备选择码输出至 I^2C 设备。I2C_DEV_SEL_W 子程序会设置 R/W 位＝0，表示是对 I^2C 设备做写入动作；I2C_DEV_SEL_R 子程序会设置 R/W 位＝1，表示是对 I^2C 设备进行读取动作。程序中会调用 WRITE_BYTE 子程序，将设备选择码以串行方式一一送出。
206~211　I2C_SET_ADRS 子程序，将读取的地址输出到 I^2C 设备。程序中除了会调用 WRITE_BYTE 子程序，将地址数据以串行方式一一送出之外，还会调用 CHECK_ACK 子程序，等待 I^2C 设备回应。
216~235　I2C_RM_READ(Random Read)与 I2C_CT_READ(Current Address Read)子程序，Random Read 与 Current Address Read 唯一的差别就在于是否送出地址通知 I^2C 设备，所以就将两个子程序集成在一起。在调用 I2C_RM_READ 子程序之前，必须先在 I2C_ADRS 寄存器中设置所要读取的地址；I2C_CT_READ 子程序则不需指定。由 I^2C 设备所读回的值会存放在 BYTE_DATA 寄存器中。
241~264　I2C_SQ_RM_READ(Sequential Random Read)与 I2C_CT_READ(Sequential Current Address Read)子程序，Sequential Random Read 与 Sequential Current Address Read 唯一的差别就在于是否送出地址通知 I^2C 设备，所以就将两个子程序集成在一起。在调用 I2C_SQ_RM_READ 子程序之前，必须先在 I2C_ADRS 寄存器中设置所要读取的起始地址；I2C_SQ_CT_READ 子程序则不需指定。至于要连续读取几个字节，必须在 PAGE_COUNT 寄存器中先予以设置。由 I^2C 设备所读回的值，会以间接寻址法存在数据存储器，用户必须在 MP0 中先设置起始地址。
268~290　I2C_BYTE_WRITE 子程序，此子程序会将一个字节的数据(BYTE_DATA 寄存器)写入用户所指定的设备地址之中(由 I2C_ADRS 寄存器指定地址)。
295~318　I2C_PAGE_WRITE 子程序，此子程序会将数个(由 PAGE_COUNT 寄存器予以指定，但 PAGE_COUNT≤16)字节的数据写入用户所指定的设备地址(由 I2C_ADRS 寄存器指定)之中。写入 I^2C 设备的数据，会以间接寻址法由 HT46x23 的数据存储器循序读出，用户必须在 MP0 中先设置数据的起始地址。
322~334　I2C_STOP 子程序，产生 STOP Condition。
338~364　WRITE_BYTE 子程序，将 Acc 寄存器的数据由 SDA 引脚循序移出(配合 SCL 信号)。
368~391　READ_BYTE 子程序，将 SDA 引脚上的信号循序移入(配合 SCL 信号)Acc 寄存器。
395~407　CHECK_ACK 子程序，等待 I^2C 设备回应 ACK 信号。
411~416　DELAY_10 子程序，延时 10 个指令周期的时间，以确保 I^2C 设备来得及反应。
420~448　SET_PASS 子程序，此程序会先调用 BUFFER 子程序，要求用户输入旧密码(Old_Password)。若密码正确(调用 CHECK 子程序检查)，会再要求用户输入新密码(New_Password, 调用 KEY_PAS 子程序)，并将新密码传送至系统密码区(由地址 PASS_WORD 开始，调用 CHA_PAS 子程序)。在 CHA_PAS 子程序中，会再调用 I2C_PAGE_WRITE 子程序将新密码存到 HT24LC16。
452~458　S_ERROR 子程序，显示 ERROR 字符串并调用 DELAY 子程序延时 1.5 s。
462~479　KEY_PAS 子程序，此程序会调用 BUFFER 子程序读回用户所输入的密码，然后检查其位数(Digit)是否介于 4~10 之间。若位数超过此范围，则返回 0 表错误；否则返回 1。
483~512　CHA_PAS 子程序，首先运用间接寻址法将新输入的密码由 KEY_BUFFER 传送至系统密码区(由地址 PASS_WORD 开始)。接下来分别调用 I2C_BYTE_WRITE 与 I2C_PAGE_WRITE 子程序将 POWER_CODE、密码长度(PWD_COUNT)以及用户输入的新密码存到 HT24LC16。其中写入 POWER_CODE 的主要目的，是让 HT46x23 判断系统是否是第一次使用，若是则由程序存储器(ROM)读取系统密码；否则，就从 HT24LC16 中读取用户最后一次输入的自设密码。
516~538　CHECK 子程序，运用间接寻址法检查用户输入的密码(KEY_BUFFER)与系统密码(PASS_WORD)是否相同。如果相同则返回 1；否则返回 0 表密码错误。

542~580 BUFFER 子程序，负责读取用户输入的密码(由地址 KEY_BUFFER 开始存放)，并返回密码的长度(DIG_COUNT)。若输入 Clear 键，则清除之前所有输入的按键值并将 DIG_COUNT 归零；若输入 Enter 键，表示密码输入结束。另外，若输入的按键值不是 0~9，将不予理会。

584~601 READ_KEY_PRESSED 子程序，此子程序会调用 READ_KEY 子程序读取按键值，但必须确定用户按下按键，且等到按键放开后才会跳出此子程序。此举的主要目的，是避免 LCD 受 4×4 Key Pad 的干扰(因为 KEY_PORT 与 LCD_DATA Port 是共用的)。

605~643 PHONE 子程序，显示 "Call 110…" 字符串并发出警报声。待按下 Clear 键时，重新启动系统。PHONE 子程序会启动 Timer/Event Counter 开始计数，通过 TMRH、TMRL 计数值的改变配合 Timer/Event Counter 中断服务子程序产生音调的变化。

647~660 PRINT 子程序，此子程序负责将定义好的字符串依序显示在 LCD 上。在调用此子程序之前，除了必须先设置好 LCD 的位置之外，还需先在 Acc 寄存器中指定字符串的起始地址(字符串必须存放在最末程序页 Last Page)，并请于字符串的最后一个字符后插入 STR_END，代表字符串结束。

664~688 WLCMD 与 WLCMC 子程序，请参考 5-5 节中的说明。

693~720 READ_KEY 子程序，请参考 4-7 节中的说明。

724~737 DELAY 子程序，延时时间的计算请参考 4-1 节中的说明。

740~741 系统原始密码存放区，请记得以 END_CODE 代表密码结束。

742~750 字符串定义区，请注意每一个字符串必须以 STR_END 作为结束。

在第82~87行是以POWER_CODE的值作为是否是第一次使用系统的判断依据，当调用I2C_RM_READ子程序(第83行)时，HT24C16的第0个地址的内容会读至Acc寄存器之中。如果用户曾经修改过密码，会在存储密码到HT24LC16的同时(CHA_PAS子程序；程序第483~512行)，在第0个地址存入55h，所以程序中就以此作为判断的依据。至于操作的方式与专题二相同，请读者参考该专题的内容。

请读者注意：由于E^2PROM的写入速度较慢，所以在HT24LC16内部有一块存取速度较快的RAM Buffer，外界写入的数据都是先暂时存放在Buffer内，待进入STOP Condition后，HT24LC16再把Buffer内的数据复制到E^2PROM上。这样的作法是为了提高与外部传输的速度，但是因为内部的Buffer只有16字节，所以外界一次最多只能连续写入16字节的数据就必须进入STOP Condition，否则将会造成数据错误。另外在写入数据时也要注意时间的间距，请参考图6-3-19的时序图，其中t_{WR}是所谓的 "Write Cycle Time"，其代表写入数据后的STOP Condition与下一个动作的STOP Condition的最短间隔时间。如前所述，当写入数据到HT24LC16时，HT24LC16先将数据暂存在内部的RAM中，待进入STOP Condition后才将数据复制到E^2PROM上，所以当写入数据到HT24LC16后必须保留一段时间让其完成内部数据复制的工作。如果在数据还未完成复制前又执行写入的动作，将造成HT24LC16的动作不正常。HT24LC16的t_{WR}约为5 ms。

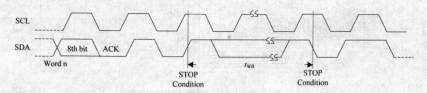

图6-3-19　HT24LC16 STOP Condition的间距

6-3-5 动动脑+动动手

- 使用本专题的密码锁时，一旦重新设置过密码，系统将永远以 HT24LC16 存储器值当作密码。万一用户忘了自设的密码，除了换一个 HT24LC16 再以系统密码重新进入设置之外，好像别无他途。请读者写一小段程序将 HT24LC16 第 0 个地址填入 55h 以外的任意数值，以省去拔换 IC 的过程。
- 在写入密码到 HT24LC16 的同时，除了 POWER_CODE 与密码长度也一并被存储之外，请将密码的检查和(Check SUM；1字节或2字节)也一并写入，以增加系统的可靠性。在由 HT24LC16 读回密码时，也应重新计算检查和并与存储的值做比较，若两者不相等则改由系统自设密码(0123)重新进入系统。

6-4 专题四：24小时时钟

6-4-1 专题功能概述

本专题为24小时的时钟设计，除了计时的基本功能之外，还可调整时间并设置闹铃(闹铃时间需精确至秒)等功能。强调的重点是：当用户设置闹铃时间时，时钟主体仍须继续计时，以避免因设置闹铃而产生时间误差。

6-4-2 学习重点

本专题的实现电路，可说是几个实验的综合，包含了LCD显示、4×4键盘扫描以及蜂鸣器发声控制，在计时方面则运用HT46x23的Timer/Event Counter中断功能，使得当用户输入闹铃时间时，HT46x23依然能够持续计时的动作，不受按键输入的影响。

6-4-3 电路图（见图6-4-1）

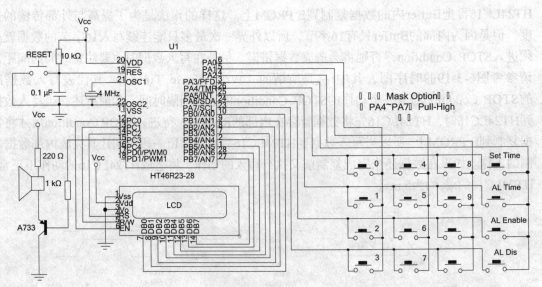

图6-4-1 24小时时钟控制电路

首先说明按键的功能：
0~9：   时间输入按键。
Set Time：   时间设置键，按下此键表示要更改目前的时间。
AL Time：   按下此键表示要更改闹铃的时间。
AL Enable：   启动闹铃功能。
AL Dise：   关闭闹铃功能，或进入闹铃时关闭喇叭的叫声。

谈到时钟的设计，想必读者脑海中必定浮现一个既定的程序模型，那就是找一个一秒的延时子程序，然后逐次依时将代表秒、分、时的寄存器加一，即可达到计时的效果，如图6-4-2的流程图所示。

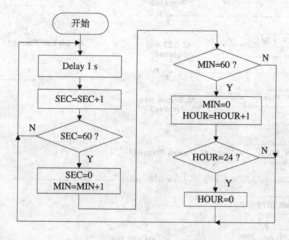

图6-4-2　时钟设计上的概念

以上概念都正确，但是若要做一个误差小的时钟可不容易，如果图6-4-2中的1 s延时子程序采用软件来设计，那时钟的准确性肯定遭人质疑，因为在延时的过程中还要兼具显示时间、用户的按键输入以及判断是否到达闹铃时间等处理过程，所以要做得很准确实在不是件容易的事。因为HT46x23已配备有Timer/Event Counter计数器，所以用硬件的方式来解决应该不会增加成本，再加上其Timer/Event Counter具有中断功能，所以即使是在用户进入设置时间的模式时仍旧可以继续计数，不影响时间的准确性。既然如此，还是请读者先把第2章与第4章有关Timer/Event Counter以及中断相关的说明及实验拿出来复习一下，写程序时才能够一气呵成。

接下来仍是需要产生一个1 s的延时，让HT46x23每隔1 s就自动更新相关的计时寄存器。由于振荡器的频率为4 MHz，再配合分频器的运用，Timer/Event Counter的计时溢位时间最长可达128×65536 μs＝2.097 s，要达到1 s的延时的确相当容易。但是考虑到用户按下按键时，必须立即反应(例如在LCD上显示相关消息)，所以在程序处理上是每隔20 ms就让Timer/Event Counter溢位一次，50次中断之后再处理相关计时寄存器的更新。因此必须再外加一个寄存器(程序中为TMR_COUNT)，每当Timer/Event Counter溢位时就减一，如果TMR_COUNT已减至0，则表示1 s已到。问题大概解决了，其他程序请读者参考流程图以及程序中的说明。

6-4-4 流程图及程序

1. 流程图(见图6-4-3)

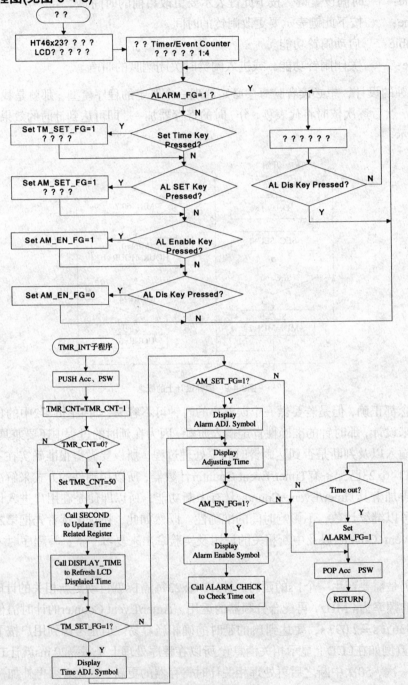

图6-4-3 流程图

2. 程序 6-4 24 小时时钟

```
1    ;PROGRAM : 6-4.ASM  (6-4.PRJ)                    BY STEVEN
2    ;FUNCTION: 24 HOUR CLOCK/ALARM DEMO PROGRAM      2002.DEC.07.
3    #INCLUDE HT46R23.INC
4             .CHIP    HT46R23
5    ;-----------------------------------------------------------------
6    MY_DATA   .SECTION 'DATA'           ;== DATA SECTION ==
7    DEL1           DB   ?               ;DELAY LOOP COUNT 1
8    DEL2           DB   ?               ;DELAY LOOP COUNT 2
9    DEL3           DB   ?               ;DELAY LOOP COUNT 3
10   DC_FLAG        DBIT                 ;LCD DATA/COMMAND FLAG
11   TM_SET_FG      DBIT                 ;TIME UNSET/SET FLAG
12   AM_SET_FG      DBIT                 ;ALARM UNSET/SET FLAG
13   AM_EN_FG       DBIT                 ;ALARM ENABLE FLAG
14   ALARM_FG       DBIT                 ;ALARM FLAG
15   LCD_CUR_ADRS   DB   ?               ;LCD CURSOR ADDRESS
16   HOR_H          DB   ?               ;HIGH DIGIT HOUR BUFFER
17   HOR_L          DB   ?               ;LOW  DIGIT HOUR BUFFER
18   MIN_H          DB   ?               ;HIGH DIGIT MIN. BUFFER
19   MIN_L          DB   ?               ;LOW  DIGIT MIN. BUFFER
20   SEC_H          DB   ?               ;HIGH DIGITSEC. BUFFER
21   SEC_L          DB   ?               ;LOW  DIGITSEC. BUFFER
22   AM_HOR_H       DB   6 DUP(?)        ;HIGH DIGIT HOUR BUFFER
23   ADJ_TIME       DB   6 DUP(?)        ;HIGH DIGIT HOUR BUFFER
24   TMR_COUNT      DB   ?               ;TMR COUNT FOR 1SEC
25   COUNT          DB   ?               ;UNIVERSAL COUNTER
26   KEY_COUNT      DB   ?               ;COUNT FOR READ_KEY ROUTINE
27   KEY            DB   ?               ;KEY CODE REGISTER
28   KEY_PS         DB   ?               ;KEY CODE REGISTER
29   STACK_A        DB   ?               ;STACK FOR ACC
30   STACK_MP0      DB   ?               ;STACK FOR MP0
31   STACK_MP1      DB   ?               ;STACK FOR MP1
32   STACK_PSW      DB   ?               ;STACK FOR STATUS
33   ;-----------------------------------------------------------------
34   LCD_CONTR    EQU  PC                ;DEFINE LCD CONTROL PORT
35   LCD_CONTRC   EQU  PCC               ;DEFINE LCD CONTROL PORT CON. REG.
36   LCD_DATA     EQU  PB                ;DEFINE LCD DATA PORT
37   LCD_DATAC    EQU  PBC               ;DEFINE LCD DATA PORT CON. REG.
38   LCD_EN       EQU  LCD_CONTR.0       ;DEFINE EN CONTROL PIN
39   LCD_RW       EQU  LCD_CONTR.1       ;DEFINE RW CONTROL PIN
40   LCD_RS       EQU  LCD_CONTR.2       ;DEFINE RS CONTROL PIN
41   LCD_READY    EQU  LCD_DATA.7        ;DEFINE READY BIT OF LCD
42   KEY_PORT     EQU  PA                ;DEFINE KEY PORT
43   KEY_PORTC    EQU  PAC               ;DEFINR KEY PORT CONTROL REG.
44   SPK_PORT     EQU  PC                ;DEFINE BUZZER PORT
45   MY_CODE  .SECTION AT 0 'CODE'       ;== PROGRAM SECTION ==
46          ORG      00H                 ;HT-46RXX RESET VECTOR
47          JMP      START
48          ORG      08H                 ;HT-46RXX TMR INTERRUPT VECTOR
49          JMP      TMR_INT
50   START:
51          CLR      LCD_CONTRC          ;CONFIG LCD_CONTR AS OUTPUT MODE
52          MOV      A,00000101B         ;ENABLE GLOBAL AND TMR INTERRUPT
53          MOV      INTC0,A
54          MOV      A,10000010B         ;CONFIG TMR 0 IN MODE 2(TIMER MODE)
55          MOV      TMRC,A              ;fINT=fSYS(4MHz)/4
56          MOV      A,HIGH (65536-20000) ;SET 20ms INTERRUTP TIME CONSTANT
57          MOV      TMRH,A
58          MOV      A,LOW  (65536-20000) ;SET 20ms INTERRUTP TIME CONSTANT
59          MOV      TMRL,A
60          MOV      A,50
61          MOV      TMR_COUNT,A         ;INITIAL TMR_COUNT 20mSx50=1s.
62          MOV      A,3*6
63          MOV      COUNT,A
64          MOV      A,OFFSET HOR_H      ;INITIAL ALL TIME BUFFER TO ZERO
65          MOV      MP0,A
```

```
66  BUF_CLEAR:
67        CLR       IAR0
68        INC       MP0
69        SDZ       COUNT
70        JMP       BUF_CLEAR
71        CLR       TM_SET_FG              ;RESET TIME SETTING FLAG
72        CLR       AM_SET_FG              ;RESET ALARM TIME SETTING FLAG
73        CLR       AM_EN_FG               ;RESET ALARM ENABLE FLAG
74        CLR       ALARM_FG               ;RESET ALARM FLAG
75        MOV       A,38H                  ;FUNCTION SET: 8-BIT,2-LINE,5X10 DOTS
76        CALL      WLCMC
77        MOV       A,0CH                  ;ON/OFF CONTR: DISPLAY ON,CURSOR BLINKING OFF
78        CALL      WLCMC
79        MOV       A,06H                  ;ENTRY MODE : INCREMENT,DISPLAY NOT SHIFT
80        CALL      WLCMC
81        MOV       A,01H                  ;CLEAR DISPLAY
82        CALL      WLCMC
83        SET       TON                    ;START TIMER
84  WAIT_KEY:
85        SZ        ALARM_FG               ;IS ALARM TIME REACH?
86        CALL      ALARMING               ;YES, ALARMING
87        CALL      READ_KEY               ;SCAN KEY
88        MOV       A,KEY
89        XOR       A,0CH
90        SZ        Z                      ;IS "SET"(C) KEY PRESSED?
91        JMP       TIME_SET               ;YES, JUMP TO SET_TIME
92        MOV       A,KEY
93        XOR       A,0DH
94        SZ        Z                      ;IS "ALRAM/SET" (D)KEY PRESSED?
95        JMP       ALARM_SET              ;YES, JUMP TO SET TIME
96        MOV       A,KEY
97        XOR       A,0EH
98        SZ        Z                      ;IS "ALRAM/ENABLE" (E)KEY PRESSED?
99        SET       AM_EN_FG               ;YES, SET ALARM_ENABLE FLAG
100       MOV       A,KEY
101       XOR       A,0FH
102       SNZ       Z                      ;IS "ALRAM/CLR" (F)KEY PRESSED?
103       JMP       WAIT_KEY               ;NO, ILLEGAL KEY, RE-READ.
104       CLR       AM_EN_FG               ;DISABLE AM_ENG_FG
105       CLR       ALARM_FG               ;DISABLE ALARMING FLAG
106       JMP       WAIT_KEY
107 ALARM_SET:
108       SET       AM_SET_FG              ;SET ALARM TIME SEETING FLAG
109       MOV       A,OFFSET AM_HOR_H      ;GET ALARM TIME BUFFER ADDRESS
110       MOV       MP0,A
111       CALL      SET_TIME               ;SET TIME
112       CLR       AM_SET_FG              ;CLEAR ALARM TIME SEETING FLAG
113       JMP       WAIT_KEY
114 TIME_SET:
115       SET       TM_SET_FG              ;SET TIME SEETING FLAG
116       MOV       A,OFFSET HOR_H         ;GET TIME BUFFER ADDRESS
117       MOV       MP0,A
118       CALL      SET_TIME               ;SET TIME
119       CLR       TM_SET_FG              ;CLEAR TIME SEETING FLAG
120       JMP       WAIT_KEY
121 ;*********************************************************************
122 ;           ALARM/SYSTEM TIME SETTING PROCEDURE
123 ;*********************************************************************
124 SET_TIME   PROC
125       MOV       A,OFFSET ADJ_TIME
126       MOV       MP1,A
127       MOV       A,0C4H
128       MOV       LCD_CUR_ADRS,A         ;INITIAL LCD CURSOR ADDRESS
129 SET:
130       CALL      READ_KEY_PRESSED       ;READ A KEY PRESSED
131       MOV       A,KEY_PS
132       SUB       A,3
```

```
133         SZ       C                            ;IS KEY <=2 ?
134         JMP      SET_TIME                     ;NO, INVALID KEY. RE-READ.
135         MOV      A,KEY_PS
136         MOV      R1,A                         ;SET AM/HOR_H.
137         INC      MP1
138         INC      LCD_CUR_ADRS                 ;INCREASE CURSOR ADDRESS BY 1
139         SUB      A,02
140         SZ       ACC
141         JMP      HORH_LESS_2
142 SET_TIME_0:
143         CALL     READ_KEY_PRESSED             ;READ A KEY PRESSED
144         MOV      A,KEY_PS
145         SUB      A,4
146         SZ       C                            ;IS KEY <=3 ?
147         JMP      SET_TIME_0                   ;NO, INVALID KEY. RE-READ.
148         MOV      A,KEY_PS
149         MOV      IAR1,A                       ;SET AM/HOR_L.
150         INC      MP1
151         JMP      SET_TIME_1
152 HORH_LESS_2:
153         CALL     READ_KEY_PRESSED             ;READ A KEY PRESSED
154         MOV      A,KEY_PS
155         SUB      A,10
156         SZ       C                            ;IS KEY <=3 ?
157         JMP      HORH_LESS_2                  ;NO, INVALID KEY. RE-READ.
158         MOV      A,KEY_PS
159         MOV      R1,A                         ;SET AM/HOR_L.
160         INC      MP1
161 SET_TIME_1:
162         INC      LCD_CUR_ADRS                 ;INCREASE CURSOR ADDRESS BY 1
163         INC      LCD_CUR_ADRS                 ;INCREASE CURSOR ADDRESS BY 1
164         CALL     READ_KEY_PRESSED             ;READ A KEY PRESSED
165         MOV      A,KEY_PS
166         SUB      A,6
167         SZ       C                            ;IS KEY <=5 ?
168         JMP      SET_TIME_1                   ;NO, INVALID KEY. RE-READ.
169         MOV      A,KEY_PS
170         MOV      IAR1,A                       ;SET AM/MIN_H.
171         INC      MP1
172         INC      LCD_CUR_ADRS                 ;INCREASE CURSOR ADDRESS BY 1
173 SET_TIME_2:
174         CALL     READ_KEY_PRESSED             ;READ A KEY PRESSED
175         MOV      A,KEY_PS
176         SUB      A,10
177         SZ       C                            ;IS KEY <=5 ?
178         JMP      SET_TIME_2                   ;NO, INVALID KEY. RE-READ.
179         MOV      A,KEY_PS
180         MOV      IAR1,A                       ;SET AM/MIN_L.
181         INC      MP1
182         INC      LCD_CUR_ADRS                 ;INCREASE CURSOR ADDRESS BY 1
183         INC      LCD_CUR_ADRS                 ;INCREASE CURSOR ADDRESS BY 1
184 SET_TIME_3:
185         CALL     READ_KEY_PRESSED             ;READ A KEY PRESSED
186         MOV      A,KEY_PS
187         SUB      A,5
188         SZ       C                            ;IS KEY <=5 ?
189         JMP      SET_TIME_3                   ;NO, INVALID KEY. RE-READ.
190         MOV      A,KEY_PS
191         MOV      IAR1,A                       ;SET AM/SEC_H.
192         INC      MP1
193         INC      LCD_CUR_ADRS                 ;INCREASE CURSOR ADDRESS BY 1
194 SET_TIME_4:
195         CALL     READ_KEY_PRESSED             ;READ A KEY PRESSED
196         MOV      A,KEY_PS
197         SUB      A,10
198         SZ       C                            ;IS KEY <=5 ?
199         JMP      SET_TIME_4                   ;NO, INVALID KEY. RE-READ.
```

```
200         MOV         A,KEY_PS
201         MOV         IAR1,A                  ;SET AM/SEC_L.
202         MOV         A,6
203         MOV         KEY,A
204         MOV         A,OFFSET ADJ_TIME       ;MOVE TIME TO TIME/ALARM BUFFER
205         MOV         MP1,A
206 SET_TIME_5:
207         MOV         A,IAR1
208         MOV         IAR0,A
209         INC         MP0
210         INC         MP1
211         SDZ         KEY
212         JMP         SET_TIME_5
213         RET
214 SET_TIME    ENDP
215 ;****************************************************************
216 ;              TIMER INTERRUPT SERVICE ROUTINE
217 ;****************************************************************
218 TMR_INT     PROC
219         MOV         STACK_A,A               ;PUSH A
220         MOV         A,STATUS
221         MOV         STACK_PSW,A             ;PUSH STATUS
222         MOV         A,MP1
223         MOV         STACK_MP1,A             ;PUSH MP1
224         MOV         A,MP0
225         MOV         STACK_MP0,A             ;PUSH MP0
226         SDZ         TMR_COUNT               ;IS 1s REACH?
227         JMP         TMR_INT_0               ;NO, JUMP TO TMR_INT_0
228         MOV         A,50
229         MOV         TMR_COUNT,A             ;RE-LOAD TMR_COUNT (50x20mS=1s.)
230         MOV         A,01H                   ;CLEAR LCD
231         CALL        WLCMC
232         CALL        SECOND                  ;INCREASE 1s
233         MOV         A,OFFSET HOR_H          ;GET TIME BUFFER START ADDRESS
234         MOV         MP0,A
235         MOV         A,84H                   ;SET LCD LINE 1, POSITION 4
236         CALL        WLCMC
237         CALL        DISPLAY_TIME            ;DISPLAY TIME
238 TMR_INT_0:
239         SNZ         TM_SET_FG               ;IS TIME SEETING ?
240         JMP         TMR_INT_1               ;NO.
241         MOV         A,0C2H                  ;SET LINE1, POSITION 0
242         CALL        WLCMC
243         MOV         A,40H                   ;PRINT ADJUST TIME SYMBOL
244         CALL        WLCMD
245         JMP         TMR_INT_2
246 TMR_INT_1:
247         SNZ         AM_SET_FG               ;IS SETTING ALARM SETTING?
248         JMP         TMR_INT_3               ;NO,
249         MOV         A,0C2H                  ;SET LINE1, POSITION 0
250         CALL        WLCMC
251         MOV         A,0EFH                  ;PRINT ADJUST ALARM SYMBOL
252         CALL        WLCMD
253 TMR_INT_2:
254         MOV         A,0FH                   ;ON/OFF CONTR: DISPLAY ON,CURSOR BLINKING ON
255         CALL        WLCMC
256         MOV         A,OFFSET ADJ_TIME       ;GET TIME BUFFER START ADDRESS
257         MOV         MP0,A
258         MOV         A,0C4H                  ;SET LCD LINE 2, POSITION 4
259         CALL        WLCMC
260         CALL        DISPLAY_TIME            ;DISPLAY TIME
261         MOV         A,LCD_CUR_ADRS          ;SET CURSOR ADDRESS
262         CALL        WLCMC
263         JMP         TMR_INT_4
264 TMR_INT_3:
265         MOV         A,0CH                   ;ON/OFF CONTR: DISPLAY ON,CURSOR BLINKING OFF
266         CALL        WLCMC
```

```
267          SNZ      AM_EN_FG            ;IS ALARM ENABLE?
268          JMP      TMR_INT_4           ;NO.
269          MOV      A,8DH               ;SET LCD LINE 2, POSITION 13
270          CALL     WLCMC
271          MOV      A,0EFH              ;DISPLAY ALARM SYMBOL
272          CALL     WLCMD
273          CALL     ALARM_CHECK         ;RETURN WITH A=1 FOR TIME=ALARM TIME!!
274          SZ       ACC                 ;NO,
275          SET      ALARM_FG            ;SET FLAG FOR ALARMING
276 TMR_INT_4:
277          MOV      A,STACK_MP0         ;POP MP0
278          MOV      MP0,A
279          MOV      A,STACK_MP1         ;POP MP1
280          MOV      MP1,A
281          MOV      A,STACK_PSW         ;POP STATUS
282          MOV      STATUS,A
283          MOV      A,STACK_A
284          RETI                         ;POP Acc
285 TMR_INT  ENDP
286 ;************************************************************************
287 ;                    INCREASE TIME BUFFER WITH 1SECOND
288 ;************************************************************************
289 SECOND   PROC
290          INC      SEC_L               ;SECOND+1
291          MOV      A,SEC_L
292          SUB      A,10
293          SZ       ACC                 ;IS OVER 10S.?
294          RET                          ;NO.
295          CLR      SEC_L               ;SEC_L=0
296          INC      SEC_H               ;SEC_H=SEC_H+1
297          MOV      A,SEC_H
298          SUB      A,6
299          SZ       ACC                 ;IS OVER 60s.?
300          RET                          ;NO.
301          CLR      SEC_H               ;YES,SEC_H=0
302          INC      MIN_L               ;MINUTE+1
303          MOV      A,MIN_L
304          SUB      A,10
305          SZ       ACC                 ;IS OVER 10 MIN.?
306          RET                          ;NO.
307          CLR      MIN_L               ;YES, MIN_L=0
308          INC      MIN_H               ;MIN_H=MIN_H+1
309          MOV      A,MIN_H
310          SUB      A,6
311          SZ       ACC                 ;IS OVER 60 MIN.?
312          RET                          ;NO.
313          CLR      MIN_H               ;YES, MIN_H=0
314          MOV      A,HOR_H
315          SUB      A,02H
316          SZ       ACC                 ;IS HOR_H = 20?
317          JMP      LESS_20             ;NO, GOTO LESS_20
318          INC      HOR_L               ;YES, HOR_L+1
319          MOV      A,HOR_L
320          SUB      A,04H
321          SZ       ACC                 ;IS 24 HOURS?
322          RET                          ;NO.
323          CLR      HOR_L               ;YES, HOR_L=0
324          CLR      HOR_H               ;HOR_H=0
325          RET
326 LESS_20:
327          INC      HOR_L               ;HOR_L+1
328          MOV      HOR_L,A
329          SUB      A,10
330          SZ       ACC                 ;IS HOR_L > 10?
331          RET                          ;NO.
332          CLR      HOR_L               ;HOR_L=0
333          INC      HOR_H               ;HOR_H+1
```

```
334         RET
335 SECOND      ENDP
336 ;**********************************************************************
337 ;           CHECK ALARM TIME IS EQUAL TO CURRENT TIME OR NOT
338 ;**********************************************************************
339 ALARM_CHECK    PROC
340         MOV     A,6
341         MOV     COUNT,A                 ;SET COMAPRE COUNT hh:mm:ss
342         MOV     A,OFFSET HOR_H          ;GET TIME BUFFER START ADDRESS
343         MOV     MP0,A
344         MOV     A,OFFSET AM_HOR_H       ;GET ALARM TIME START ADDRESS
345         MOV     MP1,A
346 ALARM_CHK_0:
347         MOV     A,IAR0
348         SUB     A,IAR1
349         SZ      ACC                     ;IS REACH ALARM TIME?
350         RET     A,0                     ;NO, RETURN WITH A=0!
351         INC     MP0
352         INC     MP1
353         SDZ     COUNT                   ;COUNT-1 = 0?
354         JMP     ALARM_CHK_0
355         RET     A,1                     ;YES, RETURN WITH A=1
356 ALARM_CHECK    ENDP
357 ;**********************************************************************
358 ;                 DISPLAY TIME SPECIFIED BY MP0 ON LCD
359 ;**********************************************************************
360 DISPLAY_TIME    PROC
361         MOV     A,2
362         MOV     COUNT,A
363 DISPLAY_1:
364         MOV     A,IAR0                  ;HOUR/MIN HIGH-DIGIT
365         ADD     A,30H                   ;CONVERT TO ASCII CODE
366         CALL    WLCMD
367         INC     MP0
368         MOV     A,IAR0                  ;HOUR/MIN LOW-DIGIT
369         ADD     A,30H                   ;CONVERT TO ASCII CODE
370         CALL    WLCMD
371         INC     MP0
372         MOV     A,3AH                   ;":"
373         CALL    WLCMD
374         SDZ     COUNT
375         JMP     DISPLAY_1
376         MOV     A,IAR0                  ;SEC. HIGH-DIGIT
377         ADD     A,30H                   ;CONVERT TO ASCII CODE
378         CALL    WLCMD
379         INC     MP0
380         MOV     A,IAR0                  ;SEC. LOW-DIGIT
381         ADD     A,30H                   ;CONVERT TO ASCII CODE
382         CALL    WLCMD
383         RET
384 DISPLAY_TIME    ENDP
385 ;**********************************************************************
386 ;         GENERATING ALARM AND WAIT UNTIL ALM/CLR KEY PRESSED
387 ;**********************************************************************
388 ALARMING    PROC
389         MOV     A,100
390         MOV     DEL1,A                  ;SET BUZZER TIME CONSTANT
391 ALARM_1:
392         MOV     A,20
393         SUB     A,DEL1
394         SZ      Z
395         MOV     A,100
396         MOV     DEL1,A
397         MOV     DEL2,A
398         MOV     A,00001000B
399         XORM    A,SPK_PORT              ;TOGGLE BUZZER
400 ALARM_2:
```

```
401         SDZ     DEL2                    ;DEL2-1 = 0?
402         JMP     ALARM_2                 ;NO.
403         CALL    READ_KEY                ;YES, SCAN KEYPAD
404         MOV     A,0FH
405         XOR     A,KEY
406         SZ      ACC                     ;IS ALM/CLR KEY PRESSED?
407         JMP     ALARM_1                 ;NO, ALARMING
408         CLR     ALARM_FG                ;YES, RESET ALARM FLAG
409         CALL    READ_KEY_PRESSED        ;WAIT KEY RELEASE
410         RET                             ;RETURN
411 ALARMING    ENDP
412 ;*********************************************************************
413 ;            LCD DATA/COMMAND WRITE PROCEDURE
414 ;*********************************************************************
415 WLCMD PROC
416         SET     DC_FLAG                 ;SET DC_FLAG=1 FOR DATA WRITE
417         JMP     WLCM
418 WLCMC:
419         CLR     DC_FLAG                 ;SET DC_FLAG=0 FOR COMMAND WRITE
420 WLCM:
421         SET     LCD_DATAC               ;CONFIG LCD_DATA AS INPUT MODE
422         CLR     LCD_CONTR               ;CLEAR ALL LCD CONTROL SIGNAL
423         SET     LCD_RW                  ;SET RW SIGNAL (READ)
424         NOP                             ;FOR TAS
425         SET     LCD_EN                  ;SET EN HIGH
426         NOP                             ;FOR TDDR
427 WF:
428         SZ      LCD_READY               ;IS LCD BUSY?
429         JMP     WF                      ;YES, JUMP TO WAIT
430         CLR     LCD_DATAC               ;NO, CONFIG LCD_DATA AS OUTPUT MODE
431         MOV     LCD_DATA,A              ;LATCH DATA/COMMAND ON PB(LCD DATA BUS)
432         CLR     LCD_CONTR               ;CLEAR ALL LCD CONTROL SIGNAL
433         SZ      DC_FLAG                 ;IS COMMAND WRITE?
434         SET     LCD_RS                  ;NO, SET RS HIGH
435         SET     LCD_EN                  ;SET EN HIGH
436         NOP
437         CLR     LCD_EN                  ;SET EN LOW
438         RET
439 WLCMD ENDP
440 ;*********************************************************************
441 ;              READ KEYPAD AND WAIT UNTIL KEY IS PRESSED
442 ;*********************************************************************
443 READ_KEY_PRESSED    PROC
444         CALL    READ_KEY                ;SCAN KEYPAD
445         MOV     A,16
446         XOR     A,KEY
447         SZ      Z                       ;IS KEY BEEN PRESSED?
448         JMP     READ_KEY_PRESSED        ;NO, RE-SCAN KEYPAD
449         MOV     A,KEY
450         MOV     KEY_PS,A                ;RESERVED KEY CODE
451 WAIT_KEY_RELEASED:
452         CALL    READ_KEY                ;SCAN KEYPAD
453         MOV     A,16
454         XOR     A,KEY
455         SZ      ACC                     ;IS KEY BEEN RELEASED?
456         JMP     WAIT_KEY_RELEASED       ;NO, RE-SCAN KEYPAD
457         MOV     A,4
458         CALL    DELAY                   ;DELAY 40ms
459         RET
460 READ_KEY_PRESSED    ENDP
461 ;*********************************************************************
462 ;   SCAN 4x4 MATRIX ON KEY PORT AND RETURN THE CODE IN KEY REGISTER
463 ;      IF NO KEY BEEN PRESSED, KEY=16.
464 ;*********************************************************************
465 READ_KEY    PROC
466         MOV     A,11110000B
467         MOV     KEY_PORTC,A             ;CONFIG KEY_PORT
```

```
468         SET     KEY_PORT                ;INITIAL KEY PORT
469         CLR     KEY                     ;INITIAL KEY REGISTER
470         MOV     A,04
471         MOV     KEY_COUNT,A             ;SET ROW COUNTER
472         CLR     C                       ;CLEAR CARRY FLAG
473 SCAN_KEY:
474         RLC     KEY_PORT                ;ROTATE SCANNING BIT
475         SET     C                       ;MAKE SURE C=1
476         SNZ     KEY_PORT.4              ;COLUMN 0 PRESSED?
477         JMP     END_KEY                 ;YES.
478         INC     KEY                     ;NO, INCREASE KEY CODE.
479         SNZ     KEY_PORT.5              ;COLUMN 1 PRESSED?
480         JMP     END_KEY                 ;YES.
481         INC     KEY                     ;NO, INCREASE KEY CODE.
482         SNZ     KEY_PORT.6              ;COLUMN 2 PRESSED?
483         JMP     END_KEY                 ;YES.
484         INC     KEY                     ;NO, INCREASE KEY CODE.
485         SNZ     KEY_PORT.7              ;COLUMN 3 PRESSED?
486         JMP     END_KEY                 ;YES.
487         INC     KEY                     ;NO, INCREASE KEY CODE.
488         SDZ     KEY_COUNT               ;HAVE ALL ROWs BEEN CHECKED?
489         JMP     SCAN_KEY                ;NO, NEXT ROW.
490 END_KEY:
491         RET
492 READ_KEY    ENDP
493 ;*********************************************************************
494 ;                       Delay about DEL1*10ms
495 ;*********************************************************************
496 DELAY       PROC
497         MOV     DEL1,A                  ;SET DEL1 COUNTER
498 DEL_1:
499         MOV     A,30
500         MOV     DEL2,A                  ;SET DEL2 COUNTER
501 DEL_2:
502         MOV     A,110
503         MOV     DEL3,A                  ;SET DEL3 COUNTER
504 DEL_3:
505         SDZ     DEL3                    ;DEL3 DOWN COUNT
506         JMP     DEL_3
507         SDZ     DEL2                    ;DEL2 DOWN COUNT
508         JMP     DEL_2
509         SDZ     DEL1                    ;DEL1 DOWN COUNT
510         JMP     DEL_1
511         RET
512 DELAY       ENDP
513         END
```

程序说明

7~32   依序定义变量地址。

34~37   定义 LCD_DATA Port 与 LCD_CONTR Port 分别为 PB 与 PC。

38~41   定义 LCD 的控制信号引脚。

42~43   定义 KEY_PORT 为 PA。

44   定义 SPK_PORT 为 PC。

47   声明存储器地址由 000h 开始。

49   声明存储器地址由 008h 开始。

52   定义 LCD_CONTR Port 为输出模式。

53~54   Global Interrupt 与 Timer/Event Counter Interrupt 有效。

第 6 章 实践应用篇

行号	说明
55~56	定义 TMRC 控制寄存器,Timer/Event Counter 为 Timer Mode、预分频比例=1:4($f_{INT}=f_{SYS}/4$)。
57~60	设置 TMRH 与 TMRL,使 Timer/Event Counter 的计数溢位时间为 20 ms。
61~62	设置 TRM_COUNT=50,在 Timer/Event Counter 中断服务子程序中,会将 TRM_COUNT−1,当其递减至 0 时即代表 1 s(20 ms×50)。
65~71	运用间接寻址法清除计时相关的寄存器。
72~75	清除程序中用来作为判断的相关标志位: TM_SET_FG:用户是否按下 Set Time 按键的判断标志位; AM_SET_FG:用户是否按下 AL Time 按键的判断标志位; AM_EN_FG:是否启动闹铃功能的判断标志位; ALARM_FG:是否进入闹铃程序的判断标志位;
76~77	将 LCD 设置为双行显示(N = 1)、使用 8 位(DB7 ~ DB0)控制模式(DL = 1)、5×7 点矩阵字形(F = 0)。
78~79	将 LCD 设置为显示所有数据(D = 1)、显示光标(C = 1)、光标所在位置的字不会闪烁(B = 0)。
80~81	将 LCD 的地址标志位(AC)设置为递加(I/D = 1)、显示器画面不因读/写数据而移动(S = 0)。
82~83	将 LCD 整个显示器清空。
84	启动 Timer/Event Counter 开始计数。
86~87	判断是否已到达闹铃时间(ALARM_FG=1),若是则调用 ALARMING 子程序,进入闹铃程序。
88	调用 READ_KEY 子程序,扫描键盘。
89~92	判断是否按下 Set Time 按键,若是则进入时间设置过程(JMP TIME_SET)。
93~96	判断是否按下 AL Time 按键,若是则进入闹铃时间设置过程(JMP ALARM_SET)。
97~100	判断是否按下 AL Enable 按键,若是则设置 AM_EN_FG=1。
101~106	判断是否按下 AL Dis 按键,若是则清除 AM_EN_FG 与 ALARM_FG。
107	重新扫描键盘(JMP WAIT_KEY)。
108~114	闹铃时间设置过程,会调用 SET_TIME 子程序将用户设置的闹铃时间存入 AM_HOR_H(6 字节)寄存器。因为 SET_TIME 子程序用间接寻址法将输入的时间存放在 MP0 所指定的地址,所以在第 110~111 行先将 MP0 指向 AM_HOR_H 寄存器的起始地址。而第 109 行将 AM_SET_FG 设置为 1 的目的,是为了控制 Timer/Event Counter 中断服务子程序做相关的显示。用户输入完毕之后,会再重新进行扫描键盘的动作(JMP WAIT_KEY)。
115~121	时间设置程序,会调用 SET_TIME 子程序将用户设置的时间存入 HOR_H(6 字节)寄存器。因为 SET_TIME 子程序是用间接寻址法将输入的时间存放在 MP0 所指定的地址,所以在第 117~118 行先将 MP0 指向 HOR_H 寄存器的起始地址。而第 116 行将 TM_SET_FG 设置为 1 的目的,是为了控制 Timer/Event Counter 中断服务子程序做相关的显示。用户输入完毕之后,会再重新进行扫描键盘的动作(JMP WAIT_KEY)。
125~215	SET_TIME 子程序,此段程序的主要功能是将用户输入的时间(目前时间或闹铃时间)存放在 MP0 所指定的地址。为了在输入过程中,让 LCD 仍然正常显示计时的时间,所以先将用户输入的时间存放在 ADJ_TIME(6 字节)寄存器,待完成输入之后再将其复写到指定的地址。程序中有许多判断的指令,其目的只是在确定用户输入的格式是否正确(小时、分、秒的个位数是否小于 10;分、秒的十位数是否小于 6;小时的十位数是否小于 3)。
219~286	Timer/Event Counter 中断服务子程序,配合 TMRH、TMRL 的初值设置,每隔 20 ms 就会进入中断服务子程序,此段程序有几个主要功能:

① 将TMR_COUNT-1，若为0表示已经计时1 s，首先将TMR_COUNT重新设为50，为下一秒钟的计时工作做好准备；其次调用SENCOND子程序进行计时相关寄存器的更新过程，然后调用DISPLA_TIME子程序将更新后的时间显示在LCD上。以上是第227~238行程序所完成的工作。

② 检查是否进入时间调整过程(TM_SET_FG=1)。若是，则在LCD上显示"☺"符号以供用户识别，并跳到TMR_INT_2(第254行)显示调整时间的相关消息。以上是第240~246行程序所完成的工作。

③ 检查是否进入闹铃时间调整过程(AM_SET_FG=1)。若是，则在LCD上显示"Ö"符号以供用户识别(第248~253行)，并调用DISPLAY_TIME子程序显示调整闹铃时间的相关消息(第255~264行)。

④ 检查是否启动闹铃功能。若是，则在LCD上显示"Ö"符号以供用户识别(第266~273行)，其次调用ALARM_CHECK子程序分析是否已到达闹铃时间，如果已经到达闹铃时间，就设置ALARM_FG=1(第274~276行)。

290~336 SENCOND子程序，更新小时(HOR_H、HOR_L)、分(MIN_H、MIN_L)、秒(SEC_H、SEC_L)计时相关寄存器的值，若已计时至23:59:59则将上述寄存器的值归0。

340~357 ALARM_CHECK子程序，运用间接寻址法比较闹铃时间与目前时间是否相同，若相同则返回Acc=1；否则返回Acc=0。

361~385 DISPLAY_TIME子程序，将指定的计时相关寄存器数据显示在LCD上。其中运用了间接寻址法的技巧，所以在调用之前必须先将MP0指向寄存器的起始地址。

389~412 ALARMING子程序，运用DELAY的技巧让蜂鸣器发声，期间会调用READ_KEY子程序，检查用户是否按下AL Dis按键，若是则会清除ALARM_FG；否则就一直持续闹铃的过程。在第410行调用READ_KEY_PRESSED子程序目的，是要确认用户已经放开按键。

416~440 WLCMD与WLCMC子程序，请参考5-5节中的说明。

444~461 READ_KEY_PRESSED子程序，此子程序会调用READ_KEY子程序读取按键值，但必须确定用户按下按键，且等到按键放开后才会跳出此子程序。

466~493 READ_KEY子程序，请参考4-7节中的说明。

497~510 DELAY子程序，延时时间的计算请参考4-1节中的说明。

以下介绍24小时时钟的操作方式，程序执行时，LCD会显现如下的画面(计时模式)：

			0	0	:	0	0	:	0	0		

并且开始计时，此刻用户可以按下Time Set键输入正确的时间，画面如下(时间设置模式)：

			0	0	:	0	0	:	0	3		
	☺		0	0	:	0	0	:	0	0		

此时LCD第一排代表目前的时间，仍旧持续计时。第二排代表输入的时间(以"☺"符号表示)，正确地输入时、分、秒之后，会回到计时模式继续显示时间。如果输入的时间格式错误(例如输入35:00:00)，除不予显示之外也不影响原来的计时值。

按下AL Time按键之后，用户可以设置闹铃时间，此时LCD将显示如下的画面：

				0	0	:	0	0	:	1	3		
	Ö			0	0	:	0	0	:	0	0		

第一排代表目前的时间，仍旧持续计时。第二排代表输入的时间(以"Ö"符号表示)，正确地输入时、分、秒之后，会回到计时模式继续显示时间。如果输入的时间格式错误(例如输入35:00:00)，除不予显示之外也不影响原来的计时值。

用户可以按下AL Enable键启动闹铃功能，此时LCD的显示如下：

				0	0	:	0	0	:	1	9		Ö

如果想要取消闹铃功能，只要按下AL Dis按键就可以。此时代表闹铃功能的"Ö"符号会消失，表示回到正常的计时模式。当到达闹铃时间而且闹铃功能又被启动时，蜂鸣器会发出声响提醒用户，此时显示的时间仍旧持续计时。请按下AL Dis键停止闹铃，系统会回到计时模式继续显示时间。读者应该很容易发现，因为计时的动作是在Timer/Event Counter中断中进行的，因此所有的按键输入动作都不会影响时间的准确度。

6-4-5　动动脑＋动动手

  📖 当进入闹铃的过程时，除非用户按下 AL Dis 键，否则蜂鸣器会一直叫个不停。试着改写程序，如果 3 min 后用户仍未按下 AL Dis 按键，就自动停止蜂鸣器发出声音。

6-5　专题五：猜数字游戏机

6-5-1　专题功能概述

  本游戏机的游戏规则是：由HT46x23产生一组4位数的随机数(wxyz)，让用户猜。若用户所猜的数值(w'y'x'z')中有n个位数与随机数数值、位置均相同，有m个位数与随机数数值相同但位置不同，则呈现nAmB的消息通知用户目前猜对4个号码，但有2个数值与随机数字置不相同，让用户归纳这些消息之后能顺利地猜出正确的数值。其中随机数的规则是w、x、y、z各个位数的范围为0~9，而且彼此之间不能重复。

6-5-2　学习重点

  本专题的实现电路，可说是几个实验的综合，在硬件方面对读者而言应该是驾轻就熟，因此本专题的重点乃着重于程序技巧的磨练。

6-5-3　电路图（见图6-5-1）

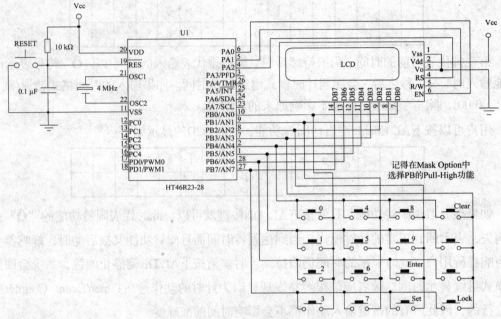

图6-5-1　猜数字游戏机控制电路

此专题可分为三大部分来说明：随机数产生规则、对比并判断nAmB以及人机接口。首先探讨随机数的产生方式，在5-9节与5-10节中，以Timer/Event Counter来产生随机数的方式在此再度被应用，但是必须加上4个位数不能重复的限制。请读者参考流程图(见图6-5-2)及程序。

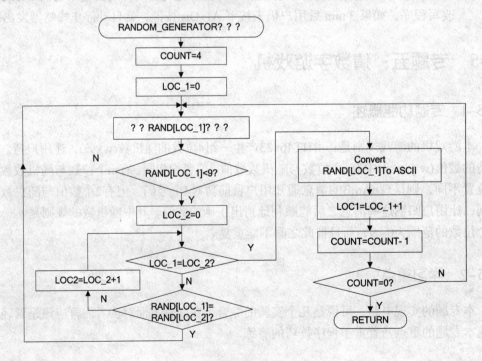

图6-5-2　流程图

```
1   ;*********************************************************************
2   ;           GENERATE 4 UNDUPILCATED 0~9 DIGIT
3   ;*********************************************************************
4   RANDOM_GENERATOR    PROC
5           MOV     A,OFFSET RANDOM     ;INITIAL DATA POINTER TO RANDOM
6           MOV     MP0,A
7           MOV     A,04
8           MOV     COUNT,A             ;SET COUNT=4
9   GET_RANDOM:
10          MOV     A,TMRH              ;GET TMRH
11          XOR     A,TMRL              ;XOR WITH TRML
12          AND     A,0FH               ;MASK HIGH BYTE
13          MOV     IAR0,A
14          SUB     A,10
15          SZ      C                   ;A<10?
16          JMP     GET_RANDOM          ;NO, RE-READ
17          MOV     A,MP0                ;YES,DECREASE POINTER BY 1
18          MOV     MP1,A
19  CHECK_DUPLICATE:
20          DEC     MP1                 ;DECREASE POINTER BY 1
21          MOV     A,OFFSET RANDOM-1
22          XOR     A,MP1
23          SZ      Z                   ;IS THE FIRST RANDOM DIGIT?
24          JMP     NO_DUPLICATE        ;YES, GOTO NO_DUPLICATE
25          MOV     A,IAR0
26          ADD     A,30H               ;CONVERT TO ASCII CODE
27          XOR     A,IAR1
28          SZ      Z                   ;DOES THE NUMBER DUPLICATE?
29          JMP     GET_RANDOM          ;YES, RE-READ
30          JMP     CHECK_DUPLICATE
31  NO_DUPLICATE:
32          MOV     A,30H
33          ADDM    A,IAR0              ;CONVERT TO ASCII CODE
34          INC     MP0                 ;INCREASE DATA POINTER BY 1
35          XOR     A,TMRL              ;DISTURB A
36  DISTURB_ACC:
37          SDZ     ACC                 ;FOR TMR GENERATE RANDOM
38          JMP     DISTURB_ACC
39          SDZ     COUNT               ;COUNT-1 = 0?
40          JMP     GET_RANDOM          ;NO, NEXT RANDOM DIGIT
41          RET                         ;YES, RETURN
42  RANDOM_GENERATOR    ENDP
```

程序中利用间接寻址法(IRA0与MP0、IRA1与MP1)来存放产生随机数以及进行数字是否重复的对比过程，结合流程图的说明读者应该可以理解。

比较特别的是在第35~38行的延时循环，虽然第一个数字的产生与用户按下按键的时间有关，但由于每次按下按键的时间相同的机率不大，因此很难产生两组一样的随机数。但是其他数字的产生只与程序执行的流程有关，如果让其直接产生，用户将很容易归纳出数字间的关系，所以在此故意将随机数与TMRL互斥或之后的数值作为延时的小循环，让随机数之间的关系更为凌乱。

至于对比的过程与密码锁中的方式也极为类似，但现在不是光提供"对"或"错"的答案而已，而是必须提供用户有几个位数的数值、位置均相同(nA)，有几个位数仅数值一样但位置不相同的信息(mB)，请读者参考流程图(见图6-5-3)及程序。

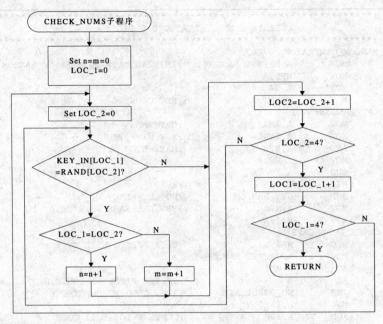

图6-5-3 流程图

其中的KEY_IN[]代表用户输入的数列，RAND[]代表HT46x23所产生的随机数数列，此段流程可用如下的C语言来描述：

```c
for (LOC_1=0;LOC_1<4;LOC_1++)
{    for (LOC_2=0;LOC_2<4;LOC_2++)
     {    if (KEY_IN[LOC_1]=RAND[LOC_2])
          {    if (LOC_1=LOC_2)
                    n=n+1;
               else
                    m=m+1;
          }
     }
}
```

写成HT46x23的汇编语言，就变成以下的形式了：

```
1     ;********************************************************
2     ;    COMPARE 4 INPUT DIGITS WITH RANDOM AND RESULTING m,n
3     ;********************************************************
4     CHECK_NUMS   PROC
5          CLR     N                      ;SET n=0
6          CLR     M                      ;SET m=0
7          MOV     A,OFFSET RANDOM
8          MOV     MP0,A                  ;INITIAL DATA POINTER TO RANDOM BUFFER
9     CHECK_1:
10         MOV     A,OFFSET KEYIN_BUF     ;INITIAL DATA POINTER TO KEYIN_BUF
11         MOV     MP1,A
12    CHECK_2:
13         MOV     A,IAR0
14         XOR     A,IAR1
15         SZ      ACC                    ;EQUAL?
16         JMP     CHECK_3                ;NO.
17         MOV     A,MP0                  ;YES.
18         ADD     A,04
19         XOR     A,MP1
20         SZ      ACC                    ;THE SAME POSITION?
```

```
21          JMP     B_INC                   ;NO.
22  A_INC:
23          INC     N                       ;YES,n=n+1
24          JMP     CHECK_3
25  B_INC:
26          INC     M                       ;m=m+1
27  CHECK_3:
28          INC     MP1                     ;INCREASE DATA POINTER BY 1
29          MOV     A,MP1
30          XOR     A,OFFSET KEYIN_BUF+4
31          SZ      ACC                     ;END OF KEYIN_BUF?
32          JMP     CHECK_2                 ;NO, NEXT
33          INC     MP0                     ;YES,INCREASE DATA POINTER BY 1
34          MOV     A,MP0
35          XOR     A,OFFSET RANDOM+4
36          SZ      ACC                     ;END OF RANDOM BUFFER?
37          JMP     CHECK_1                 ;NO, NEXT
38          RET                             ;YES, RETURN
39  CHECK_NUMS  ENDP
```

最后,关于人机接口的部分,读者由电路图中可以发现,在此再次以LCD与4×4 Key Pad作为输入/输出的接口,这部分应该可以不需要再多做说明。

6-5-4 流程图及程序

1. 流程图(见图 6-5-4)

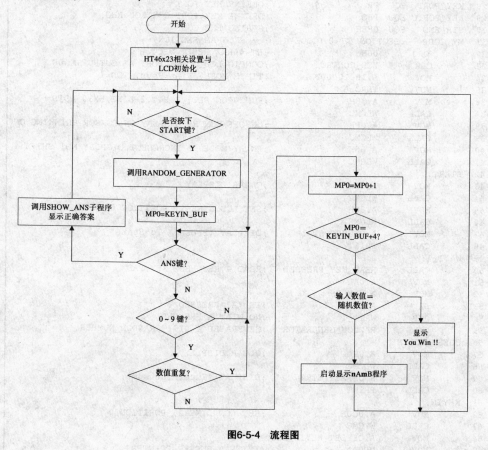

图6-5-4 流程图

2. 程序6-5 猜数字游戏机

```
1   ;PROGRAM : 6-5.ASM  (6-5.PRJ)              BY STEVEN
2   ;FUNCTION: RANDOM GAME DEMO PROGRAM        2002.DEC.07.
3   #INCLUDE   HT46R23.INC
4              .CHIP    HT46R23
5   ;------------------------------------------------------------------
6   MY_DATA    .SECTION 'DATA'         ;== DATA SECTION ==
7   DEL1       DB   ?                  ;DELAY LOOP COUNT 1
8   DEL2       DB   ?                  ;DELAY LOOP COUNT 2
9   DEL3       DB   ?                  ;DELAY LOOP COUNT 3
10  DC_FLAG    DBIT                    ;LCD DATA/COMMAND FLAG
11  COUNT      DB   ?                  ;UNIVERSAL COUNT
12  KEY_COUNT  DB   ?                  ;ROW COUNT USED IN READ_KEY
13  KEY        DB   ?                  ;REGISTER FOR READ_KEY
14  KEY_PS     DB   ?                  ;BUFFER FOR READ_A_KEY
15  M          DB   ?                  ;mA RESULT REGISTER
16  N          DB   ?                  ;nB RESULT REGISTER
17  RANDOM     DB   4 DUP(?)           ;BUFFER FOR RANDOM NUMBER
18  KEYIN_BUF  DB   4 DUP(?)           ;BUFFER FOR USER KEY-IN
19  ;------------------------------------------------------------------
20  LCD_CONTR  EQU  PA                 ;DEFINE LCD CONTROL PORT
21  LCD_CONTRC EQU  PAC                ;DEFINE LCD CONTROL PORT CON. REG.
22  LCD_DATA   EQU  PB                 ;DEFINE LCD DATA PORT
23  LCD_DATAC  EQU  PBC                ;DEFINE LCD DATA PORT CON. REG.
24  LCD_EN     EQU  LCD_CONTR.0        ;DEFINE EN CONTROL PIN
25  LCD_RW     EQU  LCD_CONTR.1        ;DEFINE RW CONTROL PIN
26  LCD_RS     EQU  LCD_CONTR.2        ;DEFINE RS CONTROL PIN
27  LCD_READY  EQU  LCD_DATA.7         ;DEFINE READY BIT OF LCD
28  KEY_PORT   EQU  PB                 ;DEFINE KEY PORT
29  KEY_PORTC  EQU  PBC                ;DEFINE KEY PORT CONTROL REG.
30  STR_END    EQU  0FEH               ;STRING END CODE
31  MY_CODE    .SECTION AT 0 'CODE'    ;== PROGRAM SECTION ==
32         ORG     00H                 ;HT-46RXX RESET VECTOR
33         CLR     LCD_CONTRC          ;CONFIG LCD_CONTR PORT AS OUTPUT MODE
34         MOV     A,10010000B         ;TIMER MODE, Fint=Fsys, TON=1
35         MOV     TMRC,A
36         MOV     A,38H               ;FUNCTION SET: 8-BIT,2-LINE,5X10 DOTS
37         CALL    WLCMC
38         MOV     A,0FH               ;ON/OFF CONTR: DISPLAY ON,CURSOR BLINKING ON
39         CALL    WLCMC
40         MOV     A,06H               ;ENTRY MODE  : INCREMENT,DISPLAY NOT SHIFT
41         CALL    WLCMC
42  START:
43         MOV     A,01H               ;CLEAR DISPLAY
44         CALL    WLCMC
45         MOV     A,80H               ;SET LCD LINE 1, POSITION 0
46         CALL    WLCMC
47         MOV     A,OFFSET STR1       ;DISPLAY "Press A To Start"
48         CALL    PRINT
49  WAIT_1:
50         CALL    READ_KEY_PRESSED    ;READ A KEY FROM KEY PAD
51         MOV     A,0AH
52         XOR     A,KEY_PS
53         SZ      ACC                 ;IS "A" PRESSED ?
54         JMP     WAIT_1              ;NO,WIAT A PRESSED
55         CALL    RANDOM_GENERATOR    ;GENERATED 4 DIGIT RANDOM NUMBER
56  GUESS_AGAIN:
57         MOV     A,01                ;CLEAR DISPLAY
58         CALL    WLCMC
59         MOV     A,OFFSET STR2       ;DISPLAY "Enter The Nums:"
60         CALL    PRINT
61  KEYIN:
62         MOV     A,0C0H              ;SET LCD LINE 2, POSITION 0
63         CALL    WLCMC
64         MOV     A,OFFSET KEYIN_BUF
65         MOV     MP0,A               ;INITIAL BUFFER POINTER
```

```
66  GET_KEY:
67       CALL    READ_KEY_PRESSED    ;READ KEY
68       MOV     A,0BH
69       XOR     A,KEY_PS
70       SZ      ACC                 ;IS "ANS(B)" BEEN PRESSED?
71       JMP     NEXT                ;NO, JUMP TO NEXT
72       CALL    SHOW_ANS            ;YES, JUMP TO SHOW ANSWER
73       JMP     WAIT                ;GOTO WAIT KEY "A" PRESSED
74  NEXT:
75       MOV     A,KEY_PS
76       SUB     A,10
77       SZ      C                   ;IS KEY_PS > 9?
78       JMP     GET_KEY             ;YES,RE-READ THE KEY PAD
79       MOV     A,30H               ;NO, CONVERT TO ASCII CODE
80       ADDM    A,KEY_PS            ;SAVE IN KEY_PS
81  CHECK:
82       DECA    MP0                 ;LOAD PREVIOUS DIGIT POINTER TO Acc
83       MOV     MP1,A
84       XOR     A,OFFSET KEYIN_BUF-1
85       SZ      Z                   ;IS THE FIRST KEY-IN NUMBER?
86       JMP     DISPLAY             ;YES,JUMP FOR DISPLAYING
87  CHK_LL:
88       MOV     A,KEY_PS
89       XOR     A,IAR1
90       SZ      Z                   ;IS THE NUMBER SUPLICATE WITH PREVIOUS?
91       JMP     GET_KEY             ;YES,RE-READ
92       DEC     MP1                 ;NEXT BUFFER
93       MOV     A,OFFSET KEYIN_BUF-1
94       XOR     A,MP1
95       SZ      ACC                 ;HAS ALL KEY IN NUMBER BEEN CHECKED?
96       JMP     CHK_LL              ;NO,CHECK NEXT INPUT NUMs
97  DISPLAY:
98       MOV     A,KEY_PS
99       MOV     IAR0,A              ;SAVE KEY IN DIGIT(ASCII) TO BUFFER
100      CALL    WLCMD               ;DSIPLAY INPUT DATA
101      INC     MP0                 ;INCREASE DATA POINTER
102      MOV     A,OFFSET KEYIN_BUF+4
103      XOR     A,MP0
104      SZ      ACC                 ;DOES 4 DIGITS INPUT OK?
105      JMP     GET_KEY             ;NO, READ NEXT INPUT DIGIT
106      CALL    CHECK_NUMS          ;YES, CHECK NUMBERS
107 OUTPUT_RESULT:
108      MOV     A,0C8H              ;SET LCD LIN 2, POSITION 8
109      CALL    WLCMC               ;DISPLAY THE nAmB RESULT
110      MOV     A,N                 ;GET n
111      ADD     A,30H               ;CONVERT TO ASCII CODE
112      CALL    WLCMD
113      MOV     A,41H               ;PRINT 'A'
114      CALL    WLCMD
115      MOV     A,M                 ;GET m
116      ADD     A,30H               ;CONVERT TO ASCII CODE
117      CALL    WLCMD
118      MOV     A,42H               ;PRINT 'B'
119      CALL    WLCMD
120      MOV     A,N                 ;GET n
121      XOR     A,4
122      SZ      Z                   ;IS KEYIN_BUF = RANDOM?
123      JMP     GUESS_RIGHT         ;YES, JUMP TO GUESS_RIGHT
124 WAIT_KEY_A:
125      CALL    READ_KEY_PRESSED    ;NO, WAIT "A" KEY TO GUESS AGAIN!
126      MOV     A,KEY_PS
127      XOR     A,0AH
128      SZ      ACC                 ;IS "A" PRESSED?
129      JMP     WAIT_KEY_A          ;NO, RE-READ KEYPAD
130 WAIT_3:
131      MOV     A,01H               ;YES,CLEAR DISPLAY
132      CALL    WLCMC
```

```
133         JMP       GUESS_AGAIN
134 GUESS_RIGHT:
135         MOV       A,80H               ;SET LCD LINE 1, POSITION 0
136         CALL      WLCMC
137         MOV       A,OFFSET STR3       ;DISPLAY "You WIN !!!"
138         CALL      PRINT
139 WAIT:
140         CALL      READ_KEY_PRESSED    ;NO, WAIT "A" KEY TO RESTART!
141         MOV       A,KEY_PS
142         XOR       A,0AH
143         SZ        ACC                 ;IS "A" PRESSED?
144         JMP       WAIT                ;NO, RE-READ KEYPAD
145         JMP       START               ;YES, RE-START THE GAME
146 ;**********************************************************************
147 ;        DISPLAY THE RANDOM NUMBER GENERATED BY PROGRAM
148 ;**********************************************************************
149 SHOW_ANS   PROC
150         MOV       A,01                ;CLEAR DISPLAY
151         CALL      WLCMC
152         MOV       A,STR4              ;DISPLAY "The Answer Is:"
153         CALL      PRINT
154         MOV       A,0C0H              ;SET LCD LINE 2, POSITION 0
155         CALL      WLCMC
156         MOV       A,4
157         MOV       COUNT,A             ;SET COUNT=4
158         MOV       A, OFFSET RANDOM
159         MOV       MP0,A               ;INITIAL DATA POINTER
160 SHOW_1:
161         MOV       A,IAR0              ;DISPLAY DIGIT
162         CALL      WLCMD
163         INC       MP0                 ;INCREASE DATA POINTER
164         SDZ       COUNT               ;COUNT-1 =0?
165         JMP       SHOW_1              ;NO, DISPLAY NEXT DIGIT
166         RET                           ;YES,RETURN
167 SHOW_ANS   ENDP
168 ;**********************************************************************
169 ;     COMPARE 4 INPUT DIGITS WITH RANDOM AND RESULTING m,n
170 ;**********************************************************************
171 CHECK_NUMS    PROC
172         CLR       N                   ;SET n=0
173         CLR       M                   ;SET m=0
174         MOV       A,OFFSET RANDOM
175         MOV       MP0,A               ;INITIAL DATA POINTER TO RANDOM BUFFER
176 CHECK_1:
177         MOV       A,OFFSET KEYIN_BUF  ;INITIAL DATA POINTER TO KEYIN_BUF
178         MOV       MP1,A
179 CHECK_2:
180         MOV       A,IAR0
181         XOR       A,IAR1
182         SZ        ACC                 ;EQUAL?
183         JMP       CHECK_3             ;NO.
184         MOV       A,MP0               ;YES.
185         ADD       A,04
186         XOR       A,MP1
187         SZ        ACC                 ;THE SAME POSITION?
188         JMP       B_INC               ;NO.
189 A_INC:
190         INC       N                   ;YES,n=n+1
191         JMP       CHECK_3
192 B_INC:
193         INC       M                   ;m=m+1
194 CHECK_3:
195         INC       MP1                 ;INCREASE DATA POINTER BY 1
196         MOV       A,MP1
197         XOR       A,OFFSET KEYIN_BUF+4
198         SZ        ACC                 ;END OF KEYIN_BUF?
199         JMP       CHECK_2             ;NO, NEXT
```

```
200         INC         MP0                     ;YES,;INCREASE DATA POINTER BY 1
201         MOV         A,MP0
202         XOR         A,OFFSET RANDOM+4
203         SZ          ACC                     ;END OF RANDOM BUFFER?
204         JMP         CHECK_1                 ;NO, NEXT
205         RET                                 ;YES, RETURN
206 CHECK_NUMS      ENDP
207 ;**********************************************************************
208 ;              GENERATE 4 UNDUPILCATED 0~9 DIGIT
209 ;**********************************************************************
210 RANDOM_GENERATOR    PROC
211         MOV         A,OFFSET RANDOM         ;INITIAL DATA POINTER TO RANDOM
212         MOV         MP0,A
213         MOV         A,04
214         MOV         COUNT,A                 ;SET COUNT=4
215 GET_RANDOM:
216         MOV         A,TMRH                  ;GET TMRH
217         XOR         A,TMRL                  ;XOR WITH TRML
218         AND         A,0FH                   ;MASK HIGH BYTE
219         MOV         IAR0,A
220         SUB         A,10
221         SZ          C                       ;A<10?
222         JMP         GET_RANDOM              ;NO, RE-READ
223         MOV         A,MP0                   ;YES,DECREASE POINTER BY 1
224         MOV         MP1,A
225 CHECK_DUPLICATE:
226         DEC         MP1                     ;DECREASE POINTER BY 1
227         MOV         A,OFFSET RANDOM-1
228         XOR         A,MP1
229         SZ          Z                       ;IS THE FIRST RANDOM DIGIT?
230         JMP         NO_DUPLICATE            ;YES, GOTO NO_DUPLICATTE
231         MOV         A,IAR0
232         ADD         A,30H                   ;CONVERT TO ASCII CODE
233         XOR         A,IAR1
234         SZ          Z                       ;DOES THE NUMBER DUPLICATE?
235         JMP         GET_RANDOM              ;YES, RE-READ
236         JMP         CHECK_DUPLICATE
237 NO_DUPLICATE:
238         MOV         A,30H
239         ADDM        A,IAR0                  ;CONVERT TO ASCII CODE
240         INC         MP0                     ;INCREASE DATA POINTER BY 1
241         XOR         A,TMRL                  ;DISTURB A
242 DISTURB_ACC:
243         SDZ         ACC                     ;FOR TMR GENERATE RANDOM
244         JMP         DISTURB_ACC
245         SDZ         COUNT                   ;COUNT-1 = 0?
246         JMP         GET_RANDOM              ;NO, NEXT RANDOM DIGIT
247         RET                                 ;YES, RETURN
248 RANDOM_GENERATOR    ENDP
249 ;**********************************************************************
250 ;       LCD PRINT PROCEDURE (START ADDRS=TBLP)
251 ;**********************************************************************
252 PRINT PROC
253         MOV         TBLP,A                  ;LOAD TABLE POINTER
254 PRINT_1:
255         TABRDL      ACC                     ;READ STRING
256         XOR         A,STR_END
257         SZ          Z                       ;END OF STRING CHARACTER?
258         JMP         END_PRINT               ;YES, STOP PRINT
259         TABRDL      ACC                     ;NO, RE-READ
260         CALL        WLCMD                   ;SEND TO LCD
261         INC         TBLP                    ;INCREASE DATA POINTER
262         JMP         PRINT_1                 ;NEXT CHARACTER
263 END_PRINT:
264         RET                                 ;YES, RETURN
265 PRINT ENDP
266 ;**********************************************************************
```

```
267 ;              LCD DATA/COMMAND WRITE PROCEDURE
268 ;****************************************************************
269 WLCMD   PROC
270         SET     DC_FLAG         ;SET DC_FLAG=1 FOR DATA WRITE
271         JMP     WLCM
272 WLCMC:
273         CLR     DC_FLAG         ;SET DC_FLAG=0 FOR COMMAND WRITE
274 WLCM:
275         SET     LCD_DATAC       ;CONFIG LCD_DATA AS INPUT MODE
276         CLR     LCD_CONTR       ;CLEAR ALL LCD CONTROL SIGNAL
277         SET     LCD_RW          ;SET RW SIGNAL (READ)
278         NOP                     ;FOR TAS
279         SET     LCD_EN          ;SET EN HIGH
280         NOP                     ;FOR TDDR
281 WF:
282         SZ      LCD_READY       ;IS LCD BUSY?
283         JMP     WF              ;YES, JUMP TO WAIT
284         CLR     LCD_DATAC       ;NO, CONFIG LCD_DATA AS OUTPUT MODE
285         MOV     LCD_DATA,A      ;LATCH DATA/COMMAND ON PB(LCD DATA BUS)
286         CLR     LCD_CONTR       ;CLEAR ALL LCD CONTROL SIGNAL
287         SZ      DC_FLAG         ;IS COMMAND WRITE?
288         SET     LCD_RS          ;NO, SET RS HIGH
289         SET     LCD_EN          ;SET EN HIGH
290         NOP
291         CLR     LCD_EN          ;SET EN LOW
292         RET
293 WLCMD   ENDP
294 ;****************************************************************
295 ;       READ KEYPAD AND WAIT UNTIL KEY IS PRESSED THEN RELEASED
296 ;****************************************************************
297 READ_KEY_PRESSED    PROC
298         CALL    READ_KEY        ;SCAN KEYPAD
299         MOV     A,16
300         XOR     A,KEY
301         SZ      Z               ;IS KEY BEEN PRESSED?
302         JMP     READ_KEY_PRESSED ;NO, RE-SCAN KEYPAD
303         MOV     A,KEY
304         MOV     KEY_PS,A        ;RESERVED KEY CODE
305 WAIT_KEY_RELEASED:
306         CALL    READ_KEY        ;SCAN KEYPAD
307         MOV     A,16
308         XOR     A,KEY
309         SZ      ACC             ;IS KEY BEEN RELEASED?
310         JMP     WAIT_KEY_RELEASED ;NO, RE-SCAN KEYPAD
311         MOV     A,4
312         CALL    DELAY           ;DELAY 40ms
313         RET
314 READ_KEY_PRESSED    ENDP
315 ;****************************************************************
316 ; SCAN 4x4 MATRIX ON KEY PORT AND RETURN THE CODE IN KEY REGISTER
317 ; IF NO KEY BEEN PRESSED, KEY=16.
318 ;****************************************************************
319 READ_KEY    PROC
320         MOV     A,11110000B
321         MOV     KEY_PORTC,A     ;CONFIG KEY_PORT
322         SET     KEY_PORT        ;INITIAL KEY PORT
323         CLR     KEY             ;INITIAL KEY REGISTER
324         MOV     A,04
325         MOV     KEY_COUNT,A     ;SET ROW COUNTER
326         CLR     C               ;CLEAR CARRY FLAG
327 SCAN_KEY:
328         RLC     KEY_PORT        ;ROTATE SCANNING BIT
329         SET     C               ;MAKE SURE C=1
330         SNZ     KEY_PORT.4      ;COLUMN 0 PRESSED?
331         JMP     END_KEY         ;YES.
332         INC     KEY             ;NO, INCREASE KEY CODE.
333         SNZ     KEY_PORT.5      ;COLUMN 1 PRESSED?
334         JMP     END_KEY         ;YES.
335         INC     KEY             ;NO, INCREASE KEY CODE.
```

```
336         SNZ       KEY_PORT.6            ;COLUMN 2 PRESSED?
337         JMP       END_KEY               ;YES.
338         INC       KEY                   ;NO, INCREASE KEY CODE.
339         SNZ       KEY_PORT.7            ;COLUMN 3 PRESSED?
340         JMP       END_KEY               ;YES.
341         INC       KEY                   ;NO, INCREASE KEY CODE.
342         SDZ       KEY_COUNT             ;HAVE ALL ROWs BEEN CHECKED?
343         JMP       SCAN_KEY              ;NO, NEXT ROW.
344 END_KEY:
345         RET
346 READ_KEY    ENDP
347 ;****************************************************************
348 ;                   Delay about Acc*10ms
349 ;****************************************************************
350 DELAY       PROC
351         MOV       DEL1,A                ;SET DEL1 COUNTER
352 DEL_1:
353         MOV       A,30
354         MOV       DEL2,A                ;SET DEL2 COUNTER
355 DEL_2:
356         MOV       A,110
357         MOV       DEL3,A                ;SET DEL3 COUNTER
358 DEL_3:
359         SDZ       DEL3                  ;DEL3 DOWN COUNT
360         JMP       DEL_3
361         SDZ       DEL2                  ;DEL2 DOWN COUNT
362         JMP       DEL_2
363         SDZ       DEL1                  ;DEL1 DOWN COUNT
364         JMP       DEL_1
365         RET
366 DELAY ENDP
367         ORG       LASTPAGE
368 STR1: DC          'Press A To Start',STR_END
369 STR2: DC          'Enter The Nums :',STR_END
370 STR3: DC          'You Win The Game',STR_END
371 STR4: DC          'The Answer Is:',STR_END
372         END
```

程序说明

7~18   依序定义变量地址。

20~23   定义 LCD_DATA Port 与 LCD_CONTR Port 分别为 PB 与 PA。

24~27   定义 LCD 的控制信号引脚。

28~29   定义 KEY_PORT 为 PB。

30   定义常量 STR_END 为 0FEh，作为字符串的结束字符。

33   声明存储器地址由 000h 开始。

34   定义 LCD_CONTR Port 为输出模式。

35~36   定义 TMRC：Timer/Event Counter 为 Timer Mode、预分频比例为 1:1($f_{INT}=f_{SYS}$)、TON=1(启动 Timer/Event Counter 开始计数)。

37~38   将 LCD 设置为双行显示(N = 1)、使用 8 位(DB7 ~ DB0)控制模式(DL = 1)、5×7 点矩阵字形(F = 0)。

39~40   将 LCD 设置为显示所有数据(D = 1)、显示光标(C = 1)、光标所在位置的字会闪烁(B = 1)。

41~42   将 LCD 的地址标志位(AC)设置为递加(I/D = 1)、显示器画面不因读/写数据而移动(S = 0)。

44~45   将 LCD 整个显示器清空。

46~47   将光标移到第一行的第一个位置。

48~49   在 LCD 上显示 "Press START Key!" 字符串。

50~55   调用 READ_KEY_PRESSED 子程序，等待用户按下按键；此时唯有按下 START 按键，程序才会继续往下执行。

56   调用 READ_KEY_PRESSED 子程序，产生一组 4 位数随机数。

58~61   在 LCD 上显示 "Enter The Nums : " 字符串。

行号	说明
63~66	设置 LCD 光标地址，并设置 MP0=KEYIN_BUF 准备开始存放用户输入的数字。
68	调用 READ_KEY_PRESSED 子程序，等待用户按下按键。
69~74	检查用户是否按下 ANS 按键，如果是则调用 SHOW_ANS 子程序显示系统所产生的随机数值，然后跳到 WAIT 准备进行下一场游戏(JMP WAIT)。
76~79	检查用户是否按下 0~9 按键，如果不是则重新读取按键值(JMP GET_KEY)。
80~81	将用户输入的数值转换成 ASCII 码。
82~97	检查用户输入的数值是否重复。因为用户总共必须输入 4 个 0~9 数字，而且 4 个数字彼此不能重复。因此，利用间接寻址法将目前输入的数字与之前输入的数字进行对比，若发现数字重复的情形就舍弃目前输入的数字，重新读取按键值(JMP GET_KEY)。
99~101	用户目前输入的数字并没有重复的情形，先利用间接寻址法将数字存放在 KEYIN_BUF 缓冲区，并调用 WLCMD 子程序予以显示。
102~107	检查用户是否已经输入 4 个数字，若是则调用 CHECK_NUMS 子程序进行对比工作；否则继续读取下一个按键值(JMP GET_KEY)。判断用户是否已经输入 4 个数字的方式，是以 MP0 是否已经到达 KEYIN_BUF+4 为依据，由于 MP0 一开始被设置为 KEYIN_BUF 的起始地址(第 65~66 行)，而且每输入一个数字就将其递增(第 102 行)，所以当 MP0=KEYIN_BUF+4 时即代表用户已经输入 4 个数字。
109~120	显示 CHECK_NUMS 子程序所返回的对比结果(nAmB)。
121~134	判断用户是否 4 个数字都猜对(包含数字大小及位置)，如果是(N=4)就跳到 GUESS_RIGHT 显示猜对的消息；否则就等待用户按下 START 按键，以进行下一笔数字的输入(JMP GUESS_AGAIN)。
136~146	显示 "You Win The Game" 消息，并等待用户按下 START 按键，以进行下一场游戏(JMP START)。
150~168	SHOW_ANS 子程序，显示系统所产生的随机数值。首先显示 "The Answer Is: " 消息，其次运用间接寻址法将系统产生的随机数值一一显示在 LCD 上。
172~207	CHECK_NUMS 子程序，运用间接寻址法将用户输入的数字与系统所产生的随机数值进行对比，对比的结果会存放在 N、M 两个寄存器。N 寄存器代表数值与位置均正确的数字个数，M 寄存器则代表数值正确但位置不正确的数字个数。
211~249	RANDOM_GENERATOR 子程序，负责产生 4 个不重复的 0~9 数值。首先由 TMRH 与 TMRL 产生一个随机数(第 217~219 行)，并判断其是否大于 10，若是则舍弃该数值重新由 TMRH 与 TMRL 产生随机数(第 221~223 行)。其次是判断是否有重复的数值产生，若是则舍弃该数值重新由 TMRH 与 TMRL 产生随机数(第 224~237 行)；若数值没有重复，先将其转换成 ASCII 码后再存放在 RANDOM 缓冲区内(第 239~240 行)。第 246~248 行检查是否已经产生 4 个数值，若是则返回主程序；否则继续产生下一个位数。第 242~245 行其实是一个延时循环，其延时时间因产生的随机数值与 TMRL 而变(Acc⊕TMRL)，主要的用意是希望随机数之间相关性不要太高。
253~266	PRINT 子程序，负责将定义好的字符串依序显示在 LCD 上。在调用此子程序之前，除了必须先设置好 LCD 的位置之外，还需先在 Acc 寄存器中指定字符串的起始地址(字符串必须存放在最末程序页 Last Page)，并请在字符串的最后一个字符后插入 STR_END，代表字符串结束。
270~294	WLCMD 与 WLCMC 子程序，请参考 5-5 节中的说明。
298~315	READ_KEY_PRESSED 子程序，此子程序会调用 READ_KEY 子程序读取按键值，但必须确定用户按下按键，且等到按键放开后才会跳出此子程序。此举的主要目的，是避免 LCD 受 4×4 Key Pad 的干扰(因为 KEY_PORT 与 LCD_DATA Port 是共用的)。
320~347	READ_KEY 子程序，请参考 4-7 节中的说明。
351~364	DELAY 子程序，延时时间的计算请参考 4-1 节中的说明。
366~369	字符串定义区，请注意每一个字符串必须以 STR_END 作为结束。

以下介绍猜数字游戏机的操作方式，程序执行时，LCD 会显现如下的画面：

P	r	e	s	s		S	T	A	R	T		K	e	y	!

用户可以按下START键进入游戏,此时呈现的画面如下(此为数值输入模式):

E	n	t	e	r		T	h	e		N	u	m	s		:

这时系统内部的随机数已经产生,等待用户输入所猜的数值。请注意此时只有输入0~9的数值才有效,而且如果有重复的位数出现,系统不会接受。当输入4个不重复的0~9数值后,系统出现如下的画面:

E	n	t	e	r		T	h	e		N	u	m	s		:
0	1	2	3							0	A	2	B		

此时用户可以按下START键,回到数值输入模式继续猜。

如果在数值输入模式下按ANS键,此时系统将会显示答案。当按下ANS键时的画面如下:

T	h	e			A	n	s	w	e	r		i	s	:	
2	0	4	5												

此时请按下START键重新进入游戏。

当用户输入正确的答案时,系统显示以下的消息:

Y	o	u		W	i	n	!	!							

此时请按下START键,开始新的游戏。

6-5-5 动动脑 + 动动手

 📖 修改程序,当用户输入正确数值时,除了显示画面之外,随便播放一段音效。

 📖 修改程序,在画面上显示目前用户已经猜过几次的消息,当猜错次数超过10次时,在画面上显示"Yor Lost !!"的消息,表示用户输了这场比赛。

6-6 专题六:逻辑测试笔

6-6-1 专题功能概述

在数字系统的设计与制作或数字电路的检测过程中,在手边没有示波器、逻辑分析仪等重型设备(重型=昂贵)的状况下,一支可用来测量频率、周期与Duty Cycle的逻辑笔是不可获缺的工具。然而市售的逻辑笔单价都不低,价格由数百到数千元不等。本专题以HT46x23的Timer/Event Counter的脉宽调制模式(Pulse Witdh Measurement Mode)配合程序的设计,制作一个可以测量数字信号的周期(Period)、频率(Frequency)、工作周期(Duty Cycle)以及状态(High或Low)的简易逻辑笔,并将测量的数据显示在LCD上。电路元件中只有LCD的单价较高,虽然也考虑用七段显示器来降低成本，不过顾及到操作上的方便性与显示数值的准确性，最后仍选用LCD作为显示的设备。

6-6-2 学习重点

通过本专题实验，读者应了解HT46x23 Timer/Event Counter的脉宽调制模式工作原理，并且能充分运用其特性。另外，对于在程序中需使用到的除法及测试结果的显示技巧(如Hz、kHz、ns、ms)，读者应确实明了其原理。

6-6-3 电路图（见图6-1-1）

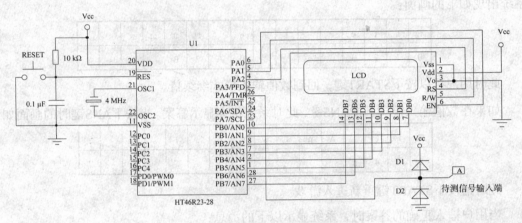

图6-6-1 逻辑测试笔控制电路

要测量一个脉冲的周期，不外乎就是计算输入脉冲电压准位由Low→High(或High→Low)到下一次准位由Low→High(或High→Low)之间的时间，如图6-6-2所示。但由于本专题还要求能够显示工作周期，因此必须将输入脉冲维持在High(T_{On})与Low(T_{Off})的时间分开计算，方能达到此要求。这对具有脉宽调制模式的HT46x23单片机而言，可谓是如鱼得水，相当的容易。

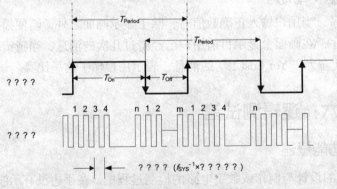

图6-6-2 待测脉冲与计数脉冲的关系

在此先将HT46x23的脉宽调制模式稍做复习。

图6-6-3是HT46x23 Timer/Event Counter内部电路结构，计数的信号来源、计数倍率，可以由TMRC寄存器加以控制，有关TMRC控制寄存器(见表6-6-1)的说明，请读者参考2-5节。

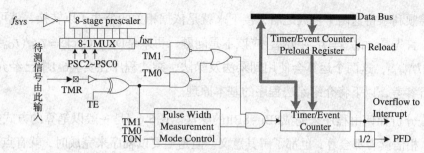

图6-6-3 HT46x23 Timer/Event Counter内部电路结构

表6-6-1 HT46x23 TMRC控制寄存器

TM1	TM0	—	TON	TE	PSC2	PSC1	PSC0

脉宽调制模式(Pulse Width Measurement Mode)：此模式主要适用来测量由TMR引脚所输入的脉冲宽度，在TON与TE位均为1的情况下，当TMR引脚有Low→High的电位变化时即启动计时器开始计数(计数的时钟脉冲来源为f_{INT})，一直到TMR引脚回复到Low电位时才停止，此时TON位会自动清除成0；也就是说只对脉冲宽度进行一次测量。在TON＝1、TE＝0的情况下，当TMR引脚有High→Low的电位变化时即启动计时器开始计数，一直到TMR引脚回复到High电位时才停止计数，此时TON位亦会自动清除为0。若在脉冲宽度的测量过程中发生计数器溢位时，除了会将TF(INT0.5)位设置为1之外，并会同时由Preload Register载入起始数值继续计数。如果定时器/计数器中断有效的话，在计数器溢位时也将产生中断要求的信号来要求CPU服务。

由上述的说明可知，只要分别让TE＝0与1各进行一次脉冲测量，就可以求得图6-6-2中的T_{On}与T_{Off}时间，再由：

$$T_{Period}=T_{On}+T_{Off}$$
$$\text{Duty cycle}=T_{On}/(T_{On}+T_{Off})\times100\%$$
$$f=1/T_{Period}$$

换算出周期、工作周期以及频率等各项参数。至于测量数据的分辨率，则取决于计数时钟脉冲周期的大小，以预分频比例为1:1(PSC2~0＝000)且f_{SYS}＝4 MHz的情况为例，分辨率为250 ns，然而局限于Timer/Event Counter只有16位，所以此时可以测量的最大脉冲宽度约只有16.4 ms。因此为了提高周期的测量范围，必须配合预分频比例的适当调整，但是随着预分频比例的增加，测量的分辨率也将随之变差。在程序中想到的作法是：先将预分频器的预分频比例调至最低(1:1)，当计数退出之后若发现Timer/Event Counter产生溢位(TF＝1)，则将预分频比例调高一级后再重新测量一遍，直到不再产生计数溢位的情况为止。然后再将TMRH与TMRL寄存器内的计数脉冲数转换为时间即可，其关系如下：

T_{On}＝计数脉冲个数×计数脉冲周期×预分频比例(TE＝1时)

T_{Off}＝计数脉冲个数×计数脉冲周期×预分频比例(TE＝0时)

能够顺利求得这两个参数之后，接下来就是依据相关的公式予以转换，就可以求得所要的数据。首先是 $T_{Period}=T_{On}+T_{Off}$，应该不是问题；再来是Duty Cycle＝$T_{On}/(T_{On}+T_{Off})×100\%$以及$f=1/T_{Period}$，这两个运算会使用到乘法及除法，乘法程序在实验6-1中已经介绍过了，请读者自行参考，以下将介绍除法程序的基本原理。

首先分析如何由程序来完成8Bit÷8Bit的除法。图6-6-4是一般以笔算的方式完成8位除法的过程，相信读者都会算，也都了解其意义。但是当要以程序来完成时，就有点不好入手了。在此尝试以写程序的逻辑观点来解释此运算过程，希望能对读者在程序的了解上有所帮助。先看STEP-1，因为被除数的Bit7＜除数(Divider)，因此将商的Bit7设置为0；在SETP-2中，被除数的Bit7~Bit6仍旧小于除数，所以商的Bit6设置为0；在SETP-3中，被除数的Bit7~Bit5≥除数，故将商的Bit5设置为1，并且将被除数Bit7~Bit5减去商(或者称此减完的结果为"部分余数")；在STEP-4中，"部分余数"与被除数Bit4的合并值＜除数，因此将商的Bit4设置为0；余依此类推，逐一依据合并值("部分余数"与被除数字符合并后的值)与除数间的大小关系，决定商字符应该为0或1。由上述的分析发现，除法的过程其实只是一连串的大小判断与减法运算而已，如果将图6-6-4的笔算范例转换为寄存器间的关系来表达，对于初次接触除法程序的读者应该会有所帮助。在STEP-1中，判断被除数的Bit7与除数的关系(一个位与字节比较)；而在STEP-2中判断被除数的Bit7~Bit6与除数的关系(两个位与字节比较)。这在程序中要如何达成呢？

首先使用一个寄存器(TEMP)，然后将被除数的最高位移位到TEMP寄存器的最低位，然后再根据TEMP寄存器与除数的大小关系设置商字符的值，不是就好像完成图6-6-4中的STEP-1了吗？大小关系的判断可以利用减法指令配合进位标志位的判断来完成，而移位与位设置的动作，读者应该都知道可以用循环指令、SET、CLR指令来完成。

图6-6-5是依据上述的概念，将图6-6-4的运算过程加以转换，请读者参考。寄存器中的"？"符号代表未知数，而在每一个STEP中，当判断完TEMP寄存器与除数的大小关系之后，都是将1(TEMP≥除数)或0(TEMP＜除数)写至商寄存器的最低位，不过在进入下一个STEP之前要将商寄存器左移一位。这样完成8个STEP之后，商就求出来了，而TEMP寄存器的最终数值即为余数。在此8Bit÷8Bit的除法图解范例中，使用了4个寄存器，可是读者应该很清楚地看到，被除数寄存器移位之后好像空出了一个位，而且与逐一设置的商字符不相冲突。因此可以将商寄存器与被除数寄存器进一步合并，进而达到节省数据存储器的目的。至于为什么需要8个STEP才能完成除法呢？这主要是因为在8Bit÷8Bit除法中，商可能为8位，因此有8个商位要一一判断。

了解以上的说明之后，只要能巧用HT46xx的SUB、RLC与位判断等指令，写成程序应该就不是难事了。请读者参考图6-6-6 8Bit÷8Bit除法的流程图及程序(DIV_8_8子程序)。

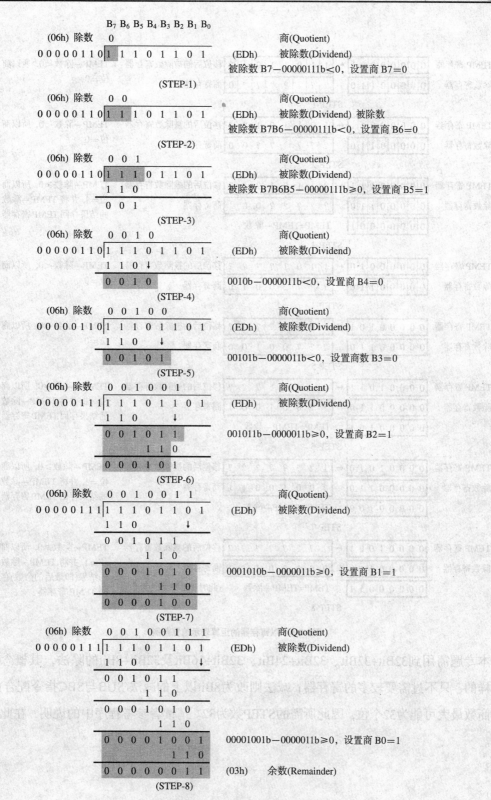

图6-6-4　8Bit÷8Bit的除法运算范例

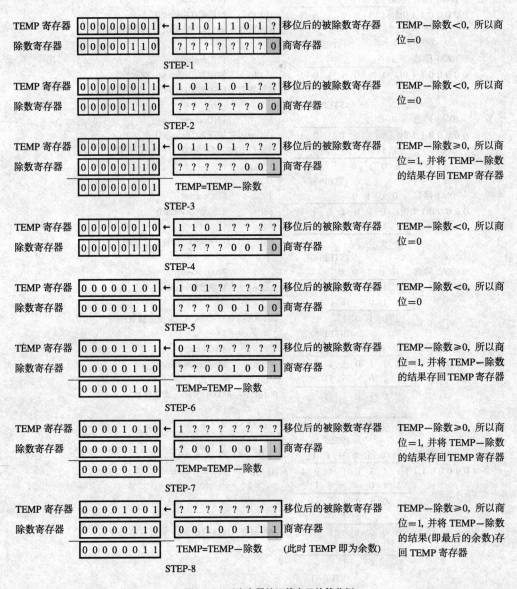

图6-6-5 以寄存器的运算表示笔算范例

本专题需用到32Bit÷32Bit、32Bit÷24Bit、32Bit÷16Bit及32Bit÷8Bit的除法,其概念其实是一样的,只不过需要较多的寄存器;减法则改为8Bit以上的减法(SUB与SBC指令配合),而由于商数最大可能为32个位,因此所需的STEP数为32。请读者参考程序中的说明,在此不再赘述。

第 6 章 实践应用篇

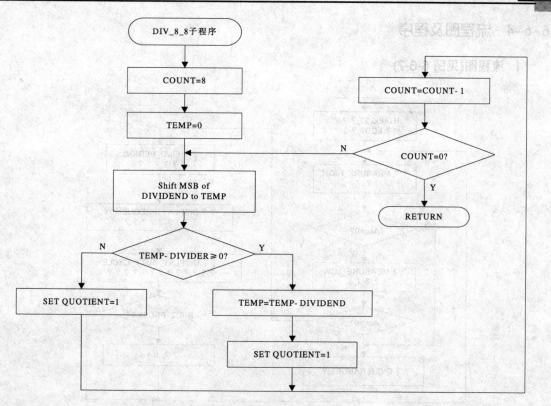

图6-6-6 DIV_8_8子程序流程图

【DIV_8_8 子程序】

```
1   ;**************************************************************************
2   ;                        8-BIT/8-BIT DIVISION ROUTINE
3   ; NOTE:   MP0 POINTING TO DIVIDER REGISTER & MP1 POINTING TO DIVIDEND REGISTER
4   ;         THE QUOTIENT AND REMAINDER WILL BE GIVEN IN DIVIDEND
5   ;         AND TEMP REGISTER RESPECTIVELY.
6   ;**************************************************************************
7   DIV_8_8     PROC
8       CLR     TEMP                    ;CLEAR TEMP BUFFER
9       MOV     A,8
10      MOV     COUNT,A                 ;SET 8 STEPS
11  NEXT_BIT:
12      CLR     C
13      RLC     IAR1                    ;SET QUOTIENT BIT TO 0
14      RLC     TEMP
15      MOV     A,TEMP
16      SUB     A,IAR0                  ;A=TEMP-DIVIDER
17      SNZ     C                       ;IS A < 0?
18      JMP     $+3                     ;YES, JUMP
19      MOV     TEMP,A                  ;NO, UPDATE TEMP WITH TEMP-DIVIDER
20      SET     IAR1.0                  ;SET QUOTIENT BIT TO 1
21      SDZ     COUNT                   ;8 STEPS OVER?
22      JMP     NEXT_BIT                ;NO, NEXT QUOTIENT BIT
23      RET
24  DIV_8_8     ENDP
```

6-6-4 流程图及程序

1. 流程图(见图6-6-7)

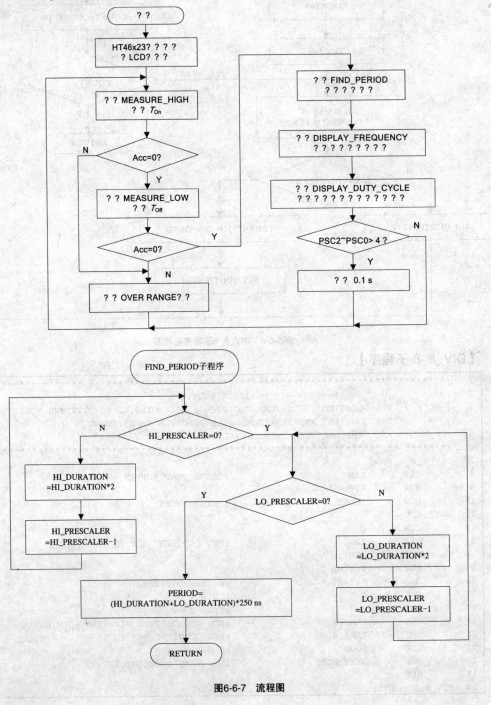

图6-6-7 流程图

第6章 实践应用篇

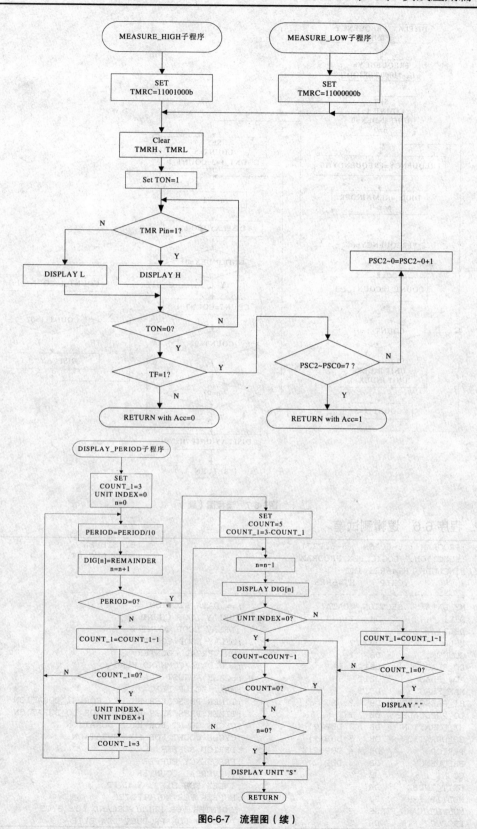

图6-6-7 流程图（续）

HT46xx 单片机原理与实践

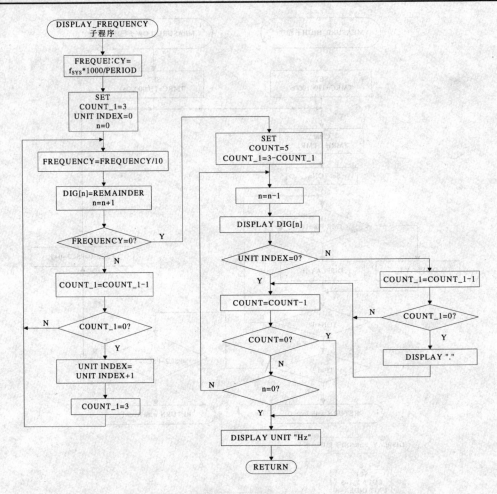

图6-6-7 流程图（续）

2. 程序 6-6 逻辑测试笔

```
1     ;PROGRAM : 6-6.ASM  (6-6.PRJ)              BY STEVEN
2     ;FUNCTION: LOGIC PEN PROGRAM                2002.DEC.07.
3     #INCLUDE HT46R23.INC
4              .CHIP   HT46R23
5     ;------------------------------------------------------------------
6     MY_DATA       .SECTION 'DATA'          ;== DATA SECTION ==
7     DEL1          DB    ?                  ;DELAY LOOP COUNT 1
8     DEL2          DB    ?                  ;DELAY LOOP COUNT 2
9     DEL3          DB    ?                  ;DELAY LOOP COUNT 3
10    COUNT         DB    ?                  ;UNIVERSAL COUNTER
11    DC_FLAG       DBIT                     ;LCD DATA/COMMAND FLAG
12    ADJ_FLAG      DBIT                     ;SCALE ADJUST FLAG
13    MAX_FLAG      DBIT                     ;MAX. SCALE FLAG
14    HI_PRESCALER  DB    ?                  ;TIMER PRESCALER BUFFER FOR HI_DURATION
15    LO_PRESCALER  DB    ?                  ;TIMER PRESCALER BUFFER FOR LO_DURATION
16    HI_DURATION   DB    3 DUP(?)           ;CYCLE COUNT FOR HIGH DURATION
17    LO_DURATION   DB    3 DUP(?)           ;CYCLE COUNT FOR LOW DURATION
18    PERIOD        DB    4 DUP(?)           ;PERIOD BUFFER
19    FREQUENCY     DB    3 DUP(?)           ;FREQUENCY BUFFER
20    COUNT_1       DB    ?                  ;UNIVERSAL COUNTER
21    UNIT_INDEX    DB    ?                  ;INDEX FOR DISPLAY UNIT
22    MULTIPLIER    DB    ?                  ;BUFFER FOR MULTIPLIER
23    MULTIPLICAND_PTR DB ?                  ;POINTER FOR MULTIPLICAND LOW BYTE
24    PRODUCT_PTR   DB    ?                  ;POINTER FOR PRODUCT LOW BYTE
```

```
25      TEMP            DB      ?
26      DIVIDEND        DB      4 DUP(?)                ;BUFFER FOR DIVIDEND
27      DIVIDER         DB      4 DUP(?)                ;BUFFER FOR DIVIDER
28      REMAINDER       DB      4 DUP(?)                ;BUFFER FOR REMAINDER
29      DIG             DB      10 DUP(?)               ;DECIMAL DIGIT BUFFER
30      ;----------------------------------------------------------------
31      LCD_CONTR       EQU     PA                      ;DEFINE LCD CONTROL PORT
32      LCD_CONTRC      EQU     PAC                     ;DEFINE LCD CONTROL PORT CON. REG.
33      LCD_DATA        EQU     PB                      ;DEFINE LCD DATA PORT
34      LCD_DATAC       EQU     PBC                     ;DEFINE LCD DATA PORT CON. REG.
35      LCD_EN          EQU     LCD_CONTR.0             ;DEFINE EN CONTROL PIN
36      LCD_RW          EQU     LCD_CONTR.1             ;DEFINE RW CONTROL PIN
37      LCD_RS          EQU     LCD_CONTR.2             ;DEFINE RS CONTROL PIN
38      LCD_READY       EQU     LCD_DATA.7              ;DEFINE READY BIT OF LCD
39      TEST_PIN        EQU     PA.4
40      STR_END         EQU     0FEH                    ;STRING END CODE
41      MY_CODE         .SECTION AT 0 'CODE'            ;== PROGRAM SECTION ==
42              ORG     00H                             ;HT-46RXX RESET VECTOR
43      START:
44              CLR     LCD_CONTRC                      ;CONFIG LCD_CONTR AS OUTPUT MODE
45              MOV     A,38H                           ;FUNCTION SET: 8-BIT,2-LINE,5X10 DOTS
46              CALL    WLCMC
47              MOV     A,0CH                           ;ON/OFF CONTR: DISPLAY ON,CURSOR BLINKING OFF
48              CALL    WLCMC
49              MOV     A,06H                           ;ENTRY MODE  : INCREMENT,DISPLAY NOT SHIFT
50              CALL    WLCMC
51              MOV     A,01H                           ;CLEAR DISPLAY
52              CALL    WLCMC
53      RE_START:
54              CALL    MEASURE_HIGH                    ;MEASURE TOn
55              SZ      ACC                             ;OUT OF RANGE?
56              JMP     OVER_RANGE                      ;YES, JUMP TO ERROR_CONDITION
57              MOV     A,TMRC                          ;NO, RESTORE PRESCALER
58              AND     A,00000111B
59              MOV     HI_PRESCALER,A
60              MOV     A,TMRH                          ;RESTORE HIGH DURATION
61              MOV     HI_DURATION[1],A
62              MOV     A,TMRL
63              MOV     HI_DURATION[0],A
64              CALL    MEASURE_LOW                     ;MEASURE TOff
65              SZ      ACC                             ;OUT OF RANGE?
66              JMP     OVER_RANGE                      ;YES, JUMP TO ERROR_CONDITION
67              MOV     A,TMRC                          ;NO, RESTORE PRESCALER
68              AND     A,00000111B
69              MOV     LO_PRESCALER,A
70              MOV     A,TMRH                          ;RESTORE LOW DURATION
71              MOV     LO_DURATION[1],A
72              MOV     A,TMRL
73              MOV     LO_DURATION[0],A
74              CALL    FIND_PERIOD                     ;GET PERIOD IN ns UNIT
75              CALL    DISPLAY_PERIOD
76              CALL    DISPLAY_FREQUENCY
77              CALL    DISPLAY_DUTY_CYCLE
78              MOV     A,TMRC
79              AND     A,00000111B
80              SUB     A,4
81              SZ      C
82              JMP     RE_START
83              MOV     A,25                            ;DELAY 250ms
84              CALL    DELAY
85              JMP     RE_START
86      OVER_RANGE:
87              MOV     A,01H                           ;CLEAR LCD DISPLAY
88              CALL    WLCMC
89              MOV     A,0C0H                          ;SET LCD DDRAM ADDRESS
90              CALL    WLCMC
91              MOV     A,STR1                          ;DISPLAY "OVER RANGE" MESSAGE
92              CALL    PRINT
93              JMP     RE_START
94      ;****************************************************************
95      ;                       DURATION MEASURE ROUTINE
```

```
96   ;********************************************************************
97   MEASURE_HIGH   PROC
98       MOV        A,11001000B         ;CONFIG TMR IN MODE3,TE=1,PSC=000
99       JMP        MEASURE
100  MEASURE_LOW:
101      MOV        A,11000000B         ;CONFIG TMR IN MODE3,TE=0,PSC=000
102  MEASURE:
103      MOV        TMRC,A
104  RE_MEASURE:
105      CLR        TMRL
106      CLR        TMRH
107      CLR        TF
108      SET        TON                 ;START TO MEASURE DURATION
109      MOV        A,8CH
110      CALL       WLCMC
111      MOV        A,48H               ;'H'
112      SNZ        TEST_PIN
113      MOV        A,4CH               ;'L'
114      CALL       WLCMD
115      SZ         TON
116      JMP        $-7
117      SNZ        TF                  ;IS TIMER OVERFLOW?
118      RET        A,0                 ;MEASURE IS OK, RETURN WITH 0
119      MOV        A,TMRC
120      AND        A,00000111B
121      SUB        A,4
122      SZ         C                   ;IS PRESCALER REACHED 1:16?
123      RET        A,1                 ;YES, OVERRANGE, RETURN WITH A=1
124      INC        TMRC                ;NO, INCREASE PRESCALER
125      JMP        RE_MEASURE          ;REMEASURE DURATION AGAIN
126  MEASURE_HIGH   ENDP
127  ;********************************************************************
128  ;                   FINDOUT PERIOD ROUTINE
129  ;********************************************************************
130  FIND_PERIOD    PROC
131      CLR        HI_DURATION[2]
132      CLR        ACC
133      XOR        A,HI_PRESCALER
134      SZ         Z                   ;PRESCALER = 0?
135      JMP        FIND_LOW            ;YES.
136  SHIFT_HI:
137      CLR        C                   ;NO, CONVERT TO 1/fSYS UNIT
138      RLC        HI_DURATION[0]
139      RLC        HI_DURATION[1]
140      RLC        HI_DURATION[2]
141      SDZ        ACC
142      JMP        SHIFT_HI
143  FIND_LOW:
144      CLR        LO_DURATION[2]
145      CLR        ACC
146      XOR        A,LO_PRESCALER
147      SZ         Z                   ;PRESCALER = 0?
148      JMP        FIND_HI_LO          ;YES.
149  SHIFT_LO:
150      CLR        C                   ;NO, CONVERT TO 1/fSYS UNIT
151      RLC        LO_DURATION[0]
152      RLC        LO_DURATION[1]
153      RLC        LO_DURATION[2]
154      SDZ        ACC
155      JMP        SHIFT_LO
156  FIND_HI_LO:
157      MOV        A,HI_DURATION[0]
158      ADD        A,LO_DURATION[0]
159      MOV        DIVIDEND[0],A
160      MOV        FREQUENCY[0],A
161      MOV        A,HI_DURATION[1]
162      ADC        A,LO_DURATION[1]
```

```
163         MOV         DIVIDEND[1],A
164         MOV         FREQUENCY[1],A
165         MOV         A,HI_DURATION[2]
166         ADC         A,LO_DURATION[2]
167         MOV         DIVIDEND[2],A           ;PERIOD IS PERIOD IN 1/fsys UNIT
168         MOV         FREQUENCY[2],A
169         MOV         A,OFFSET DIVIDEND
170         MOV         MULTIPLICAND_PTR,A
171         MOV         A,250
172         MOV         MULTIPLIER,A
173         MOV         A,OFFSET PERIOD
174         MOV         PRODUCT_PTR,A
175         CALL        MUL_24X8                ;PERIOD IS PERIOD IN ns UNIT
176         RET
177 FIND_PERIOD ENDP
178 ;****************************************************************************
179 ;                   DISPLAY PERIOD SUBROUTINE
180 ;****************************************************************************
181 DISPLAY_PERIOD PROC
182         MOV         A,PERIOD[3]             ;RELAOD PERIOD TO DIVIDEND BUFFER
183         MOV         DIVIDEND[3],A
184         MOV         A,PERIOD[2]
185         MOV         DIVIDEND[2],A
186         MOV         A,PERIOD[1]
187         MOV         DIVIDEND[1],A
188         MOV         A,PERIOD[0]
189         MOV         DIVIDEND[0],A
190         MOV         A,10                    ;SET DIVIDER=10
191         MOV         DIVIDER[0],A
192         CLR         UNIT_INDEX              ;RESET UINT INDEX
193         MOV         A,OFFSET DIG            ;GET ADDRES OF DIG BUFFER
194         MOV         MP0,A
195 DISP_PERIOD_1:
196         MOV         A,3
197         MOV         COUNT_1,A               ;SET COUNT_1=3
198 DISP_PERIOD_2:
199         CALL        DIV_32_8                ;DIVIDED BY 10
200         MOV         A,REMAINDER[0]          ;GET REMAINDER AS DIGIT
201         MOV         IAR0,A
202         INC         MP0                     ;INCREASE BUFFER POINTER
203         MOV         A,DIVIDEND[3]
204         OR          A,DIVIDEND[2]
205         OR          A,DIVIDEND[1]
206         OR          A,DIVIDEND[0]
207         SZ          Z                       ;DIVIDEND REACH 0?
208         JMP         DISP_PERIOD_3           ;YES.
209         SDZ         COUNT_1                 ;COUNT1-1 = 0?
210         JMP         DISP_PERIOD_2           ;NO.
211         INC         UNIT_INDEX              ;YES, INCREASE UNIT INDEX
212         JMP         DISP_PERIOD_1           ;RELOAD COUNT_1=3 & DIVIDE BY 10 AGAIN
213 DISP_PERIOD_3:
214         MOV         A,3
215         SUBM        A,COUNT_1               ;CONVERT COUNT_1 TO '.' POSITION
216         MOV         A,80H                   ;SET LCD DDRAM ADDRESS
217         CALL        WLCMC
218         MOV         A,5                     ;SET COUNT=5 TO DISPLAY THE MOST 5 DIGITS
219         MOV         COUNT,A
220 DISP_PERIOD_4:
221         DEC         MP0                     ;DECREASE BUFFER POINTER BY 1
222         MOV         A,IAR0                  ;GET DATA
223         ADD         A,30H                   ;CONVERT TO ASCII CODE
224         CALL        WLCMD                   ;DISPLAY DATA
225         SZ          UNIT_INDEX              ;'nS' DEGREED?
226         JMP         $+2                     ;YES
227         JMP         DISP_PERIOD_5           ;NO, NO '.' PROCESS
228         SZ          COUNT_1                 ;REACH '.' POSITION?
229         JMP         DISP_PERIOD_5           ;NO.
```

```
230         MOV     A,2EH                   ;YES, PRINT '.'
231         CALL    WLCMD
232 DISP_PERIOD_5:
233         MOV     A,OFFSET DIG            ;REACH BUFFER START ADDRESS CHECKING
234         XOR     A,MP0
235         SZ      Z                       ;THE LAST DIGIT BEEN DISPLAIED?
236         JMP     DISP_PERIOD_6           ;YES.
237         DEC     COUNT_1                 ;NO, DECREASE '.' POSITION COUNT BY 1
238         SDZ     COUNT                   ;5 DIGITS BEEN DISPLAYED?
239         JMP     DISP_PERIOD_4           ;NO,DISPLAY NEXT DIGIT
240 DISP_PERIOD_6:
241         MOV     A,UNIT_INDEX            ;GET UNIT INDEX
242         CALL    GET_PERIOD_UNIT
243         CALL    WLCMD                   ;DISPALY UNIT
244         MOV     A,53H
245         CALL    WLCMD                   ;DISPLAY "S"
246         MOV     A,20H
247         CALL    WLCMD                   ;CLEAR LCD DDRAM DATA
248         CALL    RLCMA                   ;READ LCD DDRAM ADDRESS
249         XOR     A,10
250         SNZ     Z
251         JMP     $-5
252         RET
253 DISPLAY_PERIOD ENDP
254 ;***********************************************************************
255 ;                GET UNIT OF PERIOD
256 ;***********************************************************************
257 GET_PERIOD_UNIT    PROC
258         ADDM    A,PCL
259         RET     A,6EH                   ;"n"
260         RET     A,0E4H                  ;"u"
261         RET     A,6DH                   ;"m"
262         RET     A,20H                   ;" "
263 GET_PERIOD_UNIT    ENDP
264 ;***********************************************************************
265 ;               DISPLAY FREQUENCY SUBROUTINE
266 ;***********************************************************************
267 DISPLAY_FREQUENCY  PROC
268         MOV     A,(1000*4000000/65536) >> 8
269         MOV     DIVIDEND[3],A           ;SCALE FREQUENCY BY 1000
270         MOV     A,(1000*4000000)/65536
271         MOV     DIVIDEND[2],A
272         MOV     A,HIGH (1000*4000000%65536)
273         MOV     DIVIDEND[1],A
274         MOV     A,LOW  (1000*4000000%65536)
275         MOV     DIVIDEND[0],A
276         MOV     A,FREQUENCY[2]
277         MOV     DIVIDER[2],A
278         MOV     A,FREQUENCY[1]
279         MOV     DIVIDER[1],A
280         MOV     A,FREQUENCY[0]
281         MOV     DIVIDER[0],A
282         CALL    DIV_32_24               ;CONVERT PERIOD TO FREQUENCY
283         MOV     A,DIVIDEND[2]           ;FREQUENCY UNIT IN Hz
284         MOV     FREQUENCY[2],A
285         MOV     A,DIVIDEND[1]
286         MOV     FREQUENCY[1],A
287         MOV     A,DIVIDEND[0]
288         MOV     FREQUENCY[0],A
289         MOV     A,10
290         MOV     DIVIDER[0],A
291         CLR     UNIT_INDEX              ;RESET UINT INDEX
292         MOV     A,OFFSET DIG            ;GET ADDRES OF DIG BUFFER
293         MOV     MP0,A
294 DISP_FREQUENCY_1:
295         MOV     A,3
296         MOV     COUNT_1,A               ;SET COUNT_1=3
```

```
297 DISP_FREQUENCY_2:
298       CALL      DIV_32_8                ;DIVIDED BY 10
299       MOV       A,REMAINDER[0]          ;GET REMAINDER AS DIGIT
300       MOV       IAR0,A
301       INC       MP0                     ;INCREASE BUFFER POINTER
302       MOV       A,DIVIDEND[3]
303       OR        A,DIVIDEND[2]
304       OR        A,DIVIDEND[1]
305       OR        A,DIVIDEND[0]
306       SZ        Z                       ;DIVIDEND REACH 0?
307       JMP       DISP_FREQUENCY_3        ;YES.
308       SDZ       COUNT_1                 ;COUNT1-1 = 0?
309       JMP       DISP_FREQUENCY_2        ;NO.
310       INC       UNIT_INDEX              ;YES, INCREASE UNIT INDEX
311       JMP       DISP_FREQUENCY_1        ;RELOAD COUNT_1=3 & DIVIDE BY 10 AGAIN
312 DISP_FREQUENCY_3:
313       MOV       A,3
314       SUBM      A,COUNT_1               ;CONVERT COUNT_1 TO '.' POSITION
315       MOV       A,0C0H                  ;SET LCD DDRAM ADDRESS
316       CALL      WLCMC
317       MOV       A,5                     ;COUNT=5 TO DISPLAY THE MOST 5 DIGITS
318       MOV       COUNT,A
319 DISP_FREQUENCY_4:
320       DEC       MP0                     ;DECREASE BUFFER POINTER BY 1
321       MOV       A,IAR0                  ;GET DATA
322       ADD       A,30H                   ;CONVERT TO ASCII CODE
323       CALL      WLCMD                   ;DISPLAY DATA
324       SZ        UNIT_INDEX              ;SEC DEGREED?
325       JMP       $+2                     ;YES
326       JMP       DISP_FREQUENCY_5        ;NO, NO '.' PROCESS
327       SZ        COUNT_1                 ;REACH '.' POSITION?
328       JMP       DISP_FREQUENCY_5        ;NO.
329       MOV       A,2EH                   ;YES, PRINT '.'
330       CALL      WLCMD
331 DISP_FREQUENCY_5:
332       MOV       A,OFFSET DIG            ;REACH BUFFER START ADDRESS CHECKING
333       XOR       A,MP0
334       SZ        Z                       ;THE LAST DIGIT BEEN DISPLAIED?
335       JMP       DISP_FREQUENCY_6        ;YES.
336       DEC       COUNT_1                 ;NO, DECREASE '.' POSITION COUNT BY 1
337       SDZ       COUNT                   ;5 DIGITS BEEN DISPLAIED?
338       JMP       DISP_FREQUENCY_4        ;NO,DISPLAY NEXT DIGIT
339 DISP_FREQUENCY_6:
340       MOV       A,UNIT_INDEX            ;GET UNIT INDEX
341       CALL      GET_FREQUENCY_UNIT
342       CALL      WLCMD                   ;DISPALY UNIT
343       MOV       A,48H
344       CALL      WLCMD                   ;DISPLAY "H"
345       MOV       A,7AH
346       CALL      WLCMD                   ;DISPLAY "z"
347       MOV       A,20H
348       CALL      WLCMD                   ;CLEAR LCD DDRAM DATA
349       CALL      RLCMA                   ;READ LCD DDRAM ADDRESS
350       XOR       A,40H+10
351       SNZ       Z
352       JMP       $-5
353       RET
354 DISPLAY_FREQUENCY  ENDP
355 ;*********************************************************************
356 ;             GET UNIT OF FREQUENCY
357 ;*********************************************************************
358 GET_FREQUENCY_UNIT PROC
359       DEC       ACC
360       ADDM      A,PCL
361       RET       A,20H                   ;" "
362       RET       A,4BH                   ;"K"
363       RET       A,4DH                   ;"M"
```

```
364 GET_FREQUENCY_UNIT ENDP
365 ;*********************************************************************
366 ;                    DISPLAY DUTY CYCLE SUBROUTINE
367 ;*********************************************************************
368 DISPLAY_DUTY_CYCLE PROC
369       MOV        A,HI_DURATION[0]
370       ADD        A,LO_DURATION[0]          ;HI_DURATION+LO_DURATION
371       MOV        DIVIDER[0],A
372       MOV        A,HI_DURATION[1]
373       ADC        A,LO_DURATION[1]
374       MOV        DIVIDER[1],A
375       MOV        A,HI_DURATION[2]
376       ADC        A,LO_DURATION[2]
377       MOV        DIVIDER[2],A
378       CLR        C                         ;HI_DURATION=HI_DURATION*4
379       RLC        HI_DURATION[0]
380       RLC        HI_DURATION[1]
381       RLC        HI_DURATION[2]
382       CLR        C
383       RLC        HI_DURATION[0]
384       RLC        HI_DURATION[1]
385       RLC        HI_DURATION[2]
386       MOV        A,OFFSET DIVIDEND
387       MOV        PRODUCT_PTR,A
388       MOV        A,OFFSET HI_DURATION
389       MOV        MULTIPLICAND_PTR,A
390       MOV        A,250
391       MOV        MULTIPLIER,A
392       CALL       MUL_24X8                  ;HI_DURATION*250
393       CALL       DIV_32_24                 ;HI_DURATION*1000/HI_ +LO_DURATION
394       MOV        A,10
395       MOV        DIVIDER[0],A
396       MOV        A,4
397       MOV        COUNT_1,A
398       MOV        A,OFFSET DIG              ;GET ADDRES OF DIG BUFFER
399       MOV        MP0,A
400 DISP_DUTY_1:
401       CALL       DIV_32_8                  ;DIVIDED BY 10
402       MOV        A,REMAINDER[0]            ;GET REMAINDER AS DIGIT
403       MOV        IAR0,A
404       INC        MP0                       ;INCREASE BUFFER POINTER
405       SDZ        COUNT_1                   ;COUNT_1-1 = 0?
406       JMP        DISP_DUTY_1               ;NO, DICIDED BY 10 AGAIN
407       MOV        A,0CBH
408       CALL       WLCMC                     ;SET LCD DDRAM ADDRESS
409       MOV        A,DIG[2]                  ;GET DIGIT 2
410       XOR        A,00H
411       SNZ        Z                         ;DIGIT 2 = 0?
412       ADD        A,30H-20H                 ;NO, SPACE FOR LEADIND ZERO
413       ADD        A,20H
414       CALL       WLCMD
415       MOV        A,DIG[1]                  ;GET DIGIT 1
416       ADD        A,30H                     ;CONVERT TO ASCII CODE
417       CALL       WLCMD
418       MOV        A,2EH                     ;'.'
419       CALL       WLCMD
420       MOV        A,DIG[1]                  ;GET DIGIT 0
421       ADD        A,30H                     ;CONVERT TO ASCII CODE
422       CALL       WLCMD
423       MOV        A,25H                     ;'%'
424       CALL       WLCMD
425       RET
426 DISPLAY_DUTY_CYCLE ENDP
427 ;************************************************* *******************
428 ;                  24-BIT X 8-BIT MULTIPLIER ROUTINE
429 ; NOTE:
430 ;       MULTIPILICAND_PTR = MULTIPILICAND LO-BYTE ADRS (OFFSET MULTIPILICAND)
431 ;       MULTIPLIER = 8-BIT MULTIPLIER
432 ;       PRODUCT_PTR = PRODUCT LO-BYTE ADRS (OFFSET PRODUCT)
```

```
433 ;************************************** ************************
434 MUL_24X8    PROC
435     MOV     A,PRODUCT_PTR
436     MOV     MP0,A
437     MOV     A,4
438     MOV     COUNT_1,A              ;CLEAR RESULT BUFFER(4 BYTES)
439     CLR     IAR0
440     INC     MP0
441     SDZ     COUNT_1
442     JMP     $-3
443     CLR     TEMP                   ;CLEAR TEMP BUFFER
444     MOV     A,8                    ;SET COUNTER
445     MOV     COUNT,A                ;SET DIG3 AS COUNTER
446 LOOP:
447     RRC     MULTIPLIER             ;CHECK THE LSB OF MULITPLIER
448     SNZ     C                      ;SKIP IF BIT=1
449     JMP     EQU_0                  ;JUMP IF BIT=0
450     MOV     A,PRODUCT_PTR
451     MOV     MP0,A
452     MOV     A,MULTIPLICAND_PTR
453     MOV     MP1,A
454     MOV     A,IAR1                 ;ADDITION
455     ADDM    A,IAR0                 ;ADD TO RESULT BUFFER BYTE 0
456     INC     MP0
457     INC     MP1
458     MOV     A,IAR1                 ;ADDITION
459     ADCM    A,IAR0                 ;ADD TO RESULT BUFFER BYTE 1
460     INC     MP0
461     INC     MP1
462     MOV     A,IAR1                 ;ADDITION
463     ADCM    A,IAR0                 ;ADD TO RESULT BUFFER BYTE 2
464     INC     MP0
465     MOV     A,TEMP
466     ADCM    A,IAR0
467 EQU_0:
468     MOV     A,MULTIPLICAND_PTR
469     MOV     MP0,A
470     MOV     A,3
471     MOV     COUNT_1,A              ;MULTIPLICANT SHIFT LEFT 1 BIT
472     RLC     IAR0
473     INC     MP0
474     SDZ     COUNT_1
475     JMP     $-3
476     RLC     TEMP
477     SDZ     COUNT                  ;SKIP IF COUNTER = 0
478     JMP     LOOP
479     RET
480 MUL_24X8    ENDP
481 ;********************************************************** ******************
482 ;                32-BIT/8~32-BIT DIVISION ROUTINE
483 ; NOTE: DIVIDEND [0~3]   :THE QUOTIENT WILL RETURN TO DIVIDEND
484 ;       DIVIDER [0~3]    :UNCHANGED
485 ;       REMAINDER[0~3])
486 ;*******************************************************
   ********************************
487 DIV_32_8    PROC
488     CLR     DIVIDER[1]             ;CLEAR 2-ND BYTTE BUFFER
489 DIV_32_16:
490     CLR     DIVIDER[2]             ;CLEAR 3-RD BYTTE BUFFER
491 DIV_32_24:
492     CLR     DIVIDER[3]             ;CLEAR 4-TH BYTTE BUFFER
493 DIV_32_32:
494     CLR     REMAINDER[0]
495     CLR     REMAINDER[1]
496     CLR     REMAINDER[2]
497     CLR     REMAINDER[3]
498     MOV     A,32
```

```
499         MOV     COUNT,A
500 NEXT_BIT:
501         CLR     C                       ;PRESET QUOTIENT BIT TO 0
502         RLC     DIVIDEND[0]
503         RLC     DIVIDEND[1]
504         RLC     DIVIDEND[2]
505         RLC     DIVIDEND[3]
506         RLC     REMAINDER[0]
507         RLC     REMAINDER[1]
508         RLC     REMAINDER[2]
509         RLC     REMAINDER[3]
510         MOV     A,REMAINDER[0]
511         SUB     A,DIVIDER[0]
512         MOV     REMAINDER[0],A
513         MOV     A,REMAINDER[1]
514         SBC     A,DIVIDER[1]
515         MOV     REMAINDER[1],A
516         MOV     A,REMAINDER[2]
517         SBC     A,DIVIDER[2]
518         MOV     REMAINDER[2],A
519         MOV     A,REMAINDER[3]
520         SBC     A,DIVIDER[3]
521         MOV     REMAINDER[3],A
522         SNZ     C                       ;TEMP-DIVIDER >= 0?
523         JMP     $+3                     ;NO, RESTORE TEMP
524         SET     DIVIDEND[0].0           ;SET QUOTIENT BIT TO 1
525         JMP     $+9
526         MOV     A,DIVIDER[0]            ;RESTORE VALUE OF TEMP
527         ADDM    A,REMAINDER[0]
528         MOV     A,DIVIDER[1]
529         ADCM    A,REMAINDER[1]
530         MOV     A,DIVIDER[2]
531         ADCM    A,REMAINDER[2]
532         MOV     A,DIVIDER[3]
533         ADCM    A,REMAINDER[3]
534 DIV_32_16_2:
535         SDZ     COUNT
536         JMP     NEXT_BIT
537         RET
538 DIV_32_8    ENDP
539 ;************************************************************************
540 ;       LCD PRINT PROCEDURE (START ADDRS=TBLP)
541 ;************************************************************************
542 PRINT PROC
543         MOV     TBLP,A                  ;LOAD TABLE POINTER
544 PRINT_1:
545         TABRDL  ACC                     ;READ STRING
546         XOR     A,STR_END
547         SZ      Z                       ;END OF STRING CHARACTER?
548         JMP     END_PRINT               ;YES, STOP PRINT
549         TABRDL  ACC                     ;NO, RE-READ
550         CALL    WLCMD                   ;SEND TO LCD
551         INC     TBLP                    ;INCREASE DATA POINTER
552         JMP     PRINT_1                 ;NEXT CHARACTER
553 END_PRINT:
554         RET                             ;YES, RETURN
555 PRINT ENDP
556 ;************************************************************************
557 ;       LCD DATA/ADDRESS READ PROCEDURE
558 ;************************************************************************
559 RLCMD PROC
560         SET     DC_FLAG                 ;SET DC_FLAG=1 FOR DATA READ
561         JMP     RLCM
562 RLCMA:
563         CLR     DC_FLAG                 ;SET DC_FLAG=0 FOR ADDRESS READ
564 RLCM:
565         SET     LCD_DATAC               ;CONFIG LCD_DATA AS INPUT MODE
566         CLR     LCD_CONTR               ;CLEAR ALL LCD CONTROL SIGNAL
567         SET     LCD_RW                  ;SET RW=1
568         NOP                             ;FOR TAS
569         SET     LCD_EN                  ;SET EN=1
```

```
570         NOP                                 ;FOR TDDR
571 RWF:
572         SZ      LCD_READY                   ;IS LCD BUSY?
573         JMP     RWF                         ;YES,WAIT
574         MOV     A,LCD_DATA
575         SNZ     DC_FLAG                     ;ADDRESS READ?
576         RET                                 ;YES.
577         CLR     LCD_CONTR                   ;CLEAR ALL LCD CONTROL SIGNAL
578         SET     LCD_RW                      ;SET LCD_RW
579         SET     LCD_RS                      ;SET RS=1 (DATA READ)
580         NOP
581         SET     LCD_EN                      ;SET EN
582         NOP
583         MOV     A,LCD_DATA                  ;READ LCD DATA
584         CLR     LCD_EN                      ;CLEAR EN
585         RET
586 RLCMD ENDP
587 ;************************************************************************
588 ;              LCD DATA/COMMAND WRITE PROCEDURE
589 ;************************************************************************
590 WLCMD PROC
591         SET     DC_FLAG                     ;SET DC_FLAG=1 FOR DATA WRITE
592         JMP     WLCM
593 WLCMC:
594         CLR     DC_FLAG                     ;SET DC_FLAG=0 FOR COMMAND WRITE
595 WLCM:
596         SET     LCD_DATAC                   ;CONFIG LCD_DATA AS INPUT MODE
597         CLR     LCD_CONTR                   ;CLEAR ALL LCD CONTROL SIGNAL
598         SET     LCD_RW                      ;SET RW SIGNAL (READ)
599         NOP                                 ;FOR TAS
600         SET     LCD_EN                      ;SET EN HIGH
601         NOP                                 ;FOR TDDR
602 WF:
603         SZ      LCD_READY                   ;IS LCD BUSY?
604         JMP     WF                          ;YES, JUMP TO WAIT
605         CLR     LCD_DATAC                   ;NO, CONFIG LCD_DATA AS OUTPUT MODE
606         MOV     LCD_DATA,A                  ;LATCH DATA/COMMAND ON PB(LCD DATA BUS)
607         CLR     LCD_CONTR                   ;CLEAR ALL LCD CONTROL SIGNAL
608         SZ      DC_FLAG                     ;IS COMMAND WRITE?
609         SET     LCD_RS                      ;NO, SET RS HIGH
610         SET     LCD_EN                      ;SET EN HIGH
611         NOP
612         CLR     LCD_EN                      ;SET EN LOW
613         RET
614 WLCMD ENDP
615 ;************************************************************************
616 ;                    Delay about DEL1*10ms
617 ;************************************************************************
618 DELAY   PROC
619         MOV     DEL1,A                      ;SET DEL1 COUNTER
620 DEL_1:
621         MOV     A,30
622         MOV     DEL2,A                      ;SET DEL2 COUNTER
623 DEL_2:
624         MOV     A,110
625         MOV     DEL3,A                      ;SET DEL3 COUNTER
626 DEL_3:
627         SDZ     DEL3                        ;DEL3 DOWN COUNT
628         JMP     DEL_3
629         SDZ     DEL2                        ;DEL2 DOWN COUNT
630         JMP     DEL_2
631         SDZ     DEL1                        ;DEL1 DOWN COUNT
632         JMP     DEL_1
633         RET
634 DELAYENDP
635         ORG     LASTPAGE
636 STR1:   DC      '!_OVER RANGE_!',STR_END
637         END
```

程序说明

行号	说明
7~29	依序定义变量地址。
31~34	定义 LCD_DATA Port 与 LCD_CONTR Port 分别为 PA 与 PC。
35~38	定义 LCD 的控制信号引脚。
39	定义待测信号输入引脚。
40	定义字符串结束码 STR_END=FEh。
43	声明存储器地址由 000h 开始(HT46xx Reset Vector)。
43	将 LCD_CONTR Port(PA)定义为输出模式。
46~47	将 LCD 设置为双行显示(N = 1)、使用 8 位(DB7 ~ DB0)控制模式(DL = 1)、5×7 点矩阵字形(F = 0)。
48~49	将 LCD 设置为显示所有数据(D = 1)、显示光标(C = 1)、光标所在位置的字不会闪烁(B = 0)。
50~51	将 LCD 的地址标志位(AC)设置为递加(I/D = 1)、显示器画面不因读/写数据而移动(S = 0)。
52~53	将 LCD 整个显示器清空。
55~57	调用 MEASURE_HIGH 子程序,测量 T_{On}。该子程序会自动切换 Timer/Event counter 的预分频比例,如果切换至最大预分频比例时仍发生计数器溢位的现象,则代表待测信号的脉冲宽度超过范围,此时会返回 Acc=0。因此第 56、57 两行程序是根据 Acc 判断是否为超过范围的情况,若是则跳至 OVER_RANGE 显示相关消息;否则继续往下执行。
58~64	将预分频比例(PSC2~0)与测量结果(TMRH、TMRL)存储在相关寄存器中(HI_PRESCALER、HI_DURATION)。
65~67	调用 MEASURE_LOW 子程序,测量 T_{Off}。该子程序会自动切换 Timer/Event counter 的预分频比例,如果切换至最大预分频比例时仍发生计数器溢位的现象,则代表待测信号的脉冲宽度超过范围,此时会返回 Acc=0。因此第 66、57 两行程序是根据 Acc 判断是否为超过范围的情况,若是则跳至 OVER_RANGE 显示相关消息;否则继续往下执行。
68~74	将预分频比例(PSC2~0)与测量结果(TMRH、TMRL)存储在相关寄存器中(LO_PRESCALER、LO_DURATION)。
75	调用 FIND_PERIOD 子程序,根据 MEASURE_HIGH、MEASURE_LOW 子程序的测量结果计算出待测信号的周期。
76	调用 DISPLAY_PERIOD 子程序,根据 FIND_PERIOD 子程序的计算结果,将待测信号的周期以适当格式在 LCD 上显示。
77	调用 DISPLAY_FREQUENCY 子程序,根据 MEASURE_HIGH、MEASURE_LOW 子程序的测量结果计算出待测信号的频率,并以适当格式在 LCD 上显示。
78	调用 DISPLAY_DUTY_CYCLE 子程序,根据 MEASURE_HIGH、MEASURE_LOW 子程序的测量结果计算出待测信号的工作周期,并以适当格式在 LCD 上显示。
80~85	由于脉冲的测量时间会受待测信号的周期影响,周期越长测量所需时间就越长,而测量结果的显示时间也越久;但当待测信号频率很高时,测量结果会因为显示的时间太短(因为更新速度太快)而导致看不清楚数值。这几行程序首先依据预分频比例判断是否为高频的信号,以决定是否要调用 DELAY 子程序(延时 250 ms),以避免出现高频时测量结果看不清楚的情形。
86	重新开始测量。
88~94	调用 PRINT 子程序,在 LCD 上显示 OVER RANGE 消息。
98~127	MEASURE_HIGH 与 MEASURE_LOW 子程序,由于测量 T_{On} 与 T_{Off} 的唯一差异就是 TE=1 或 0,因此将这两个子程序集成在一起。
99	MEASURE_HIGH 子程序进入点,定义 Timer/Event Counter 的工作模式:Pulse Width Measurement Mode、计数比例 1:1($f_{INT}=f_{SYS}$)、TE=1。

101　MEASURE_LOW 子程序进入点，定义 Timer/Event Counter 的工作模式：Pulse Width Measurement Mode、计数比例 1:1($f_{INT}=f_{SYS}$)、TE=0。

106~108　清除 TMRL、TMRH 寄存器，并设置 TF=0。

109　设置 TON=1 开始测量脉冲宽度。

110~115　在脉冲测量过程中将待测信号的准位显示在 LCD 上。

116~117　检查是否测量退出，若不是则继续等待。

118~119　检查是否产生定时器计数溢位的现象，若没有计数溢位，则返回主程序(此时 Acc=0，表示测量完成)。

120~126　检查是否已经到达最大预分频比例(PSC2~PSC0=111)，若是则返回主程序(此时 Acc=1，表示脉冲宽度超过容许范围)；否则将预分频比例再往上调升一级并重新测量。

131~178　FIND_PERIOD 子程序，根据 MEASURE_HIGH、MEASURE_LOW 子程序的测量结果计算出待测信号的周期。其所完成的运算为：

$$(HI\_DURATION \times HI\_PRESCALER + O\_DURATION \times LO\_PRESCALER) \times f_{SYS}^{-1}$$

132~143　将预分频比例(HI_PRESCALER)与 HI_DURATION 相乘计算出 T_{On} 时的计数脉冲数(其单位为 $1/f_{SYS}$)；由于预分频比例为 2^N(N=0~7)，在程序中是以左移的方式来完成乘法的动作(左移一位相当于乘2)。由于 Timer/Event Counter 为 16Bit(2^{16})、最大预分频比率为 128(2^7)，因此乘积最大为 23Bit。

144~156　将预分频比例(LO_PRESCALER)与 LO_DURATION 相乘计算出 T_{Off} 时的计数脉冲数(其单位为 $1/f_{SYS}$)；由于预分频比例为 2^N(N=0~7)，在程序中是以左移的方式来完成乘法的动作(左移一位相当于乘 2)。由于 Timer/Event Counter 为 16Bit(2^{16})、最大预分频比率为 128(2^7)，因此乘积最大为 23 位。

157~177　将计数脉冲数换算为周期时间，即将(HI_DURATION×HI_PRESCALER)与(LO_DURATION×LO_PRESCALER)的和乘上 f_{SYS}^{-1}(为 250 ns @f_{SYS}=4 MHz)，计算出待测脉冲的周期(此时的单位为ns)。请注意：此时计数时间最大值为 31Bit($2^{23} \times 2 \times 2^7$)。在此同时也将(HI_DURATION×HI_PRESCALER)与(LO_DURATION×LO_PRESCALER)的和复制一份存放在 FREQUENCY 寄存器，以便在计算频率时用。

182~254　DISPLAY_PERIOD 子程序，根据 FIND_PERIOD 子程序求得的值将待测信号的周期显示在 LCD 上。通过 FIND_PERIOD 子程序所求得的周期最大为(2^{31}-1) ns，基于显示器的显示字数与实际应用上的考虑，只显示最高 5 个不为 0 的十进位数值。第 197~213 行的程序，是先将十六进制的周期时间以连续除十的方式转成十进制，并以间接寻址法依序将余数存放在 DIG 缓冲区。为了能够以科学记数方式来显示测试结果，在此同时也一并分析出其单位(ns、μs、ms 或 s)以及小数点位置。程序第 215~240 行则是依据上述的运算显示 5 位数的测量结果，并在适当位置插入小数点。最后再调用 GET_PERIOD_UNIT 子程序，显示适当的周期单位(ns、μs、ms 或 s)。

258~264　GET_PERIOD_UNIT 子程序，依据 Acc 的值回传适当的周期单位。

268~355　DISPLAY_FREQUENCY 子程序，其主要功能是将 FIND_PERIOD 子程序求得的待测信号周期转换为频率之后，显示在 LCD 上。在 FIND_PERIOD 子程序中，已经将(HI_DURATION×HI_PRESCALER)与(LO_DURATION× LO_PRESCALER)的和复制到 FREQUENCY 寄存器，因此只要完成 f_{SYS}÷ FREQUENCY 寄存器的运算就可以了。但是如果这样做就只能显示"？？？Hz"，而没有小数点以下的位数，为了显示测量值的准确度，实际上是执行 1000×f_{SYS}/(FREQUENCY)的运算，此即程序第 269~283 行所完成的任务。基于显示器的显示字数与实际应用上的考虑，只显示最高 5 个不为 0 的十进位数值。第 292~312 行的程序，是先将十六进制的周期时间以连续除十的方式转成十进制，并以间接寻址法依序将余数存放在 DIG 缓冲区。为了能够以科学记数方式来显示测试结果，在此同时也一并分析出其单位(Hz、kHz、MHz)以及小数点位置。程序第 314~339 行则是依据上述的运算显示 5 位数的测量结果，并在适当位置插入小数点。最后再调用 GET_FREQUENCY_UNIT 子程序，显示适当的频率单位(Hz、kHz、MHz)。

359~365　GET_FREQUENCY_UNIT 子程序，依据 Acc 的值回传适当的频率单位。

369~427 DISPLAY_DUTY_CYCLE 子程序,依据 HI_DURATION 与 LO_DURATION 计算出待测信号的工作周期,并显示在 LCD 上。其实计算工作周期只需套用以下的公式即可:

Duty Cycle＝$T_{On}/(T_{Off}+T_{Off}) \times 100\%$

但是为了能够显示小数点以下的数值,实际上是完成以下的运算:

Duty Cycle＝$T_{On}/(T_{Off}+T_{Off}) \times 1000\%$

也就是先将 $T_{On} \times 1000$ 之后再除以$(T_{Off}+T_{Off})$,为避免再多写一个 24Bit×16Bit 的乘法子程序,在第 379~386 行先将 T_{On} 左移 2 次(相当于乘 4),然后再调用 MUL_24X8 子程序乘上 250,如此便完成 $T_{On} \times 1000$ 的运算。在第 394 行调用 DIV32_24 子程序后即计算出工作周期。第 395~426 行则负责将结果转换为十进制显示,并在适当位置插入小数点。

435~481 MUL_24X8 子程序,其原理请参考 6-1 节中相关说明。

488~539 DIV32_32、DIV32_24、DIV32_16 与 DIV32_8 子程序,有关除法程序的原理请参考 6-6-2 小节的说明。在被除数均为 32 位的情况下,其实运算方式可以予以集成。

543~556 PRINT 子程序,负责将定义好的字符串依序显示在 LCD 上。在调用此子程序之前,除了必须先设置好 LCD 的位置之外,还需先在 Acc 寄存器中指定字符串的起始地址(字符串必须存放在最末程序页 Last Page),并请于字符串的最后一个字符后插入 STR_END,代表字符串结束。

560~587 RLCMD 与 RLCMA 子程序,请参考 5-5 节中的说明及 LCD READ 时序图。

591~615 WLCMD 与 WLCMC 子程序,请参考 5-5 节的说明。

619~632 DELAY 子程序,延时时间的计算请参考 4-1 节中的说明。

634 字符串定义区。

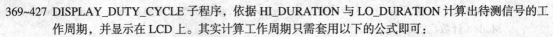

在 f_{SYS}＝4 MHz 的工作频率下,本专题所实现的逻辑测试笔的分辨率最大约为 250 ns,而且随着待测频率的降低其准确度随之增加。在 1 Hz~700 kHz 左右笔者实际测试均有正确的测量结果,读者可以试着提高 f_{SYS} 来增加其准确度与测量范围。不过本专题的最大要求,是期望读者能借此机会将 HT46xx 的脉宽调制模式彻底了解清楚,同时对除法运算的原理也能有所认识,以后若在其他型号的单片机应用上需使用到除法时,可以驾轻就熟地以汇编语言来完成程序的编写。另外有关小数点的处理以及科学记数的显示方式,也在其他单片机相关书籍中较少涉及,但实际应用上却又是经常使用到的技巧,希望通过本专题的介绍能让读者洞悉其中的原理。

6-6-5 动动脑+动动手

- 修改程序,使频率与周期的显示改为 7 位数字。
- 修改程序,当测试信号的周期小于 1μs 时,LCD 即显示 "UNDER_RANGE" 的消息。

6-7 专题七:频率计数器(Counter)的制作

6-7-1 专题功能概述

以 HT46x23 Timer/Event Counter 的事件计数模式(Event Count Mode),设计一个可以测量数字信号频率(Frequency)的简易频率计数器,并将测量的数据显示在 LCD 上。

6-7-2 学习重点

通过本专题实验，读者应了解HT46x23 Timer/Event Counter的事件计数模式(Event Count Mode)工作原理，并且能充分运用其特性。另外，在程序中需使用到除法及测试结果的显示技巧(如Hz、kHz、MHz)，读者应确实了解其原理。

6-7-3 电路图（见图6-7-1）

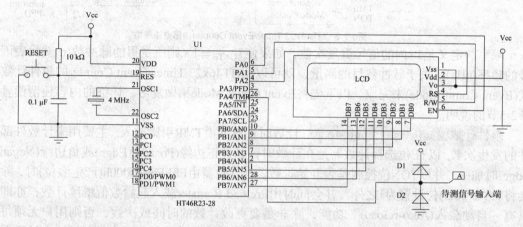

图6-7-1 频率计数器控制电路

在6-6节中，读者应该已经了解测量信号频率的基本概念，但是因为在该实验中要求同时显示工作周期(Duty Cycle)，因此就以脉宽调制模式来完成所需的参数测量。然而，在此模式之下测试频率就无法太高，而且准确度在频率超过1 MHz时就有待商议。在某些应用场合，如果只需测量频率，就可以采用Timer/Event Counter的事件计数模式来提高频率的测试范围。其概念也相当简单，如图6-7-2所示，是以单位时间内(T_{GATE})所取得的脉冲数来换算为频率值，其公式如下：

$$f = (T_{GATE}/n)^{-1} = n/T_{GATE}$$

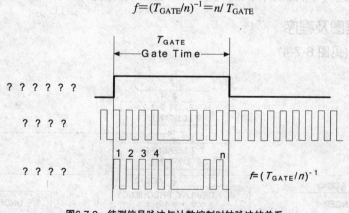

图6-7-2 待测信号脉冲与计数控制时钟脉冲的关系

图6-7-3为HT46x23 Timer/Event Counter内部电路结构。

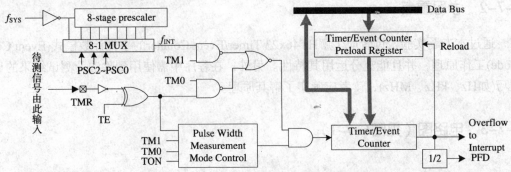

图6-7-3 HT46x23 Timer/Event Counter内部电路结构

频率的定义是1 s内的信号重复次数，如果使$T_{GATE}=1$ s,则所量得的脉冲数(n)就代表该信号的频率(nHz)。关于脉冲数目的测量，就可以用HT46x23 Timer/Event Counter的事件计数模式(Event Count Mode)来完成。以下先将Event Count Mode稍做复习，详细的内容还请读者参考2-5节的说明。

事件计数模式(Event Count Mode)：计数时钟脉冲由TMR引脚输入，主要用来计数外部事件的发生次数，以TE位选择计数器是在计数时钟脉冲正边缘(Positive Edge)或负边缘(Negative Edge)时加一，并以TON位控制是否开始计数。当计数器由FFFFh→0000h(产生溢位)时，除了会将TF(INT0.5)位设置为1之外，并会同时由Preload Register载入起始数值继续计数；亦即其具有"自动载入(Auto-reload)"功能，除非是要更改计数的时间或次数，否则用户无须每次重新载入起始数值，以避免无谓的Overhead。如果定时器/计数器中断有效的话，在计数器溢位时也将产生中断要求信号来要求CPU服务。

如果直接以$T_{GATE}=1$ s来设计，其实是最简单的。只要取得1 s内TMRH与TMRL的增量，就可得知输入信号的频率，但是由于TMRH与TMRL为16位的计数长度，所以在$T_{GATE}=1$ s的情况下，充其量也只能测到65.535 kHz的频率。为了扩大频率的测试范围，程序中必须加上自动切换T_{GATE}的功能，当发现计数溢位的情形时，就自动缩短T_{GATE}的时间(如$T_{GATE}=1$ ms)，最后再依$f=n/T_{GATE}$算出频率。另外，为提高准确度，最好是让计数器在TE=0(正边缘加一)与TE=1(负边缘加一)时各进行一次测量，然后再求其平均值，以避免不必要的误差。

6-7-4 流程图及程序

1. 流程图(见图 6-7-4)

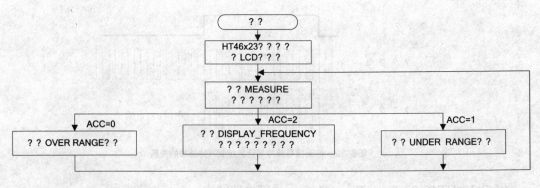

图6-7-4 流程图

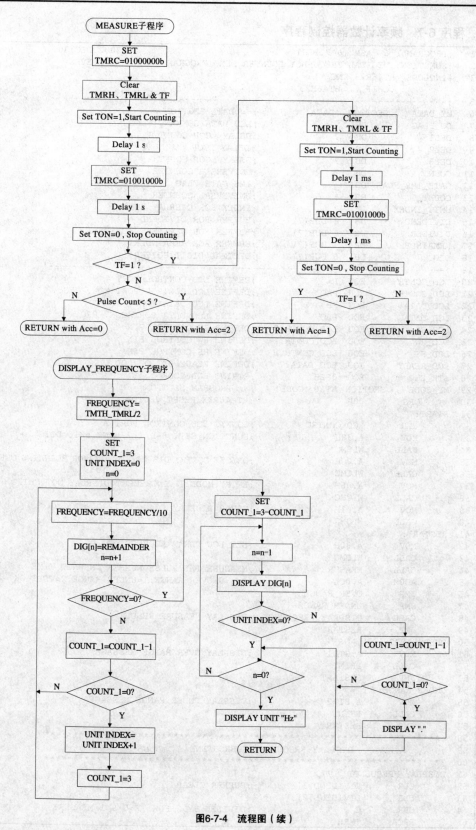

图6-7-4 流程图（续）

2. 程序 6-7 频率计数器控制程序

```
1    ;PROGRAM : 6-7ASM (6-7.PRJ)              2002.DEC.30.
2    ;FUNCTION: DIGITAL FREQUENCY COUNTER DEMO PROGRAM    BY STEVEN
3    #INCLUDE   HT46R23.INC
4              .CHIP    HT46R23
5    ;----------------------------------------------------------------
6    MY_DATA    .SECTION 'DATA'         ;== DATA SECTION ==
7    DC_FLAG         DBIT               ;LCD DATA/COMMAND FLAG
8    DEL1            DB    ?            ;DELAY LOOP COUNT 1
9    DEL2            DB    ?            ;DELAY LOOP COUNT 2
10   DEL3            DB    ?            ;DELAY LOOP COUNT 3
11   COUNT           DB    ?            ;UNIVERSAL COUNTER
12   GATE_1MS_FLAG   DBIT               ;1MS GATE FLAG
13   COUNT_1         DB    ?            ;UNIVERSAL COUNTER
14   UNIT_INDEX      DB    ?            ;INDEX FOR DISPLAY UNIT
15   DIVIDEND        DB    4 DUP(?)     ;BUFFER FOR DIVIDEND
16   DIVIDER         DB    4 DUP(?)     ;BUFFER FOR DIVIDER
17   REMAINDER       DB    4 DUP(?)     ;BUFFER FOR REMAINDER
18   DIG             DB    5 DUP(?)     ;DECIMAL DIGIT BUFFER
19   ;----------------------------------------------------------------
20   LCD_CONTR       EQU   PA           ;DEFINE LCD CONTROL PORT
21   LCD_CONTRC      EQU   PAC          ;DEFINE LCD CONTROL PORT CON. REG.
22   LCD_DATA        EQU   PB           ;DEFINE LCD DATA PORT
23   LCD_DATAC       EQU   PBC          ;DEFINE LCD DATA PORT CON. REG.
24   LCD_EN          EQU   LCD_CONTR.0  ;DEFINE EN CONTROL PIN
25   LCD_RW          EQU   LCD_CONTR.1  ;DEFINE RW CONTROL PIN
26   LCD_RS          EQU   LCD_CONTR.2  ;DEFINE RS CONTROL PIN
27   LCD_READY       EQU   LCD_DATA.7   ;DEFINE READY BIT OF LCD
28   STR_END         EQU   0FEH         ;STRING END CODE
29   MY_CODE   .SECTION AT 0 'CODE'     ;== PROGRAM SECTION==
30         ORG       00H                ;HT-46RXX RESET VECTOR
31   START:
32         CLR       LCD_CONTRC         ;CONFIG LCD CONTROL PORT A
33         MOV       A,38H              ;FUNCTION SET: 8-BIT,2-LINE,5X10 DOTS
34         CALL      WLCMC
35         MOV       A,0CH              ;ON/OFF CONTR: DISPLAY ON,CURSOR BLINKING OFF
36         CALL      WLCMC
37         MOV       A,06H              ;ENTRY MODE : INCREMENT,DISPLAY NOT SHIFT
38         CALL      WLCMC
39         MOV       A,01H              ;CLEAR DISPLAY
40         CALL      WLCMC
41   RESTART:
42         MOV       A,80H              ;SET LCD DDRAM ADDRESS
43         CALL      WLCMC
44         CALL      MEASURE            ;MEASURE THE PULSE NUMBER IN UNIT TIME
45         ADDM      A,PCL              ;Acc=0:OVER RANGE;1=UNDER RANGE;2=TEST OK
46         JMP       OVER_RANGE
47         JMP       UNDER_RANGE
48         CALL      DISPLAY_FREQUENCY  ;DISPLAY MEASURE RESULT
49         JMP       RESTART
50   OVER_RANGE:
51         MOV       A,STR1             ;DISPLAY OVER RANGE MESSAGE
52         CALL      PRINT
53         JMP       RESTART
54   UNDER_RANGE:
55         MOV       A,STR2             ;DISPLAY UNDER RANGE MESSAGE
56         CALL      PRINT
57         JMP       RESTART
58   ;****************************************************************
59   ;              DISPLAY FREQUENCY SUBROUTINE
60   ;****************************************************************
61   DISPLAY_FREQUENCY   PROC
62         CLR       DIVIDEND[3]        ;BUFFER CLEAR
63         CLR       DIVIDEND[2]
64         CLR       C                  ;DIVIDED BY 2 FOR TE=0 & 1 MEASURE
65         RRCA      TMRH
```

```
66              MOV     DIVIDEND[1],A
67              RRCA    TMRL
68              MOV     DIVIDEND[0],A
69              MOV     A,10
70              MOV     DIVIDER[0],A
71              CLR     UNIT_INDEX              ;RESET UINT INDEX
72              MOV     A,OFFSET DIG            ;GET ADDRES OF DIG BUFFER
73              MOV     MP0,A
74      DISP_FREQUENCY_1:
75              MOV     A,3
76              MOV     COUNT_1,A               ;SET COUNT_1=3
77      DISP_FREQUENCY_2:
78              CALL    DIV_32_8                ;DIVIDED BY 10
79              MOV     A,REMAINDER[0]          ;GET REMAINDER AS DIGIT
80              MOV     IAR0,A
81              INC     MP0                     ;INCREASE BUFFER POINTER
82              MOV     A,DIVIDEND[1]
83              OR      A,DIVIDEND[0]
84              SZ      Z                       ;DIVIDEND REACH 0?
85              JMP     DISP_FREQUENCY_3        ;YES.
86              SDZ     COUNT_1                 ;COUNT1-1 = 0?
87              JMP     DISP_FREQUENCY_2        ;NO.
88              INC     UNIT_INDEX              ;YES, INCREASE UNIT INDEX
89              JMP     DISP_FREQUENCY_1        ;RELOAD COUNT_1=3 & DIVIDE BY 10 AGAIN
90      DISP_FREQUENCY_3:
91              MOV     A,3
92              SUBM    A,COUNT_1               ;CONVERT COUNT_1 TO '.' POSITION
93      DISP_FREQUENCY_4:
94              DEC     MP0                     ;DECREASE BUFFER POINTER BY 1
95              MOV     A,IAR0                  ;GET DATA
96              ADD     A,30H                   ;CONVERT TO ASCII CODE
97              CALL    WLCMD                   ;DISPLAY DATA
98              SZ      UNIT_INDEX              ;SEC DEGREED?
99              JMP     $+2                     ;YES
100             JMP     DISP_FREQUENCY_5        ;NO, NO '.' PROCESS
101             SZ      COUNT_1                 ;REACH '.' POSITION?
102             JMP     DISP_FREQUENCY_5        ;NO.
103             MOV     A,2EH                   ;YES, PRINT '.'
104             CALL    WLCMD
105     DISP_FREQUENCY_5:
106             DEC     COUNT_1                 ;NO, DECREASE '.' POSITION COUNT BY 1
107             MOV     A,OFFSET DIG            ;REACH BUFFER START ADDRESS CHECKING
108             XOR     A,MP0
109             SNZ     Z                       ;THE LAST DIGIT BEEN DISPLAIED?
110             MOV     A,UNIT_INDEX            ;GET UNIT INDEX
111             SZ      GATE_1MS_FLAG           ;GATE DELAY = 1MS?
112             INC     ACC                     ;YES, INCREASE DRGREED BY 1
113             CALL    GET_FREQUENCY_UNIT
114             CALL    WLCMD                   ;DISPALY UNIT
115             MOV     A,48H
116             CALL    WLCMD                   ;DISPLAY "H"
117             MOV     A,7AH
118             CALL    WLCMD                   ;DISPLAY "z"
119             MOV     A,20H
120             CALL    WLCMD                   ;CLEAR LCD DDRAM DATA
121             CALL    RLCMA                   ;READ LCM DDRAM ADDRESS
122             XOR     A,40H+10                ;CLEAR 10 REST DDRAM
123             SNZ     Z
124             JMP     $-5
125             RET
126     DISPLAY_FREQUENCY   ENDP
127     ;********************************************************************
128     ;               GET UNIT OF FREQUENCY
129     ;********************************************************************
130     GET_FREQUENCY_UNIT PROC
131             ADDM    A,PCL
132             RET     A,20H                   ;" "
```

```
133         RET     A,4BH           ;"K"
134         RET     A,4DH           ;"M"
135 GET_FREQUENCY_UNIT ENDP
136 ;****************************************************************
137 ;                   MEASURE SUBROUTINE
138 ;****************************************************************
139 MEASURE   PROC
140         CLR     GATE_1MS_FLAG   ;CLEAR 1MS GET FLAG
141         CLR     TF              ;CLEAR TIMER/EVENT COUNTER OVERFOLW FLAG
142         CLR     TMRL            ;CLEAR PULSE COUNTER
143         CLR     TMRH
144         MOV     A,01000000B     ;CONFIG TMR IN MODE 1 WITH TE=0(L TO H ACTIVE)
145         MOV     TMRC,A          ;fINT COMES FROM TMR PIN WITHOUT PRESCALER
146         SET     TON
147         CALL    DELAY           ;1SEC GATE TIME
148         CLR     TON
149         MOV     A,01001000B     ;CONFIG TMR IN MODE 1 WITH TE=1(H TO L ACTIVE)
150         MOV     TMRC,A          ;fINT COMES FROM TMR PIN WITHOUT PRESCALER
151         SET     TON
152         CALL    DELAY           ;1SEC GATE TIME
153         CLR     TON
154         SZ      TF
155         JMP     MEASURE_1MS
156         SZA     TMRH            ;TMRH = 0?
157         RET     A,2             ;NO, RETURN WITH OK
158         MOV     A,TMRL
159         SUB     A,5
160         SZ      C               ;TMRH_TMRL > 5?
161         RET     A,2             ;YES, RETURN WITH OK
162         RET     A,1             ;NO,RETURN WITH UNDER RANGE
163 MEASURE_1MS:
164         SET     GATE_1MS_FLAG   ;SET 1MS GET FLAG
165         CLR     TF              ;CLEAR TIMER/EVENT COUNTER OVERFOLW FLAG
166         CLR     TMRL            ;CLEAR PULSE COUNTER
167         CLR     TMRH
168         MOV     A,01000000B     ;CONFIG TMR IN MODE 1 WITH TE=0(L TO H ACTIVE)
169         MOV     TMRC,A          ;fINT COMES FROM TMR PIN WITHOUT PRESCALER
170         SET     TON
171         CALL    DELAY_1MS       ;1 mSEC GATE TIME
172         CLR     TON
173         MOV     A,01001000B     ;CONFIG TMR IN MODE 1 WITH TE=1(H TO L ACTIVE)
174         MOV     TMRC,A          ;fINT COMES FROM TMR PIN WITHOUT PRESCALER
175         SET     TON
176         CALL    DELAY_1MS       ;1 mSEC GATE TIME
177         CLR     TON
178         SNZ     TF              ;EVENT COUNTER OVERFLOW?
179         RET     A,2             ;NO, RETURN WITH OK
180         RET     A,0             ;YES,RETURN WITH OVER RANGE
181 MEASURE   ENDP
182 ;****************************************************************
183 ;                     Delay 1 ms
184 ;****************************************************************
185 DELAY_1MS  PROC
186         MOV     A,5
187         MOV     DEL2,A          ;SET DEL2 COUNTER
188         MOV     A,65
189         MOV     DEL3,A          ;SET DEL3 COUNTER
190         SDZ     DEL3            ;DEL3 DOWN COUNT
191         JMP     $-1
192         SDZ     DEL2            ;DEL2 DOWN COUNT
193         JMP     $-5
194         RET
195 DELAY_1mS ENDP
196 ;****************************************************************
197 ;              32-BIT/8~32-BIT DIVISION ROUTINE
198 ; NOTE: DIVIDEND [0~3] :THE QUOTIENT WILL RETURN TO DIVIDEND
199 ;         DIVIDER [0~3]   :UNCHANGED
```

```
200 ;              REMAINDER[0~3])
201 ;**********************************************************************
202 DIV_32_8    PROC
203      CLR       DIVIDER[1]          ;CLEAR 2-ND BYTTE BUFFER
204 DIV_32_16:
205      CLR       DIVIDER[2]          ;CLEAR 3-RD BYTTE BUFFER
206 DIV_32_24:
207      CLR       DIVIDER[3]          ;CLEAR 4-TH BYTTE BUFFER
208 DIV_32_32:
209      CLR       REMAINDER[0]
210      CLR       REMAINDER[1]
211      CLR       REMAINDER[2]
212      CLR       REMAINDER[3]
213      MOV       A,32
214      MOV       COUNT,A
215 NEXT_BIT:
216      CLR       C                   ;PRESET QUOTIENT BIT TO 0
217      RLC       DIVIDEND[0]
218      RLC       DIVIDEND[1]
219      RLC       DIVIDEND[2]
220      RLC       DIVIDEND[3]
221      RLC       REMAINDER[0]        ;TEMP
222      RLC       REMAINDER[1]
223      RLC       REMAINDER[2]
224      RLC       REMAINDER[3]
225      MOV       A,REMAINDER[0]
226      SUB       A,DIVIDER[0]
227      MOV       REMAINDER[0],A
228      MOV       A,REMAINDER[1]
229      SBC       A,DIVIDER[1]
230      MOV       REMAINDER[1],A
231      MOV       A,REMAINDER[2]
232      SBC       A,DIVIDER[2]
233      MOV       REMAINDER[2],A
234      MOV       A,REMAINDER[3]
235      SBC       A,DIVIDER[3]
236      MOV       REMAINDER[3],A
237      SNZ       C                   ;TEMP-DIVIDER >= 0?
238      JMP       $+3                 ;NO, RESTORE TEMP
239      SET       DIVIDEND[0].0       ;SET QUOTIENT BIT TO 1
240      JMP       $+9
241      MOV       A,DIVIDER[0]        ;RESTORE VALUE OF TEMP
242      ADDM      A,REMAINDER[0]
243      MOV       A,DIVIDER[1]
244      ADCM      A,REMAINDER[1]
245      MOV       A,DIVIDER[2]
246      ADCM      A,REMAINDER[2]
247      MOV       A,DIVIDER[3]
248      ADCM      A,REMAINDER[3]
249 DIV_32_16_2:
250      SDZ       COUNT
251      JMP       NEXT_BIT
252      RET
253 DIV_32_8    ENDP
254 ;**********************************************************************
255 ;          LCD PRINT PROCEDURE (START ADDRS=TBLP)
256 ;**********************************************************************
257 PRINT PROC
258      MOV       TBLP,A              ;LOAD TABLE POINTER
259 PRINT_1:
260      TABRDL    ACC                 ;READ STRING
261      XOR       A,STR_END
262      SZ        Z                   ;END OF STRING CHARACTER?
263      JMP       END_PRINT           ;YES, STOP PRINT
264      TABRDL    ACC                 ;NO, RE-READ
265      CALL      WLCMD               ;SEND TO LCD
266      INC       TBLP                ;INCREASE DATA POINTER
267      JMP       PRINT_1             ;NEXT CHARACTER
268 END_PRINT:
```

```
269         RET                         ;YES, RETURN
270 PRINT ENDP
271 ;***********************************************************************
272 ;         LCD DATA/ADDRESS READ PROCEDURE
273 ;***********************************************************************
274 RLCMD PROC
275         SET     DC_FLAG             ;SET DC_FLAG=1 FOR DATA READ
276         JMP     RLCM
277 RLCMA:
278         CLR     DC_FLAG             ;SET DC_FLAG=0 FOR ADDRESS READ
279 RLCM:
280         SET     LCD_DATAC           ;CONFIG LCD_DATA AS INPUT MODE
281         CLR     LCD_CONTR           ;CLEAR ALL LCD CONTROL SIGNAL
282         SET     LCD_RW              ;SET RW=1
283         NOP                         ;FOR TAS
284         SET     LCD_EN              ;SET EN=1
285         NOP                         ;FOR TDDR
286 RWF:
287         SZ      LCD_READY           ;IS LCD BUSY?
288         JMP     RWF                 ;YES,WAIT
289         MOV     A,LCD_DATA
290         SNZ     DC_FLAG             ;ADDRESS READ?
291         RET                         ;YES.
292         CLR     LCD_CONTR           ;CLEAR ALL LCD CONTROL SIGNAL
293         SET     LCD_RW              ;SET LCD_RW
294         SET     LCD_RS              ;SET RS=1 (DATA READ)
295         NOP
296         SET     LCD_EN              ;SET EN
297         NOP
298         MOV     A,LCD_DATA          ;READ LCD DATA
299         CLR     LCD_EN              ;CLEAR EN
300         RET
301 RLCMD ENDP
302 ;***********************************************************************
303 ;         LCD DATA/COMMAND WRITE PROCEDURE
304 ;***********************************************************************
305 WLCMD PROC
306         SET     DC_FLAG             ;SET DC_FLAG=1 FOR DATA WRITE
307         JMP     WLCM
308 WLCMC:
309         CLR     DC_FLAG             ;SET DC_FLAG=0 FOR COMMAND WRITE
310 WLCM:
311         SET     LCD_DATAC           ;CONFIG LCD_DATA AS INPUT MODE
312         CLR     LCD_CONTR           ;CLEAR ALL LCD CONTROL SIGNAL
313         SET     LCD_RW              ;SET RW SIGNAL (READ)
314         NOP                         ;FOR TAS
315         SET     LCD_EN              ;SET EN HIGH
316         NOP                         ;FOR TDDR
317 WF:
318         SZ      LCD_READY           ;IS LCD BUSY?
319         JMP     WF                  ;YES, JUMP TO WAIT
320         CLR     LCD_DATAC           ;NO, CONFIG LCD_DATA AS OUTPUT MODE
321         MOV     LCD_DATA,A          ;LATCH DATA/COMMAND ON PB(LCD DATA BUS)
322         CLR     LCD_CONTR           ;CLEAR ALL LCD CONTROL SIGNAL
323         SZ      DC_FLAG             ;IS COMMAND WRITE?
324         SET     LCD_RS              ;NO, SET RS HIGH
325         SET     LCD_EN              ;SET EN HIGH
326         NOP
327         CLR     LCD_EN              ;SET EN LOW
328         RET
329 WLCMD ENDP
330 ;***********************************************************************
331 ;              Delay about 999937us
332 ;***********************************************************************
333 DELAY   PROC
334         MOV     A,95
335         MOV     DEL1,A              ;SET DEL1 COUNTER
336 DEL_1:
337         MOV     A,21
338         MOV     DEL2,A              ;SET DEL2 COUNTER
339 DEL_2:
340         MOV     A,157
```

```
341             MOV         DEL3,A                  ;SET DEL3 COUNTER
342 DEL_3:
343             SDZ         DEL3                    ;DEL3 DOWN COUNT
344             JMP         DEL_3
345             SDZ         DEL2                    ;DEL2 DOWN COUNT
346             JMP         DEL_2
347             SDZ         DEL1                    ;DEL1 DOWN COUNT
348             JMP         DEL_1
349             RET
350 DELAY ENDP
351             ORG         LASTPAGE
352 STR1: DC                '!_OVER RANGE_!',STR_END
353 STR2: DC                '_UNDER_RANGER',STR_END
352             END
```

程序说明

- **7~18** 依序定义变量地址。
- **20~23** 定义 LCD_DATA Port 与 LCD_CONTR Port 分别为 PA 与 PC。
- **24~27** 定义 LCD 的控制信号引脚。
- **28** 定义字符串结束码 STR_END=FEh。
- **31** 声明存储器地址由 000h 开始(HT46xx Reset Vector)。
- **33** 将 LCD_CONTR Port(PA)定义为输出模式。
- **34~35** 将 LCD 设置为双行显示(N = 1)、使用 8 位(DB7 ~ DB0)控制模式(DL = 1)、5×7 点矩阵字形(F = 0)。
- **36~37** 将 LCD 设置为显示所有数据(D = 1)、显示光标(C = 1)、光标所在位置的字不会闪烁(B = 0)。
- **38~39** 将 LCD 的地址标志位(AC)设置为递加(I/D = 1)、显示器画面不因读/写数据而移动(S = 0)。
- **40~41** 将 LCD 整个显示器清空。
- **43~44** 设置 LCD DDRAM 地址。
- **45** 调用 MEASURE 子程序，测量 T_{On} 与 T_{Off} 的脉冲数。该子程序会自动切换 T_{GATE} 的时间，如果切换至 $T_{GATE}=1$ ms 时仍发生计数器溢位的现象，则代表待测信号的频率超过范围，此时会返回 Acc=0(Over Range)。如果切换至 $T_{GATE}=1$ s 时，计数器计数结果小于 5，则代表待测信号的频率过低，此时会返回 Acc=1(Under Range)；反之则返回 Acc=2，表示完成脉冲数的测量。
- **46~48** 根据 MEASURE 子程序所返回的值，分别跳至对应的位置执行以下的工作：在 LCD 显示 "OVER RANGE"消息(ACC=0)、在 LCD 显示 "UNDER RANGE"消息(ACC=1)、调用 DISPLAY_FREQUENCY 子程序显示待测信号的频率。
- **52~54** 调用 PRINT 子程序，在 LCD 显示 "UNDER RANGE"消息，并重新进行测量。
- **56~58** 调用 PRINT 子程序，在 LCD 显示 "OVER RANGE"消息，并重新进行测量。
- **62~126** DISPLAY_FREQUENCY 子程序，其主要功能是将 MEASURE 子程序在 T_{GATE} 时间求得的待测信号计数脉冲数转换为频率之后，显示在 LCD 上。为了获得较高的准确度，在 MEASURE 子程序中分别让 TE=0 与 1，共进行了 2 次测量，所以第 65~68 行的程序，是先将测量除以 2 求得其平均值，其次将十六进制的测量结果以连续除以 10 的方式转成十进制，并以间接寻址法依序将余数存放在 DIG 缓冲区。为了能够以科学记数方式来显示测试结果，在此同时也一并分析出其单位(Hz、kHz、MHz)以及小数点位置。程序第 91~110 行则是依据上述的运算显示测量的结果，并在适当位置插入小数点。最后再调用 GET_FREQUENCY_UNIT 子程序，显示适当的频率单位(Hz、kHz、MHz)。
- **131~135** GET_FREQUENCY_UNIT 子程序，依据 Acc 的值回传适当的频率单位。
- **140~182** MEASURE 子程序，测量 T_{On} 与 T_{Off} 的脉冲数。此子程序会自动切换 T_{GATE} 的时间，如果切换至 $T_{GATE}=1$ ms 时仍发生计数器溢位的现象，则代表待测信号的频率超过范围，此时会返回 Acc=0(Over Range)。如果切换到 $T_{GATE}=1$ s 时，计数器计数结果小于 5，则代表待测信号的频率过低，此时会返回 Acc=1(Under Range)。反之则返回 Acc=2，表示完成脉冲数的测量。

141~144 清除 GATE_1MS_FLAG，此标志位值为 1 代表脉冲数目是在 $T_{GATE}=1$ ms 时的测量结果；若 GATE_1MS_FLAG=0，则代表脉冲数目是在 $T_{GATE}=1$ s 时的测量结果。在 DISPLAY_FREQUENCY 子程序中，会依据此标志位设置适当的频率单位；将 TF 与 TMRL、TMRH 清除为 0，准备测量。

145~146 设置 Timer/Event Counter 为 Event Count 工作模式、TE=0(正边缘加一)。

147~152 设置 TON=1 开始测量脉冲数目，调用 DELAY 子程序延时 1 s 后将 TE 设置为 1，并重新进行一次测量。

153~154 调用 DELAY 子程序延时 1 s 后，设置 TON=0 停止测量。

155~163 检查是否产生定时器计数溢位的现象，若没有计数溢位，则判断计数脉冲数是否大于 5，分别返回 ACC=2(表示测量 OK)与 ACC=0(表示测量信号频率过低)。反之，如果发生定时器计数溢位的现象，则将 T_{GATE} 设为 1 ms，重新再测量一次。

165 设置 GATE_1MS_FLAG=1，表示是在 $T_{GATE}=1$ ms 下进行测量。

166~168 将 TF 与 TMRL、TMRH 清除为 0，准备进行测量。

169~170 设置 Timer/Event Counter 为 Event Count 工作模式、TE=0(正边缘加一)。

171~176 设置 TON=1 开始测量脉冲数目，调用 DELAY_1MS 子程序延时 1 ms 后将 TE 设置为 1，并重新进行一次测量。

177~178 调用 DELAY_1MS 子程序延时 1 ms 后，设置 TON=0 停止测量。

178~181 检查是否产生定时器计数溢位的现象，若没有计数溢位，则返回 ACC=2(表示测量 OK)；反之，则返回 ACC=1(表示测量信号频率过高)。

186~196 DELAY_1MS 子程序，延时时间的计算请参考 4-1 节中的说明。

203~254 DIV32_32、DIV32_24、DIV32_16 与 DIV32_8 子程序，有关除法程序的原理请参考 6-6 节的说明。在被除数均为 32 位的情况下，其实运算方式可以予以集成。

258~271 PRINT 子程序，此子程序负责将定义好的字符串依序显示在 LCD 上。在调用此子程序之前，除了必须先设置好 LCD 的位置之外，还需先在 Acc 寄存器中指定字符串的起始地址(字符串必须存放在最末程序页 Last Page)，并请在字符串的最后一个字符后插入 STR_END，代表字符串结束。

275~302 RLCMD 与 RLCMA 子程序，请参考 5-5 节中的说明及 LCD READ 时序图。

306~330 WLCMD 与 WLCMC 子程序，请参考 5-5 节的说明。

334~348 DELAY 子程序，延时时间的计算请参考 4-1 节中的说明。

350~351 字符串定义区。

此频率计数器的准确度主要取决于 T_{GATE} 时间的精准度，以 DELAY_1MS 子程序为例，细心的读者或许已经发现，其延时时间其实只有 998 μs，但请注意再加上 "CALL DELAY_1MS" 指令的执行时间后恰好就等于 1ms。至于测量频率的最高范围就受限于 TMR 引脚输入信号的 Rising 与 Falling Time 限制，这在 HT46x23 数据手册中并无明确的规格，只是规定了在 $V_{DD}=5$ V 的工作电压下，f_{TIMER}(Timer Input Frequency)的上限为 8 000 kHz。由于笔者手边的信号发生器(Function Generator)的输出频率最高只能到达 5.5 MHz，而本专题所设计的频率计数器在此范围内的测试结果都十分正确。

6-7-5 动动脑+动动手

📖 程序 6-7 中的 DELAY 子程序的正确延时时间其实是 999 937 μs 再加上 "CALL DELAY" 的指令执行时间，顶多也是只有 999 939 μs 延时而已，这可能导致测量时造成误差。请写一个延时时间恰为 1s 的延时子程序取代 DELAY 子程序，以提高频率测量的准确度。

6-8 专题八：简易信号发生器的制作

6-8-1 专题功能概述

本专题拟以HT46x23的PWM输出接口，制作一个可以输出正弦波、方波、锯齿波与方波的信号发生器，而且其频率、直流偏压值都可调整。

6-8-2 学习重点

本专题的实现电路，只是PWM输出接口的再应用，读者应不会陌生才是。以PWM或R-2R Ladder作为D/A转换，是在许多具有模拟输出控制的玩具上经常使用的技巧，然而一般单片机相关书籍较少探讨以PWM来实现D/A转换的功能。希望通过本专题的介绍，读者能够将此技巧运用自如。

6-8-3 电路图（见图6-8-1）

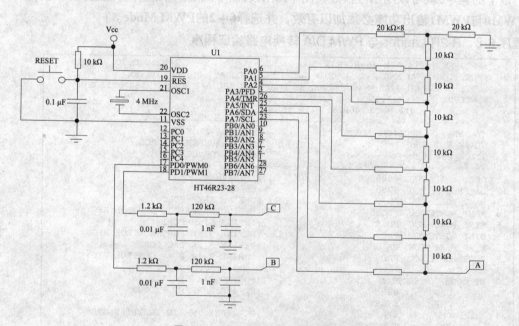

图6-8-1　以R-2R Ladder与PWM完成D/A转换控制电路

在许多单片机应用场合，经常都需要产生模拟输出信号，虽然目前市面上有各式各样的D/A转换IC可供使用，但基于成本的考虑，通常会利用R-2R梯状电阻(R-2R Ladder)或PWM的控制来完成。图6-8-1中的PA直接连接到R-2R梯状电阻网路，在A点可以测得输出的模拟电压。若考虑成本，此种D/A转换方式的成本并不高，而且具有低电磁辐射干扰(EMI)的特性，高频谐波(Harmonics)成分也较少。因此，在多数应用中都不需要再设计低通滤波电路来滤除高频谐波成分。然而R-2R梯状电阻式D/A转换器最大的缺点就是I/O引脚的需要量较大，以8位的分辨率为例(见图6-8-1)，就需使用到8个I/O引脚，对于I/O资源弥足珍贵的小型单片机而言，无疑是一种浪费。所以一般对于D/A转换速率不需太快的场合,还是以PWM来实现D/A转换的做法较为可行(关于HT46x23的PWM接口的控制,请读者参阅2-7节),只要一个I/O引脚就可以完成一个Channel的D/A转换器。当然,在其输出端必须再加上滤除高频谐波成分的低通滤波器(LPF)才行。一般基于成本的考虑,只需简单的RC滤波即可,如图6-8-2所示。

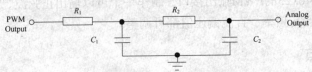

图 6-8-1 RC 滤波

其中截止频率$F_C=(2\pi RC)^{-1}$,而$R_1C_1=R_2C_2=RC$。F_C应该大于模拟输出信号的带宽(Bandwidth),以避免信号衰减;而F_C也须远小于PWM信号的频率,以免产生谐波失真。一般此类二阶RC滤波器(2-nd Order Low Pass Filter)的设计,在$R_2 \gg R_1$时可拥有较佳的响应(Response),因此在图6-8-1中将R_2与R_1刻意设计成100倍的差距。

为了验证这两种电路设计方式都可达到D/A转换的功能,请读者根据图6-8-1连接电路,并执行以下的程序,就可以用示波器观察图中A、B及C点的输出模拟波形。(**注意:Code Option 中的PWM0与PWM1输出功能必须加以有效,并选择6+2的PWM Mode。**)

程序 6-8   R-2R Ladder 与 PWM D/A 转换电路验证程序

```
1     ;PROGRAM : 6-8.ASM (6-8.PRJ)              2002.DEC.30.
2     ;FUNCTION: PWM AND R-2R LADDER DAC DEMO PROGRAM    BY STEVEN
3     ;NOTE    : PWM1 AND PMW0 SHOULD BE ENABLED, SEL 6+2 MODE (MASK OPTION).
4     #INCLUDE  HT46R23.INC
5               .CHIP    HT46R23
6     ;--------------------------------------------------------------
7     MY_DATA   .SECTION 'DATA'          ;== DATA SECTION ==
8     STACK_A   DB  ?                    ;STACK BUFFER FOR ACC
9     STACK_PSW DB  ?                    ;STACK FOR STATUS
10    ;--------------------------------------------------------------
11    OSC       EQU  4000000             ;fSYS=4MHz
12    DA_FREQ   EQU  8000                ;fDAC=8KHz
13    RAMP_PORT EQU  PD.0                ;DEFINE RAMP PORT
14    RAMP_PORTC EQU PDC.0               ;DEFINE RAMP PORT CONTROL REG.
15    SINE_PORT EQU  PD.1                ;DEFINE SINE PORT
16    SINE_PORTC EQU PDC.1               ;DEFINE SINE PORT CONTROL REG.
17    R_2R_PORT EQU  PA                  ;DEFINE R_2R DAC PORT
18    R_2R_PORTC EQU PAC                 ;DEFINE R_2R PORT CONTROL REG.
19
20    MY_CODE   .SECTION AT 0 'CODE'     ;== PROGRAM SECTION ==
21        ORG     00H                    ;HT-46RXX RESET VECTOR
22        JMP     START
23        ORG     08H                    ;HT-46RXX TMR INTERRUPT VECTOR
24        JMP     TMR_INT
25    START:
26        MOV     A,00000101B            ;ENABLE GLOBAL & TMR INTERRUPT
27        MOV     INTC0,A
28        MOV     A,10000000B            ;CONFIG TMR 0 IN TIMER MODE
29        MOV     TMRC,A                 ;fINT=fSYS(4MHz/1).
30        CLR     RAMP_PORTC             ;CONFIG PD0 AS OUTPUT MODE
31        CLR     SINE_PORTC             ;CONFIG PD0 AS OUTPUT MODE
32        CLR     R_2R_PORTC             ;CONFIG R_2R PORT AS OUTPUT MODE
33        MOV     A,LOW (65536-OSC/DA_FREQ)
34        MOV     TMRL,A
35        MOV     A,HIGH (65536-OSC/DA_FREQ)
36        MOV     TMRH,A
37        MOV     A,OFFSET SINE_TABLE
38        MOV     TBLP,A                 ;INITIAL SINE TABLE POINTER
39        CLR     PWM0                   ;RESET PWM0
```

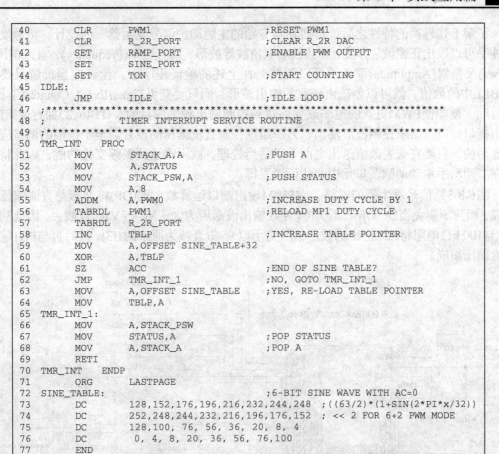

此验证程序是以Timer/Event Counter为主体，配合中断的控制让PWM0、PWM1与PB输出每隔125μs更新一次，更新的值分别为逐次递增的锯齿波与查表所得的正弦波数值。若以程序6-8为例，应该在A、C两点测得正弦波输出信号，而在B点测得锯齿波输出信号。笔者为了比较R-2R Ladder与PWM D/A转换器的转换效果，曾将程序6-8稍加修改后，分别测得锯齿波与正弦波的输出波形供读者参考。图6-8-3与图6-8-4中的CH1为R-2R Ladder式D/A转换器的输出波形，CH2则为PWM式D/A转换器的输出波形。读者应该可以看出其实两者的差异并不大，PWM式的输出的确是谐波成分多了一些。

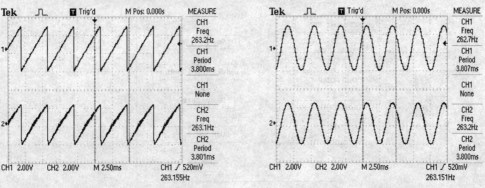

图6-8-3　R-2R Ladder与PWM D/A转换器输出锯齿波的情形　　图6-8-4　R-2R Ladder与PWM D/A转换器输出正弦波的情形

了解上述转换的特性之后，再回到本专题的主题即"信号发生器"的设计，信号发生器不外乎可以输出正弦波、方波、锯齿波、三角波等波形，而且频率(Frequency)、直流偏压(DC Offset)及振幅(Amplitude)都必须可以调整。在上述的验证程序中，读者应当理解只要改变TABLE中的数值，就可以创造出任意的输出波形，而只要变更Timer/Event Counter的计数溢位时间，便能使PWM更新的速率改变而达到控制输出频率的目的。HT46x23拥有两组PWM输出接口，以一组来控制波形输出，另一组则控制直流偏压的设置。至于振幅的调整应该是最容易的，只要在查表数值送出之前予以适当处理，就不难达到振幅变化功能，因此将其保留在"动动手＋动动脑"单元中，让读者发挥。

图6-8-5是信号发生器的电路，在PWM输出端以运算放大器(OP)将波形与直流偏压予以集成，而为了避免当DC Offset不为0 V导致输出波形因为大于V_{CC}(5 V)而被截去，因此利用33 kΩ与100 kΩ电阻将PD0输出的波形信号先予以衰减(衰减为原来的1/3)，然后再与PD1输出的直流偏压集成。

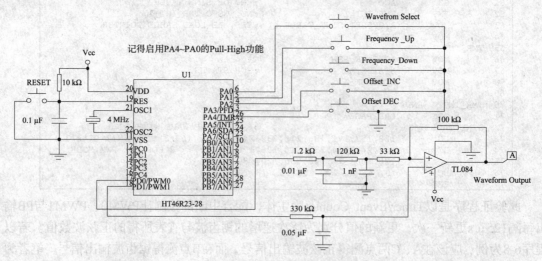

图6-8-5　简易信号发生器控制电路

6-8-4　流程图及程序
1. 流程图(见图6-8-6)

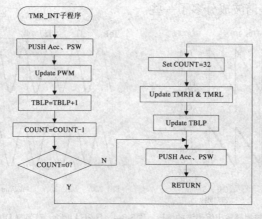

图6-8-6　流程图

第6章 实践应用篇

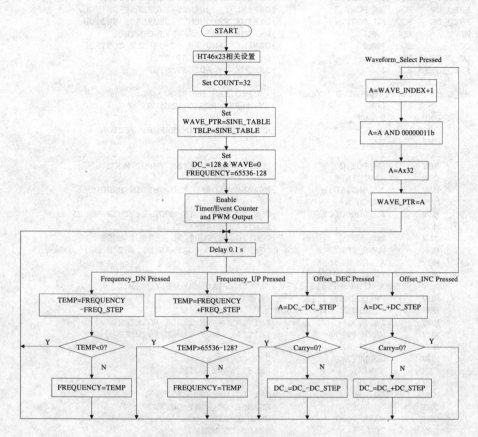

图6-8-6 流程图(续)

2. 程序 6-8-1 简易信号发生器

```
1   ;PROGRAM  : 6-8-1.ASM 6-8-1.PRJ)                          2002.DEC.30.
2   ;FUNCTION: DIGITAL FUNCTION GENERATOR DEMO PROGRAM        BY STEVEN
3   ;NOTE    : PWM1 AND PMW0 SHOULD BE ENABLED, SEL 6+2 MODE (MASK OPTION).
4   #INCLUDE  HT46R23.INC
5             .CHIP     HT46R23
6   ;----------------------------------------------------------------------
7   MY_DATA    .SECTION 'DATA'       ;== DATA SECTION ==
8   DEL1       DB     ?              ;DELAY LOOP COUNT 1
9   DEL2       DB     ?              ;DELAY LOOP COUNT 2
10  DEL3       DB     ?              ;DELAY LOOP COUNT 3
11  WAVE_INDEXDB      ?              ;WAVEFORM STYLE INDEX
12  WAVE_PTR   DB     ?              ;WAVEFORM DATA POINTER
13  COUNT      DB     ?
14  FREQUENCY  DB     2 DUP(?)       ;FREQUENCY CONTROL REGISTER
15  TEMP       DB     2 DUP(?)
16  STACK_A    DB     ?              ;STACK BUFFER FOR ACC
17  STACK_PSW  DB     ?              ;STACK FOR STATUS
18  ;----------------------------------------------------------------------
19  FSYS       EQU    4000000        ;fSYS=4MHz
20  DC_PORT    EQU    PD.0           ;DEFINE DC OFFSET PORT
21  DC_PORTC   EQU    PDC.0          ;DEFINE DC OFFSET CONTROL REG.
22  WAVE_PORT  EQU    PD.1           ;DEFINE WAVE PORT
23  WAVE_PORTCEQU     PDC.1          ;DEFINE WAVE PORT CONTROL REG.
24  SW_PORT    EQU    PA             ;DEFINE SWITCH PORT
25  SW_PORTC   EQU    PAC            ;DEFINE ALL SWITCH CONTROL REG.
26  WAVE_SW    EQU    SW_PORT.0      ;DEFINE WAVEFORM SELECT SWITCH.
27  OFFSET_INCEQU     SW_PORT.1      ;DEFINE OFFSET INCREASE SWITCH
```

```
28  OFFSET_DEC EQU  SW_PORT.2           ;DEFINE OFFSET DECREASE SWITCH
29  FREQ_UP    EQU  SW_PORT.3           ;DEFINE FREQUENCY INCREASE SWITCH
30  FREQ_DN    EQU  SW_PORT.4           ;DEFINE FREQUENCY DECREASE SWITCH
31  WAVE       EQU  PWM1                ;DEFINE PWM1 AS WAVE CONTROL
32  DC_        EQU  PWM0                ;DEFINE PWM0 AS DC OFFSET CONTROL
33  DC_STEP    EQU  4                   ;DC OFFSET ADJUST STEP
34  FREQ_STEP  EQU  8                   ;FREQUENCY ADJUST STEP
35
36  MY_CODE    .SECTION AT 0 'CODE'     ;== PROGRAM SECTION ==
37         ORG  00H                     ;HT-46RXX RESET VECTOR
38         JMP  START
39         ORG  08H                     ;HT-46RXX TMR INTERRUPT VECTOR
40         JMP  TMR_INT
41  START:
42         SET  SW_PORTC                ;CONFIG SW_PORT AS INPUT MODE
43         CLR  WAVE_INDEX
44         MOV  A,00000101B             ;ENABLE GLOBAL AND TMR INTERRUPT
45         MOV  INTC0,A
46         MOV  A,10000000B             ;CONFIG TMR 0 IN MODE 2(TIMER MODE)
47         MOV  TMRC,A                  ;fINT=fSYS(4MHz/1).
48         CLR  DC_PORTC                ;CONFIG DC_PORT AS OUTPUT MODE
49         CLR  WAVE_PORTC              ;CONFIG WAVE PORT AS OUTPUT MODE
50         MOV  A,LOW (65536-128)
51         MOV  FREQUENCY[0],A
52         MOV  A,HIGH (65536-128)      ;SET INITIAL MAX. FREQUENCY
53         MOV  FREQUENCY[1],A
54         MOV  A,OFFSET SINE_TABLE
55         MOV  TBLP,A                  ;INITIAL WAVE TABLE POINTER
56         MOV  WAVE_PTR,A              ;INITIAL WAVE TABLE POINTER
57         MOV  A,32
58         MOV  COUNT,A                 ;SET COUNT
59         CLR  WAVE                    ;RESET WAVE OUTPUT
60         MOV  A,128
61         MOV  DC_,A                   ;SET DC OFFSET
62         SET  DC_PORT                 ;ENABLE DC_PORT OUTPUT
63         SET  WAVE_PORT               ;ENABLE SINE_PORT OUTPUT
64         SET  TON                     ;START COUNTING
65  WAIT_KEY:
66         CALL DELAY
67         SNZ  WAVE_SW                 ;WAVE_SW PRESSED?
68         JMP  WAVEFORM_CHANGE         ;YES.
69         SNZ  OFFSET_INC              ;OFFSET_INC PRESSED?
70         JMP  OFFSET_INCREASE         ;YES.
71         SNZ  OFFSET_DEC              ;OFFSET_DEC PRESSED?
72         JMP  OFFSET_DECREASE         ;YES.
73         SNZ  FREQ_UP                 ;FREQ_UP PRESSED?
74         JMP  FREQUENCY_UP            ;YES.
75         SNZ  FREQ_DN                 ;FREQ_DN PRESSED?
76         JMP  FREQUENCY_DOWN          ;YES.
77         JMP  WAIT_KEY                ;WAIT KEY INPUT
78  FREQUENCY_DOWN:
79         MOV  A,FREQUENCY[0]
80         ADD  A,-FREQ_STEP
81         MOV  TEMP[0],A
82         SET  ACC
83         ADC  A,FREQUENCY[1]
84         MOV  TEMP[1],A
85         SNZ  C                       ;REACH MIN. FREQUENCY ?
86         JMP  WAIT_KEY                ;YES.
87         MOV  A,TEMP[0]               ;NO, INCREASE FREQUENCY
88         MOV  FREQUENCY[0],A
89         MOV  A,TEMP[1]
90         MOV  FREQUENCY[1],A
91         JMP  WAIT_KEY
92  FREQUENCY_UP:
93         MOV  A,FREQUENCY[0]
94         ADD  A,FREQ_STEP
```

```
 95        MOV     TEMP[0],A
 96        CLR     ACC
 97        ADC     A,FREQUENCY[1]
 98        MOV     TEMP[1],A
 99        MOV     A,LOW  -(65536-128)
100        ADD     A,TEMP[0]
101        MOV     A,HIGH -(65536-128)
102        ADC     A,TEMP[1]
103        SZ      C                    ;REACH MAX. FREQUENCY ?
104        JMP     WAIT_KEY             ;YES.
105        MOV     A,TEMP[0]            ;NO, INCREASE FREQUENCY
106        MOV     FREQUENCY[0],A
107        MOV     A,TEMP[1]
108        MOV     FREQUENCY[1],A
109        JMP     WAIT_KEY
110 OFFSET_DECREASE:
111        MOV     A,DC_
112        SUB     A,DC_STEP
113        SZ      C                    ;REACH MIN. DC OFFSET?
114        MOV     DC_,A                ;NO, DECREASE DC OFFSET
115        JMP     WAIT_KEY
116 OFFSET_INCREASE:
117        MOV     A,DC_
118        ADD     A,DC_STEP
119        SNZ     C                    ;REACH MAX. DC OFFSET?
120        MOV     DC_,A                ;NO, INCREASE DC OFFSET
121        JMP     WAIT_KEY
122 WAVEFORM_CHANGE:
123        INCA    WAVE_INDEX           ;ACC=WAVE_INDEX+1
124        AND     A,00000011B          ;MASK 6 MSB BIT TO ENSURE Acc < 4
125        MOV     WAVE_INDEX,A         ;UPDATA WAVE STYLE INDEX
126        RL      ACC                  ;CHANGE WAVE INDEX TO TABLE ADDRESS
127        RL      ACC                  ;BY *32 (EASCH WAVEFORM CONTAIN 32 DATA)
128        RL      ACC                  ;ACHIVE *32 BY SHIFT LEFT 5 TIMES
129        RL      ACC
130        RL      ACC
131        MOV     WAVE_PTR,A           ;UPDATE TABLE POINTER
132        JMP     WAIT_KEY
133 ;************************************************************************
134 ;             TIMER INTERRUPT SERVICE ROUTINE
135 ;************************************************************************
136 TMR_INT    PROC
137        MOV     STACK_A,A            ;PUSH A
138        MOV     A,STATUS
139        MOV     STACK_PSW,A          ;PUSH STATUS
140        TABRDL  WAVE                 ;RELOAD WAVE DUTY CYCLE
141        INC     TBLP                 ;INCREASE TABLE POINTER
142        SDZ     COUNT                ;TABLE END?
143        JMP     TMR_INT_1            ;NO.
144        MOV     A,32                 ;YES, RELOAD COUNTER
145        MOV     COUNT,A
146        MOV     A,WAVE_PTR
147        MOV     TBLP,A               ;RE-LOAD TABLE POINTER
148        MOV     A,FREQUENCY[0]       ;RE-LOAD TMRH & TMRL
149        MOV     TMRL,A
150        MOV     A,FREQUENCY[1]
151        MOV     TMRH,A
152 TMR_INT_1:
153        MOV     A,STACK_PSW
154        MOV     STATUS,A             ;POP STATUS
155        MOV     A,STACK_A            ;POP A
156        RETI
157 TMR_INT    ENDP
158 ;************************************************************************
159 ;             Delay about DEL1*10ms
160 ;************************************************************************
161 DELAY      PROC
```

```
162         MOV        A,10
163         MOV        DEL1,A              ;SET DEL1 COUNTER
164 DEL_1:
165         MOV        A,30
166         MOV        DEL2,A              ;SET DEL2 COUNTER
167 DEL_2:
168         MOV        A,110
169         MOV        DEL3,A              ;SET DEL3 COUNTER
170 DEL_3:
171         SDZ        DEL3                ;DEL3 DOWN COUNT
172         JMP        DEL_3
173         SDZ        DEL2                ;DEL2 DOWN COUNT
174         JMP        DEL_2
175         SDZ        DEL1                ;DEL1 DOWN COUNT
176         JMP        DEL_1
177         RET
178 DELAY ENDP
179         ORG        LASTPAGE
180 SINE_TABLE:                            ;6-BIT SINE WAVE WITH AC=0
181         DC         128,152,176,196,216,232,244,248  ;((63/2)*(1+SIN(2*PI*x/32))
182         DC         252,248,244,232,216,196,176,152  ; << 2 FOR 6+2 PWM MODE
183         DC         128,100, 76, 56, 36, 20,  8,  4
184         DC           0,  4,  8, 20, 36, 56, 76,100
185 TRI_TABLE:                             ;6-BIT TRI-ANGLE WAVE WITH AC=0
186         DC           0, 16, 32, 48, 64, 80, 96,112  ;(63*(x/16)) << 2
187         DC         128,140,156,172,188,204,220,236  ; IF x <16 FOR 6+2 PWM MODE
188         DC         252,236,220,204,188,172,156,140  ;(63-63*(x-16)/16)) << 2
189         DC         128,112, 96, 80, 64, 48, 32, 16  ;IF x >=16 FOR 6+2 PWM MODE
190 SAW_TABLE:                             ;6-BIT SAWTOOTH WAVE WITH AC=0
191         DC           0,  8, 16, 24, 32, 40, 48, 56  ;(63*x/32) << 2 FOR 6+2 PWM MODE
192         DC          64, 72, 80, 88, 96,104,112,120  ;
193         DC         128,132,140,148,156,164,172,180
194         DC         188,196,204,212,220,228,236,244
195 RET_TABLE:                             ;6-BIT RETANGULAR WAVE WITH AC=0
196         DC           0,  0,  0,  0,  0,  0,  0,  0  ;
197         DC           0,  0,  0,  0,  0,  0,  0,  0
198         DC         252,252,252,252,252,252,252,252
199         DC         252,252,252,252,252,252,252,252
200         END
```

程序说明

8~17   依序定义变量地址。

19   定义 HT46x23 的工作时钟脉冲为 4 MHz。

20~23   定义 DC_PORT 与 WAVE_PORT 分别为 PD.0 与 PD.1。

24~30   定义 SW_PORT 为 PA，并依序定义各按键开关的控制引脚。

31~32   定义 WAVE 与 DC_分别为 PWM1 与 PWM0。

33~34   定义频率与直流偏压值的变化量。

37   声明存储器地址由 000h 开始，此为 HT46x23 的 Reset Vector。

39   声明存储器地址由 008h 开始，此为 HT46x23 的 Timer/Event Counter Interrupt Vector。

42~43   将 SW_PORT 定义为输入模式，并将 WAVE_INDEX 清除为零，亦即本信号发生器 Default 输出正弦波的波形。

44~45   Global Interrupt 与 Timer/Event Counter Interrupt 有效。

46~47   将 Timer/Event Counter 定义如下：Timer Mode(Mode 2)、预分频比例 1:1($f_{INT}=f_{SYS}$)。

48~49   将 DC_PORT 与 WAVE_PORT 定义为输出模式。

50~53   设置 Timer/Event Counter 的中断时间，此时间的长短决定了输出频率的高低。

54~56　设置表格数据的索引值 WAVE_PTR=SINE_TABLE，WAVE_PTR 的内容决定了该提取表格中的哪一笔数据作为 WAVE 的更新值。

57~58　将 COUNT 设置为 32，此值代表每一组波形的数据为 32 笔。

59~61　清除 WAVE，并将 DC_设置为 128，此时的值零偏压值约为 2.5 V。

62~63　PWM 开始输出有效。

64　　　启动 Timer/Event Counter 开始计数；在此之后会依 FREQUENCY 寄存器的内容大小，每隔一段时间 Timer/Event Counter 就中断一次。

66~77　依序检查各个按键是否被按下，若是则跳至各按键的处理程序；否则就持续按键检查的动作。第 66 行调用 DELAY 子程序延时 0.1 s，其目的是为了避免开关的弹跳现象所引发的误动作。

78~91　按下 Frequency_Down 按键的处理程序，会对 FREQUENCY 寄存器进行减量(减去 FREQ_STEP)的动作，以延长 Timer/Event Counter 计数溢位的时间，进而达到降低输出信号频率的目的。这其中还必须注意是：是否 FREQUENCY 已递减到零以下，若是则不能再进行更新的动作，要让其维持原值。

92~109　按下 Frequency_Up 按键的处理程序，会对 FREQUENCY 寄存器进行增量(加上 FREQ_STEP)的动作，以缩短 Timer/Event Counter 计数溢位的时间，进而达到提高输出信号频率的目的。这其中还必须注意是：是否 FREQUENCY 已递增到最大值，若是则不能再进行更新的动作，要让其维持原值。

110~115　按下 Offset_DEC 按键的处理程序，会对 DC_寄存器进行减量(减去 DC_STEP)的动作，以降低输出信号的直流准位。这其中还必须注意 DC_寄存器是否已递减到零以下，若是则不能再进行更新的动作，要让其维持原值。

116~121　按下 Offset_INC 按键的处理过程，会对 DC_寄存器进行增量(加上 DC_STEP)的动作，以增加输出信号的直流准位。这其中还必须注意 DC_寄存器的递增是否产生进位，若是则不能再进行更新的动作，要让其维持原值。

122~132　按下 Waveform_Select 按键的处理过程，会对 WAVE_INDEX 寄存器进行加一的动作，以改变输出波形。因为本例中只建了 4 种波形的数据表，因此必须让 WAVE_INDEX 的值在 0~3 之间变化。另外因为每一组波形都有 32 笔数据，因此只要将 WAVE_INDEX 乘上 32 就可获得该组波形的数据起始地址，程序第 126~130 行就是以连续左移 5 次的方式来完成乘 32 的目的。

136~157　TMR_INT 中断服务子程序，根据 TBLP 寄存器的值提取波形数据更新 WAVE 值；并将 COUNT−1 判断其是否为零，如为零表示已处理至波形的最后一笔数据，必须将 TBLP 指向该波形数据的起始地址，在此同时也将 TMRL 与 TMRH 依 FREQUENCY 寄存器的值更新，若频率有变化的话会在此时发生作用。

161~175　DELAY 子程序，延时时间的计算请参考 4-1 节中的说明。

177~181　正弦波数据定义区，其每笔数据依下式的公式求得：
$$x[n]=\left[\frac{63}{2}\times(1+\sin(\frac{2\pi n}{32}))\right]<<2,\ n=0\sim31$$

182~186　三角波数据定义区，其每笔数据依下式的公式求得：
$$x[n]=\left\lfloor 63\times(\frac{n}{16})\right\rfloor<<2,\ n=0\sim15$$
$$x[n]=\left\lfloor 63-63\times(\frac{n-16}{16})\right\rfloor<<2,\ n=16\sim31$$

187~191　锯齿波数据定义区，其每笔数据依下式的公式求得：
$$x[n]=\left[63\times\frac{n}{32}\right]<<2,\ n=0\sim31$$

192~196　方波数据定义区。

图6-8-7是本专题各式波形在DC Offset＝0 V的输出结果,由于强调单电源的操作,因此OP-AMP的工作电压使用与HT46x23相同的5 V,这限制了电压变动的范围。读者可以视实际的需要改变OP-AMP的工作电压,以增加输出信号的动态变化范围。如前所述,Timer/Event Counter计数溢位的时间决定了输出信号的频率,然而计数溢位的时间是否可以无限地缩短呢？请读者注意,TMR_INT中断服务子程序所需的执行时间大约在20~24μs左右,如果中断的时间过短,将导致主程序分配执行的时间过短,而使按键的检测变得极不灵敏。再夸张一点,如果让Timer/Event Counter计数溢位的时间小于20μs,会使得CPU跳出中断服务子程序后又立刻进入(即在TMR_INT中断服务子程序执行过程中又发生计数溢位的状况),此时就无法再由按键来选择输出波形、频率以及直流偏压值了。

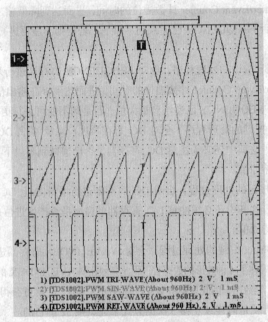

图6-8-7 简易信号发生器的4种输出波形

表6-8-1是各种实现D/A转换方式的比较表,供读者参考。

表6-8-1 各种实现D/A转换方式的比较

项　目	软件 PWM	硬件 PWM	R-2R Ladder	外加 DAC 晶片
成本	低	中	低	高
外接零件数目	少	少	视分辨率而定 $2n$	少
所需 I/O Pin	1	1	n	2(串行式)/n(并列式)
准确度	佳	佳	佳	佳
频宽	低	中	高	高
谐波失真	高	中	低	低

图6-8-8~图6-8-13是不同频率及直流偏压值下的输出波形,特将其列出以供读者参考。

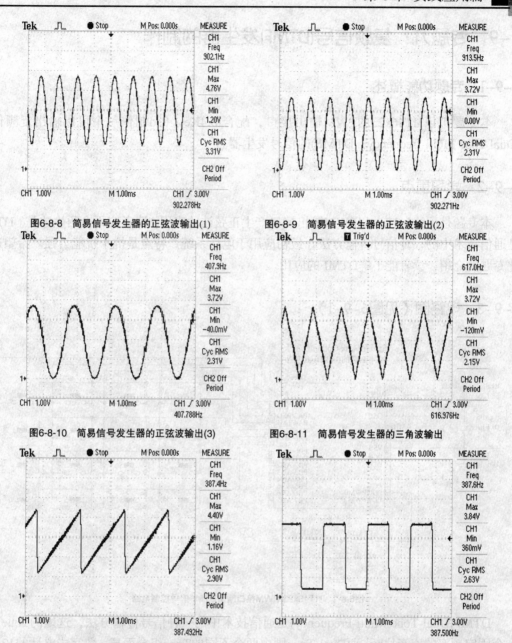

图6-8-8 简易信号发生器的正弦波输出(1)　　图6-8-9 简易信号发生器的正弦波输出(2)

图6-8-10 简易信号发生器的正弦波输出(3)　　图6-8-11 简易信号发生器的三角波输出

图6-8-12 简易信号发生器的锯齿波输出　　图6-8-13 简易信号发生器的方波输出

6-8-5 动动脑+动动手

- 在 PA.5(Ampitude_INC)与 PA.6(Ampitude_DEC)再加上两个按键开关，并修改程序使输出波形的峰对峰值(Peak to Peak)也可由用户加以调整。
- 为此简易信号发生器多增加几组输出波形，如指数波、阶梯波等。

6-9 专题九：复频信号(DTMF)发生器的制作

6-9-1 专题功能概述

本专题拟以HT46x23的PWM输出接口，配合4×4Key Pad制作一个可以输出复频信号(Dual Tone Multi-frequency；DTMF)的信号发生器。

6-9-2 学习重点

本专题是承续6-8节，以PWM输出接口产生正弦波的应用，读者应不会陌生才是。DTMF是通信技术中的一种信号传输方法，最初应用于电话系统，特点是抗干扰能力强。希望通过此专题的介绍，读者能了解DTMF的应用。

6-9-3 电路图（见图6-9-1）

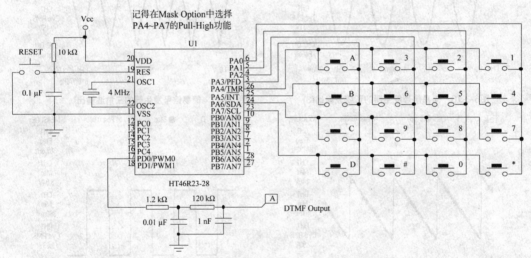

图6-9-1　HT46x23 PWM接口输出DTMF信号控制电路

DTMF(Dual-Tone Multi-Frequency)是通信技术中的一种信号传输方法，这里的Tone代表一个固定频率的声音片断，而Dual-Tone则由两个不同的Tone组合而成。数字式电话的12个按键分别代表了12种不同的复频组合，借助于对频率组合的检测，DTMF Decoder可分辨出所按的是哪一个键，从而达到与电话另一端的用户互动控制的目的。DTMF一般用于电话的拨号脉冲，电话的拨号脉冲可分为脉冲信号和复频信号两种。

脉冲信号：以电话机为例，拨号盘截断直流回路控制交换机，依照电流中断次数而接线。每中断一次称为一个脉冲(Dialing or Pulsing)。标准的脉冲速度为每秒10次，每次100 ms。标准脉冲比又称为断续比(Break-Mark Ratio)，是指断脉冲对通脉冲的比例，应该为 $66\frac{2}{3}:33\frac{1}{3}$ 或 2:1。

复频信号：由高频群(Column 1~4)1 209 Hz、1 336 Hz、1 477 Hz、1 633 Hz及低频群(Row 1~4)697 Hz、770 Hz、852 Hz、941 Hz组成，频率与电话按键码的关系如图6-9-2所示，一般话机只使用0~9、*以及#共12个按键，但为其方便性，本专题拟产生表中的16组复频信号。

	1209Hz	1336Hz	1477Hz	1633Hz	
697Hz	1	2	3	A	Row 1
770Hz	4	5	6	B	Row 2
852Hz	7	8	9	C	Row 3
941Hz	*	0	#	D	Row 4
	Column 1	Column 2	Column 3	Column 4	

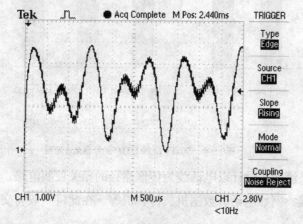

图6-9-2　DTMF频率与按码关系

以按下电话按键9为例，其DTMF信号由低频群的852 Hz正弦波信号与高频群的1 477 Hz正弦波信号所组成，如图6-9-3所示。

图6-9-3　电话按键9的DTMF信号

在CCITT Q.23的相关规范中，还有表6-9-1的规定：

表6-9-1　CCITT Q.23规范

Frequency Tolerance	Operation:	<=1.5%
	Non-Operation:	>=3.5%
Signal Duration	Operation:	40ms(MAX)
	Non-Operation:	23ms(MIN)
Signal Interruption		10ms(MAX)

Signal Duration是指DTMF信号所维持的时间，如果信号时间小于23 ms，则DTMF接收端不会检测此信号；若信号时间介于23~40 ms，即使DTMF接收端检测到信号也将予以忽略。所以如果所设计的DTMF发生器要为一般DTMF接收器所接收，其Signal Duration至少需维持40 ms以上。而Signal Interruption是指两个DTMF信号的间隔时间，至少要10 ms以上。

也就是说频率的误差允许范围在1.5%以内(见图6-9-4),如果DTMF发生器所产生的频率误差超过3.5%,则DTMF接收端不会检测此信号,而如果信号的频率误差在1.5%~3.5%之间的话,即使DTMF接收端检测到信号也将予以忽略。所以如果所设计的DTMF发生器要为一般DTMF接收器所接收,其误差范围一定要控制在1.5%以内。另外由于高频信号的衰减情形比低频来得严重,因此在发送端通常会加上预强调(Pre-Emphasis)处理,将高频群(Column Frequency)的信号放大2 dB。

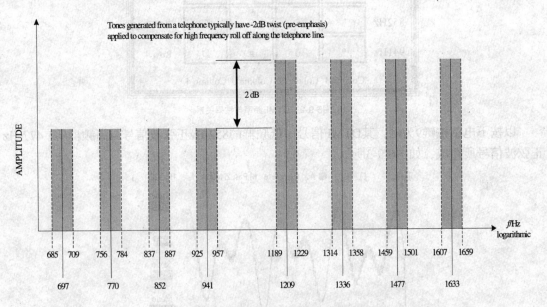

图6-9-4 CCITT Q.23 DTMF的频率允许范围

除此之外,如果读者想设计以电话线为传输途径的相关控制电路,通常都必须使用一些信号与交换机作连接,例如电话、数据机、传真机等。在此将一般与交换机作连接的信号大致介绍一下:

① 拨号音:是指交换机准备开始接收发送端所发出的信号。常用的频率有33、160、400、420、480、500 Hz,最常用的是400 Hz。
② 回铃音:所谓的回铃音是指交换机通知发送信号端所送出的信号。
③ 振铃信号:接收信号端由交换机接收振铃信号。
④ 忙音:表示线路忙碌。

由6-8节读者应该已经了解如何以HT46x23的PWM输出接口来产生正弦波,并通过PWM寄存器的更新速率来控制其频率。但是HT46x23的Timer/Event Counter只有一组,如何同时产生两个不同频率的正弦波输出呢? 这表示不能再以Timer/Event Counter计数溢位的时间控制输出频率,如果在Timer/Event Counter计数溢位时间固定的情况之下,更改正弦波波形数据的点数,其实也可以达到频率变化的目的。

如图6-9-5所示,假设采样周期(Sample Period,即PWM寄存器的更新速率)为t_{Sample},则

SINE 1与SINE 2的频率分别为$(m \times t_{Sample})^{-1}$ Hz及$(n \times t_{Sample})^{-1}$ Hz。所以在t_{Sample}为固定的情况之下，是可以用一个周期正弦波的波形数据点数(N)来决定其频率。例如$t_{Sample}=100$ μs，且取$N=32$时，正弦波的频率为312.5 Hz；若$N=16$，频率为625 Hz。而波形数据分别为：

$$x[n] = \left[\frac{63}{2} \times (1+\sin(\frac{2n\pi}{32}))\right] << 2 ， n=1\sim32 \quad \text{for } N=32 (312.5\text{ Hz})$$

$$x[n] = \left[\frac{63}{2} \times (1+\sin(\frac{2n\pi}{16}))\right] << 2 ， n=1\sim16 \quad \text{for } N=16 (625\text{ Hz})$$

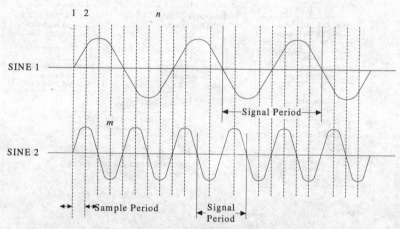

图6-9-4 取样频率、数据点数与正弦波频率的关系示意图

式中的"[ ]"符号代表取整数之意，而左移2位(<<2)的主要原因是要采用HT46x23的(6+2)PWM Mode，而且AC=00(AC Part为零)。当然，在相同的正弦波输出频率之下，波形数据点数越多(即N越大)，那么输出的波形更加细腻，可是采样周期(t_{Sample})就必须更短才能维持相同的输出频率。所以当t_{Sample}越高，将使得输出的正弦波波形更为细致，但是因为笔者打算以Timer/Event Counter中断来进行PWM寄存器数据的更新，万一t_{Sample}小于Timer/Event Counter中断服务子程序的处理时间，又会造成递归式(Recursive)中断的窘境(也就是在中断服务子程序的执行过程中，又发生Timer/Event Counter计数溢位的情形，所以当执行完中断服务子程序后又立即进入)，导致CPU一直在中断服务子程序里打转。但是如果t_{Sample}太大，又会使得波形数据点数(N值)太少而造成输出波形过于粗糙。所以结论是：t_{Sample}必须越小越好，但是又必须满足t_{Sample}小于Timer/Event Counter中断服务子程序的处理时间。其实笔者是先大致估算中断服务子程序执行所需的时间(t_{INT}；约为38μs @ $f_{SYS}=4$ MHz)，然后再推导出各频率所需的波形数据点数。其公式如下：

$$N = (t_{INT} \times f_{FREQUENCY})^{-1}$$

式中的$f_{FREQUENCY}$代表所要产生的正弦波频率。求出波形的点数(N)后，再依下式求得各点的波形数据：

$$x[n] = \left[\frac{63}{2} \times (1+\sin(\frac{2n\pi}{N}))\right] << 2 , n=1\sim N$$

6-9-4 流程图及程序

1. 流程图(见图6-9-6)

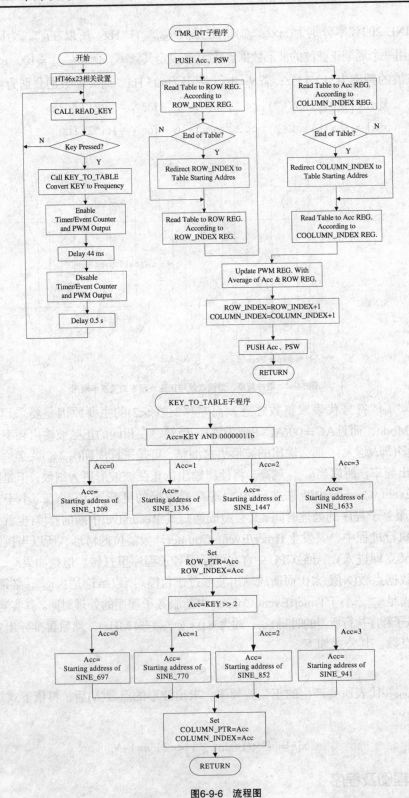

图6-9-6 流程图

注意：必须使能 Code Option 中的 PWM0 输出功能，并选择 6+2 的 PWM Mode。

2. 程序6-9 复频(DTMF)信号发生器

```asm
1     ;PROGRAM : 6-9.ASM (6-9.PRJ)                2002.DEC.30.
2     ;FUNCTION: DTMF GENERATOR DEMO PROGRAM         BY STEVEN
3     ;NOTE    : PMW0 SHOULD BE ENABLED, SEL 6+2 MODE (MASK OPTION).
4     #INCLUDE   HT46R23.INC
5                .CHIP    HT46R23
6     ;-------------------------------------------------------------
7     MY_DATA    .SECTION  'DATA'         ;== DATA SECTION ==
8     DEL1          DB    ?               ;DELAY LOOP COUNT 1
9     DEL2          DB    ?               ;DELAY LOOP COUNT 2
10    DEL3          DB    ?               ;DELAY LOOP COUNT 3
11    STACK_A       DB    ?               ;STACK BUFFER FOR ACC
12    STACK_PSW     DB    ?               ;STACK FOR STATUS
13    ROW           DB    ?               ;BUFFER FOR ROW WAVE DATA
14    ROW_PTR       DB    ?               ;ROW TABLE POINTER
15    ROW_INDEX     DB    ?               ;ROW TABLE INDEX
16    COLUMN_PTR    DB    ?               ;COLUMN TABLE POINTER
17    COLUMN_INDEX  DB    ?               ;COLUMN TABLE INDEX
18    COUNT         DB    ?               ;COUNTER
19    KEY           DB    ?               ;KEY INDEX
20    ;-------------------------------------------------------------
21    KEY_PORT      EQU   PA              ;DEFINE KEYPAD PORT
22    KEY_PORTC     EQU   PAC             ;DEFINE KEY PORT VONTROL REG.
23    DTMF_PORT     EQU   PD.0            ;DEFINE SINE PORT
24    DTMF_PORTC    EQU   PDC.0           ;DEFINE SINE PORT CONTROL REG.
25    TABLE_END     EQU   1
26
27    MY_CODE       .SECTION AT 0 'CODE'  ;== PROGRAM SECTION ==
28            ORG   00H                   ;HT-46RXX RESET VECTOR
29            JMP   START
30            ORG   08H                   ;HT-46RXX TMR INTERRUPT VECTOR
31            JMP   TMR_INT
32    START:
33            MOV   A,00000101B
34            MOV   INTC0,A               ;ENABLE GLOBAL AND TMR INTERRUPT
35            MOV   A,10000010B           ;CONFIG TMR 0 IN MODE 2(TIMER MODE)
36            MOV   TMRC,A                ;fINT=fSYS/4(4MHz/4).
37            CLR   DTMF_PORTC            ;CONFIG PD0 AS OUTPUT MODE
38            MOV   A,LOW (65536-38)
39            MOV   TMRL,A
40            MOV   A,HIGH (65536-38)
41            MOV   TMRH,A                ;SET DELAY FOR SAMPLE POINTS
42    WAIT_KEY:
43            CALL  READ_KEY              ;READ KEY PAD
44            MOV   A,KEY
45            XOR   A,16
46            SZ    Z
47            JMP   WAIT_KEY              ;KEY PRESSED?
48            CALL  KEY_TO_TABLE          ;NO.
49            CLR   PWM0                  ;COVERT KEY TO TABLE ADDRESS
50            SET   DTMF_PORT             ;RESET PWM0
51            SET   TON                   ;START PWM OUTPUT
52            MOV   A,7                   ;START DTMF WAVE OUTPUT
53            CALL  DELAY                 ;DELAY 44mS for DTMF SIGNAL DURATION
54            CLR   TON
55            CLR   DTMF_PORT             ;STOP DTMF WAVE OUTPUT
56            CLR   ACC                   ;STOP PWM OUTPUT
57            CALL  DELAY                 ;DELAY 0.13 SEC
58            CALL  DELAY                 ;DELAY 0.13 SEC
59            CALL  DELAY                 ;DELAY 0.13 SEC
60            CALL  DELAY                 ;DELAY 0.13 SEC
61            JMP   WAIT_KEY
62    ;****************************************************************
63    ;             TIMER INTERRUPT SERVICE ROUTINE
64    ;****************************************************************
65    TMR_INT    PROC
```

```
66          MOV     STACK_A,A               ;PUSH A
67          MOV     A,STATUS
68          MOV     STACK_PSW,A             ;PUSH STATUS
69          MOV     A,ROW_INDEX             ;GET ROW WAVE INDEX
70          MOV     TBLP,A
71          TABRDL  ROW                     ;GET ROW WAVE DATA
72          SNZ     ROW.0                   ;TABLE ENDDING?
73          JMP     TONE_1                  ;NO.
74          MOV     A,ROW_PTR               ;RELOAD ROW INDEX VALUE
75          MOV     ROW_INDEX,A
76          MOV     TBLP,A
77  TONE_1:
78          TABRDL  ROW                     ;GET ROW WAVE DATA
79          MOV     A,COLUMN_INDEX          ;GET COLUMN WAVE INDEX
80          MOV     TBLP,A
81          TABRDL  ACC                     ;GET ROW WAVE DATA
82          SNZ     ACC.0                   ;TABLE ENDDING?
83          JMP     TONE_2                  ;NO.
84          MOV     A,COLUMN_PTR            ;RELOAD COLUMN INDEX VALUE
85          MOV     COLUMN_INDEX,A
86          MOV     TBLP,A
87  TONE_2:
88          TABRDL  ACC                     ;GET COLUMN WAVE DATA
89          ADD     A,ROW                   ;COMBINE ROW & COULMN WAVE DATA
90          RRC     ACC                     ;DIVIDED BY 2
91          MOV     PWM0,A                  ;UPDATE PWM
92          INC     ROW_INDEX               ;INCREASE ROW INDEX
93          INC     COLUMN_INDEX            ;INCREASE COLUMN INDEX
94          MOV     A,STACK_PSW
95          MOV     STATUS,A                ;POP STATUS
96          MOV     A,STACK_A               ;POP A
97          RETI
98  TMR_INT     ENDP
99  ;**********************************************************************
100 ;               CONVERT KEY TO TABLE ADDRESS ROUTINE
101 ;**********************************************************************
102 KEY_TO_TABLE    PROC
103         MOV     A,KEY                   ;GET KEY VALUE
104         AND     A,00000011B             ;GET LAST 2 BIT FOR ROW
105         CALL    TRANS_ROW
106         MOV     COLUMN_PTR,A            ;SET COLUMN WAVE DATA POINTER
107         MOV     COLUMN_INDEX,A
108         MOV     A,KEY                   ;GET KEY VALUE
109         AND     A,00001100B             ;GET LAST 2 BIT FOR COLUMN
110         RRC     ACC                     ;DIVIDE BY 4
111         RRC     ACC
112         CALL    TRANS_COLUMN
113         MOV     ROW_PTR,A               ;SET ROW WAVE DATA POINTER
114         MOV     ROW_INDEX,A
115         RET
116 TRANS_COLUMN:
117         ADDM    A,PCL
118         RET     A,SINE_1209
119         RET     A,SINE_1336
120         RET     A,SINE_1477
121         RET     A,SINE_1633
122 TRANS_ROW:
123         ADDM    A,PCL
124         RET     A,SINE_697
125         RET     A,SINE_770
126         RET     A,SINE_852
127         RET     A,SINE_941
128 KEY_TO_TABLE    ENDP
129 ;**********************************************************************
130 ;   SCAN 4x4 MATRIX ON KEY PORT AND RETURN THE CODE IN KEY REGISTER
131 ;       IF NO KEY BEEN PRESSED, KEY=16.
132 ;**********************************************************************
```

```
133 READ_KEY    PROC
134     MOV     A,11110000B
135     MOV     KEY_PORTC,A         ;CONFIG PORT B
136     SET     KEY_PORT            ;INITIAL PORT B
137     CLR     KEY                 ;INITIAL KEY REGISTER
138     MOV     A,04
139     MOV     COUNT,A             ;SET ROW COUNTER
140     CLR     C                   ;CLEAR CARRY FLAG
141 SCAN_KEY:
142     RLC     KEY_PORT            ;ROTATE SCANNING BIT
143     SET     C                   ;MAKE SURE C=1
144     SNZ     KEY_PORT.4          ;COLUMN 0 PRESSED?
145     JMP     END_KEY             ;YES.
146     INC     KEY                 ;NO, INCREASE KEY CODE.
147     SNZ     KEY_PORT.5          ;COLUMN 1 PRESSED?
148     JMP     END_KEY             ;YES.
149     INC     KEY                 ;NO, INCREASE KEY CODE.
150     SNZ     KEY_PORT.6          ;COLUMN 2 PRESSED?
151     JMP     END_KEY             ;YES.
152     INC     KEY                 ;NO, INCREASE KEY CODE.
153     SNZ     KEY_PORT.7          ;COLUMN 3 PRESSED?
154     JMP     END_KEY             ;YES.
155     INC     KEY                 ;NO, INCREASE KEY CODE.
156     SDZ     COUNT               ;HAVE ALL ROWs BEEN CHECKED?
157     JMP     SCAN_KEY            ;NO, NEXT ROW.
158 END_KEY:
159     RET
160 READ_KEY    ENDP
161 ;************************************************************************
162 ;     Delay about DEL1*511+2 uS WITHOUT TIMER INTERRUPT
163 ;************************************************************************
164 DELAY   PROC
165     MOV     DEL1,A              ;SET DEL1 COUNTER
166 DEL_1:
167     MOV     A,3
168     MOV     DEL2,A              ;SET DEL2 COUNTER
169 DEL_2:
170     MOV     A,55
171     MOV     DEL3,A              ;SET DEL3 COUNTER
172 DEL_3:
173     SDZ     DEL3                ;DEL3 DOWN COUNT
174     JMP     DEL_3
175     SDZ     DEL2                ;DEL2 DOWN COUNT
176     JMP     DEL_2
177     SDZ     DEL1                ;DEL1 DOWN COUNT
178     JMP     DEL_1
179     RET
180 DELAY ENDP
181     ORG     LASTPAGE
182 SINE_697:                       ;6-BIT SINE WAVE WITH AC=0
183     DC      148,168,184,204,220,232,240,248 ;((63/2)*(1+SIN(2*PI*x/N))
184     DC      252,252,248,240,232,220,204,184 ; << 2 FOR 6+2 PWM MODE
185     DC      168,148,128,104, 84, 68, 48, 32
186     DC       20, 12,  4,  0,  0,  4, 12, 20
187     DC       32, 48, 68, 84,104,124
188     DC      TABLE_END
189 SINE_770:                       ;6-BIT SINE WAVE WITH AC=0
190     DC      148,172,192,212,228,240,248,252 ;((63/2)*(1+SIN(2*PI*x/N))
191     DC      252,248,240,228,212,192,172,148 ; << 2 FOR 6+2 PWM MODE
192     DC      128,104, 80, 60, 40, 24, 12,  4
193     DC        0,  0,  4, 12, 24, 40, 60, 80
194     DC      104,124
195     DC      TABLE_END
196 SINE_852:                       ;6-BIT SINE WAVE WITH AC=0
197     DC      152,176,196,216,232,244,252,252 ;((63/2)*(1+SIN(2*PI*x/N))
198     DC      248,240,224,208,188,164,140,112 ;<< 2 FOR 6+2 PWM MODE
199     DC       88, 64, 44, 28, 12,  4,  0,  0
```

```
200         DC          8, 20, 36, 56, 76,100,124
201         DC          TABLE_END
202 SINE_941:                                      ;6-BIT SINE WAVE WITH AC=0
203         DC          156,180,204,224,240,248,252,248  ;((63/2)*(1+SIN(2*PI*x/N))
204         DC          240,224,204,180,156,128, 96, 72  ; << 2 FOR 6+2 PWM MODE
205         DC          48, 28, 12,  4,  0,  4, 12, 28
206         DC          48, 72, 96,124
207         DC          TABLE_END
208 SINE_1209:                                     ;6-BIT SINE WAVE WITH AC=0
209         DC          160,196,220,240,252,252,240,220  ;((63/2)*(1+SIN(2*PI*x/N))
210         DC          196,160,128, 92, 56, 32, 12,  0  ; << 2 FOR 6+2 PWM MODE
211         DC           0, 12, 32, 56, 92,124
212         DC          TABLE_END
213 SINE_1336:                                     ;6-BIT SINE WAVE WITH AC=0
214         DC          164,200,228,244,252,244,228,200  ;((63/2)*(1+SIN(2*PI*x/N))
215         DC          164,128, 88, 52, 24,  8,  0,  8  ; << 2 FOR 6+2 PWM MODE
216         DC          24, 52, 88,124
217         DC          TABLE_END
218 SINE_1477:                                     ;6-BIT SINE WAVE WITH AC=0
219         DC          168,208,236,252,252,236,208,168  ;((63/2)*(1+SIN(2*PI*x/N))
220         DC          128, 84, 44, 16,  0,  0, 16, 44  ; << 2 FOR 6+2 PWM MODE
221         DC           84,124
222         DC          TABLE_END
223 SINE_1633:                                     ;6-BIT SINE WAVE WITH AC=0
224         DC          176,216,244,252,244,216,176,128  ;((63/2)*(1+SIN(2*PI*x/N))
225         DC          76, 36,  8,  0,  8, 36, 76,124   ; << 2 FOR 6+2 PWM MODE
226         DC          TABLE_END
227         END
```

程序说明

行号	说明
8~19	依序定义变量地址。
21~22	定义 KEY_PORT 为 PA。
23~24	定义 DTMF_PORT 为 PD.0。
25	定义各组数据表格的结束代码 TABLE_END=1;由于各频率正弦波的数据长度不一,因此以 TABLE_END 作为结束的代码,以方便程序判断是否该重新由数据起点开始提取数据。
28	声明存储器地址由 000h 开始,此为 HT46x23 的 Reset Vector。
30	声明存储器地址由 008h 开始,此为 HT46x23 的 Timer/Event Counter Interrupt Vector。
33~34	Global Interrupt 与 Timer/Event Counter Interrupt 有效。
35~36	将 Timer/Event Counter 定义如下:Timer Mode(Mode 2)、预分频比例设置为 1:4(f_{INT}=f_{SYS}/4)。
37	将 DTMF_PORT 定义为输出模式。
38~41	设置 Timer/Event Counter 的中断时间,在此设置为每 38μs 中断一次。
42~47	调用 READ_KEY 子程序读取按键,并判断是否有按键被按下,如没有则持续检查按键(JMP WAIT_KEY)。
48	调用 KEY_TO_TABLE 子程序,将用户按下的按键值转换为 ROW Frequency 与 Column Frequency 的波形起始地址,并分别存放在 ROW_PTR(ROW_INDEX)与 COLUMN_PTR(COLUMN_INDEX) 寄存器中。
49~50	清除 PWM0 并启动其开始输出 PWM 信号。
51	启动 Timer/Event Counter 开始计数。
52~53	调用 DELAY 子程序延时约 44 ms,请参考后续的说明。
54~55	停止 Timer/Event Counter 的计数与 PWM0 的输出。
56~61	延时约 0.5 s 后重新读取按键(JMP WAIT_KEY)。
65~98	TMR_INT 中断服务子程序,其主要的功能是根据 ROW、COLUMN 寄存器的值分别提取 Row Frequency 与 Column Frequency 波形数据,并以两者的平均值更新 WAVE 值(即 PWM 输出值)。由于各

个频率的波形数据长度不一致，所以在每一组频率数据的最后一笔插入 TABLE_END 作为结束的判别码；又因为各笔数据都是以 AC=0 来设计，所以当发现查表所得的数据最低位(LSB)不为零时，就是代表已经到达该频率波形数据的最后一笔，必须更新其指针值(ROW_INDEX、COLUMN_INDEX)至波形数据的起点(ROW_PTR、COLUMN_PTR)，以便输出下一周期的波形。

69~78　依据 ROW_INDEX 寄存器的值查取波形数据(存放在 ROW 寄存器)，若所查得数据的LSB=1，表示已到达该频率波形的最尾端，此时需将 ROW_INDEX 依 ROW_PTR 寄存器的值更新，指向该频率波形数据的起始地址，并重新读取数据存放在 ROW 寄存器。

79~88　依据 COLUMN_INDEX 寄存器的值查取波形数据(存放在 ACC 寄存器)，若所查得数据的LSB=1，表示已到达该频率波形的最尾端，此时需将 COLUMN_INDEX 依 COLUMN_PTR 寄存器的值更新，指向该频率波形数据的起始地址，并重新读取数据存放在 ACC 寄存器。

89~91　求取 ROW 与 ACC 的平均值后，更新 PWM0 的输出。

92~93　将 ROW_INDEX 与 COLUMN_INDEX 寄存器加一，分别指向取 Row Frequency 与 Column Frequency 的下一笔波形数据。

102~128　KEY_TO_TABLE 子程序，将 KEY 寄存器的按键值转换为对应的 ROW Frequency 与 Column Frequency 波形起始地址，并分别存放在 ROW_PTR、ROW_INDEX 与 COLUMN_PTR、COLUMN_INDEX 寄存器中。

133~160　READ_KEY 子程序，请参考 4-7 节中的说明。

164~177　DELAY 子程序，延时时间为(ACC × 511+2)μs(@f_{SYS}=4 MHz)，详细的计算请参考 4-1 节中的说明。

179~185　697 Hz 正弦波数据定义区，其每笔数据依下式的公式求得：
$$x[n]=\left[\frac{63}{2} \times (1+\sin(\frac{2n\pi}{38}))\right]<<2, \quad n=1\sim38$$

186~192　770 Hz 正弦波数据定义区，其每笔数据依下式的公式求得：
$$x[n]=\left[\frac{63}{2} \times (1+\sin(\frac{2n\pi}{36}))\right]<<2, \quad n=1\sim36$$

193~198　852 Hz 正弦波数据定义区，其每笔数据依下式的公式求得：
$$x[n]=\left[\frac{63}{2} \times (1+\sin(\frac{2n\pi}{31}))\right]<<2, \quad n=1\sim31$$

199~204　941 Hz 正弦波数据定义区，其每笔数据依下式的公式求得：
$$x[n]=\left[\frac{63}{2} \times (1+\sin(\frac{2n\pi}{28}))\right]<<2, \quad n=1\sim28$$

205~209　1 209 Hz 正弦波数据定义区，其每笔数据依下式的公式求得：
$$x[n]=\left[\frac{63}{2} \times (1+\sin(\frac{2n\pi}{22}))\right]<<2, \quad n=1\sim22$$

210~214　1 336 Hz 正弦波数据定义区，其每笔数据依下式的公式求得：
$$x[n]=\left[\frac{63}{2} \times (1+\sin(\frac{2n\pi}{20}))\right]<<2, \quad n=1\sim20$$

215~219　1 477 Hz 正弦波数据定义区，其每笔数据依下式的公式求得：
$$x[n]=\left[\frac{63}{2} \times (1+\sin(\frac{2n\pi}{18}))\right]<<2, \quad n=1\sim18$$

220~223　1 633 Hz 正弦波数据定义区，其每笔数据依下式的公式求得：
$$x[n]=\left[\frac{63}{2} \times (1+\sin(\frac{2n\pi}{16}))\right]<<2, \quad n=1\sim16$$

　　笔者认为在此必须提出说明的是程序第52~53行，明明应该只是延时(7 × 511+2) μs=3.579 ms，可是却可以输出大约44 ms左右的复频信号，为什么呢？请读者注意在程序第51行已将TON设置为1，此时Timer/Event Counter开始计数，而且每隔38 μs就会中断一次。然而中断服务子程序的执行时间约在31~36 μs(因为有"条件式判断指令"，所以会有所差异)，也就是说每38

μs之中，实际上只有2~7μs是在执行DELAY子程序中的指令，所以在Timer/Event Counter中断持续发生的情况下，DELAY子程序所造成的延时时间应该是：

$$T_{MIN} = \frac{7 \times 511 + 2}{7} \times 38 \text{ μs} \approx 19.4 \text{ ms}$$

$$T_{MAX} = \frac{7 \times 511 + 2}{7} \times 38 \text{ μs} \approx 68.0 \text{ ms}$$

取其平均值大约为43.7 ms，恰好符合CCITT DTMF Signal Duration 至少须维持40 ms的规定范围。而在第56~60行程序中，因为在第54行已将TON设置为0，此时Timer/Event Counte不再计数，故中断不再发生。所以第56~60行所造成的延时时间为$(256 \times 511 + 2) \times 4$ μs ≈ 0.5 s。

表6-9-2将本专题的Row Frequency与Column Frequency频率值列出与CCITT DTMF信号规范相比较，依其规定各频率的误差范围需在1.5%以内，而本专题产生的频率均在规定之内。

表6-9-2 本专题的Row Frequency与Column Frequency频率值

Type	频率/ Hz	最小频率/Hz	最大频率/Hz	本专题 DTMF 输出频率/Hz
ROW	697	686.55	707.46	692.52
	770	758.45	781.85	773.99
	852	839.22	864.78	848.90
	941	926.89	955.12	939.85
COLUMN	1 209	1 190.87	1 227.14	1 196.17
	1 336	1 315.96	1 356.04	1 316.10
	1 477	1 454.85	1 499.16	1 461.99
	1 633	1 608.51	1 657.50	1 644.74

图6-9-7~图6-9-9为本专题实际输出的DTMF波形，在此将其列出供读者参考。

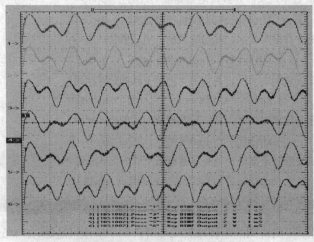

图6-9-7 按键1~6的DTMF输出波形

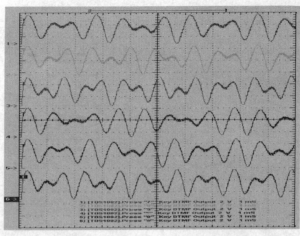

图6-9-8 按键7~0、*、#的DTMF输出波形

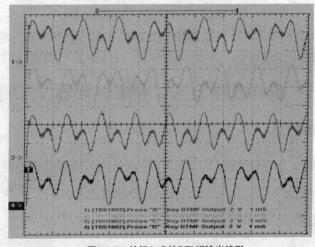

图6-9-9 按键A~D的DTMF输出波形

6-9-5 动动脑 + 动动手

  读者可以试着将 PWM 的输出经功率放大之后直接驱动喇叭(见图 6-1-10)，按下按键后可以听到打电话时按按键的声音。如果喇叭的响应不太差，可以拿起电话听筒，以其产生的 DTMF 信号达到拨号的功能。(注：HT82V733 为 Holtek 半导体公司在放大器系列(Amplifier Series)的产品，详细数据请参考随书所附的光碟。其中的/CE 引脚＝1 时，该芯片即进入 Power Off 模式，此时约只消耗 1 μA 的电流。图 6-9-8 是让 HT82V733 持续工作于正常模式。读者可以视实际需要以 HT46x23 的一根 I/O 引脚并配合程序的运作，达到省电的目的。)

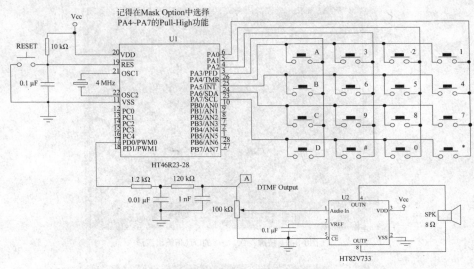

图6-9-8　DTMF加上Speaker推动电路

📖 本程序是每按下按键之后就立即输出 DTMF 信号，请将本程序修改成当按完 10 个数字之后再依序送出 DTMF 信号，而且每个 DTMF 信号间需延时 20 ms。

6-10　专题十：简易低频电压－频率转换器(VCO)的制作

6-10-1　专题功能概述

本专题拟以HT46x23的ADC接口，制作一个可由输入模拟电压控制输出方波频率与振幅的压控振荡器(Voltage Control Oscillator,VCO)，或称为电压—频率转换电路。

6-10-2　学习重点

本专题是承续实验6-8，以R-2R梯状电阻网路实现DAC的功能并产生正弦波的应用，通过HT46x23 ADC接口输入的电压值，控制正弦波的频率及振幅，读者应该不会陌生。

6-10-3　电路图（见图6-10-1）

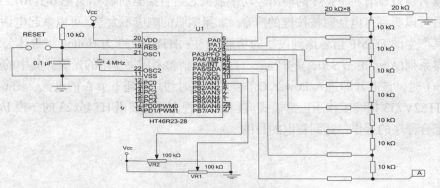

图6-10-1　VCO控制电路

电压—频率的转换器(VCO)是通信电路中常用的元件,一般作为收发机可调控频道的本地振荡源或发射机的调变元件。本专题拟以HT46x23的ADC转换接口与R-2R Ladder DAC转换的特性,并配合程序的控制完成以电压控制输出方波频率的目的。由于HT46x23提供了8个通道的A/D转换器,本实验利用其中两个通道分别作为输出方波的频率(A0,由VR1控制)及振幅(A1,由VR2控制)控制之用。在程序中将使用到Timer/Event Counter中断以及ADC转换接口,其基本原理及控制方式请读者自行参考各相关的章节内容(见表6-10-1)。

6-10-4 流程图及程序

1. 流程图(见图6-10-2)

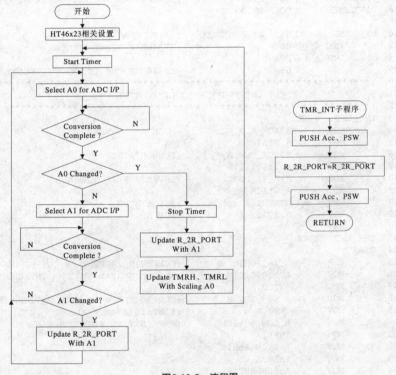

图6-10-2 流程图

表6-10-1 本实验的相关控制寄存器与参考章节

特殊功能寄存器	Bit7	Bit6	Bit5	Bit4	Bit3	Bit2	Bit1	Bit0	参考章节
INTC0	—	ADF	TF	EIF	EADI	ETI	EEI	EMI	2-4
TMRC	TM1	TM0	—	TON	TE	PSC2	PSC1	PSC0	2-5
ADSR	TEST	—	—	—	—	—	ADCS1	ADCS0	2-9
ADCR	START	EOC	PCR2	PCR1	PCR0	ACS0	ACS1	ACS0	2-9

2. 程序 6-10 简易 VCO 控制器

```
1    ;PROGRAM : 6-10.ASM (6-10.PRJ)            2002.DEC.30.
2    ;FUNCTION: VCO AND VCA  DEMO PROGRAM      BY STEVEN
3    #INCLUDE   HT46R23.INC
4              .CHIP     HT46R23
5    ;----------------------------------------------------------------
6    MY_DATA   .SECTION  'DATA'        ;== DATA SECTION ==
7    PRODUCT        DB   3 DUP(?)      ;BUFFER FOR RPRODUCT
```

```
8       MULTIPLIER      DB      ?               ;BUFFER FOR MULTIPLIER
9       MULTIPLICAND    DB      2 DUP(?)        ;BUFFER FOR MULTIPLICAND
10      REMAINDER       DB      3 DUP(?)        ;BUFFER FOR REMAINDER
11      DIVIDER         DB      3 DUP(?)        ;BUFFER FOR DIVIDER
12      TEMP            DB      ?
13      COUNT           DB      ?
14      FREQ_REG        DB      ?               ;BUFFER FOR FREQUENCY
15      AMP_REG         DB      ?               ;BUFFER FOR AMPLITUDE
16      STACK_A         DB      ?               ;STACK BUFFER FOR ACC
17      STACK_PSW       DB      ?               ;STACK FOR STATUS
18      ;----------------------------------------------------------------
19      OSC             EQU     4000000         ;fSYS=4MHz
20      R_2R_PORT       EQU     PA              ;DEFINE R_2R DAC PORT
21      R_2R_PORTC      EQU     PAC             ;DEFINE R_2R DAC PORT CONTROL REG.
22      EOC             EQU     ADCR.6          ;END OF CONVERSION
23      SADC            EQU     ADCR.7          ;DEFINE SADC AS ADC START BIT
24      DIVIDEND        EQU     PRODUCT
25
26      MY_CODE         .SECTION AT 0 'CODE'    ;== PROGRAM SECTION ==
27              ORG     00H                     ;HT-46RXX RESET VECTOR
28              JMP     START
29              ORG     08H                     ;HT-46RXX TMR INTERRUPT VECTOR
30      ;****************************************************************
31      ;               TIMER INTERRUPT SERVICE ROUTINE
32      ;****************************************************************
33      TMR_INT PROC
34              MOV     STACK_A,A               ;PUSH A
35              MOV     A,STATUS
36              MOV     STACK_PSW,A             ;PUSH STATUS
37              MOV     A,11111111B
38              XORM    A,R_2R_PORT             ;INVERSE R-2R OUTPUT
39              MOV     A,STACK_PSW
40              MOV     STATUS,A                ;POP STATUS
41              MOV     A,STACK_A               ;POP A
42              RETI
43      TMR_INT ENDP
44      START:
45              CLR     R_2R_PORTC              ;CONFIG R_2R PORT AS OUTPUT MODE
46              MOV     A,00000101B             ;ENABLE GLOBAL AND TMR INTERRUPT
47              MOV     INTC0,A
48              MOV     A,10000001B             ;CONFIG TMR 0 IN MODE 2(TIMER MODE)
49              MOV     TMRC,A                  ;fINT=fSYS(4MHz/2).
50              MOV     A,LOW 65510             ;SET INITIAL TIME CONSTANT
51              MOV     TMRL,A
52              MOV     A,HIGH 65510
53              MOV     TMRH,A
54              MOV     A,00000001B
55              MOV     ACSR,A                  ;= SYSTEM CLOCK/8
56              MOV     A,00010000B             ;SET PB0 & PB1 AS A/D CHANNEL
57              MOV     ADCR,A                  ;&SELECT A0 FOR ADC
58      RESTART:
59              SET     TON                     ;ENABLE TABLE COUNT
60      IDLE:
61              MOV     A,00010000B             ;SET PB0 & PB1 AS A/D CHANNEL
62              MOV     ADCR,A                  ;&SELECT A0 FOR ADC
63              SET     SADC                    ;START A0 ADC CONVERSION
64              CLR     SADC
65              SZ      EOC                     ;END OF CONVERSION?
66              JMP     $-1                     ;NO,WAIT FOR ADC
67              MOV     A,ADRH
68              XOR     A,FREQ_REG
69              SNZ     Z                       ;VCO I/P CHANGE?
70              JMP     CHANGE_FREQUENCY        ;YES
71              MOV     A,00010001B             ;SET PB0 & PB1 AS A/D CHANNEL
72              MOV     ADCR,A                  ; & SELECT A1 FOR ADC
73              SET     SADC                    ;START A1 ADC CONVERSION
74              CLR     SADC
```

```
75          SZ      EOC                     ;END OF CONVERSION?
76          JMP     $-1                     ;NO,WAIT FOR ADC
77          MOV     A,ADRH
78          XOR     A,AMP_REG
79          SZ      Z                       ;AMPLITUDE CHANGE?
80          JMP     IDLE                    ;NO.
81          MOV     A,ADRH                  ;YES.
82          MOV     AMP_REG,A
83          MOV     R_2R_PORT,A             ;UPDATE R-2R OUTPUT
84          JMP     IDLE
85  CHANGE_FREQUENCY:
86          CLR     TON
87          MOV     A,ADRH
88          MOV     FREQ_REG,A              ;UPDATE FREQUENCY REGISTER
89          MOV     MULTIPLIER,A
90          MOV     A,HIGH 65510
91          MOV     MULTIPLICAND[1],A       ;SCALING FREQ_REG BY 65510
92          MOV     A,LOW 65510
93          MOV     MULTIPLICAND[0],A
94          CALL    MUL_16x8
95          MOV     A,255
96          MOV     DIVIDER[0],A
97          CALL    DIV_24_8
98          MOV     A,PRODUCT[0]            ;RELOAD TIMER CONDTANT
99          MOV     TMRL,A
100         MOV     A,PRODUCT[1]
101         MOV     TMRH,A
102         JMP     RESTART                 ;IDLE LOOP
103 ;**********************************************************************
104 ;                 16-BIT X 8-BIT MULTIPLIER ROUTINE
105 ; NOTE:
106 ;   MULTIPILICAND = MULTIPILICAND
107 ;       MULTIPLIER = 8-BIT MULTIPLIER
108 ;           PRODUCT = PRODUCT
109 ;**********************************************************************
110 MUL_16X8    PROC
111         CLR     PRODUCT[2]
112         CLR     PRODUCT[1]
113         CLR     PRODUCT[0]
114         CLR     TEMP                    ;CLEAR TEMP BUFFER
115         MOV     A,8                     ;SET COUNTER
116         MOV     COUNT,A                 ;SET DIG3 AS COUNTER
117 LOOP:
118         RRC     MULTIPLIER              ;CHECK THE LSB OF MULITPLIER
119         SNZ     C                       ;SKIP IF BIT=1
120         JMP     EQU_0                   ;JUMP IF BIT=0
121         MOV     A,MULTIPLICAND[0]
122         ADDM    A,PRODUCT[0]
123         MOV     A,MULTIPLICAND[1]
124         ADCM    A,PRODUCT[1]
125         MOV     A,TEMP
126         ADCM    A,PRODUCT[2]
127 EQU_0:
128         RLC     MULTIPLICAND[0]
129         RLC     MULTIPLICAND[1]
130         RLC     TEMP
131         SDZ     COUNT                   ;SKIP IF COUNTER = 0
132         JMP     LOOP
133         RET
134 MUL_16X8    ENDP
135 ;**********************************************************************
136 ;               24-BIT/8~24-BIT DIVISION ROUTINE
137 ; NOTE:     DIVIDER[0~2]    :UNCHANGED
138 ;           DIVIDEND[0~2]   :THE QUOTIENT WILL RETURN TO DIVIDEND
139 ;           REMAINDER[0~2])
140 ;**********************************************************************
141 DIV_24_8    PROC
```

```
142         CLR     DIVIDER[1]          ;CLEAR 2-ND BYTTE BUFFER
143 DIV_24_16:
144         CLR     DIVIDER[2]          ;CLEAR 3-RD BYTTE BUFFER
145 DIV_24_24:
146         CLR     REMAINDER[0]
147         CLR     REMAINDER[1]
148         CLR     REMAINDER[2]
149         MOV     A,24
150         MOV     COUNT,A
151 NEXT_BIT:
152         CLR     C                   ;PRESET QUOTIENT BIT TO 0
153         RLC     DIVIDEND[0]
154         RLC     DIVIDEND[1]
155         RLC     DIVIDEND[2]
156         RLC     REMAINDER[0]        ;TEMP
157         RLC     REMAINDER[1]
158         RLC     REMAINDER[2]
159         MOV     A,REMAINDER[0]
160         SUB     A,DIVIDER[0]
161         MOV     REMAINDER[0],A
162         MOV     A,REMAINDER[1]
163         SBC     A,DIVIDER[1]
164         MOV     REMAINDER[1],A
165         MOV     A,REMAINDER[2]
166         SBC     A,DIVIDER[2]
167         MOV     REMAINDER[2],A
168         SNZ     C                   ;TEMP-DIVIDER >= 0?
169         JMP     $+3                 ;NO, RESTORE TEMP
170         SET     DIVIDEND[0].0       ;SET QUOTIENT BIT TO 1
171         JMP     $+7
172         MOV     A,DIVIDER[0]        ;RESTORE VALUE OF TEMP
173         ADDM    A,REMAINDER[0]
174         MOV     A,DIVIDER[1]
175         ADCM    A,REMAINDER[1]
176         MOV     A,DIVIDER[2]
177         ADCM    A,REMAINDER[2]
178 DIV_24_16_2:
179         SDZ     COUNT
180         JMP     NEXT_BIT
181         RET
182 DIV_24_8 ENDP
183         END
```

程序说明

- 7~17 依序定义变量地址。
- 20~21 定义 R_2R_PORT 为 PA，此即为 R-2R Ladder D/A 转换器的输出端口。
- 22~23 定义 EOC 与 SADC 分别为 ADCR.6 与 ADCR.7，由于这两个位在 HT46R23.INC 定义文件中并未定义，为了方便程序的编写与易读性，在此先予以定义。
- 24~25 定义 DIVIDEND 与 PRODUCT 是指向相同的数据存储器位置。
- 27   声明存储器地址由 000h 开始，此为 HT46x23 的 Reset Vector。
- 29   声明存储器地址由 008h 开始，此为 HT46x23 的 Timer/Event Counter Interrupt Vector。
- 33~42 TMR_INT 中断服务子程序，由于本专题只是很单纯地输出方波波形，因此中断子程序只负责将 R_2R_PORT 上的输出数据予以反向而已。
- 45   将 R_2R_PORT 定义为输出模式。
- 46~47 Global Interrupt 与 Timer/Event Counter Interrupt 有效。
- 48~49 将 Timer/Event Counter 定义如下：Timer Mode(Mode 2)、预分频比例设置为 1:2($f_{INT}=f_{SYS}/2$)。
- 50~53 设置 Timer/Event Counter 的中断时间，在此设置为每 13 μs([65 536－65 510]×0.5 μs)中断一次。

行号	说明
54~55	定义 ACSR 控制寄存器：ADC Clock=f_{SYS}/8。
56~57	定义 ADCR 控制寄存器：定义 PB0＝A0、PB0＝A1，并指定 A0 为 ADC 转换的信号来源。
59	启动 Timer/Event Counter 开始计数。
61~62	定义 ADCR 控制寄存器：定义 PB0＝A0、PB0＝A1，并指定 A0 为 ADC 转换的信号来源。
63~64	启动 HT46x23 ADC 接口开始进行转换。
65~66	等待 ADC 接口完成转换。
67~70	检查 A0 的输入电压是否改变，若产生变化则代表必须重新设置输出波形的频率，此时程序跳到 CHANGE_FREQUENCY 进行频率调整的程序。
71~72	定义 ADCR 控制寄存器：定义 PB0＝A0、PB0＝A1，并指定 A1 为 ADC 转换的信号来源。
73~74	启动 HT46x23 ADC 接口开始进行转换。
75~75	等待 ADC 接口完成转换。
77~80	检查 A1 的输入电压是否改变，若产生变化则代表必须重新设置输出波形的振幅，此时将 ADC 转换的结果送到 R_2R_PORT 以更新输出波形的电压。
85~102	频率调整的程序。
86	停止 Timer/Event Counter 计数功能。
88~87	依 A0 的转换结果将 FREQ_REG 寄存器的值予以更新。
89~97	依据转换的数值计算出 TMRL 与 TMRH 的时间常量；由于转换的结果为 8 位，如果直接以此数值来更新 TMRL 或 TMRH，会有输出频率变化量不够细腻与范围太小的情形，所以程序中采取的做法是依据下式计算出 TMRL 与 TMRH 的数值： $$TRM=\frac{FREQ\_REG \times 65\,510}{255}$$ 因为只取 8 位的 ADC 转换结果，所以上式可以保证中断间隔的时间最短为 13μs。至于为什么中断间隔时间要限定在 13μs 以上，请参考文章的后续说明。
98~102	依据计算的结果更新 TMRL 与 TMRH，并重新执行程序(JMP RESATRT)。
110~134	MUL_16X8 子程序，请参考 6-1 节中乘法原理相关的说明。
141~182	DIV_24_8 子程序，请参考 6-6 节中除法原理相关的说明。
141~182	DIV_24_8 子程序，请参考 6-6 节中除法原理相关的说明。

笔者认为在此唯一必须再提出说明的是："为什么中断间隔时间要限定在13μs以上？"。请读者观察一下TMR_INT中断服务子程序，可以估算一下这段程序的执行大约是10个指令周期(在f_{SYS}＝4 MHz时，为10μs)。如果为了提高输出频率而使中断间隔时间小于此数值，将会造成单片机持续执行中断程序的窘境，导致主程序停滞不前。

读者可以试着故意缩短Timer/Event Counter的计时溢位时间(例如让TMRL_TMRH＝65 530)，读者应当可以发现虽然仍有波形输出，但此时不管如何调整VR1、VR2的数值，都无法再改变其频率与振幅。

由于中断间隔时间最短为13μs，所以本专题的最高输出频率为$(2\times13\mu s)^{-1}\approx38.5$ kHz，最低输出频率为$(2\times65\,536\times0.5\mu s)^{-1}\approx15.3$ Hz，频率最小变化量约为151 Hz。图6-10-3~图6-10-5为本专题实际输出的波形，在此将其列出供读者参考。

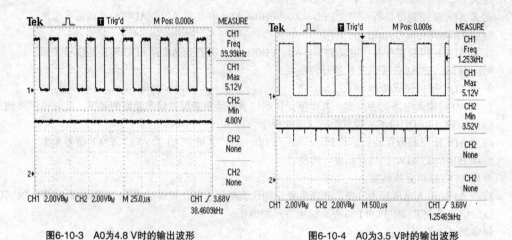

图6-10-3　A0为4.8 V时的输出波形　　　　图6-10-4　A0为3.5 V时的输出波形

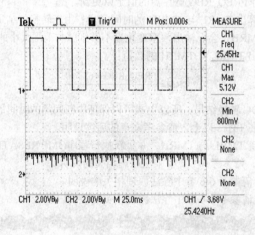

图6-10-5　A0为0.8 V时的输出波形

6-10-5　动动脑+动动手

  本专题的实用性虽不高，但若能与其他实验相结合，其实也可以获得相当有趣的结果。例如读者可以将6-1节的温度传感器电路输出作为本实验的A0与A1的输入，再将R-2R DAC的输出通过功率放大之后驱动喇叭(见图6-10-6)，就可以制作一个随温度变化而改变叫声高低与音量的警报器。当然，还得配合程序的适当控制，这就当成读者自我磨练的机会。不过提醒读者，人类的听觉最高频率约在20 kHz，感觉最刺耳的音频范围约在3~4 kHz。根据这些听觉特性来修改程序6-10，应该会有令人兴奋的成果与收获。

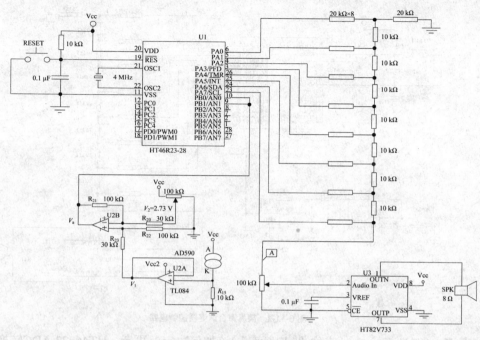

图6-10-6 温控音量与频率电路

6-11 专题十一：简易声音调变器的制作

6-11-1 专题功能概述

本专题拟以HT46x23的ADC接口，制作一个简易的变声器，将麦克风输入的信号乘上三角波的波形之后再由R-2R Ladder D/A接口输出，达到简单的声音变化效果。当不再有语音信号输入时，就让HT46x23进入HALT Mode，以达到省电的目的。

6-11-2 学习重点

本专题是承续实验6-8、实验6-10，以R-2R梯状电阻网路实现DAC的功能并结合乘法子程序的运用。语音信号通过HT46x23 ADC接口输入之后，通过HT46x23的运算与处理再予以输出。希望通过此专题让读者了解数字信号处理的基本观念。

6-11-3 电路图（见图6-11-1）

本专题的功能要求其实相当容易完成，只要把模拟输入信号由HT46x23 单片机ADC接口转换为数字数值后再与事先建好的调变波振幅值相乘就完成了。实际上只要将之前实验用过的子程序予以适当的排列组合，就可达到最终的目的。不过笔者想介绍一些数字信号处理的基本概念。首先是为何在放大电路之后要加上低通滤波电路(Low Pass Filter，LPF)，其截止频

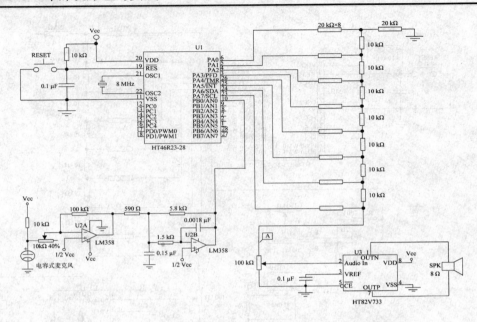

图 6-11-1 简易声音调变器控制电路

率(Cut-Off Frequency，或称"3dB频率")该如何决定？再者，HT46x23 ADC转换接口的采样频率又必须为多少才合理呢？

众所皆知，取样频率应该是越高越好，但相对的就必须采用转换速度更快的ADC与执行速度更高的单片机，方能在采样时间之内完成所需的运算。采样频率必须无止境地提高吗？其实不然，在数字信号处理领域中有个很重要的理论——**"采样定理(Sampling Theorem)：若信号所含的最高频率为F_{MAX}，则为保持采样后的信号不失真，采样频率F_{SAMPLE}必须满足$F_{SAMPLE} \geq 2F_{MAX}$"**。这实际上可通过数学推导加以证实，不过在此笔者并不想引入太多的数学公式，仅以简单的图示让读者了解基本的概念，请读者参考图6-11-2。

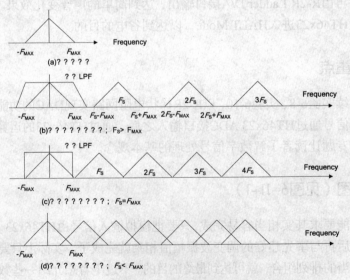

图6-11-2 原信号、采样频率与采样后的信号频谱关系

图6-11-2(a)中所示为一个信号的频谱，其最高频率为F_{MAX}。其实采样在频域(Frequency

Domain)上所造成的结果是信号频谱的复制,在采样频率(F_S)的N倍处会产生一个与原信号相同的频谱($N=-\infty \sim +\infty$),所以如果要将采样后的信号还原,理论上只须利用滤波器将$N=0$以外的频谱滤除即可。图6-11-2(b)是当$F_{SAMPLE} > 2F_{MAX}$的情形;图6-11-2(c)则是$F_{SAMPLE} = 2F_{MAX}$的情形,很明显此时若想将信号还原,就必须使用阶数(Oreder)相当高的滤波器(理论上为无限多阶)方能实现;图6-11-2(d)则是$F_{SAMPLE} < 2F_{MAX}$的情形,此时频谱已经发生重叠的现象(Aliasing),要想将信号还原是一件非常棘手的苦差事。

或许读者已经知道CD唱片的信号采样频率为44.1 kHz,这是因为人类的听觉系统最高大约只能听到20 kHz左右的频率,所以就设置$F_{MAX}=20$ kHz。这样$F_{SAMPLE}=44.1$ kHz恰好符合"采样频率F_{SAMPLE}必须满足$F_{SAMPLE} \geq 2F_{MAX}$"的需求。本专题以语音信号为输入,所以采样频率无须像CD唱片那么高。像电话线路的带宽大约只有3.4 kHz左右,就已经可以听到不错的通话音质(当然,如果想通过电话线直接传送音响的效果,那就另当别论了),因此在此设置本专题的取样频率为8 kHz。

至于为何在放大电路之后要加上低通滤波器呢?因为既然已经决定采样频率为8 kHz,那么输入信号的最高频率就必须小于4 kHz,否则就会产生如图6-11-2(d)所示的频谱重叠现象(Aliasing)。所以在输入信号进入A0采样之前,先由低通滤波器滤除会造成重叠的高频成分,因此此滤波器一般称为"Anti-Aliasing Filter"。

另外为了方便,电路以单电源方式设计,所以运算放大器的非反相输入端被偏压于$V_{cc}/2$;也就是当静音情况时,A点的电压为$V_{cc}/2$,通过HT46x23 ADC接口转换后的数值约为80H。

6–11–4 流程图及程序

1. 流程图(见图 6-11-3)

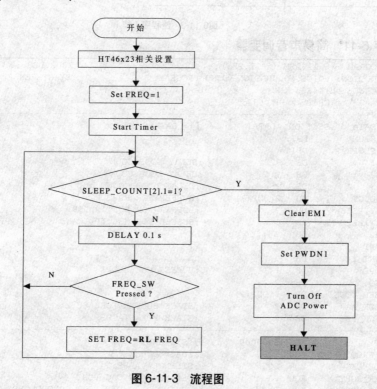

图 6-11-3 流程图

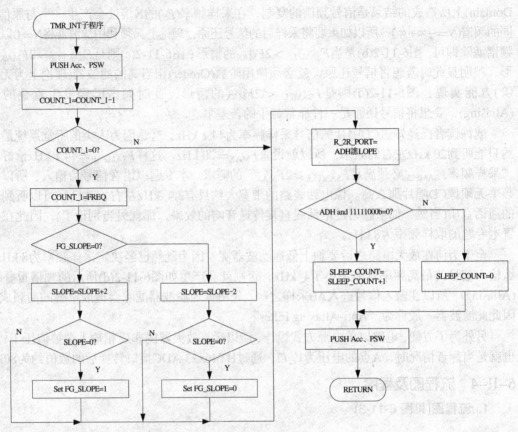

图6-11-3 流程图（续）

2. 程序 6-11 简易声音调变器

```
1   ;PROGRAM : 6-11.ASM (6-11.PRJ)              2002.DEC.30.
2   ;FUNCTION: VOICE MODULATION  DEMO PROGRAM   BY STEVEN
3       #INCLUDE  HT46R23.INC
4               .CHIP     HT46R23
5   ;-----------------------------------------------------------------
6   MY_DATA     .SECTION 'DATA'         ;== DATA SECTION ==
7   FG_SIGN         DBIT
8   FG_SLOPE        DBIT
9   DEL1            DB      ?           ;DELAY LOOP COUNT 1
10  DEL2            DB      ?           ;DELAY LOOP COUNT 2
11  DEL3            DB      ?           ;DELAY LOOP COUNT 3
12  PRODUCT         DB      2 DUP(?)    ;BUFFER FOR RPRODUCT
13  MULTIPLIER      DB      ?           ;BUFFER FOR MULTIPLIER
14  MULTIPLICAND    DB      ?           ;BUFFER FOR MULTIPLICAND
15  TEMP            DB      ?
16  COUNT           DB      ?
17  COUNT_1         DB      ?
18  SLOPE           DB      ?           ;BUFFER FOR SLOPE
19  FREQ            DB      ?
20  SLEEP_COUNT     DB      3 DUP(?)
21  STACK_A         DB      ?           ;STACK BUFFER FOR ACC
22  STACK_PSW       DB      ?           ;STACK FOR STATUS
23  ;-----------------------------------------------------------------
24  OSC             EQU     8000000     ;fSYS=8MHz
25  DA_FREQ         EQU     8000        ;fDAC=8000Hz
26  SILENCE         EQU     80H         ;SILENCE VALUE
27  R_2R_PORT       EQU     PA          ;DEFINE R_2R DAC PORT
28  R_2R_PORTC      EQU     PAC         ;DEFINE R_2R DAC PORT CONTROL REG.
```

```
29      EOC             EQU     ADCR.6          ;END OF CONVERSION
30      SADC            EQU     ADCR.7          ;DEFINE SADC AS ADC START BIT
31      SW_PORT         EQU     PB              ;DEFINE SWITCH PORT
32      SW_PORTC        EQU     PBC             ;DEFINE ALL SWITCH CONTROL REG.
33      FREQ_SW         EQU     SW_PORT.1       ;DEFINE FREQUENCY CONTROL SWITCH
34      PWDN1           EQU     SW_PORT.3       ;DEFINE HT82V733 /CE CONTROL PIN
35      MY_CODE         .SECTION AT 0 'CODE'    ;== PROGRAM SECTION ==
36              ORG     00H                     ;HT-46RXX RESET VECTOR
37              JMP     START
38              ORG     08H                     ;HT-46RXX TMR INTERRUPT VECTOR
39              JMP     TMR_INT
40      START:
41              MOV     A,00000111B
42              MOV     SW_PORTC,A
43              CLR     PWDN1
44              MOV     A,00000101B             ;ENABLE GLOBAL AND TMR INTERRUPT
45              MOV     INTC0,A
46              MOV     A,10000000B             ;CONFIG TMR 0 IN MODE 2(TIMER MODE)
47              MOV     TMRC,A                  ;fINT=fSYS(8MHz/1).
48              MOV     A,LOW (65536-OSC/DA_FREQ)
49              MOV     TMRL,A
50              MOV     A,HIGH (65536-OSC/DA_FREQ)
51              MOV     TMRH,A
52              MOV     A,00000001B
53              MOV     ACSR,A                  ;SELECT CONVERSION LOCK = SYSTEM CLOCK/8
54              MOV     A,00001000B             ;SET PB0 AS A/D CHANNEL & SELECT A0 FOR ADC
55              MOV     ADCR,A
56              CLR     R_2R_PORTC              ;CONFIG R_2R PORT AS OUTPUT MODE
57              CLR     SLOPE                   ;INITIAL SLOOP
58              CLR     SLEEP_COUNT[0]          ;INITIAL SLEP COUNT
59              CLR     SLEEP_COUNT[1]
60              CLR     SLEEP_COUNT[2]
61              SET     SADC                    ;RESET ADC
62              CLR     SADC                    ;START NEXT CONVERSION
63              MOV     A,1
64              MOV     FREQ,A
65              SET     TON                     ;ENABLE TABLE COUNT
66      WAIT_KEY:
67              SZ      SLEEP_COUNT[2].1        ;7.3 SEC*2^N INVOKE POWER DOWN
68              JMP     SLEEP
69              CALL    DELAY
70              SNZ     FREQ_SW                 ;FREQ_SW PRESSED?
71              RL      FREQ                    ;YES
72              JMP     WAIT_KEY                ;NO.
73      SLEEP:
74              CLR     EMI                     ;DISABLE INTERRUPT
75              SET     PWDN1                   ;RESET /CE
76              SET     R_2R_PORTC              ;DISABLE D/A OUTPUT
77              MOV     A,00000000B             ;SET PCR2~0=000 TO POWER OFF ADC
78              MOV     ADCR,A
79              HALT                            ;INVOKE POWER DOWN
80              JMP     $
81      ;***********************************************************************
82      ;               TIMER INTERRUPT SERVICE ROUTINE
83      ;***********************************************************************
84      TMR_INT PROC
85              MOV     STACK_A,A               ;PUSH A
86              MOV     A,STATUS
87              MOV     STACK_PSW,A             ;PUSH STATUS
88              MOV     A,ADRH                  ;GET PCM SAMPLE VALUE
89              SET     SADC                    ;RESET ADC
90              CLR     SADC                    ;START NEXT CONVERSION
91              CLR     FG_SIGN                 ;CLEAR SIGN FLAG
92              SUB     A,SILENCE               ;SUBTRACT SILENCE VALUE
93              SNZ     ACC.7                   ;IS RESULT > 0?
94              JMP     TMR_INT_0               ;YES.
95              SET     FG_SIGN                 ;NO, SET SIGN FLAG
```

```
96          XOR     A,11111111B         ;TAKE ABS(Acc)
97          INC     ACC
98   TMR_INT_0:
99          MOV     MULTIPLIER,A
100         SDZ     COUNT_1             ;COUNT_1-1 = 0?
101         JMP     TMR_INT_2           ;NO.
102         MOV     A,FREQ              ;YES, RELOAD COUNT_1
103         MOV     COUNT_1,A
104         SZ      FG_SLOPE            ;POSITIVE SLOPE?
105         JMP     TMR_INT_1           ;NO.
106         INC     SLOPE               ;INCREAE SLOPE BY 2
107         INC     SLOPE
108         SZ      Z                   ;REACH 0?
109         SET     FG_SLOPE            ;YES, SET NEGATIVE SLOPE
110         JMP     TMR_INT_2
111  TMR_INT_1:
112         DEC     SLOPE               ;DECREASE SLOPE BY 2
113         DEC     SLOPE
114         SZ      Z                   ;REACH 0 ?
115         CLR     FG_SLOPE            ;YES, SET POSITIVE SLOPE
116  TMR_INT_2:
117         MOV     A,SLOPE
118         MOV     MULTIPLICAND,A
119         CALL    MUL_8X8             ;MULTIPLY INPUT VALUE BY SLOPE
120         SNZ     FG_SIGN             ;IS NEGATIVE NUMBER?
121         JMP     TMR_INT_3           ;NO.
122         SET     ACC                 ;TAKE 2'S COMPLEMENT OF RESULT
123         XORM    A,PRODUCT[1]
124         XOR     A,PRODUCT[0]
125         ADD     A,1
126         CLR     ACC
127         ADCM    A,PRODUCT[1]
128  TMR_INT_3:
129         MOV     A,PRODUCT[1]
130         SUB     A,SILENCE           ;RESTORE SILENCE VALUE
131         MOV     R_2R_PORT,A         ;OUTPUT TO R-2R DAC
132         ADD     A,SILENCE           ;COVERT TO 2'S COMPLEMENT
133         SNZ     ACC.7               ;TAKE ABS(Acc)
134         JMP     $+3
135         CPL     ACC
136         INCA    ACC
137         AND     A,11111000B         ;MASK LOW NIBLE
138         SZ      Z                   ;IS SILENCE?
139         JMP     SILENCE_            ;YES.
140         CLR     SLEEP_COUNT[0]      ;CLEAR SLEEP COUNT
141         CLR     SLEEP_COUNT[1]      ;CLEAR SLEEP COUNT
142         CLR     SLEEP_COUNT[2]      ;CLEAR SLEEP COUNT
143         JMP     TMR_INT_4
144  SILENCE_:
145         MOV     A,1                 ;INCREASE SLEEP COUNT BY 1
146         ADDM    A,SLEEP_COUNT[0]
147         CLR     ACC
148         ADCM    A,SLEEP_COUNT[1]
149         ADCM    A,SLEEP_COUNT[2]
150  TMR_INT_4:
151         MOV     A,STACK_PSW
152         MOV     STATUS,A            ;POP STATUS
153         MOV     A,STACK_A           ;POP A
154         RETI
155  TMR_INT    ENDP
156  ;****************************************************************
157  ;              8-BIT X 8-BIT MULTIPLIER ROUTINE
158  ; NOTE:
159  ;       MULTIPILICAND = MULTIPILICAND
160  ;           MULTIPLIER = 8-BIT MULTIPLIER
161  ;              PRODUCT = PRODUCT
162  ;****************************************************************
```

```
163 MUL_8X8    PROC
164     CLR     PRODUCT[1]
165     CLR     PRODUCT[0]
166     CLR     TEMP                ;CLEAR TEMP BUFFER
167     MOV     A,8                 ;SET COUNTER
168     MOV     COUNT,A             ;SET DIG3 AS COUNTER
169 LOOP:
170     RRC     MULTIPLIER          ;CHECK THE LSB OF MULITPLIER
171     SNZ     C                   ;SKIP IF BIT=1
172     JMP     EQU_0               ;JUMP IF BIT=0
173     MOV     A,MULTIPLICAND
174     ADDM    A,PRODUCT[0]
175     MOV     A,TEMP
176     ADCM    A,PRODUCT[1]
177 EQU_0:
178     RLC     MULTIPLICAND
179     RLC     TEMP
180     SDZ     COUNT               ;SKIP IF COUNTER = 0
181     JMP     LOOP
182     RET
183 MUL_8X8    ENDP
184 ;**********************************************************************
185 ;                 Delay about DEL1*10ms
186 ;**********************************************************************
187 DELAY      PROC
188     MOV     A,10
189     MOV     DEL1,A              ;SET DEL1 COUNTER
190 DEL_1:
191     MOV     A,30
192     MOV     DEL2,A              ;SET DEL2 COUNTER
193 DEL_2:
194     MOV     A,110
195     MOV     DEL3,A              ;SET DEL3 COUNTER
196 DEL_3:
197     SDZ     DEL3                ;DEL3 DOWN COUNT
198     JMP     DEL_3
199     SDZ     DEL2                ;DEL2 DOWN COUNT
200     JMP     DEL_2
201     SDZ     DEL1                ;DEL1 DOWN COUNT
202     JMP     DEL_1
203     RET
202 DELAY ENDP
203     END
```

程序说明

7~22 依序定义变量地址。

24   定义 OSC=8 MHz(即系统工作时钟脉冲f_{SYS})，因为本专题以语音信号作为输入，为了保持采样后的信号不失真，所以必须将采样频率维持在 8 kHz。也就是说 A/D 转换必须在 125μs 内完成。为达此目的必须将f_{SYS}提升到 8 MHz，请参考后续的说明。

25   定义采样频率为 8 kHz。

26   定义静音时的信号准位 80H。

27~28 定义 R_2R_PORT 为 PA，此即为 R-2R Ladder D/A 转换器的输出端口。

29~30 定义 EOC 与 SADC 分别为 ADCR.6 与 ADCR.7，由于这两个位在 HT46R23.INC 定义文件中并未定义，为了方便程序的编写与易读性，在此先予以定义。

31~34 定义控制调变频率的输入开关为(FREQ_SW)PB.1；控制 HT82V733 是否进入省电模式(Standby Mode)的输出引脚为 PB.3。

37   声明存储器地址由 000h 开始，此为 HT46x23 的 Reset Vector。

40   声明存储器地址由 008h 开始，此为 HT46x23 的 Timer/Event Counter Interrupt Vector。

行号	说明
42~42	将 PB.1 定义为输入模式、PB.3 定义为输出模式。
43	将 PWDN1 设置为 0；此引脚用来控制 HT82V733 的 $\overline{CE}$ 信号，将其设置为 0，表示 HT82V733 进入一般的工作模式(Normal Mode)。
45~46	Global Interrupt 与 Timer/Event Counter Interrupt 有效。
47~48	将 Timer/Event Counter 定义如下：Timer Mode(Mode 2)、预分频比例设置为 1:1($f_{INT}=f_{SYS}$)。
49~52	设置 Timer/Event Counter 的计数溢位时间，在此设置为 125μs。
53~54	定义 ACSR 控制寄存器：ADC Clock=f_{SYS}/8；此时转换完成所需时间约为 76μs。
55~56	定义 ADCR 控制寄存器：定义 PB0=A0，并指定 A0 为 A/D 转换的信号来源。
57	将 R_2R_PORT 定义为输出模式。
58	清除 SLOPE 寄存器。
59~61	清除 SLEEP_COUNT 寄存器(为 3 字节的长度)；此寄存器用来统计究竟连续采样了几个振幅小于 7 的信号(由于只采集了 HT46x23 A/D 转换后的高 8 位信号，所以严格来说是统计连续采样了几个振幅小于 31 的信号)。
62~63	启动 HT46x23 ADC 接口开始进行转换。
64~65	设置 FREQ 寄存器的初值为 1；FREQ 寄存器的值决定调变波(三角波)的频率高低。
66	启动 Timer/Event Counter 开始计数。
68~69	因为进入 TMR_INT 中断服务子程序后，会依转换后的信号振幅是否大于 7 判断有无语音信号输入，若无语音信号输入就将 SLEEP_COUNT 值加一。由于之前已经定义 Timer/Event Counter 的中断时间为 125μs，所以当 SLEEP_COUNT[2].1 为 1 时，表示已经大约 16s($2^{18-1}×125μs$)没有语音信号输入了，此时就准备进入省电模式的相关程序(JMP SLEEP)。
70~73	检查 FREQ_SW 是否被按下；若有则将 FREQ 寄存器带进位位左移，用以达到调整三角波频率的目的(请参考 TMR_INT 中断服务子程序中的说明)。
74~80	省电模式的相关程序。
75	HT46x23 的中断功能禁止。
76	将 PWDN1 设置为 1(即 HT82V733 的 $\overline{CE}$ 被设置为 1，进入省电模式(Standby Mode)。
78~79	关闭 HT46x23 的 ADC 接口电源。
80	HT46x23 进入 HALT Mode。
85~156	TMR_INT 中断服务子程序，此子程序的主要功能是将经 ADC 转换的输入数值乘上调变波(三角波)的值后再由 R-2R Ladder DAC 转换接口输出。同时若输入的信号是小于设置值(7)，就将 SLEEP_COUNT 加一，否则将 SLEEP_COUNT 清除为零，以作为主程序决定是否进入省电模式的判断依据。
89	取得 ADC 转换值。
90~91	再次启动 ADC 转换。
93~98	因为麦克风放大及滤波电路工作在单电源(V_{cc})状态下，所以运算放大器的非反相输入端被偏压于 V_{cc}/2；也就是在静音情况时，A 点的电压为 V_{cc}/2，经 ADC 转换后的数值约为 80H。所以在第 93 行先减去静音的值后再取其绝对值(第 94~98 行)，就可取得目前信号的振幅。
100	将信号的振幅存于乘数寄存器(MULTIPLIER)。
101~116	COUNT_1-1 是否为零，若是则将调变波振幅值加二(或减二，根据目前调变波的斜率为正或负)；否则维持原振幅。由于 COUNT_1 的值来自 FREQ 寄存器，而 FREQ 寄存器在主程序中又受 FREQ_SW 开关的控制，因此读者可以通过 FREQ_SW 开关控制调变波(三角波)频率，达到不同的声音输出效果。
118~120	将信号的振幅与调变波振幅相乘。
121~132	将调变后的信号还原为原信号的极性，并取其高 8 位(High Byte)由 R-2R Ladder DAC 接口输出。
138~150	判断目前输入的信号是否小于设置值(7)，若是就将 SLEEP_COUNT 加一，否则将 SLEEP_COUNT 清除为零，以作为主程序决定是否进入省电模式的判断依据。

164~184 MUL_8X8 子程序，请参考6-1节中乘法原理相关的说明。
188~202 DELAY 子程序，请参考4-1节中的相关说明。

图6-11-4描述了将正弦波由放大器输入端灌入的情形，读者可以看出的确达到调变的目的。图6-11-4(a)是FREQ=1的情形；图6-11-4(b)则是FREQ=8的情形，足见FREQ_SW开关也可以控制调变波(三角波)的频率。

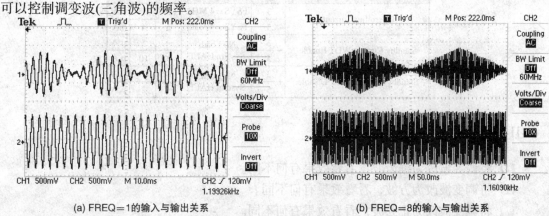

(a) FREQ=1的输入与输出关系　　　　　　　(b) FREQ=8的输入与输出关系

图6-11-4　输入正弦波时的输出波形

如果实际以声音输入，读者可以听到原本讲话的声音变成忽而大声、忽而小声的变化，如果三角波的频率控制得宜，将可造成抖音的特效。图6-11-5是语音信号"单片机"由放大器输入端输入的情形。图6-11-6为将"控"字音展开的波形。

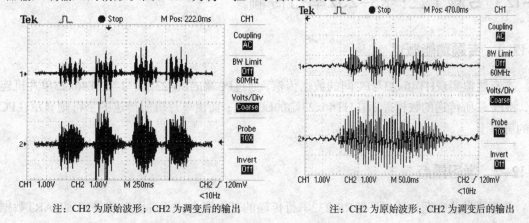

注：CH2为原始波形；CH2为调变后的输出　　　注：CH2为原始波形；CH2为调变后的输出

图6-11-5　输入"单片机"且FREQ=2时的输出波形　　图6-11-6　将"控"字音展开的波形

本专题的另外一个目的是展现如何进入省电模式。由于大多数单片机的应用场合都以电池为主要的供电设备，为了达到省电的目的，HT46xx提供了省电模式功能，一旦系统进入省电模式，其大约只消耗数μA的电流，请参考表6-11-1。但究竟何时该进入省电模式呢，就视不同的应用场合而定。以本专题为例，既然是以语音的输入为主，笔者就以麦克风输入信号的能量小于设置值做判断，但是语音信号的变化本来就是时大时小，如果光是以一个采样值作为判断依据，必然会有误判的情形发生。所以在程序中是以连续2^{18-1}个采样值都小于设置值，才判断用户已经不再使用，而让单片机进入省电模式，同时也控制负责功率放大的HT82V733进入"Standby Mode"，达到省电的效果。

表6-11-1　HT46R23省电模式的电流消耗比较

工作模式	V_{DD}/V	测试条件	Typ.	Max.	Unit
Operating Current	3	FSYS=4 MHz	0.6	1.5	mA
	5	FSYS=4 MHz	2	4	mA
Standby Current WDT Enable	3	No Load, System HALT	—	5	μA
	5		—	10	μA
Standby Current WDT Disable	3	No Load, System HALT	—	1	μA
	5		—	2	μA

6-11-5　动动脑 + 动动手

- 将调变波改为锯齿波，看看效果有何不同。
- 将调变波改为方波，看看效果有何不同。
- 将调变波改为正弦波，看看效果有何不同。
- 如果想将进入省电模式的时间由原来的16 s延长为32 s，该修改哪一行指令？
- 如果想将进入省电模式的时间由原来的16 s缩短为8 s，该修改哪一行指令？

6-12　专题十二：RS-232串行传输

6-12-1　专题功能概述

本专题拟实现HT46x23与PC间的数据传输，通过PC端的RS-232接口与HT46x23单片机连线，将PC端所传送的数据显示在HT46x23端的LCD上；而由单片机所传送的字符则显示在PC端的屏幕上。

6-12-2　学习重点

希望通过此专题让读者了解RS-232串行传输的基本概念，并进而以软件实现UART数据传输的功能。

6-12-3　电路图（见图6-12-1）

并行传输是与单片机最容易的数据传输方式，但是其最大的缺点是需要许多的I/O引脚，而且传输的距离最多也只有几米而已。基于成本的考虑，对于长距离的传输而言一般采用串行传输的方式。串行传输又可分为同步(Synchronous)与异步(Asynchronous)两种。同步传输（见图 6-12-2(a)），是发送端(Transmitter, TX)与接收端(Receiver, RX)间共同使用一个时钟脉

第6章 实践应用篇

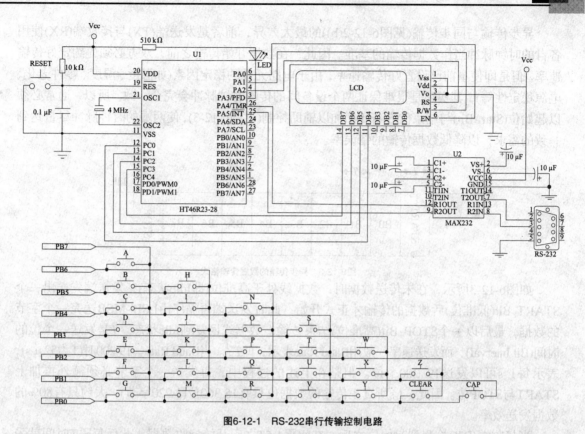

图6-12-1 RS-232串行传输控制电路

冲(Clock)当作数据传输的基准,每一个数据位都需参考时钟脉冲来进行发送与接收。这种发送端与接收端对准时序的传输方式相当稳定,不管数据位有多少,大都可以正确地发送,唯一的缺点就是必须多使用一条控制线传送时钟脉冲信号。

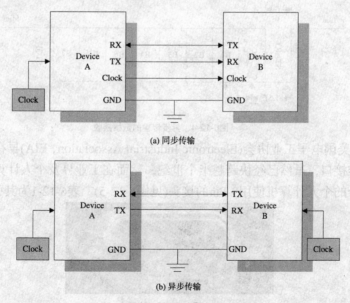

图6-12-2 同步传输与异步传输

451

异步传输与同步传输(见图6-12-2(b))的最大差异，前者是发送端(TX)与接收端(RX)使用各自的时钟脉冲当作数据传输的基准。因此，在数据开始传输之前，双方必须先约定好传输速率。但是即使事前已设置好传输速率，由于电路元件的稳定因素(如温度、湿度、零件老化、电源稳定性等)，其实还是很难保证两个设备间的传输时钟脉冲会完全一致。所以，通常必须以起始位(Start Bit)与结束位(Stop Bit)加以辅助控制(见图6-12-3)，使两端的时钟脉冲尽量达到一致的需求，以降低数据传输的错误率。

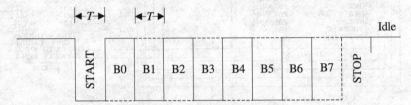

图6-12-3　异步传输的数据传输格式

如图6-12-3所示，在未传送数据时，数据线处于高准位(即Idle状态)，发送端先送出一个START Bit(低准位)后数据的传输才正式开始，此时发送端再循序由LSB~MSB送完一个字节的数据，最后以一个STOP Bit(高准位)结束传输，数据线回复到Idle状态。T是发送一个位的时间(Bit Interval)，而发送速率一般用Baud Rate表示，即T^{-1}。当Baud Rate＝19 200时(T为52μs)，表示每1 s可以发送19 200个位。但是在上述的过程中，每发送一个字节必须额外再加上START与STOP位。因此，1 s所真正传输的数据位只有15 360位(1 920字节)，大约只有80%的数据发送效率。

当接收端(RX)检测到起始位之后，在间隔1.5T之后开始接收数据，此后每隔T时间就采集一个位的数据(见图6-12-4)。间隔1.5T的主要原因，是希望能够在各位的中心位置采集数据，这样可以采集到较稳定的数据位，避免错误发生。

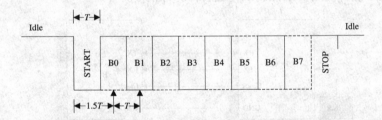

图6-12-4　异步传输的数据接收

RS-232C是美国电子工业协会(Electronic Industrial Association，EIA)早在1960年代所制订的串行传输标准接口，虽然已经快跨越半个世纪，然而在工业界及个人计算机上仍常见其踪迹。目前大部分的个人计算机使用9-Pin的接头(见图6-12-5)。表6-12-1为其引脚的定义。

图6-12-5　9针插头

表6-12-1　PC COM Port—EIA-574，RS-232/V.24引脚定义

引脚号	名　称	定　义
1	DCD	Data Carrier Detected：数据载波检测
2	RxD	Received Data：数据接收端
3	TxD	Transmitted Data：数据发送端
4	DTR	Data Terminal Ready：终端机备妥
5	SG	Signal Ground：信号接地端
6	DSR	Data Set Ready：数据已备妥
7	RTS	Ready To Send：数据发送请求
8	CTS	Clear To Send：信号已清除并准备开始传送
9	RI	Ring Indicator：回铃音指示

基于远距离传输的考虑，并避免在传输过程中受到干扰，RS-232C采用较高的电压信号传递，请参考图6-12-6。RS-232C采用负逻辑准位，也就是逻辑1以低电位(—3V~—25V)表示，一般称之为**MARK**；而逻辑0则以高电位(+3~+25 V)代表，称之为**SPACE**。因为PC的COM Port为RS-232C标准接口，因此其所送出与接收的信号必须符合此规范，因此在电路中通常是以RS-232准位转换IC(MAX232、MAX233或DS275)完成电压转换，使HT46x23与COM Port得以完成数据的传输。

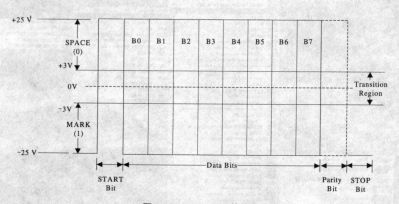

图6-12-5　RS-232C的信号准位

关于RS-232C串行数据传输方式，该如何以程序加以实现，相信读者心理已经大致上有个谱了。首先是需要一个$T(T=\text{Baud Rate}^{-1})$时间的定时器，以便每隔T时间就由RXD接收一位数据，或由TXD送出一位数据。HT46x23端何时发送数据，由HT46x23的控制程序决定，只要记得：**在发送数据之前必须先送出START Bit(0)，而在送完数据之后必须以STOP Bit(1)作为结束**。至于何时该接收数据，就必须检查RXD是否有SPACE(0)信号出现，若有，则需先延时1.5T的时间之后再每隔T时间到RXD提取一位数据(参考图6-12-4)。要注意的是当发现START Bit时(PC端有数据发送过来)，单片机要能够立即反应，否则会失去同步而导致采集数据的位置不正确。所以，用外部中断的方式来检测START Bit，应该是较适当的选择。而以HT46x23的Timer/Event Cunter来作为位时间(T或1.5T)的定时，更是恰当。在此将程序中使用到的特殊功能寄存器与其参考章节列出供读者参考。

特殊功能寄存器	Bit7	Bit6	Bit5	Bit4	Bit3	Bit2	Bit1	Bit0	参考章节
INTC0	—	ADF	TF	EIF	EADI	ETI	EEI	EMI	2-4
TMRC	TM1	TM0	—	TON	TE	PSC2	PSC1	PSC0	2-5
TMRH									2-5
TMRL									2-5

由于是PC COM Port与单片机之间的数据传输，所以光有HT46x23端的控制程序还不够，必须还要有PC端的COM Port控制程序才行。不过，笔者长期沉浸于单片机汇编语言的熏陶，早就将一些高阶语言"置之度外"，虽然明知要以Visual Basic 来控制COM Port是"轻而易举"(Not for Me！)，但是还是有点……相信部分读者应该可以体会甚至拥有相同的无奈吧？就算"写完"了(注意：是"写完"不是"写好")，万一数据传输不正常，究竟是HT46x23端控制程序的问题？还是PC端的控制程序的问题？可能还得费尽周折地两头找"BUGs"。所幸Windows 9x的超级终端控制程序可以用来作为简易的COM Port传输控制，至少先利用它来确认一下HT46x23端的控制程序没问题，建立一下自信心后再考虑其他的吧。超级终端控制程序位于"附件→通讯"表(见图6-12-7)。若读者的计算机中无此控制程序，应该是在安装Windows时未选择该项组件，请先加装通讯组件中的"超级终端"功能。

进入"超级终端"功能后会出现如图6-12-8所示的画面，执行其中的**HYPERTRM**就可以开始创建新的超级终端连接。首先出现的是如图6-12-9所示的画面，读者可以自行为此新建立的连接定义名称并选择其图示(Icon)，设置完成之后单击"确定"按钮。

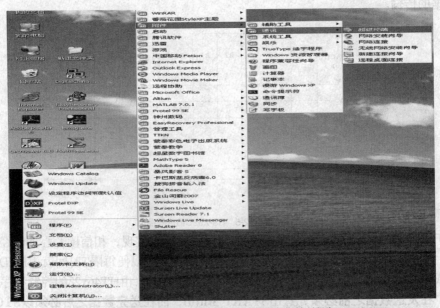

图6-12-7　Windows 9x的"超级终端"控制程序

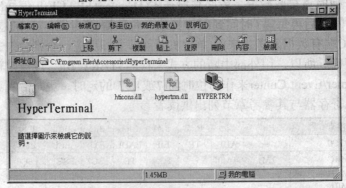

图6-12-8　进入"超级终端"功能后出现的画面

接着出现图6-12-10，在这里要设置连接的COM Port，请读者视自己计算机的COM Port使用状况加以选择(笔者是以COM 2作为连接端口，所以图6-12-1电路图中的DB-9接头就必须连接到PC COM 2端口)。

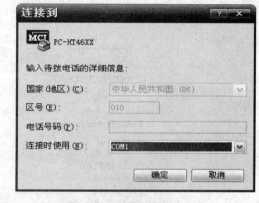

图6-12-9　定义名称并设置选择其图示　　　　图6-12-10　设置连接的COM Port

最后就是通信参数的设置了(见图6-12-11)，因为本实验的程序中Baud Rate可以有几种不同的选择(2 400、4 800、9 600与19 200)，所以请读者视实际的需要设置(在此以Baud Rate=9 600为例)，同位请选择"none"，而停止位请至少设置为"1"。单击"确定"按钮之后，就完成了PC-HT46xx的连接设置。下次再进入超级终端功能时，就会出现PC-HT46xx选项，直接点选该选项：

PC-HT46xx.ht

就可以开始进行RS-232C通信端口的数据传输。如果需要修改通信参数，只要到"文件(F)"中的"内容(R)"加以修改就可以了。

在硬件连接完成之后，读者可以先在HIDE-3000执行6-12.ASM的单片机端程序，然后再执行PC端的PC-HT46xx连接，此时在PC键盘上的按键值将会显示在LCD上，而HT46x23按下的键值则显示在PC端的PC-HT46xx超级终端窗口中（见图6-12-12）。

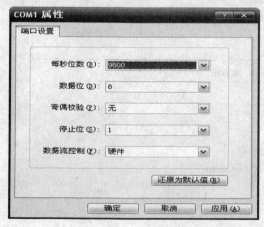

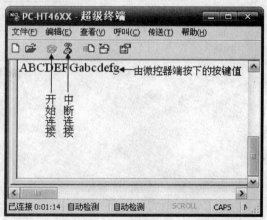

图6-12-11　通信参数的设置　　　　图6-12-12　按键值显示在**PC-HT46xx** 超级终端窗口的情形

6-12-4 流程图及程序
1. 流程图(见图6-12-13)

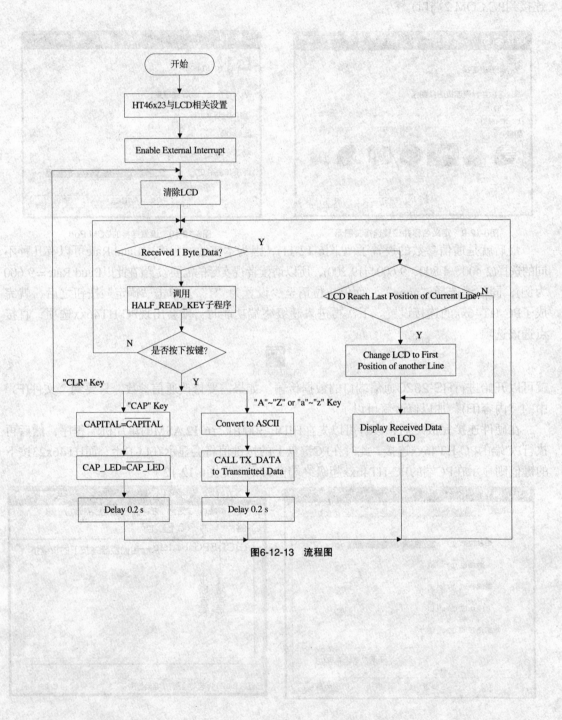

图6-12-13 流程图

第6章 实践应用篇

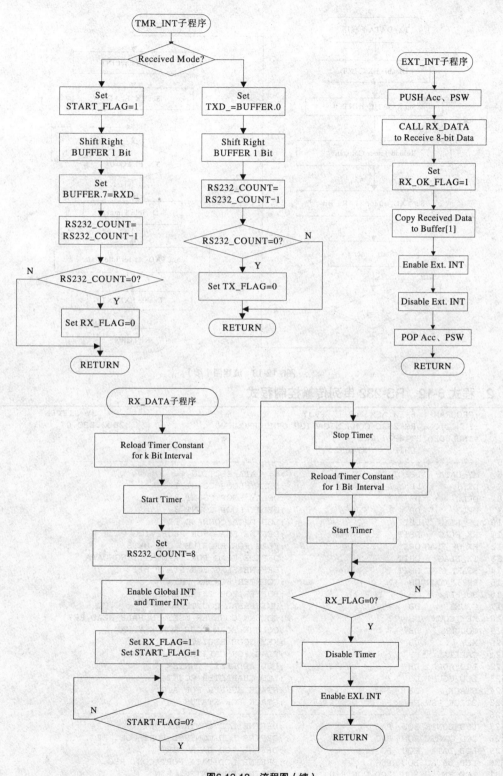

图6-12-13 流程图（续）

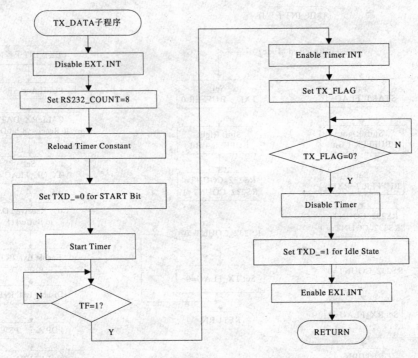

图6-12-13 流程图（续）

2. 程式 6-12 RS-232 串列传输控制程式

```
1    ;PROGRAM : 6-12.ASM  (6-12.PRJ)                          BY STEVEN
2    ;FUNCTION: RS-232C COMMUNICATION DEMO PROGRAM           2002.DEC.07.
3    #INCLUDE HT46R23.INC
4              .CHIP    HT46R23
5    ;------------------------------------------------------------------
6    MY_DATA   .SECTION 'DATA'         ;== DATA SECTION ==
7    DEL1       DB    ?                ;DELAY LOOP COUNT 1
8    DEL2       DB    ?                ;DELAY LOOP COUNT 2
9    DEL3       DB    ?                ;DELAY LOOP COUNT 3
10   DC_FLAG    DBIT                   ;LCD DATA/COMMAND FLAG
11   RX_FLAG    DBIT                   ;DEFINE FLAG FOR RECEIVING DATA
12   RX_OK_FLAG DBIT                   ;FLAG FOR RECEIVED ONE BYTE
13   TX_FLAG    DBIT                   ;DEFINE FLAG FOR TRANSMITTING DATA
14   START_FLAG DBIT                   ;DEFINE FLAG FOR START BIT
15   RS232_COUNT DB   ?                ;COUNTER FOR RX DATA
16   BUFFER     DB    2 DUP(?)         ;BUFFER FOR TX/RX DATA
17   COUNT      DB    ?                ;UNIVERSAL COUNTER
18   KEY_COUNT  DB    ?                ;COLUMN COUNTER USED IN HALF_READ_KEY
19   COLUMN     DB    ?                ;COLUMN SCAN CODE FOR KEY PAD
20   KEY        DB    ?                ;KEY CODE REGISTER
21   CAPITAL    DB    ?                ;FLAG FOR CAPITAL
22   LCD_ADRS   DB    ?                ;LCD ADDRESS COUNTER
23   LCD_COUNT  DB    ?                ;LCD CHARACTER COUNT
24   STACK_A    DB    ?                ;STACK BUFFER FOR ACC
25   STACK_PSW  DB    ?                ;STACK FOR STATUS
26   ;------------------------------------------------------------------
27   LCD_CONTR  EQU   PA               ;DEFINE LCD CONTROL PORT
28   LCD_CONTRC EQU   PAC              ;DEFINE LCD CONTROL PORT CON. REG.
29   LCD_DATA   EQU   PC               ;DEFINE LCD DATA PORT
30   LCD_DATAC  EQU   PCC              ;DEFINE LCD DATA PORT CON. REG.
31   LCD_EN     EQU   LCD_CONTR.0      ;DEFINE EN CONTROL PIN
32   LCD_RW     EQU   LCD_CONTR.1      ;DEFINE RW CONTROL PIN
33   LCD_RS     EQU   LCD_CONTR.2      ;DEFINE RS CONTROL PIN
34   CAP_LED    EQU   LCD_CONTR.3      ;CAPITAL INDICATOR
```

```
35  LCD_READY  EQU  LCD_DATA.3       ;DEFINE READY BIT OF LCD
36  TXD_       EQU  LCD_CONTR.4      ;DEFINE TX PIN
37  RXD_       EQU  LCD_CONTR.5      ;DEFINE RX PIN(PA.5 FOR EXTERNAL INT)
38  RXD_C      EQU  LCD_CONTRC.5     ;DEFINE RX PIN CONTROL REG.
39  KEY_PORT   EQU  PB               ;DEFINE KEY PORT
40  KEY_PORTC  EQU  PBC              ;DEFINE KEY PORT CONTROL REG.
41  BAUD       EQU  9600             ;DEFINE BAUD RATE
42  FSYS       EQU  4000000          ;DEFINE SYSTEM CLOCK
43  K          EQU  40*4             ;INSTRUCTION CYCLE COUNT
44
45  MY_CODE    .SECTION AT 0 'CODE'  ;== PROGRAM SECTION ==
46          ORG      00H             ;HT-46RXX RESET VECTOR
47          JMP      START
48          ORG      04H             ;HT-46RXX EXT. INTERRUPT VECTOR
49          JMP      EXT_INT
50          ORG      08H             ;HT-46RXX TIMER/EVENT COUNTER INT. VECTOR
51          JMP      TMR_INT
52  START:
53          CLR      LCD_CONTRC      ;CONFIG LCD_CONTR AS OUTPUT MODE
54          SET      RXD_C           ;SET RXD_ AS INPUT MODE
55          CLR      CAPITAL         ;INITIAL CAPITAL
56          CLR      CAP_LED         ;TURN CAPITAL LED ON
57          CLR      RX_OK_FLAG      ;INITIAL RECEIVED OK FLAG
58          SET      TXD_            ;SET IN IDLE STATE
59          CLR      START_FLAG      ;CLEAR START FLAG
60          MOV      A,28H           ;FUNCTION SET: 8-BIT,2-LINE,5X10 DOTS
61          CALL     WLCMC_4
62          MOV      A,0CH           ;ON/OFF CONTR: DISPLAY ON, NO BLINKING
63          CALL     WLCMC_4
64          MOV      A,06H           ;ENTRY MODE : INCREMENT,DISPLAY NOT SHIFT
65          CALL     WLCMC_4
66          MOV      A,10000000B     ;CONFIG TMR 0 IN MODE 2(TIMER MODE)
67          MOV      TMRC,A          ;fINT=fSYS
68          SET      EMI             ;ENABLE GLOBAL INTERRUPT
69          SET      EEI             ;EBABLE EXTERNAL INTERRUPT
70          MOV      A,20
71          MOV      LCD_COUNT,A     ;SET COUNT FOR LCD
72          MOV      A,80H
73          MOV      LCD_ADRS,A      ;SET ADDRESS FOR LCD
74  MAIN:
75          MOV      A,01H           ;CLEAR DISPLAY, LINE ONE, POSITION 0
76          CALL     WLCMC_4
77  WAIT_KEY:
78          CALL     HALF_READ_KEY   ;READ KEYPAD
79          SNZ      RX_OK_FLAG      ;HAVE DATA RECEIVED?
80          JMP      NO_RX_DATA      ;NO.
81          SDZ      LCD_COUNT       ;REACH OF LINE 1/2 LAST POSITION?
82          JMP      $+7             ;NO.
83          MOV      A,LCD_ADRS      ;YES, GET LCD POSITION 0 ADDRESS
84          XOR      A,01000000B     ;CHANE LINE 1/2 TO 2/1
85          MOV      LCD_ADRS,A      ;UPDATE LCD_ADRS BUFFER
86          CALL     WLCMC_4         ;SET ADDRESS TO LINE 1/2 POSITION 0
87          MOV      A,20
88          MOV      LCD_COUNT,A     ;RELOAD LCD POSITION COUNT
89          MOV      A,BUFFER[1]     ;PRINT DATA
90          CALL     WLCMD_4
91          CLR      RX_OK_FLAG      ;CLEAR RX_OK_FLAG
92  NO_RX_DATA:
93          MOV      A,KEY
94          SUB      A,28
95          SZ       Z               ;KEY PRESSED?
96          JMP      WAIT_KEY        ;NO, RE-READ KEYPAD
97          MOV      A,KEY
98          SUB      A,26
99          SZ       Z               ;IS "CLEAR" PRESSED?
100         JMP      MAIN            ;YES, JUMP MAIN
101         SDZ      ACC             ;IS "CAP" PRESSED?
```

```
102         JMP     ALPHA_PRESSED   ;NO. IT'S ALPHABETA KEY PRESSED
103         SET     ACC
104         XORM    A,CAPITAL       ;YES, SET CAPITAL=!CAPITAL
105         MOV     A,00001000B
106         XORM    A,LCD_CONTR     ;SET CAP_LED=!CAP_LED
107         CALL    DELAY           ;FOR DE-BOUNCING WHEN KEY PRESSED
108         JMP     WAIT_KEY
109 ALPHA_PRESSED:
110         MOV     A,KEY           ;GET KEY CODE
111         SZ      CAPITAL         ;IS CAPITAL?
112         ADD     A,20H           ;OFFSET OF ASCII "a" AND "A"
113         ADD     A,41H           ;ASCII OF "A"
114         MOV     BUFFER,A        ;SAVE KEY IN DATA IN BUFFER
115         CALL    TX_DATA         ;SERIAL DATA OUT
116         CALL    DELAY           ;DEBOUNCING
117         JMP     WAIT_KEY        ;READ NEXT KEY
118 ;**********************************************************************
119 ;           EXTERNAL INTERRUPT SERVICE ROUTINE
120 ;**********************************************************************
121 EXT_INT  PROC
122         MOV     STACK_A,A       ;PUSH ACC
123         MOV     A,STATUS
124         MOV     STACK_PSW,A     ;PUSH STATUS
125         CLR     EEI             ;DISABLE EXT. INTERRUPT
126         CALL    RX_DATA         ;RECEIVED 8 BITS DATA
127         SET     RX_OK_FLAG      ;SET RECEIVED OK FLAG
128         MOV     A,BUFFER        ;COPY DATA TO BUFFER[1]
129         MOV     BUFFER[1],A
130         CLR     EIF             ;CLEAR EIF
131         SET     EEI             ;ENABLE EXT. INTERRUPT
132         MOV     A,STACK_PSW
133         MOV     STATUS,A        ;POP PSW
134         MOV     A,STACK_A       ;POP ACC
135         RETI
136 EXT_INT  ENDP
137 ;**********************************************************************
138 ;           TIMER INTERRUPT SERVICE ROUTINE
139 ;**********************************************************************
140 TMR_INT  PROC
141         SNZ     RX_FLAG         ;RECEIVED DATA?
142         JMP     TX_MODE         ;NO. TRANSMITTED DATA.
143 RX_MODE:                        ;YES, RECEIVED DATA.
144         SET     START_FLAG      ;SET START FLAG
145         RR      BUFFER          ;ROTATE DATA
146         CLR     BUFFER.7        ;NOTE: LSB FIRST
147         SZ      RXD_            ;RX_PIN = 0?
148         SET     BUFFER.7        ;NO,SET BUFFER.0=1
149         SDZ     RS232_COUNT     ;RECEIVED 8-BIT ?
150         RETI                    ;NO.
151         CLR     RX_FLAG         ;YES, CLEAR RX_FLAG
152         RETI
153 TX_MODE:                        ;TRANSMITTED DATA.
154         CLR     TXD_
155         SZ      BUFFER.0        ;BUFFER.0 = 0?
156         SET     TXD_            ;NO, SET TXD_=1
157         RR      BUFFER
158         SDZ     RS232_COUNT     ;RECEIVED 8-BIT ?
159         RETI                    ;NO.
160         CLR     TX_FLAG         ;YES, CLEAR TX_FLAG
161         RETI
162 TMR_INT  ENDP
163 ;**********************************************************************
164 ;           TRANSMIT DATA TO RS232 VIA TXD PIN
165 ;**********************************************************************
166 TX_DATA  PROC
167         CLR     EEI             ;DISABLE RX FUNCTION
168         MOV     A,8
```

```
169        MOV      RS232_COUNT,A       ;SET TRANSMITTED COUNT=8
170        MOV      A,LOW (65536-FSYS/BAUD) ;SET 1 BIT INTERVAL TIME CONSTANT
171        MOV      TMRL,A              ;TIME CONSTANT=65536-(1/BAUD)/(1/FINT)
172        MOV      A,HIGH (65536-FSYS/BAUD);=65536-FINT/BAUD=65536-FSYS/BAUD,
173        MOV      TMRH,A              ;SINCE PRESCALER=1:1
174        CLR      TXD_                ;START BIT
175        CLR      TF                  ;CLEAR TF
176        SET      TON                 ;STAT COUNTING
177        SET      ETI                 ;ENABLE TIMER INTERRUPT
178        SET      TX_FLAG             ;SET TX_FLAG
179        SZ       TX_FLAG             ;WAIT FOR TRANSMITTED 8-BIT DATA
180        JMP      $-1
181        CLR      ETI                 ;DISABLE TIMER INTERRUPT
182        SNZ      TF                  ;LAST BIT INTERVAL
183        JMP      $-1
184        CLR      TON                 ;DISABLE TIMER COUNTING
185        SET      TXD_                ;IDLE STATE
186        SET      EEI                 ;ENABLE RX FUNCTION
187        RET
188 TX_DATA  ENDP
189 ;*********************************************************************
190 ;         RECEIVED DATA FROM RS232 VIA RXD PIN
191 ;*********************************************************************
192 RX_DATA   PROC
193        MOV      A,LOW (65536-FSYS*15/(BAUD*10)+K);SET 1.5 BIT INTERVAL for START Bit
194        MOV      TMRL,A
195        MOV      A,HIGH(65536-FSYS*15/(BAUD*10)+K)
196        MOV      TMRH,A
197        SET      TON                 ;START TIMER
198        SET      EMI                 ;ENABLE GLOBAL INTERRUPT
199        SET      ETI                 ;ENABLE TIMER INTERRUPT
200        MOV      A,8
201        MOV      RS232_COUNT,A       ;SET RECEIVED COUNT=8
202        SET      RX_FLAG             ;SET RX_FLAG
203        CLR      START_FLAG          ;CLEAR START FLAG
204        SNZ      START_FLAG          ;START INTERVAL OVER?
205        JMP      $-1                 ;NO, WAIT!
206        CLR      TON                 ;STOP TIMER
207        MOV      A,LOW (65536-FSYS/BAUD) ;SET 1 BIT INTERVAL TIME CONSTANT
208        MOV      TMRL,A              ;TIME CONSTANT=65536-(1/BAUD)/(1/FINT)
209        MOV      A,HIGH (65536-FSYS/BAUD);=65536-FINT/BAUD=65536-FSYS/BAUD,
210        MOV      TMRH,A              ;SINCE PRESCALER=1:1
211        SET      TON                 ;START TIMER
212        SZ       RX_FLAG             ;WAIT FOR RECEIVED 8-BIT DATA
213        JMP      $-1
214        CLR      ETI                 ;DISABLE TIMER UNTERRUPT
215        CLR      TON                 ;STOP TIMER
216        RET
217 RX_DATA   ENDP
218 ;*********************************************************************
219 ;   SCAN HALF MATRIX ON KEY PORT AND RETURN THE CODE IN KEY REGISTER
220 ;   IF NO KEY BEEN PRESSED, KEY=28.
221 ;*********************************************************************
222 HALF_READ_KEY PROC
223        MOV      A,01111111B
224        MOV      COLUMN,A            ;SET COLUMN FOR SCANNING
225        CLR      KEY                 ;INITIAL KEY REGISTER
226        MOV      A,7
227        MOV      KEY_COUNT,A         ;SET COUNTER=7 FOR 7 COLUMNS
228 SCAN_KEY:
229        MOV      A,COLUMN            ;GET COLUMN FOR SCANNING
230        MOV      KEY_PORTC,A         ;CONFIG KEY_PORT
231        MOV      KEY_PORT,A          ;SCANNING KEY PAD
232        MOV      A,KEY_COUNT         ;GET COLUMN COUNTER
233        ADDM     A,PCL               ;COMPUTATIONAL JUMP
234        NOP                          ;OFFSET FOR KEY_COUNT=0
235        JMP      $+24                ;KEY_COUNT=1
```

```
236         JMP      $+20              ;KEY_COUNT=2
237         JMP      $+16              ;KEY_COUNT=3
238         JMP      $+12              ;KEY_COUNT=4
239         JMP      $+8               ;KEY_COUNT=5
240         JMP      $+4               ;KEY_COUNT=6
241         SNZ      KEY_PORT.6        ;KEY_COUNT=7,ROW 1 PRESSED?
242         JMP      END_KEY           ;YES.
243         INC      KEY               ;NO, INCREASE KEY CODE
244         SNZ      KEY_PORT.5        ;ROW 2 PRESSED?
245         JMP      END_KEY           ;YES.
246         INC      KEY               ;NO, INCREASE KEY CODE
247         SNZ      KEY_PORT.4        ;ROW 3 PRESSED?
248         JMP      END_KEY           ;YES
249         INC      KEY               ;NO, INCREASE KEY CODE
250         SNZ      KEY_PORT.3        ;ROW 4 PRESSED?
251         JMP      END_KEY           ;YES.
252         INC      KEY               ;NO, INCREASE KEY CODE
253         SNZ      KEY_PORT.2        ;ROW 5 PRESSED?
254         JMP      END_KEY           ;YES.
255         INC      KEY               ;NO, INCREASE KEY CODE
256         SNZ      KEY_PORT.1        ;ROW 6 PRESSED?
257         JMP      END_KEY           ;YES.
258         INC      KEY               ;NO, INCREASE KEY CODE
259         SNZ      KEY_PORT.0        ;ROW 7 PRESSED?
260         JMP      END_KEY           ;YES.
261         INC      KEY               ;NO, INCREASE KEY CODE
262         RR       COLUMN            ;SCAN CODE FOR NEXT COLUMN
263         SDZ      KEY_COUNT         ;HAVE ALL COULMN BEEN CHECKED?
264         JMP      SCAN_KEY          ;NO, NEXT COLUMN
265 END_KEY:
266         RET
267 HALF_READ_KEY    ENDP
268 ;**********************************************************************
269 ;       4-BIT LCD DATA/COMMAND WRITE PROCEDURE
270 ;**********************************************************************
271 WLCMD_4    PROC
272         SET      DC_FLAG           ;SET DC_FLAG=1 FOR DATA WRITE
273         JMP      WLCM_4
274 WLCMC_4:
275         CLR      DC_FLAG           ;SET DC_FLAG=0 FOR COMMAND WRITE
276 WLCM_4:
277         SET      LCD_DATAC         ;CONFIG LCD_DATA AS INPUT MODE
278         CLR      LCD_EN            ;CLEAR ALL LCD CONTROL SIGNAL
279         CLR      LCD_RW
280         CLR      LCD_RS
281         SET      LCD_RW            ;SET RW SIGNAL (READ)
282         NOP                        ;FOR TAS
283         SET      LCD_EN            ;SET EN HIGH
284         NOP                        ;FOR TDDR
285 WF_4:
286         SZ       LCD_READY         ;IS LCD BUSY?
287         JMP      WF_4              ;YES, JUMP TO WAIT
288         CLR      LCD_DATAC         ;NO, CONFIG LCD_DATA AS OUTPUT MODE
289         SWAPA    ACC
290         MOV      LCD_DATA,A        ;LATCH DATA/COMMAND ON PB(LCD DATA BUS)
291         CLR      LCD_EN            ;CLEAR ALL LCD CONTROL SIGNAL
292         CLR      LCD_RW
293         CLR      LCD_RS
294         SZ       DC_FLAG           ;IS COMMAND WRITE?
295         SET      LCD_RS            ;NO, SET RS HIGH
296         SET      LCD_EN            ;SET EN HIGH
297         NOP
298         CLR      LCD_EN            ;SET EN LOW
299         SWAPA    ACC
300         SET      LCD_EN            ;SET EN HIGH
```

```
301         MOV         LCD_DATA,A
302         CLR         LCD_EN              ;SET EN LOW
303         RET
304 WLCMD_4 ENDP
305 ;********************************************************************
306 ;                   Delay about DEL1*10ms
307 ;********************************************************************
308 DELAY   PROC
309         MOV         A,20
310         MOV         DEL1,A              ;SET DEL1 COUNTER
311 DEL_1:
312         MOV         A,30
313         MOV         DEL2,A              ;SET DEL2 COUNTER
314 DEL_2:
315         MOV         A,110
316         MOV         DEL3,A              ;SET DEL3 COUNTER
317 DEL_3:
318         SDZ         DEL3                ;DEL3 DOWN COUNT
319         JMP         DEL_3
320         SDZ         DEL2                ;DEL2 DOWN COUNT
321         JMP         DEL_2
322         SDZ         DEL1                ;DEL1 DOWN COUNT
323         JMP         DEL_1
324         RET
325 DELAY ENDP
326         END
```

程序说明

行号	说明
7~25	依序定义变量地址。
27~30	定义 LCD_DATA Port 与 LCD_CONTR Port 分别为 PC 与 PA。
31~35	定义 LCD 的控制信号引脚,以及表示大小写的 LED 控制引脚。
36~38	定义 RXD_与 TXD_的引脚位置,以作为数据的传输与接收之用。
39~40	定义 KEY_PORT 为 PB。
41	定义数据传输速率 BAUD=9 600,读者可以视实际需求改变设置。
42	定义 FSYS=4 MHz(即系统工作时钟脉冲 f_{SYS})。
43	定义 K=40×4,请参考后续的说明。
46	声明存储器地址由 000h 开始,此为 HT46x23 的 Reset Vector。
48	声明存储器地址由 004h 开始,此为 HT46x23 的 External Interrupt Vector。
50	声明存储器地址由 008h 开始,此为 HT46x23 的 Timer/Event Counter Interrupt Vector。
53	将 LCD_CONTR Port(PA)定义为输出模式。
54	将 RXD_定义为输入模式,作为读取串行输入数据之用。
55~56	设置 CAPITAL=0(Default 为大写字形),并点亮代表大写字形的 LED。
57~59	清除 RX_OK_FLAG 与 START_FLAG 标志位,并将 TXD_设置为 1(Idle Staet)。
60~61	将 LCD 设置为双行显示(N = 1)、使用 4 位(DB7 ~ DB4)控制模式(DL = 1)、5×7 点矩阵字形(F = 0)。
62~63	将 LCD 设置为显示所有数据(D = 1)、显示光标(C = 1)、光标所在位置的字不会闪烁(B = 0)。
64~65	将 LCD 的地址标志位(AC)设置为递加(I/D = 1)、显示器画面不因读/写数据而移动(S = 0)。
66~67	将 Timer/Event Counter 定义如下:Timer Mode(Mode 2)、预分频比例设置为 1:1($f_{INT}=f_{SYS}$)。
68~69	Global Interrupt 与 Timer/Event Counter Interrupt 有效。
70~71	设置 LCD_COUNT=20,作为 LCD 显示位置是否已经到达该行最后一个位置的判断依据。本专题采用 20×2 的 LCD,所以设置为 20。
72~73	设置 LCD_ADRS=80h,作为字形显示在 LCD 的第 1 行与第 2 行切换之用。

75~76	将 LCD 整个显示器清空，并设置 DD RAM 地址为第 1 行第 0 个位置。
78	调用 HALF_READ_KEY 子程序，读取按键。
79~80	检查是否接收到 PC 端发送过来的数据。若无(RX_OK_FLAG=0)，则跳到 NO_RX_DATA 检查按键值；若有(RX_OK_FLAG=1)，则准备显示接收到的字节。
81~90	将 LCD_COUNT−1，判断是否已经到达该行最后一个位置。若是(LCD_COUNT=0)则将 LCD_ADRS 的 Bit 6 反相，并依据反相后的值调用 WLCMC_4 子程序切换 LCD 的行数，并将 LCD_COUNT 重设为 20。若 LCD 还未到达该行最后一个位置，则直接调用 WLCND_4 子程序显数据。
91	清除 RX_OK_FLAG 标志位，表示此笔数据已经完成显示。
93~96	根据 KEY 寄存器的值判定是否有按键被按下，若无则继续检查(JMP SCAN_KEY)。
97~100	判断是不是按下 CLEAR 键(KEY=26)，若是则清除 LCD 并重新扫描按键(JMP MAIN)。
103~106	判断是不是按下 CAP 键(KEY=27)，若是则将 CAPITAL 寄存器与 CAP_LED 反相，完成大/小写字形的切换。
107~108	调用 DELAY 子程序，延时 0.2 s 后重新扫描按键(JMP WAIT_KEY)；延时时间的计算请参考 4-1 节。
110~115	将按键值转换为 ASCII 码并调用 TX_DATA 子程序传送给 PC 端。在这之前会以 CAPITAL 寄存器的值判断目前为大写模式(CAPITAL=00h)或小写模式(CAPITAL=FFh)，以便转换为大写或小写的 ASCII 码(注：大小写英文字形的 ASCII 码相差 20h)。
116~117	调用 DELAY 子程序，延时 0.2 s 后重新扫描按键(JMP WAIT_KEY)；延时时间的计算请参考 4-1 节。
121~136	EXT_INT 子程序，由于主程序中已将外部中断有效，所以一旦 RXD_引脚出现 START Bit 信号(1→0)，就会进入此外部中断服务子程序。紧接着会调用 RX_DATA 子程序由 RXD_引脚依序读回 B0~B7 数据位，此时 RXD_不得再为中断输入引脚，而必须当成一般的 I/O 使用，所以在第 125 行先关闭 HT46x23 的外部中断功能。在接收完 8 位数据之后会将 RX_OK_FLAG 设为 1，以供主程序判断(第 79~80 行)；同时也将收到的数据由 BUFFER 寄存器复制到 BUFFER[1]作为缓冲；在返回主程序前再度将外部中断有效，准备检测下一个 START Bit 信号。
140~162	TMR_INT 子程序，每当 Timer/Event Counter 计时溢位，就依 RX_FLAG 判断是要传送数据至 PC 端(TX_MODE)，或接收由 PC 端送过来的数据(RX_MODE)。
143~152	RX_MODE：由 RXD_读取一位数据存放在接收寄存器(BUFFER)中，若已接收 8 位数据则清除 RX_FLAG，其中设置 START_FLAG=1 的目的是让 RX_DATA 知道 START Bit 时间已到，将 Timer 计时溢位时间由 1.5T 改成 T。
153~162	TX_MODE：依据传送寄存器(BUFFER)的值设置 TXD_的状态供 PC 端读取(由 LSB 先送)，若已传送完 8 位数据则清除 TX_FLAG。
166~188	TX_DATA 子程序，负责将按键值由 TXD_引脚串行输出。首先将外部中断功能关闭，以防止 PC 端此时送数据过来造成错误。接着设置 Timer 计时溢位时间为 T，将 TXD_设为 0，此目的为产生 START Bit。待经过 T 时间之后 Timer/Event Counter Interrupt 有效，每隔 T 时间循序送出一个位的数据。等所有位都完成发送之后，将 Timer/Event Counter Interrupt 关闭并重新启动外部中断功能。
192~217	RX_DATA 子程序，负责由 RXD_引脚接收一位数据。首先是设置 Timer 计时溢位时间为 1.5T(for START Bit)，并 Timer/Event Counter 中断与中断总开关(EMI)有效。由于 RX_DATA 子程序由外部中断服务子程序所调用，但是 HT46x23 单片机在进入中断之后会自动将 EMI 设为 0，因此为了能够在外部中断还未结束之前，仍能使用 Timer/Event Counter 的中断功能，所以在第 198 行需重新将其使能，并接着在第 199 行使能 Timer/Event Counter Interrupt。第 204 与 205 行是等待 1.5T 的 START Bit 时间，然后在第 207~210 行将 Timer 计时溢位时间重定为 T，以便接收剩余的 7 个数据位。由于 Timer 还在定时的情况时，HT46xx 会等到计数溢位时才会将新的计数数值载入寄存器开始计数，所以在第 206 行先停止 Timer 的动作，等新的数值载入寄存器后才在第 211 行启动其计数功能。
222~267	HALF_READ_KEY 子程序，请参考 5-12 节中的说明。

271~304 WLCMD_4 与 WLCMC 子程序，请参考 5-9 节中的说明。但不同的是由于 LCD_CONTR(PA) 除了控制 LCD 之外，还有其他用途(RXD_、TXD_、CAP_LED)，因此将原本 CLR LCD_CONTR 的动作改由 CLR LCD_RS、CLR LCD_RW 与 CLR LCD_EN 三行指令来完成。

308~322 DELAY 子程序，请参考 4-1 节中的相关说明。

  首先要说明的是第 195~198 行在 RX_DATA 子程序中的意义，其目的是载入产生 $1.5T(T(一个位时间)=BAUD^{-1})$ 所需的计数常量，以便在采样 RXD_ 引脚上的数据时恰好是在位数据的中央(见图 6-12-4)，降低错误率。既然是 1.5 倍的位时间，那么在 Timer/Event Counter 计数比例是 1:1 的情况下，在 TMRH_TMRL 寄存器所载入的数值应该是：**65 536−(f_{SYS}*1.5/BAUD)**，为何写成 **65 536-(f_{SYS}*10/(BAUD*15)+K)**？而 **K** 值又何需如程序第 33 行的定义呢？请读者仅记：本专题是以软件来实现 UART 功能，所以单片机指令的执行时间必须列入考虑，当 HT46x23 检测到 RXD_=0(即硬件中断发生)，到跳到 RX_DATA 子程序启动 Timer/Event Counter 开始计数，这期间约历经了 20~30 个指令，当 f_{SYS}=4 MHz 时约为 20~30μs，此外还必须加上 HT46x23 对中断产生反应的内部处理时间。这在 Baud Rate 低的状况下影响还不是太大(见图 6-12-14(a)、(b))，一旦 Baud Rate 超过 9 600 时就不得不注意了(见图 6-12-14(c))。请参考图 6-12-14 的波形，读者可以发现在 Baud Rate=9 600 时，采样点已经太过向右偏移。这是因为指令执行时间加上 1.5 T 时间所致，如不设法予以调整数据，错误率将会提高。因此以 **K** 作为上述时间的补偿，图 6-12-15 是补偿后的数据与采样点关系。

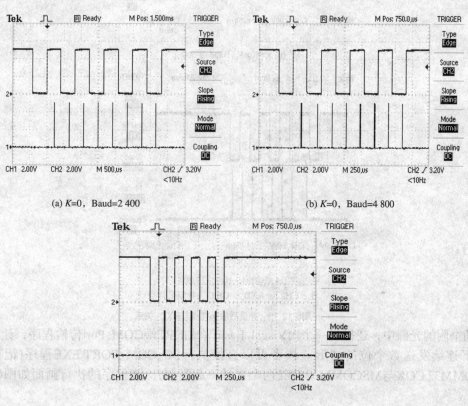

(a) K=0，Baud=2 400   (b) K=0，Baud=4 800

(c) K=0，Baud=9 600

注：CH2 for RXD_，CH1 为数据采样点

图 6-12-14   采样点与 RXD_ 关系

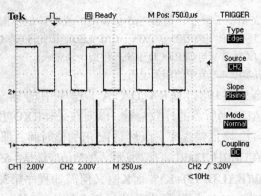

(a) $K=40\times4$,Baud=4 800

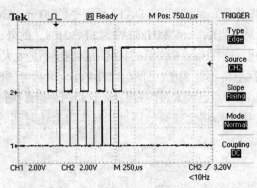

(b) $K=40\times4$,Baud=9 600

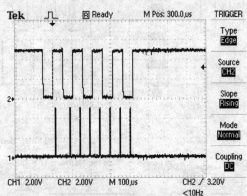

(c) $K=40\times4$,Baud=19 200

注：CH2 for RXD_，CH1 为数据采样点

图6-12-15 补偿后的采样点与RXD_关系

随书附的光碟中，还提供一个由Visual Basic写成的PC端COM Port传输程序，让读者体会一下连续发送数个位的情形。读者可以直接执行PC_COM_PORT.EXE程序(记得连同MSCOMM32.COX与MSCOMM32.DEP两个文件一起复制)，该程序的执行画面如图6-12-16所示。

第 6 章 实践应用篇

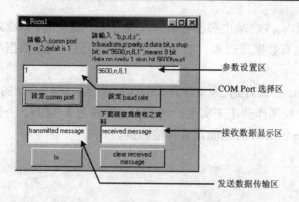

图6-12-16　PC_COM_PORT.EXE程序执行画面

　　使用方式：首先在COM Port选择区输入欲控制的COM Port，输入后单击"设置**comm port**"按钮；接着输入参数，如9600,n,8,1(Baud＝9600、No Parity Bit、8 Data Bits、1 STOP Bit)，输入后请单击"设置**baud rate**"按钮。此时完成COM Port的设置。此后，在单片机端所按下的按键会显示在"接收数据显示区"，单击"**clear revevied message**"按钮可以清除其内容。在"发送数据输入区"输入想要发送的数据，当单击"tx"按钮后，该输入区内的数据会全部被发送出去。

　　由于在HT46x23在接收一个字节之后，需要将接收到的字符显示在LCD上，因此会调用WLCMD_4子程序(如果LCD需要换行，还得先调用WLCMC_4)子程序，而LCD并非是快速的设备，写入数据或命令至少都需要40μs以上的时间，加上电路中是以4位方式来控制，所以数据写入的时间至少也在100μs以上。这在BAUD＝2 400、4 800情况下，连续传送几个字节都没有问题，但是BAUD一旦再高就来不及了(也就是刚接收的数据还未显示，就被新进的数据覆盖了)。虽然在程序中使用BUFFER[1]来做缓冲，但在BAUD＝19 200时还是来不及。读者可以通过增加PC端数据的停止位来加以改善，或是把LCD显示改以七段显示器取代就没有问题了，可是如此一来就无法看到成串的传输数据。如图6-12-17所示为使用PC_COM_PORT.EXE程序进行传输的画面。

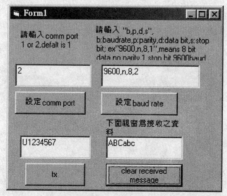

图6-12-17　使用PC_COM_PORT.EXE程序进行传输的画面

6-12-5　动动脑＋动动手

　　📖 运用6-11节的概念，如果按键历经256次的检查(KEY＝28)，用户均未按下任何键，

HT46xx 单片机原理与实践

而且此段期间内，PC 端也都没有数据发送过来，就让 HT46x23 进入 HALT Mode；但 PC 端一旦有数据送过来，单片机要能够被叫醒并恢复数据的正常传输(利用中断唤醒功能)。

📖 如果 Baud Rate 要提升到 38 400($T=26\,\mu s$)，以 $f_{SYS}=4$ MHz 的条件是很难达到的，但可以通过上拉工作频率来实现。请将 f_{SYS} 提高为 8 MHz，并适当调整 K 值，使其在 Baud Rate＝38 400 时也能正常传输数据。

附 录

附录 A   HT46xx 指令速查表

附录 B   HT46xx 家族程序存储器映射图

附录 C   HT46xx 家族数据存储器与特殊功能寄存器

附录 D   HT46xx 特殊功能寄存器速查表

附录 E   HT46xx 重置后的内部寄存器状态

附录 F   LCD 指令速查表

附录 G   本书常用子程序一览表

附录 A HT46xx 指令速查表

助忆符号		指令功能描述	指令周期	受影响标志					
				C	AC	Z	OV	PDF	TO
ADC	A,[m]	累加器 A、寄存器[m]与进位标志 C 相加,结果至累加器 A	1	■	■	■	■		
ADCM	A,[m]	寄存器[m]、累加器 A 与进位标志 C 相加,结果至[m]	1(1)	■	■	■	■		
ADD	A,[m]	累加器 A 与寄存器[m]相加,结果存至累加器 A	1	■	■	■	■		
ADD	A,x	累加器 A 与常数 x 相加,结果存至累加器 A	1	■	■	■	■		
ADDM	A,[m]	寄存器与累加器 A 相加,结果存至[m]	1(1)	■	■	■	■		
AND	A,[m]	累加器 A 与寄存器[m]执行 AND 运算,结果至累加器 A	1			■			
AND	A,x	累加器 A 与常数 x 执行 AND 运算,结果至累加器 A	1			■			
ANDM	A,[m]	寄存器与累加器 A 执行 AND 运算,结果至[m]	1(1)			■			
CALL	Addr	呼叫子程序指令(PC=Addr)	2						
CLR	[m],i	将寄存器[m]的第 i 位清除为 0 (i=0~7)	1(1)						
CLR	[m]	将寄存器[m]内容清除为 0	1(1)						
CLR	WDT	清除看门狗定时器	1					■(3)	■(3)
CLR	WDT1	看门狗定时器清除指令 1	1					■(3)	■(3)
CLR	WDT2	看门狗定时器清除指令 2	1					■(3)	■(3)
CPL	[m]	对寄存器[m]内容取补数,再将结果回存至[m]	1(1)			■			
CPLA	[m]	对寄存器[m]内容取补数,再将结果存至 A	1			■			
DAA	[m]	累加器 A 的内容转成 BCD 码后存至[m]	1(1)	■					
DEC	[m]	寄存器[m]−1,结果存至寄存器[m]	1(1)			■			
DECA	[m]	寄存器[m]−1,结果存至累加器 A	1			■			
HALT		进入省电模式	1					■	■
INC	[m]	寄存器[m]+1,结果存至[m]	1(1)			■			
INCA	[m]	寄存器[m]+1,结果存至累加器 A	1			■			
JMP	Addr	跳跃至地址 Addr(PC=Addr)	2						
MOV	A,[m]	将寄存器[m]内容放入累加器 A	1						
MOV	[m],A	将累加器 A 内容放入寄存器[m]	1(1)						
MOV	A,x	将常数 x 放入寄存器[m]	1						
NOP		不动作	1						
OR	A,[m]	累加器 A 与寄存器[m]执行 OR 运算,结果存至累加器 A	1			■			
OR	A,x	累加器 A 与常数 x 执行 OR 运算,结果存至累加器 A	1			■			
ORM	A,[m]	累加器 A 与寄存器 A 执行 OR 运算,结果存至[m]	1(1)			■			
RET		子程序返回指令(PC=Top of Stack)	2						
RET	A,x	子程序返回指令(PC=Top of Stack),并将常数 x 放入累加器 A	2						
RETI		中断子程序返回指令(PC=Top of Stack),并设置 EMI flag=1	2						
RL	[m]	寄存器[m]内容左移一个位	1(1)						
RLA	[m]	寄存器[m]内容左移一个位后,将结果存至累加器 A	1						
RLC	[m]	寄存器[m]内容连同进位标志 C 一起左移一个位	1(1)	■					
RLCA	[m]	寄存器[m]内容连同进位标志 C 一起左移一个位后,将结果存至 A	1	■					
RR	[m]	寄存器[m]内容右移一个位	1(1)						
RRA	[m]	寄存器[m]内容右移一个位后,将结果存至累加器 A	1						
RRC	[m]	寄存器[m]内容连同进位标志 C 一起右移一个位	1(1)	■					
RRCA	[m]	寄存器[m]内容连同进位标志 C 一起右移一个位后,将结果存至 A	1	■					
SBC	A,[m]	累加器 A 与寄存器[m]、进位标志 C 相减,结果存至 A	1	■	■	■	■		
SBCM	A,[m]	累加器 A 与寄存器[m]、进位标志 C 相减,结果至[m]	1(1)	■	■	■	■		
SDZ	[m]	将寄存器[m]−1 结果存至寄存器[m],若结果为 0 则跳过下一行	1(3)						
SDZA	[m]	寄存器[m]−1 结果存至累加器 A,若结果为 0 则跳过下一行	1(2)						
SET	[m],i	将寄存器[m]的第 i 位设置为 1 (i=0~7)	1(1)						
SET	[m]	将寄存器[m]内容设置为 FFh	1(1)						
SIZ	[m]	寄存器[m]+1 结果存至寄存器[m],若结果为 0 则跳过下一行	1(3)						
SIZA	[m]	寄存器[m]+1 结果存至累加器 A,若结果为 0 则跳过下一行	1(2)						
SNZ	[m],i	若寄存器[m]的第 i (i=0~7)位不为 0 则跳过下一行	1(2)						
SUB	A,x	累加器 A 与常数 x 相减,结果存至累加器 A	1	■	■	■	■		
SUB	A,[m]	累加器 A 与寄存器[m]相减,结果存至累加器 A	1	■	■	■	■		
SUBM	A,[m]	累加器 A 与寄存器[m]相减,结果至[m]	1(1)	■	■	■	■		
SWAP	[m]	将寄存器[m]的高低 4 位互换	1(1)						
SWAPA	[m]	将寄存器[m]的高低 4 位互换后之结果存至累加器 A	1						
SZ	[m]	若寄存器[m]内容为 0 则跳过下一行	1(2)						
SZ	[m],i	若寄存器[m]的第 i (i=0~7)位为 0 则跳过下一行	1(2)						
SZA	[m]	将寄存器[m]内容存至累加器 A,若为 0 则跳过下一行	1(2)						
TABRDC	[m]	依据 TBLP 读取程序存储器(目前页)之值并存放至 TBLH 与[m]	2(1)						
TABRDL	[m]	依据 TBLP 读取程序存储器(最末页)之值并存放至 TBLH 与[m]	2(1)						
XOR	A,[m]	累加器 A 与寄存器[m]执行 XOR 运算,结果存至累加器 A	1			■			
XOR	A,x	累加器 A 与常数 x 执行 XOR 运算,结果存至累加器 A	1			■			
XORM	A,[m]	累加器 A 与寄存器[m]执行 XOR 运算,结果存至[m]	1(1)			■			

i=某个位(0~7); x=8 位常数; [m]=数据存储器位置; Addr=程序存储器位置; ■=标志受影响; □=标志不受影响。
(1) 若有载入数值至 PCL 寄存器,则执行时间增加一个指令周期。
(2) 若条件成立,跳跃至下下一行指令执行时,则执行时间增加一个指令周期。
(3) 执行"CLR WDT1"与"CLR WTD2"后,TO 及 PDF 标志会被清除为「0」;若只是单独执行任何一个指令,则 TO 及 PDF 标志并不受影响。

附录B　HT46xx家族程序存储器映射图

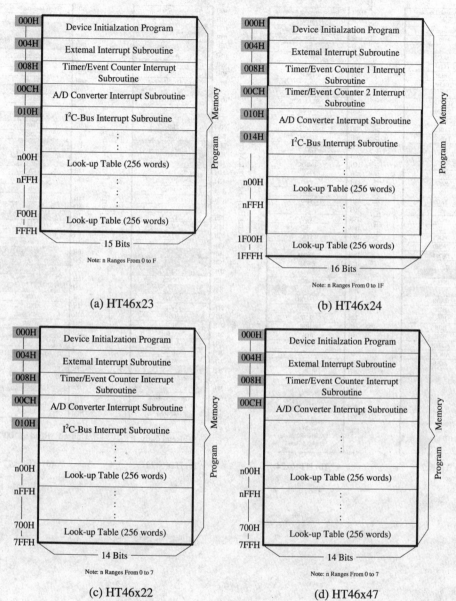

(a) HT46x23
(b) HT46x24
(c) HT46x22
(d) HT46x47

附录C　HT46xx数据存储器与特殊功能寄存器

(a) HT46x23

地址	名称
00H	Indirect Addressing Register 0
01H	MP0
02H	Indirect Addressing Register 1
03H	MP1
04H	
05H	ACC
06H	PCL
07H	TBLP
08H	TBLH*
09H	
0AH	STATUS
0BH	INTC0
0CH	TMRH
0DH	TMRL
0EH	TMRC
0FH	
10H	
11H	
12H	PA
13H	PAC
14H	PB
15H	PBC
16H	PC
17H	PCC
18H	PD
19H	PDC
1AH	PWM0
1BH	PWM1
1CH	
1DH	
1EH	INTC1
1FH	
20H	HADR
21H	HCR
22H	HSR
23H	HDR
24H	ADRL
25H	ADRH*
26H	ADCR*
27H	ACSR
28H–3FH	(Unused)
40H–FFH	General Purpose Data Memory (192 Bytes)

(b) HT46x24

地址	名称
00H	Indirect Addressing Register 0
01H	MP0
02H	Indirect Addressing Register 1
03H	MP1
04H	BP
05H	ACC
06H	PCL
07H	TBLP
08H	TBLH*
09H	
0AH	STATUS
0BH	INTC0
0CH	TMR0H
0DH	TMR0L
0EH	TMR0C
0FH	TMR1H
10H	TMR1L
11H	TMR1C
12H	PA
13H	PAC
14H	PB
15H	PBC
16H	PC
17H	PCC
18H	PD
19H	PDC
1AH	PWM0
1BH	PWM1
1CH	PWM2
1DH	PWM3
1EH	INTC1
1FH	
20H	HADR
21H	HCR
22H	HSR
23H	HDR
24H	ADRL
25H	ADRH*
26H	ADCR*
27H	ACSR
28H	PF
29H	PFC
2AH–3FH	(Unused)
40H–FFH	General Purpose Data Memory (192 Bytes×2 Banks) Bank0 & Bank1

(c) HT46x22

地址	名称
00H	Indirect Addressing Register
01H	MP
02H	
03H	
04H	
05H	ACC
06H	PCL
07H	TBLP
08H	TBLH*
09H	
0AH	STATUS
0BH	INTC0
0CH	
0DH	TMR
0EH	TMRC
0FH	
10H	
11H	
12H	PA
13H	PAC
14H	PB
15H	PBC
16H	PC
17H	PCC
18H	PD
19H	PDC
1AH	PWM
1BH	
1CH	
1DH	
1EH	INTC1
1FH	
20H	HADR
21H	HCR
22H	HSR
23H	HDR
24H	ADRL
25H	ADRH*
26H	ADCR*
27H	ACSR
28H–3FH	(Unused)
40H–7FH	General Purpose Data Memory (64Bytes)

(d) HT46x47

地址	名称
00H	Indirect Addressing Register
01H	MP
02H	
03H	
04H	
05H	ACC
06H	PCL
07H	TBLP
08H	TBLH*
09H	
0AH	STATUS
0BH	INTC
0CH	
0DH	TMR
0EH	TMRC
0FH	
10H	
11H	
12H	PA
13H	PAC
14H	PB
15H	PBC
16H	
17H	
18H	PD
19H	PDC
1AH	PWM
1BH	
1CH	
1DH	
1EH	
1FH	
20H	ADRL*
21H	ADRH*
22H	ADCR
23H	ACSR
24H–3FH	(Unused)
40H–7FH	General Purpose Data Memory (64Bytes)

* : Read Only.　　■ : Unused, Read as "00".

附录D HT46xx特殊功能寄存器速查表

特殊功能寄存器	Bit7	6	5	4	3	2	1	Bit0	参考章节	
IAR0									2-3-1	
MP0									2-3-2	
IAR1									2-3-3	
MP1									2-3-4	
BP	—	—	—	—	—	—	—	—	2-3-5	(46x24 only)
ACC									2-3-6	
PCL									2-3-7	
TBLP									2-3-8	
TBLH									2-3-9	
STATUS	—	—	TO	PDF	OV	Z	AC	C	2-3-10	
INTC0	—	ADF	TF	EIF	EADI	ETI	EEI	EMI	2-4	
INTC1	—	—	—	HIF	—	—	—	EHI	2-4	
TMRC	TM1	TM0	—	TON	TE	PSC2	PSC1	PSC0	2-5	
TMRH									2-5	
TMRL									2-5	
TMR1C	TM1	TM0	—	TON	TE	—	—	—	2-5	(46x24 only)
TMR1H									2-5	(46x24 only)
TMR1L									2-5	
PA									2-6	
PAC									2-6	
PB									2-6	
PBC									2-6	
PC									2-6	
PCC									2-6	
PD									2-6	
PDC									2-6	
PF									2-6	(46x24 only)
PFC									2-6	(46x24 only)
PWM0									2-7	
PWM1									2-7	
PWM2									2-7	(46x24 only)
PWM3									2-7	(46x24 only)
HADR				Slave Address				—	2-8	
HCR	HEN	—	—	HTX	TXAK	—	—	—	2-8	
HSR	HCF	HASS	HBB	—	—	SRW	—	RXAK	2-8	
HDR									2-8	
ADRL	D1	D0	—	—	—	—	—	—	2-9	(46x23,x24 only)
ADRH	D9	D8	D7	D6	D5	D4	D3	D2	2-9	(46x23,x24 only)
ADRL	D0	—	—	—	—	—	—	—	2-9	(46x22,x47 only)
ADRH	D8	D7	D6	D5	D4	D3	D2	D1	2-9	(46x22,x47 only)
ADSR	TEST	—	—	—	—	—	ADCS1	ADCS0	2-9	
ADCR	START	EOC	PCR2	PCR1	PCR0	ACS0	ACS1	ACS0	2-9	

附录 E HT46xx 重置后的内部寄存器状态

Register	Reset (Power On)		WDT Time-Out (Normal Operation)		RES Reset (Normal Operation)		RES Reset (HALT)		WDT Time-Out (HALT) *	
TMRL	xxxx	xxxx	xxxx	xxxx	xxxx	xxxx	xxxx	xxxx	uuuu	uuuu
TMRH	xxxx	xxxx	xxxx	xxxx	xxxx	xxxx	xxxx	xxxx	uuuu	uuuu
TMRC	00-0	1000	00-0	1000	00-0	1000	00-0	1000	uu-u	uuuu
TMR1L	xxxx	xxxx	xxxx	xxxx	xxxx	xxxx	xxxx	xxxx	uuuu	uuuu
TMR1H	xxxx	xxxx	xxxx	xxxx	xxxx	xxxx	xxxx	xxxx	uuuu	uuuu
TMR1C	00-0	1---	00-0	1---	00-0	1---	00-0	1---	uu-u	u---
Program Counter	000H		000H		000H		000H		000H	
BP	----	---0	----	---0	----	---0	----	---0	----	---u
MP0	xxxx	xxxx	uuuu	uuuu	uuuu	uuuu	uuuu	uuuu	uuuu	uuuu
MP1	xxxx	xxxx	uuuu	uuuu	uuuu	uuuu	uuuu	uuuu	uuuu	uuuu
ACC	xxxx	xxxx	uuuu	uuuu	uuuu	uuuu	uuuu	uuuu	uuuu	uuuu
TBLP	xxxx	xxxx	uuuu	uuuu	uuuu	uuuu	uuuu	uuuu	uuuu	uuuu
TBLH	-xxx	xxxx	-uuu	uuuu	-uuu	uuuu	-uuu	uuuu	-uuu	uuuu
STATUS	--00	xxxx	--1u	uuuu	--uu	uuuu	--01	uuuu	--11	uuuu
INTC0	-000	0000	-000	0000	-000	0000	-000	0000	uuuu	uuuu
INTC1	---0	---0	---0	---0	---0	---0	---0	---0	---u	---u
PA	1111	1111	1111	1111	1111	1111	1111	1111	uuuu	uuuu
PAC	1111	1111	1111	1111	1111	1111	1111	1111	uuuu	uuuu
PB	1111	1111	1111	1111	1111	1111	1111	1111	uuuu	uuuu
PBC	1111	1111	1111	1111	1111	1111	1111	1111	uuuu	uuuu
PC	---1	1111	---1	1111	---1	1111	---1	1111	---u	uuuu
PCC	---1	1111	---1	1111	---1	1111	---1	1111	---u	uuuu
PD	----	--11	----	--11	----	--11	----	--11	----	--uu
PDC	----	--11	----	--11	----	--11	----	--11	----	--uu
PF	1111	1111	1111	1111	1111	1111	1111	1111	uuuu	uuuu
PFC	1111	1111	1111	1111	1111	1111	1111	1111	uuuu	uuuu
PWM0	xxxx	xxxx	xxxx	xxxx	xxxx	xxxx	xxxx	xxxx	uuuu	uuuu
PWM1	xxxx	xxxx	xxxx	xxxx	xxxx	xxxx	xxxx	xxxx	uuuu	uuuu
PWM2	xxxx	xxxx	xxxx	xxxx	xxxx	xxxx	xxxx	xxxx	uuuu	uuuu
PWM3	xxxx	xxxx	xxxx	xxxx	xxxx	xxxx	xxxx	xxxx	uuuu	uuuu
HADR	xxxx	xxx-	xxxx	xxx-	xxxx	xxx-	xxxx	xxx-	uuuu	uuu-
HCR	0--0	0---	0--0	0---	0--0	0---	0--0	0---	u--u	u---
HSR	100-	-0-1	100-	-0-1	100-	-0-1	100-	-0-1	uuu-	-u-u
HDR	xxxx	xxxx	xxxx	xxxx	xxxx	xxxx	xxxx	xxxx	uuuu	uuuu
ADRL	xx--	----	xx--	----	xx--	----	xx--	----	uuuu	----
ADRH	xxxx	xxxx	xxxx	xxxx	xxxx	xxxx	xxxx	xxxx	uuuu	uuuu
ADCR	0100	0000	0100	0000	0100	0000	0100	0000	uuuu	uuuu
ACSR	1---	--00	1---	--00	1---	--00	1---	--00	u---	--uu

注：*：代表热重置（Warm Reset）   u：代表状态没改变   x：代表未知状态

注意：由于HT46xx家族部分寄存器位数不尽相同，故本表主要是以HT46x23为主，而表中**斜字体部分**代表仅HT46x24所拥有之寄存器，各成员之详细重置数据还是请读者参阅原厂的数据手册。

附录F　LCD指令速查表

指令	指令码										指令说明		执行时间
	RS	R/W	DB7	DB6	DB5	DB4	DB3	DB2	DB1	DB0			
清除显示器	0	0	0	0	0	0	0	0	0	1	DD RAM 里的所有地址填入空白码 20h, AC 设置为 00h, I/D 设置为「1」		1.64ms
光标归位	0	0	0	0	0	0	0	0	1	X	DD RAM 的 AC 设为 00h, 光标回到左上角第一行的第一个位置, DD RAM 内容不变		1.64ms
进入模式	0	0	0	0	0	0	0	1	I/D	S	I/D=0	CPU 写数据到 DD RAM 或读取数据之后 AC 减 1, 光标会向左移动	40μs
											I/D=1	CPU 写数据到 DD RAM 或读取数据之后 AC 加 1, 光标会向右移动	
											S=0	显示器画面不因读写数据而移动	
											S=1	CPU 写数据到 DD RAM 后, 整个显示器会向左移动(若 I/D=0)或向右移动(若 I/D=1)一个位置, 但从 DD RAM 读取数据时显示器不会移动	
显示器 ON/OFF 控制	0	0	0	0	0	0	1	D	C	B	显示器控制: D=0 所有数据不显示　D=1 显示所有数据 光标控制: C=0 不显示光标　C=1 显示光标 光标闪烁控制: B=0 不闪烁　B=1 光标所在位置的字会闪烁		40μs
光标或显示器移动	0	0	0	0	0	1	S/C	R/L	X	X	S/C=0 R/L=0 光标位置向左移(AC 值减 1) S/C=0 R/L=1 光标位置向右移(AC 值加 1) S/C=1 R/L=0 显示器与光标一起向左移 S/C=1 R/L=1 显示器与光标一起向右移		40μs
功能设置	0	0	0	0	1	DL	N	F	X	X	设置数据位长度: DL=0 使用 **4 位**(DB7~DB4)控制模式 DL=1 使用 **8 位**(DB7~DB0)控制模式 设置显示器的行数: N=0 单行显示　N=1 双行显示两行 设置字型: F=0 5×7 点矩阵字型　F=1 5×10 点矩阵字型		40μs
CG RAM 地址设置	0	0	0	1	CGRAM Address						将 CG RAM 的地址(DB5~DB0)写入 AC		40μs
DDRAM 地址设置	0	0	1	DDRAM Address							将 DD RAM 的地址(DB6~DB0)写入 AC		40μs
读取忙碌标志和地址	0	1	BF	Address Counter							BF=1, 表示目前 LCD 正忙着内部的工作, 因此无法接收外部的命令, 必须等到 BF=0 之后, 才可以接收外部的命令。在(DB6~DB0)可读出 AC 值		0
写数据到 CG RAM 或 DD RAM	1	0	Write Data								将数据写入 DD RAM 或 CG RAM		40μs
从 CG RAM 或 DD RAM 读取数据	1	1	Read Data								读取 CG RAM 或 DD RAM 数据		40μs

附录G　本书常用子程序一览表

子程序名称	子程序功能描述	参考章节
DELAY	延迟 DEL1×1ms	4-1
DELAY	延迟 DEL1×10ms	4-1 4-2
READ_KEY	读取 4×4 键盘按键值	4-7
WLCMC	将控制命令写至 LCD（8-Bit 控制模式）	5-5
WLCMD	将显示数据写至 LCD（8-Bit 控制模式）	5-5
PRINT	将字串送至 LCD 显示	5-8
RLCM	由 LCD 读回 DD RAM 之内容	5-8
WLCMC_4	将控制命令写至 LCD（4-Bit 控制模式）	5-9
WLCMD_4	将显示数据写至 LCD（4-Bit 控制模式）	5-9
HEX2ASCII	将 8 位数据转为十进制数值后,在转成其 ASCII Code 码	5-10
HALF_READ_KEY	读取半矩阵式键盘按键值	5-12
MUL_16X8	完成 16 位与 8 位之正整数乘法运算	6-1
WORD_HEX2ASCII	将 16 位数据转为十进制数值后,在转成其 ASCII Code 码	6-1
READ_KEY_PRESSED	读取 4×4 键盘按键,一直到用户按下按键并放开按键后,才传回按键值	6-2
I2C_START	产生 I²C 起始信号	6-3
I2C_DEV_SEL_W	送出 I²C Device Select Code, 同时送出 W 命令	6-3
I2C_DEV_SEL_R	送出 I²C Device Select Code, 同时送出 R 命令	6-3
I2C_SET_ADRS	送出 I²C 地址数据	6-3
I2C_RM_READ	完成 I²C Random Address Read 数据读取	6-3
I2C_CT_READ	完成 I²C Current Address Read 数据读取	6-3
I2C_SQ_RM_READ	完成 I²C Sequential Random Read 数据读取	6-3
I2C_SQ_CT_READ	完成 I²C Sequential Current Read 数据读取	6-3
I2C_BYTE_WRITE	完成 I²C Byte Write 数据写入	6-3
I2C_PAGE_WRITE	完成 I²C Page Write 数据写入	6-3
I2C_STOP	产生 I²C 结束信号	6-3
DIV_8_8	完成 8 位与 8 位之正整数除法运算	6-6
MUL_24X8	完成 24 位与 8 位之正整数乘法运算	6-6
DIV_32_8	完成 32 位与 8 位之正整数除法运算	6-6
DIV_32_16	完成 32 位与 16 位之正整数除法运算	6-6
DIV_32_24	完成 32 位与 24 位之正整数除法运算	6-6
DIV_32_32	完成 32 位与 32 位之正整数除法运算	6-6
RLCMD	由 LCD 读回 DD RAM 之内容	6-7
RLCMA	由 LCD 读回 AC（Address Counter）之值	6-7
MUL_16X8	完成 16 位与 8 位之正整数乘法运算	6-10
DIV_24_8	完成 24 位与 8 位之正整数除法运算	6-10
DIV_24_16	完成 24 位与 16 位之正整数除法运算	6-10
DIV_24_24	完成 24 位与 24 位之正整数除法运算	6-10
MUL_8X8	完成 8 位与 8 位之正整数乘法运算	6-10